Solid Fuels and Heavy Hydrocarbon Liquids

Thermal Characterization and Analysis

Elsevier Internet Homepage

http://www.elsevier.com
Consult the Elsevier homepage for full catalogue information on all books, journals and electronic products and services.

Elsevier Titles of Related Interest
The following journals related to Fuel Energy can all be found at
http://www.sciencedirect.com

Fuel
Fuel Processing Technology
Fuel and Energy Abstracts
Progress in Energy and Combustion Science
Combustion and Flame
Biomass and Bioenergy
Energy
Chemical Engineering and Processing
Journal of Petroleum Science and Engineering
International Journal of Coal Geology

To Contact the Publisher
Elsevier welcomes enquiries concerning publishing proposals: books, journal special issues, conference proceedings, etc. All formats and media can be considered. Should you have a publishing proposal you wish to discuss, please contact, without obligation, the commissioning editor responsible for Elsevier's energy science books publishing programme:

Publishing Editor
Energy
Elsevier Limited
The Boulevard, Langford Lane
Kidlington, Oxford
OX5 1GB, UK
Tel.: +44 1865 84 3000
Fax: +44 1865 84 3010
E-mail: energy@elsevier.com

General enquiries including placing orders, should be directed to Elsevier's Regional Sales Offices – please access the Elsevier internet homepage for full contact details.

Solid Fuels and Heavy Hydrocarbon Liquids

Thermal Characterization and Analysis

R. Kandiyoti
Imperial College London, UK

A. A. Herod
Imperial College London, UK

K.D. Bartle
University of Leeds, Leeds, UK

ELSEVIER

AMSTERDAM • BOSTON • HEIDELBERG • LONDON • NEW YORK • OXFORD
PARIS • SAN DIEGO • SAN FRANCISCO • SINGAPORE • SYDNEY • TOKYO

Elsevier
The Boulevard, Langford Lane, kidlington, Oxford, OX5 1GB, UK
84 Theobald's Road, London WC1X 8RR, UK
Radarweg 29, PO Box 211, 1000 AE Amsterdam, The Netherlands
525 B Street, Suite 1900, San Diego, CA 92101-4495, USA

First edition 2006

British Library Cataloguing in Publication Data
A catalogue record for this book is available from the British Library

Library of Congress Cataloging-in-Publication Data
A catalog record for this book is available from the Library of Congress

ISBN 13: 978-0-08-044486-4
ISBN 10: 0-08-044486-5

Printed and bound in Great Britain

CONTENTS

PREFACE

A little fortuitously, this book on coal, biomass and heavy hydrocarbons is approaching publication during a period of high and unstable oil and gas prices and amidst mounting uncertainty over future supplies. The underlying trend facing energy markets is one of steady, relentless rise in demand and little spare or excess capacity. The awakening, potentially giant economies of Asia are dependent on increasing energy supplies, no less than much of the already industrialized world. The USA alone expects its energy requirements to double by 2050. Many of the older industrialized countries also face a baffling set of options regarding the replacement of obsolescent electricity generating capacity.

Within this framework, how governments, industry and individuals should respond to conflicting personal, economic and environmental requirements remains to be debated. Economic growth and improved living standards require security of supply and freedom from terror attacks. However, we also desire a cleaner planet and fear the possible threat of climate change. Our objectives are at odds with each other. Much recent planning has focused on short-term solutions. Longer-term expectations have been placed on expensive and dangerous hydrogen on the one hand and technically as well as economically unrealistic CO_2-sequestration on the other. The world, the industrialized nations and the UK are all in need of reviewing feasible options that are publicly acceptable. This is a complex task that can no more be left to the market to resolve, than could be resolved by excluding it.

Overall, however, the picture we face is one of a gradually degrading environment and of a diminishing resource base. Many of the alterations suffered by the planet are of an irreversible nature. An unknown but no doubt large proportion of the readily extractable oil has already been used up. The present rush to gas can indeed delay the end of plentiful hydrocarbon supplies, but cannot avoid it. All combustion processes generate greenhouse gases as well as pollutants, such as toxic trace elements and sulfur oxides. Spillages from oil tankers and pipelines, coal mining and cleaning operations pollute the environment and are causing largely irreparable damage on grand scale.

Unfortunately, the "renewables" options do not appear as attractive, close up, as they do on first mention. The use of biomass as fuel can only assist in providing marginal amounts of energy; its large-scale use would compete directly with food production. Other renewable energy options, including wind-power suffer from high capital costs and affect the environment adversely. The inherent instabilities of wind-generated power impose a ceiling of a mere several percent contribution to the grid. In the coming decade, these difficulties will probably require bringing back the nuclear energy option.

Perhaps we have now collectively learned to have a care about decommissioning costs and those of nuclear waste storage. It would in any case be interesting to compare these costs with those of CO_2-sequestration. We will also need to live with and account for the dangers of ploughshares being turned into swords, since nuclear reactors can be used to manufacture fissile material for nuclear weapons.

Against this background of uncertain supply, a major long-term share for coal in power generation appears inevitable. Countries as diverse as the USA, China and India are all committed to increasing their use of coal-fired power generation, despite pollution from its mining, cleaning, transport and utilization and indeed despite its associated CO_2-emissions. As noted in all comparisons with oil and gas, the distribution of coals across the continents is more evenly spread. Furthermore, current and future coal use is not restricted to combustion. In albeit smaller tonnages, coal has maintained its position as a source of carbon in steel making and as raw material for high value carbon products. Furthermore, current high oil and gas prices have significantly changed the established cost calculations. The economic prospects of coal gasification, already under consideration as an efficient and less polluting alternative, and that of coal liquefaction for transport fuels, will be re-examined if oil and gas prices appear to stabilise at their current high levels.

The first stage in the processing of coal nearly always involves thermolytic reactions and one main purpose of this book is to summarize and assess the current state of knowledge about the thermochemical reactions of coals. These synoptic accounts have relied mainly on results of experimental work carried out in our laboratories at Imperial College London and the University of Leeds over the past two and a half decades. The work has been put in the context of the extensive literature on the thermochemical reactions of coals.

Several major strands run through the present manuscript. The first is the focus on the design and operation of experimental thermochemical reactors. The emphasis has been on isolating effects due to reactor design on the measured fundamental properties of solid fuels. In simple terms, this amounts to requiring the decoupling of experimental method from the properties being measured, as in any field of scientific endeavor. The clearer methodology emerging from this work has enabled the identification of similarities and differences between thermal breakdown in pyrolysis and liquefaction. The other major strand in the book has emerged from attention paid to changing product characteristics during reactor-related developments. New analytical procedures have been developed and applied to the characterization of liquid products from coal and biomass pyrolysis, coal liquefaction and petroleum derived heavy hydrocarbons.

Large molecular mass materials have been detected in coal derived liquids, in soot, in petroleum asphaltenes and vacuum residues, in solvent extracts of amber and wood, as well as in "craft" products such as Stockholm tar, used as a caulk and preservative in the "Mary Rose", famous flagship of Henry VIII. We will describe chromatographic and mass spectrometric methods developed for detecting and characterizing large molecular mass materials that have hitherto received relatively little attention. The increasing arrival of heavier crudes in oil refineries makes these developments more necessary than in the past. Controversies surround aspects of the characterization of large molecular mass materials and new insights made possible by the use of novel mass spectrometric techniques will be presented.

Little of the work described in these pages would have been possible without the untiring work of our students and associates at Imperial College and the University of Leeds. The authors take pleasure in acknowledging how much they have learned from them all, too many to name here, but whose names will appear as co-authors of many publications cited in this manuscript. We would like to thank them for their efforts and for many lasting friendships.

The authors would also like to thank the many sponsors of the work presented in these pages. Our special thanks go to the British Coal Utilisation Research Association, the Engineering and Physical Sciences Research Council, the UK Department of Trade and Industry, the Ministry of Science and Technology (China), SASOL of South Africa, CORUS and the European Community. We would like to remember the Coal Research Establishment (British Coal), which supported our laboratories with care and constancy over the years. The effective closure of British Coal by the Conservative Governments of the early 1990s caused much pain in the country, but allowed us to recruit AAH, who would like to thank them for giving him this opportunity. A special word of thanks must also go to all our friends and collaborators, spread over all five continents, to whom we humbly present the results of our work, with the hope that this book may help them in theirs.

Last but certainly not least, we would like to express our gratitude to our wives, Christine Bartle, Barbara Herod and Denise Kandiyoti, for their patience and support during the writing of this book and for putting up with us over the years.

Rafael Kandiyoti and Alan Herod, London
Keith Bartle, Leeds

CHAPTER 1

Coal and Biomass: The Study of Solid Fuels and their Utilization

This book is concerned with the thermochemical breakdown and associated thermal reactions of solid fuels. It is intended to provide an overview of the behavior of different coals and types of biomass during pyrolysis, gasification and liquefaction. Considerable space has been devoted to examining how changes during these reactions can be accurately quantified and how these measurements can be improved. The development of methods used for analyzing and characterizing gaseous and liquid products from these reactions has been described. Similar methods are applicable to petroleum derived heavy fractions. Special attention has been paid to bring together aspects of sample characterization, reactor design and the evaluation of product distributions from these reactions.

This conceptual integration is relevant to the study of solid fuels and their thermochemical reactions in a very practical sense. Put simply, most products formed during thermal breakdown are themselves reactive. Any interactions between reacting solids and reactive volatiles and their precursors have profound effects on the eventual product distribution. Thus, reactor design parameters, including local heating rates, flow rates, residence times and the design of quench zones, all affect the outcome of the experiment. The designer's task is, first, to identify the effect of each parameter on the outcome of the experiment and, then, to isolate its manipulation from that of the others. As in any other field of scientific measurement, the overall aim is to decouple the results of the measurement from the design of the experiment. We will see in Chapters 3–6 how attempts to distinguish between these two vital elements will lead us to a clearer picture of coal thermal breakdown than has been hitherto achieved. We will also show examples of how bench-scale experimental work is able to assist in the design and operation of pilot and plant scale equipment.

The second major strand running through this book is the development of analytical techniques for characterizing gaseous and liquid products from the thermochemical reactions of solid fuels. In this type of analytical work, two distinct issues loom large. The first is the complex nature of the product mixtures, with individual components only appearing in low concentrations. The second problem relates to the high molecular masses of many of the products. Many of these materials do not allow analysis using the powerful conventional methods available to the analytical chemist, such as gas chromatography coupled with mass spectroscopy, because they cannot pass through the chromatographic column. The final section of this book is devoted to describing efforts to extend the ranges of some of these powerful techniques, and to the development of alternative means of studying these large molecules. We will see that many of these techniques can be applied to the heavier petroleum derived fractions as well as solid fuel derived products.

The next few sections will present a few historical notes and quickly review the uses of coal and oil, alongside processes for converting these fossil fuels into useful products. Neither the purpose of the book nor limitations of space would allow more than a brief glance, and the reader is referred to well known works for a broader overview [Pitt & Millward, 1979; Elliott, 1981].

1.1 A Fuel for Producing Energy; a Carbon Source for Making Steel

There is still much that we do not know about this combustible sedimentary rock that has fueled so much of our modern way of life. For example, our understanding of the gelification process that transforms ancient plant debris into the vitrinite component of coals is surprisingly imprecise. Our grasp of how the solid coal matrix is held together is limited. However, mankind could not have been expected to wait until we had time to complete our studies. They had their relatively simple, dare we say, primitive needs. Coal was needed to burn as a source of heat and when technology evolved a little, for raising steam. Early on, it was also widely used as source of carbon to reduce ore and make iron and steel.

The ancients appear to have known about the properties of coal but there is not much evidence that they made extensive use of it. Perhaps they found the smell offensive. Freese [2003] has described the impact of coal, both in terms of the production of wealth in the Industrial Revolution that swept through Europe and America in the eighteenth and nineteenth centuries, and regarding its impact on the environment of industrial cities. There appear to have been banning orders against the use of coal for domestic heating and cooking as far back as the reign of Edward I (1272–1301), with the noxious fumes deemed unacceptable by the Parliament of the day [Elliott, 1981].

During several centuries coming up to the Industrial Revolution, wood and charcoal were much in evidence. The gun foundries of Henry VIII were mainly fired with charcoal from the South of England. The method is said to have caused much deforestation. It is no accident that Sweden, with its vast supply of timber, became the next great manufacturer of steel and of guns. However, the inventions and new technologies of the eighteenth century required progressively larger sources of energy. The new spinning mills of the textile trade would only come into their own when steam power was harnessed to their drive shafts. Coal was preferred to wood and charcoal, as it was denser. It gave more heat per unit weight as well as per unit volume of fuel, so it was cheaper to transport. Where it was plentiful and there was a need, it rapidly came to replace both waterpower and wood and charcoal and brought great prosperity to Europe and America. It also brought social costs associated with the now legendary poor working conditions of miners and factory workers. It brought massive environmental pollution. As late as the mid-1920s, air pollution in Manchester was being blamed for the failed salad crops of the working classes, whose window boxes did not receive enough sunlight because of the smog [Report, 1924].

Meanwhile, steam engines were being improved, powering trains and ships that made the world a smaller and more accessible place. By the time oil was discovered in the Pennsylvania of the late 1850s, coal had been king for nearly a century, occupying center stage in an industrializing world, not just as a fuel but also as a prime source of chemicals. Advances in the chemistry of aromatic compounds were driven by, and were leading to, the production of high value synthetic dyes, pharmaceuticals, militarily important trinitrotoluene (TNT) and a whole raft of products for everyday use. Later, twice in the twentieth century, chemicals and liquid fuels synthesized from coal allowed oil-starved Germany to wage war on its neighbors. Mostly, the source of these chemicals was coal tar, but the nineteenth century also saw the use of "illumination gas", first in palaces and in elegant households, then in city streets and finally in the common home.

Thus, well into the twentieth century, tar and gas were the sought after products of carbonization. Now virtually unrecognized, terms like "gasflammkohle" and "gaskohle" would conjure up some very desirable classes of coals. Gas fires for home heating may still be common, but the gas has now changed. In the lead-up to, and the aftermath of World War II, carbonization could simply not produce the quantities of feedstocks required to satisfy demand. Always hungry for cheap, abundant raw materials, the industrialized world turned to, and eventually became entirely dependent on, petroleum derived fuels and chemicals and eventually on natural gas.

In the competition between petroleum derived fuels and coal, the development of the internal combustion engine toward the end of the nineteenth century could probably be taken as a turning point. Up to that point, petroleum was being refined for lighting and heating fuels. "Oil" did not, however, greatly impinge on technologies geared to the use of coal. Nevertheless, petroleum extraction, transport and refining had already become big business before the car, the aircraft and the oil fired ship. J.D. Rockefeller, the founder of the Standard Oil Company became a very rich man, not by extracting petroleum from the ground but by establishing monopolies of the rail-transit trade that eventually led to the break-up of his empire into seven (giant) companies, nearly four decades later, in 1911.

However, the penetration of petroleum and its products was restricted by the geography of availability. After the United States, production started in a big way, first in Mexico and then in Venezuela. In Asia, Shell struck oil in Sumatra in the 1880s, whilst Baku and hapless Grozny were the early production areas in the Russian Empire. However, by 1914, production volumes in Burma, Iran and India had not caught up with the volumes coming from the Americas. Only Russia could claim a comparable share of world production whilst Europe, the industrial continent, had no oil to speak of. Coal, on the other hand could be mined on all continents and, at the time, in much greater quantities than oil. Nevertheless, by the 1920s, the industrialized world, both in and out of uniform, had become addicted to the internal combustion engine, for transport, for flying machines and not least, for fighting machines, the tank and the armored car. By then, oil companies were shifting massive quantities of crude oil from one end of the globe to the other. The relegation of coal to second place was already completed.

When it comes to considering the new rush to gas, however, a feeling of *déjà vu* seems inevitable. The urban environment has, in fact, been familiar with the use of gas for domestic illumination, heating and cooking for well over 150 years. The city of Baltimore, Maryland was the first conurbation where commercial gas lighting of residences, streets, and businesses started, as early as 1816. Soon, gas plants were being constructed by small groups of local entrepreneurs in increasing numbers of towns and cities. They mainly used standard gas works kits built in New York City and shipped west by all means available, from freight wagons and canals to rivers and rail. In the latter half of the nineteenth and early part of the twentieth century, gas works spread to towns and cities the world over. It has been estimated that in the United States alone, some 52,000 local gas plants were built over the period in most towns of 10,000 or more residents [Hatheway, http://hatheway.net/01_history.htm].

Usually fueled with coal, these gas plants were almost invariably heavily polluting installations. Some of the by-products of these processes, such as tars and effluent gases containing aromatic compounds are today considered as highly toxic. But, that is not why they eventually disappeared. In the USA, the Federal Government had invested heavily during the World War II to construct oil pipelines, to connect the Texas oil fields to the Eastern Seaboard. At the end of the war, these lines were bought by the natural gas industry and converted for gas-transmission. Coupled with new discoveries of large gas fields in Texas and Louisiana, it spelled the end of the market for manufactured gas. In Western and Northern Europe, a similar trend was repeated, with the tapping of North Sea gas and the arrival of piped Russian gas. In Europe and in the USA, the switch to natural gas was mostly completed by the 1960s and the early 1970s.

1.2 The Widening Use of Coal, Oil and Gas

Coal is rightly perceived as a source of pollution. In the industrialized world, whenever possible, it is replaced by either petroleum derived fuels or gas, nuclear power or renewable energy. However, during the past several decades, newly emerging industrial powers have, in their turn, relied on coal to power their development. Compared to oil and gas, coals have a wider natural distribution over the globe. Their increased use has provided some of the newly emerging economies with large savings in hard currency through local sourcing. Recent decades have thus witnessed a widening gap between the rush to oil and gas in the First World and increasing coal utilization in the Third. At its peak, coal consumption in China reached nearly 1.5 *billion* tpy during 1995–1996 [US DoE, 2002]. After an interlude when old, inefficient plants were scrapped, the new trend is again upward, using coal in modern power plants.

During the last stages of the Cold War, Western observers had greeted the Soviet Union's claim to surpass them in steel production with a touch of irony. *Reductions* in steel consumption have in fact become an index of post-industrial development. For a short spell, there was a similar recognition in the emerging worlds of China and India and large consumers like Poland that downward trends in coal consumption, through increased efficiency and diversification of energy production are now more desirable. However, oil prices have nearly tripled in three years and certainly doubled in less than two years.

The rush to coal by newly industrializing countries has always come at a heavy cost. Increasingly, the human and material costs of coal-derived pollution are being taken into account. "Cleaner coal" is well on its way to becoming a watchword. Sadly, many of the clean coal technologies, where available, are still expensive. Coal from deep mines is costly. In the medium-to-long term, oil and gas turn out to be cheaper, for economies that can afford them, but it is not clear where the present price surges will stabilize. In any case, even in the First World, there is still a hard backbone of coal consumption that is unlikely to disappear quickly. In the USA, the 1990s saw a 10% upward drift in consumption, which stabilized toward the end of the decade at a little over 1 billion tpy [US DoE, 2002]. This was, once again, mainly coal used for electric power generation and metallurgical coke production. The latter is a notoriously dirty and costly process, but blast-furnace technology is still today the preferred route for making iron and steel; thus, iron-makers have retained the use of coke.

Overall, coal consumption has been increasing despite its perception as a major source of global temperature rise through the emissions of the "greenhouse gas" carbon dioxide. World coal production was about 3.8 billion tonnes in 2001, rising to 4.04 billion tonnes in 2003 [World Coal Institute] and approximately 4.33 billion tonnes (or 2732 million tonnes oil equivalent) in 2004 [BP, 2005]. The production (and consumption) in 2004 was divided amongst the various regions as follows: North America 22.2% (21.7%), South and Central America 1.6% (0.7%), Europe and Eurasia 15.9% (19.3%), Middle East <0.05% (0.3%), Africa 5.1% (3.7%) and Asia Pacific 55.1% (54.2%).

Meanwhile, oil production is also expanding. Production in 2004 was a little over 3,800 million tonnes per annum or 80.26 million barrels per day [BP, 2005]. These numbers include shale oil, oil sands and natural gas liquids as well as crude oil. The production (and consumption) was divided amongst the regions in 2004 as follows: North America 17.3% (29.8%), South and Central America 8.8% (5.9%), Europe and Eurasia 22.0% (24.4%), Middle East 30.7% (6.7%), Africa 11.4% (3.3%) and Asia Pacific 9.8% (28.9%). Observers have suggested that oil will probably run out in the next 30 years or so. In fact, doomsday scenarios about depletion have been common since the end of the nineteenth century. Whilst there is no evidence that we will run out soon, there is plenty of evidence showing that oil exploration is increasingly turning to difficult terrains, from the Alaska "North Slope" to Sakhalin and the permafrost of Yakutsk in North Eastern Siberia. Furthermore, if, as it seems likely, the present surge in prices is maintained, producers will move toward processing progressively heavier crudes and exploiting increasingly unorthodox (and more expensive) sources of hydrocarbons. There are large reserves of tar sands and bitumens in both North and South America. Their present modest levels of exploitation, currently running at approximately 200,000 barrels per day for the Alberta tar sands, would be expected to increase considerably.

The third major type of fossil fuel, natural gas, is also being produced in increasing quantities. In 2004 production was at 2692 billion cubic meters or 2422 million tonnes oil equivalent [BP, 2005]. The production (and consumption) for the same regions in 2004 was as follows: North America 28.3% (29.3%), South and Central America 4.8% (4.4%), Europe and Eurasia 39.1% (41.2%), Middle East 10.4% (9.0%), Africa 5.4% (2.6%) and Asia Pacific 12.0% (13.7%).

It may be noted that these apparently unstoppable increases in fossil fuel utilization have taken place alongside raging debates about climate change, CO_2 capture and sequestration and the amazing notion of using hydrogen as a primary fuel. Furthermore, these increases do not just involve the newly industrializing countries. OECD countries have contributed to these developments at very significant levels [IEA, 2005]. In Chapter 9, we will take a closer look at matters relating to global natural resource utilization and the environment.

1.3 Processing Coals and Oils

It would be brave to attempt to cover this very wide field in the space available. However, a few pointers may be useful for those new to the field. Historically, coal pyrolysis was used in England, first to replace charcoal and then to make coke, using various types of ovens. *Coke* was widely produced in British coalfields in the seventeenth century and used in iron working by blacksmiths. Abraham Darby of Coalbrookdale in Shropshire was successful in using coke for iron smelting in the eighteenth century. During the year 1790–1791 alone, 81 blast furnaces using coke were built in the UK [Pitt and Millward, 1979]. The introduction of indirectly heated slot-ovens in the nineteenth century allowed the collection of liquid by-products from the coking of coal and led to the development of the modern coke oven. For blast furnace use, high-strength coke was necessary to withstand the pressure of the blast-furnace burden of coke, limestone and iron ore. The coke also maintained the porosity of the bed and allowed gas (laden with CO) to rise through the mixture and reduce the ore to metallic iron [Elliott, 1981]. In its essentials, this process is still the chief method of smelting iron ore. Since the reserves of prime coking coals are largely depleted, much effort goes into blending coals to produce cokes of sufficient strength for blast furnace use. The injection of pulverized coal into blast furnaces has been used as a method of reducing the quantity of expensive coke combusted and to supply heat to the blast furnace.

Historically, one major application of coal pyrolysis-gasification was the production of town gas for domestic heating and cooking. The requirement was for a suitable gas-producing coal rather than the production of strong coke. The process was in use over a period of 150 years in the UK following the first manufacture of gas from coal in the late eighteenth century. The method used air blowing to produce CO, N_2 and H_2 as well as steam blowing to produce H_2, CO and CH_4. The resulting gas that was piped to homes was highly toxic, leading to endless accidents and suicides. The weak cokes produced from this process were known as gas cokes and sold as domestic solid fuel to be combusted in covered stoves [Dainton, 1979].

Coal is still a valuable source of premium carbon products. A number of high-value carbon products originate from coal, in addition to the continuing manufacture of metallurgical coke and by-product coal-tar pitch, which finds a major application as an electrode binder. These products include graphitic carbon, activated carbons and carbon molecular sieve, carbon fibres and composites and carbons for use as catalyst support materials. This range may soon be expanded to include new products such as nanotubes [Menendez *et al.*, 2000].

Coal tars were the source of many classes of important chemicals before the age of petroleum and the increasing sophistication of petroleum refineries shifted the balance. Coal pyrolysis produces organic and aqueous liquids. The aqueous layer tends to be rich in phenolics whereas the tar contains aromatic materials. Low temperature tars (~500°C) contain much alkyl aromatic material as well as alkanes. High temperature (>1000°C) coal tar from coke ovens is distilled to produce light oils, creosote and anthracene oil. The distillation residue, involatile at 450°C, is called pitch [Owen 1979a, Elliott 1981]. Substituent alkyl groups are lost during the distillation as light alkanes and alkyl substitution found in pitches is limited mainly to methyl groups. Significant quantities of naphthalene, mainly for phthalic anhydride manufacture (~100,000 tons per year), and anthracene, the source of anthraquinone, are still produced from coal tars. Coal tars also yield smaller, but commercially important amounts of other hydrocarbons (e.g. acenaphthene, biphenyl, 1-methylnaphthalene, fluorene, phenanthrene, fluoranthene and pyrene) as well as phenols, for plastics manufacture, and nitrogen bearing compounds such as quinoline and isoquinoline [Menendez *et al.*, 2000].

The gasification of coal is an old art and an old science. Coal gasification in classical moving burden reactors is still widely practised, not least in South Africa, where SASOL's batteries of Lurgi gasifiers consume 40–50 million tons per annum. The gas produced in these huge reactors is converted by the Fischer-Tropsch synthesis route, to make transport fuels [Owen, 1979b]. This was vital during the Apartheid era, when the sale of oil products to that country was banned. At the time of writing, the gasifiers are being used to produce raw materials for plastics, chemicals and diesel fuel on a large scale. Most modern gasifiers, however, tend to be entrained bed reactors, where oxygen injection keeps temperatures well above 1,750–2,000°C and residence times are typically less than 1-second. In Europe, a slurry-fed Shell gasifier forms the core of the integrated gasification combined cycle (IGCC) plant at Buggenum (The Netherlands) while a dry-feed Koppers–Totzek reactor has been installed at the IGCC in Puerto Llano in Spain.

Coal liquefaction extracts have been produced by dissolving coal in a solvent, such as anthracene oil, supercritical toluene, or model compounds such as tetralin. Extraction at ambient temperature with strong polar solvents (1-methyl-2-pyrrolidinone) can dissolve large quantities (~60%) of selected coals. Coal liquefaction has been seen as likely to produce liquid transport fuels in the future, when the oil reserves have been used up. Several direct liquefaction routes, such as the British Coal Point of Ayr liquefaction process have been investigated (Chapter 5). However, only the indirect Fisher–Tropsch method used by SASOL in South Africa remains in commercial scale production [Owen, 1979b].

Present day petroleum refining is too well known to attempt a single paragraph description. Crude oil from the well may be mixed with water and sand, which must be removed before further treatment. This is usually done at or near wellheads in gas–oil separation plants ("GOSPs"). Refinery operations, however, are mostly carried out nearer the centers of consumption. The crude is distilled. In many refineries, the involatile fraction is distilled under vacuum at temperatures involving partial thermal cracking. However, the vacuum residue presents problems. It may be carbonized to produce hydrogen, light hydrocarbons and "petroleum coke". For the other streams, process routes in modern refineries are intricate and rather sophisticated. Middle distillates are thermally or catalytically cracked to boost automobile gasoline yields. Reforming, platforming, alkylation and bitumen blowing are performed to serve diverse needs including making fuel additives and chemical feedstocks. Dewaxing may be necessary to avoid precipitation of paraffin waxes. Extraction of aromatics may be carried out because they are either valuable commodities or detrimental to the performance of other products such as lubricating oils, and sulfur removal is generally necessary to meet environmental requirements [IP Information Service 1987].

1.4 Outline of the Book

The structure of coal: The celebrated coal chemist Van Krevelen is known for his insistence on studying coal structure, in order to learn more about the "composition of coal". His several books on the subject [Van Krevelen, 1993] have brought a measure of order into this widely dispersed field. The structure of coal is, nevertheless, poorly understood. As things are, once the "proximate" and elemental ("ultimate") analyses of coals have been completed [*e.g.* cf. Speight, 1994] the researcher is on his/her own. We are not unlike so many blind men examining the proverbial elephant. Each analytical technique offers a distinct perspective. None is able to tell the whole story.

Industrial coal consumers have developed measurements directly relevant to their particular needs. Coke-makers determine the swelling and agglomeration properties of coals and coal blends. Vitrinite reflectance measurements and scanning electron microscopy of polished test samples are routinely used in composing feedstock blends. Operators of coal burning power stations have entirely different perspectives. Flame stability, coal reactivity, slagging and fouling are some of their main concerns. Pf-grade coal injection and combustion is also relevant to blast furnace operation, where "injectant" coals are used to generate part of the heat required in the blast furnace; this reduces the direct combustion of some of the expensive coke. A number of standard – and several novel – tests give good estimates of volatile release and char reactivity, to help select coal feedstocks. The chemistry of slagging and fouling of coal ash is a science and an art on its own, where high temperature viscometry, x-ray diffraction, infrared spectroscopy and scanning electron microscopy serve as some of the primary tools.

While the power and coke-making industries remain the major coal-consumers of the present, there have been other pretenders at other times. The much-invoked hike in oil prices of the 1970s stimulated enormous research. The drive to uncover chemical routes to liquid fuels from coal re-awakened interest in some of the processes left over from WWII and others developed since 1945. During the 1970s and 1980s, solvent extraction, hydro-pyrolysis and gasification *re*-entered the vocabularies of many research laboratories. These new needs were addressed by quantum leaps in analytical sophistication. Every new technique was enthusiastically turned to examining "the structure of coal" and its products. Somewhat less frequently, these tools have been deployed to study processes taking place *during* the thermal and other transformations of coals. Throughout this book, we will attempt to shift the focus a little toward thermal transformations and to enquire whether the course of these transformations may tell us something more of the nature and structure of coals. Risking the reuse of the well-worn metaphor, we will attempt to improve our knowledge by disturbing the elephant and observing how it runs. It is bound to be a bumpy ride.

Two omissions: The reader will note that relatively little attention has been paid in this book to two important aspects of fuel science. Combustion related aspects have been left to one side, in order to concentrate on aspects more directly relevant to fuel characterization and thermal breakdown. The other major omission concerns the kinetic modeling of processes involved in pyrolysis, gasification and liquefaction. It seems clear that such models come into their own when existing knowledge about the processes involved is relatively sound. One such model has been developed for coal liquefaction (Chapter 5). It distinguishes between dissolution and mass transfer related processes taking place at low temperatures from those involving covalent bond scission, leading to the dissolution of much of the coal mass at higher temperatures. The model also takes account of mass loss during heatup to peak temperature. The assumptions underlying the model were based on observations from the novel "flowing-solvent" reactor, described in the same Chapter. The model was shown to be successful and has been described, albeit briefly. Another mathematical model constructed along similar lines, delineating pre-pyrolysis phenomena, tar generation and subsequent gas evolution has not proved successful in characterizing pyrolytic reactions (unpublished) and the reader has been spared the details. A semi-empirical correlation used in predicting pyrolysis volatile yields from their FT-infrared spectra has been described in Chapter 4. While the predictive power of the method surpassed expectations, a similar approach used for estimating gasification conversions was not nearly as successful.

Chapter 2 presents a brief review of the origins of fossil fuels and methods for their characterization. Much of what we know about this subject is generally available in the literature. It appears more productive, therefore, to point out several well-regarded works on the formation of coals and of petroleum, and on relationships between their chemical structures, morphologies and pathways of geochemical maturation. These shortcuts will allow space to evaluate methods for examining the thermal behavior of coals in detail. The brief overview on the origins of fossil fuels presented in Chapter 2 might nevertheless assist the general reader. We will briefly trace the evolution of coal and oil formation and discuss the link between coal and kerogen macerals through terrestrial and marine organic debris. We will also signal several points of debate and outline structural aspects that seem relevant to a discussion of the thermal reactions of coals.

Chapters 3–6 will focus on experimental reactors, developed for determining the thermochemical behavior of solid fuels. Results from these experiments will be carefully evaluated. As in any other area of study, the quality of data from these experiments depends largely on experiment design. One major strand running through this book will be a critique of laboratory-based techniques, developed for characterizing the thermochemical reactions of solid fuels and how these designs relate to the actual process conditions they are meant to simulate. The second major strand running through these Chapters will be a focus on the pathways of thermal breakdown. We will examine elementary processes common to pyrolysis and liquefaction and contrast outcomes of experiments, with close attention to the influence of reactor design.

Chapter 3 will thus be devoted to a review of experimental coal and biomass pyrolysis work. We will study relationships between prevailing reaction conditions and final product distributions. Clearly, the course of the initial pyrolytic step has a profound influence on the outcome of subsequent transformations during the coking, liquefaction, gasification or combustion of coals. It will also be necessary to address the greater difficulty of designing pyrolysis experiments, by attempting to free experimental results from effects relating to the designs of the experiments themselves. We will attempt to indicate reactor configurations appropriate to improving the focus and accuracy of measured product distributions and discuss their effect on resulting product quality. Several case studies will be discussed.

Chapter 4 will focus on high-pressure pyrolysis, hydropyrolysis and hydrogasification experiments, as well reactivity determinations in CO_2 and steam–air environments. Coal gasification reactivity measurements have, in the past, been largely based on thermogravimetric balances and differential thermal analyzers. However, most modern gasifiers are high heating rate, short residence time fluidized-bed or entrained flow reactors. Their requirements have substantially changed the face of solid fuel characterization. The bench-scale fuel-characterization experiments, described in Chapter 4, aim to place on a firm foundation the conditions that must be fulfilled to match these rapid processing requirements. Reactors described in this Chapter have been used to generate data to support the design and/or operation of several types of pilot plant equipment; some of these examples will be presented as case studies.

Most bench-scale liquefaction experiments are performed in batch reactors, where extracts released from the sample remain within the reaction zone until the termination of the experiment. Opportunities thus exist for products and reactants to mingle and react freely. It is difficult from the reaction mixture that is eventually recovered to deduce the reactions of the original sample. Chapter 5 will describe the configuration of and results from a liquefaction reactor conceived for decoupling the outcome of a coal liquefaction experiment from the design of the apparatus. Information gleaned from comparing results from this "flowing-solvent" reactor and a conventional batch reactor was useful in reviewing the stages of the thermal breakdown process. In Chapter 6, an attempt was made to unify observations on the thermal breakdown mechanisms of coals by juxtaposing results from pyrolysis and liquefaction experiments. The data was reviewed with the aid of results from earlier electron spin resonance experiments, to show both the differences and the significant similarities between pyrolysis and liquefaction related processes. Reference will be made to the way conditions during thermal breakdown affect the course of subsequent coking, gasification and combustion related processes.

Chapters 7 and 8 will focus on aspects relating to the analytical characterization of the gaseous and liquid products derived from the reactions or fractionation of coals, biomass materials and petroleum-derived liquids. As already described, these are complex mixtures where no individual compound is found present in significant abundance.

Because of the craft-like development of the various industries using coal and oil, many processes are not always completely understood in terms of the chemical structures of the raw materials and the changes involved in forming desirable products. There is considerable scope for achieving a better understanding of these chemical changes and possibly bringing about improvements in processing efficiency. Early analytical work in the nineteenth and early twentieth centuries on the composition of coal pyrolysis and extraction products depended on the classical procedures of distillation and crystallization to isolate and purify hundreds of single compounds. However, in the second half of the twentieth century, the development of chromatographic and spectroscopic techniques had a major impact. An early review [Bartle, 1972] has described many approaches that have since been developed. The use of Fourier transform methods in nuclear magnetic resonance and infrared spectroscopy has considerably expanded our knowledge of the chemical nature of fossil fuels and their derivatives. There has been important progress from columns for gas (capillaries) and liquid (small particles) chromatography for the characterization of smaller molecules, particularly when coupled to mass spectrometry. For the very high molecular weight constituents, modern advances in mass spectrometry (ionization methods, greatly expanded mass range) and size-exclusion chromatography have been especially valuable.

In the first of the Chapters on the analytical chemistry of fuel derived liquids, we will review established techniques as well as recent method development, intended for investigating the relatively smaller molecular mass materials, in the mass range below about 500 u. The methods surveyed include gas-, liquid- and supercritical fluid chromatography and combinations of these methods, often coupled with mass spectrometry for molecular identification. Direct mass spectrometric analyses, using different sample introduction techniques, have been considered in some detail. These methods mostly require the evaporation of the sample into the mass-spectrometer vacuum system prior to ionization.

However, not all fossil fuel and biomass derived materials can be evaporated (even) under vacuum or, for that matter, pass through a chromatographic column. There are no exact methods for determining the molecular mass distributions or for precisely identifying the structural features of these larger molecular mass materials. The techniques available for characterizing these complicated materials only allow their own particular, specialized view of the samples involved. Not unlike components of the composite eye of a spider or lizard, these measurements give possibly precise but always limited information. We are no longer able to talk of "analyzing" samples, but of "characterizing" them. Chapter 8 will present a discussion of methods developed for assessing the molecular mass distributions of coal, biomass and petroleum derived liquids and methods available for their structural characterization. These include size exclusion chromatography (SEC) and several mass spectrometric methods where ionization is thought to occur *before* analyte molecules desorb/evaporate from the sample matrix. Fractionation methods and results from bulk characterization techniques such as NMR and FT-ir spectroscopy will be described. The work is relevant to situations ranging from the modeling of volatile combustion rates in pf-combustors, to the evaluation and upgrading of heavy tars and petroleum residues.

We will attempt to pull together strands emerging from these discussions in a short concluding Chapter 9. Where possible we will place matters in the context of work performed and ideas developed in the wider scientific community. Inevitably, however, we will rely heavily on work done in our own laboratories at Imperial College and at the University of Leeds. This is because the conceptual framework of this book reflects the ideas that have guided our own experiments spanning a little over two decades. That framework has in turn been shaped and (dare we make the claim?) developed by experimental observations made in our laboratories and elsewhere.

References

Bartle, K.D. (1972) *Rev. Pure and Appl. Chem.* 22, 79.

BP (2005) Statistical Review of World Energy, June 2005 [www.bp.com/energy].

Dainton, A.D. (1979) Chapter 7 in Pitt, G.J. and Millward, G.R., *Coal and Modern Coal Processing: an Introduction*. Academic Press, London.

Elliott, M.A. (ed) (1981) *Chemistry of Coal Utilization*, 2nd supplementary volume. John Wiley & Sons, NY.

Freese, B. (2003) *Coal, a human history*, Perseus Publishing, Cambridge MA, USA.

Hatheway, A.W. *History and Chronology of Manufactured Gas*, FMGP, http://hatheway.net/01_history.htm.

IEA (2005) *Energy Statistics of OECD Countries 2002–2003*, OECD/IEA, Paris.

IP Information Service (1987), Institute of Petroleum, 61 New Cavendish Street, London W1M 8AR, UK.

Menendez, R., Bermejo, J., Figueiras, A., (2000) Chapter 5 in H. Marsh and F. Rodriguez-Reinoso eds *Science of Carbon Materials*, Universidad de Alicante, Spain.

Owen J. (1979a) Chapter 9 in Pitt, G.J. and Millward, G.R., *Coal and Modern Coal Processing: an Introduction*, Academic Press, London.

Owen J. (1979b) Chapter 8 in Pitt, G.J. and Millward, G.R., *Coal and Modern Coal Processing: an Introduction*, Academic Press, London.

Pitt, G.J. and Millward, G.R., (1979) *Coal and Modern Coal Processing: an Introduction*, Academic Press, London.

Report (1924), Report of the Smoke Abatement Conference November 4–6, 1924, Town Hall, Manchester, UK, p. 147.

Speight, J.G. (1994) *The Chemistry and Technology of Coal* (2nd Ed.) Marcel Dekker, New York.

US DoE (2002) *International Energy Annual 2002*, http://www.eia.doe.gov/emeu/iea/table14.html.

van Krevelen, D.W. (1993) *Coal* (3rd Edition), Elsevier, Amsterdam, London, New York.

World Coal Institute, Cambridge House, 180 Upper Richmond Road, Putney, London SW15 2SH, UK, http://www.worldcoal.org/.

CHAPTER 2

Fossil Fuels: Origins and Characterization Methods

2.1 Precursors and Formation of Fossil Hydrocarbons

Literature on the organic geochemistry of coals is quite wide ranging and the limitation of space would not allow an exhaustive survey. Instead, in this Chapter, we will point out several well-regarded works on the formation of coals and their macerals and on relationships between their chemical structures, morphologies and pathways of geochemical maturation. First mention would go to Stach's Textbook of Coal Petrology [Stach *et al.*, 1982] and the two Supplementary Volumes of "The Chemistry of Coal Utilisation" [Lowry, 1963; Elliott, 1981]. "Coal" by van Krevelen [1993] reviews available methods for examining the composition of coals. We will also use several ideas from Speight [1994]. Space permitting, numerous other authors, too many to name, would have deserved honorable mentions. As a personal favorite, however, the prize must go to the memorable, book-length article by the late P.H. Given [1984]. These shortcuts will allow space to evaluate methods for examining the thermal behavior of coals in greater detail. The brief overview on the origins of coals in Chapter 2 might nevertheless assist the general reader in gaining a preliminary insight. We will signal several points of debate and outline structural aspects that seem relevant to the discussion of the thermal reactions of coal, as developed in later Chapters.

2.1.1 Coal

The structure of coal is poorly understood. Generally it is considered to consist of derivatives of lignin, cellulose, tannins and seeds from plants, transformed by changes of temperature and pressure in the earth's crust over geological times [see also Chapter 2 in Francis, 1961; Mukhopadhyay, 1994]. Coal is also considered to be a sedimentary rock, formed, in the main, from plant debris, which was biotically converted to peat and then subjected to metamorphic geological changes during burial. The widely accepted view is that typical Northern Hemisphere coals were formed from peats deposited in swamps and marshes, under predominantly anaerobic and at least partly reducing conditions. After deposition, the plant material is covered by sediment or more plant-derived matter. Speight [1994] has outlined arguments favoring accumulations of *transported* material, as distinct from *in-situ* deposition. Within either model, some aquatic cover is thought necessary to preserve organic deposits from total destruction by oxidation and aerobic

Table 2.1 Elemental composition of coals[a]

Coal rank	*Carbon*	*Hydrogen*	*Oxygen*
	% w/w. dry ash free		
(Peat)	58	6	35
Lignite	71	5	23
Sub-bituminous	75	5	16
Bituminous			
high-volatile	81	6	10
low-volatile	88	4	4
Anthracite	94	3	2

[a] Representative values, excluding nitrogen and sulfur content, which show little rank dependence. After Kershaw, 1989.

bacterial action. With subsidence leading to compaction, the pressures on the deposits increase and temperatures rise with increasing depths of burial.

The burial conditions of the peat (temperature, pressure and time period under different conditions) have a marked influence on the degree of metamorphic change, which we call coalification. The severity of these conditions determines the so-called 'rank' (see Section 2.2.3) of the resulting coal.

Peat → lignite, or brown coal → sub-bituminous coal → bituminous coal → anthracite

Broadly, changes in elemental composition as a function of coalification follow the pattern shown in Table 2.1. The trends are dominated by oxygen loss and increasing aromatization during maturation and eventual coalification.

2.1.2 Fossil hydrocarbons

We may expand the definitions given above to account for the origin of a broader range of fossil hydrocarbons by considering different precursors and their molecular constituents. All biomass contains carbohydrates, consisting principally of cellulose and hemicelluloses, lignins, lipids and amino acids, along with more specialized constituents, such as chlorophyll which may be partially degraded to 'biomarkers', which are sometimes retained in the final fossil hydrocarbon mixture. These compounds are subject to anaerobic microbial degradation in, usually, aquatic domains. They mix with inorganic sediments and produce kerogen – a solid high molecular mass polymeric material defined as 'a disseminated minor organic component of inorganic sediments' [Hutton, 1995a]. Even the richest kerogen-containing sediments seldom contain >18% organic material. Kerogens are defined as organic solids insoluble in common solvents and presumed to consist of large cross-linked molecular systems. They are designated as Type I, II or III, depending on atomic H/C and O/C ratios. These parameters in turn depend on the biological origins and the geothermal history of the sediments.

Table 2.2 A simplified classification of crude oils[a] (Concentrations in wt %)

Class	*Paraffinic*	*Naphthenic*	*Aromatic*	*Wax*	*Asphaltic*
Paraffinic	46–61	22–32	12–25	1–10	0–6
Paraffinic/naphthenic	42–45	38–39	16–20	1–51	0–6
Naphthenic/paraffinic	15–26	61–76	8–13	Trace	0–6
Naphthenic/aromatic	27–35	36–47	26–33	0.05–1	0–10
Aromatic	0–8	57–78	20–25	0–0.5	0–20

[a] After Speight [1989]

The process by which lipids, proteins, carbohydrates and lignins break down to form kerogens and bitumens is known as diagenesis. Kerogens of Types I and II may subsequently be cracked and transformed by the effects of pressure and moderate temperatures (50–200°C) into crude oils and so-called sapropels. Some Type II kerogens mature to give lighter oils and natural gas. In organic geochemistry, this phase is termed catagenesis. Natural gas contains C_1–C_8 alkanes as well as variable amounts of CO_2. Crude oils are classified according to their content of paraffinic, naphthenic and aromatic hydrocarbons (Table 2.2).

A working definition of an oil shale is "a sedimentary rock containing organic matter, which when pyrolysed in a retort produces sufficient oil to generate more energy than was required to produce the oil". These are usually rocks containing more than 5 volume % of oil-producing matter [Hutton, 1995b]. Oil shale contains minor amounts of soluble bitumen, but mostly incorporates insoluble kerogen material. Most oil shales are derived from lipid-rich material in lacustrine and marine environments, rather than from terrestrial, woody plants. The catagenetic transformation of larger deposits of terrestrial plants and Type III kerogens gives rise to humic coals. Table 2.3 summarizes the stages of development for several types of biomass, leading to the formation of different types of fossil fuel deposits [Berkowitz, 1997].

Sapropels are coals formed from algae, fungal spores, pollen and marine biota in shallow marine and lake environments; they differ from humic coals particularly in their higher content of hydrogen. Humic coals, which comprise the vast majority of coals, originate from mainly terrestrial plant debris that was initially converted to peat. A detailed discussion of plant remains as fossils in coals has been given by Francis [1961]. Such coals sometimes contain fossilized but well-preserved botanical remains including leaves, woody structures and pollen [Given, 1988]. The type of original plant precursors govern the maceral composition of coals (see Section 2.2.2, below) and the degree of catagenetic change leads to the observed variation in coal rank (see Section 2.2.3).

Typical early Northern Hemisphere deposits belong to the Carboniferous period, ~350–270 million years before the present [Stach *et al.*, 1982; Given, 1984]. The initial stages of the maturation of organic deposits appear to involve bacterial and fungal activity, triggering, among other processes, the depolymerization of large biopolymers and leading to overall reductions in molecular mass (diagenesis). "The rate of decay is also greatly influenced by the inherent resistivity to bacterial attack of the various plant

Table 2.3 Development of fossil hydrocarbons from precursor biomass

Biomass material	*Algal matter*	*Plankton and simple biota*	*Cells from land plants*	*Fibres, woody & root structures*	*Reworked organic matter (charred oxidized, degraded)*
Material source	Marine, lacustrine	Marine lacustrine	Continental	Peat swamps	
Kerogen type produced by diagenesis	'Algal'	'Amorphous'	'Herbaceous'	'Woody'	'Coaly'
Kerogen classification	I, II	I, II	II	III	III
Product of catagenesis	Oils (from kerogen I) Sapropels (from kerogen II)		Sapropels	Coals	

parts, tissues, organs, and secretory products (*e.g.* resins)." [Parks, 1963]. The decay of nitrogen-containing macromolecules in the bio- and geo-sphere has been considered to contribute some nitrogen compounds to the kerogen or maceral types [Stankiewicz and van Bergen, 1998]. Nitrogen preserved in the geosphere can in some cases, be related to the biological macromolecular precursors, such as proteins, amino sugars (such as chitin) and nucleic acids.

Where magma intrusions raise deposit temperatures above otherwise expected levels, maturation is locally accelerated. The Sterling area of Scotland offers deposits where different extents of maturation may be studied in a vertical sequence. In the USA, fixed carbon contents from 59 to 85% have been found in the large Lower Kittanning seam, spread over four different States. Differences observed in the properties of coals from this seam appear to reflect primarily the conditions and extents of coalification and only to a lesser extent, the differences in original accumulated biomass [Stadnichenko, 1934]. Pressure effects due to rock folding, faulting and fracturing, normally associated with mountain building activity, are also associated with the occurrence of high rank coals, as in the South Wales (UK) coalfields [Parks, 1963]. Admixture with soil components and the adventitious ingress of inorganic ions lead to mineral matter accumulation. In marine environments, sulfates tend to increase organic and mineral sulfur contents in coals [Given, 1984].

'Bituminization' is the name given to the chemical/physical transition thought to delineate 'bituminous' coals from lignitic deposits. Much of the terminology describing these transitions appears based on microscopy, while detailed chemical sequences of coal maturation seem, at best, incomplete. Broadly, bituminization, vitrinization and gelification refer to the same catagenetic step-change thought to take place during the maturation process. Stach [1982] describes an "abrupt" transition from sub-bituminous to high volatile bituminous coals. ". . . Petroleum type hydrocarbons are formed at this stage from components of waxy leaf cuticles, pollen and spore coatings" during which oxygen groups are removed, giving rise to "a first coalification jump in the liptinites and vitrinites, which corresponds with crude oil formation in petroleum source rocks." Opinions regarding the role of coals as possible source rocks for petroleum-like liquid deposits appear divided. To migrate, as from a petroleum source rock, the newly formed bitumen would need to swamp the adsorptive capacity of "the very fine sub-microscopic pore system of the vitrinite (which) functions as a molecular sieve." It has been argued that ". . . petroleum like substances in coal are assimilated adsorptively and probably also chemically by the vitrinite with only a small part being deposited as exsudatinite in empty cavities." [Stach, 1982] Limited evidence pointing to some oil and gas deposits with origins in coal bearing sequences will be outlined below.

The progression in the degree of coalification, termed the 'rank' of a coal, correlates well with increasing organic carbon contents and with the ability of the vitrinite component to reflect light under the microscope. The proportion of oxygen (as high as 40–45% in live plant material) is reduced with maturation. Following the 'bituminization' stage, the proportion of aromatic carbon and the degree of cross-linking in bituminous coals increase with rank, thus eventually reversing the depolymerization process associated with the maturation of biomass. Hydrogen contents show less pronounced but definite downward trends in middle rank coals, usually above 85–86% elemental carbon content (dmmf basis). "On further coalification, the . . . bitumen is cracked into hydrocarbons of smaller molecular size on the one hand and into strongly reflecting, polycondensed, residual products on the other. This decomposition begins in the medium-volatile bituminous coal stage (*ca.* 28–29% VM) at the second coalification jump and while the decomposition continues, the reflectivity of the liptinites and vitrinites increases very rapidly and fluorescence is lost" [Stach, 1982].

2.2 Coal Petrography, Macerals and Rank

'Macerals' are the organic components of coals, defined in terms of their morphologies, as observed by optical microscopy. Broadly, these morphological features correspond to the structures of original plant material deposited in peat bogs. With maturation, the properties of the different biomasses tend to converge toward a narrower spectrum. Each principal maceral group (vitrinites, liptinites and inertinites) subsumes arrays of macerals classified in sub-groups. The level of possible detail is comparable with the variety and state of fossilized plant material assimilated into the original deposit [Stach, 1982; van Krevelen, 1993]. The *classification* of maceral sub-groups in coals has animated almost as much debate as the microscopic identification of these fractions.

During deposition, decay reactions are predominantly anoxic if the material is covered by water and vitrinite-rich coals are produced. Total exposure to air would lead to extinction by oxidation and bacterial action but limited exposure to oxygen is thought to give rise to inertinite-rich coals. In general, vitrinites show evidence of woody and possibly also of root tissue as starting material. In Northern Hemisphere coals, vitrinites usually represent a large proportion of the coal mass (between ~60–90%). When heated, the vitrinite component softens ('melts'). Good coking coals also swell, to impart the desired agglomerating properties to the overall mass. Liptinites usually make up less than 20% of the coal mass. They contain the more aliphatic fossilized components, chiefly sporinite (spores and pollen), cutinite (cuticles, waxes), resinite (fossil resins) and the highly paraffinic alginite [Stach, 1982; van Krevelen, 1993]. Liptinites also contain relatively higher proportions of elemental-H (up to ~8%, dmmf) and, compared to other macerals, release the largest proportions of volatile matter during pyrolysis. Most liptinites readily melt on heating but do not normally swell.

Inertinites are so-named because, mostly, they do not soften or swell during coking. They usually contain higher proportions of elemental-C and lower proportions of elemental-H compared to vitrinites and liptinites. Inertinites *usually* make up less than 20% of the coal mass in many Northern Hemisphere coals. A little surprisingly, some good coking coals (*e.g.* Cortonwood Silkstone, UK) contain nearly 40% inertinites. Within this maceral group, fusinites and semi-fusinites also show fossilized woody structure but contain less hydrogen and more carbon than vitrinites. It is often suggested that these fractions represent charred woody tissue from ancient forest fires, although opinions often diverge on such speculative matters [*e.g.* cf. Parks, 1963]. Of the three maceral groups, inertinites typically release the lowest proportion of volatiles during pyrolysis and, often, release less extract during liquefaction. However, when pyrolyzed at high temperatures (~1500°C), residual *inertinite* chars turn out to be more reactive to gasification and to combustion, compared to chars from the other two maceral groups [Cai *et al.*, 1998].

Conveniently, the microscopically identifiable morphological features of macerals correlate reasonably well with trends in measured chemical properties (*e.g.* elemental composition; aliphatic/aromatic content) and observed pyrolytic behavior. For macerals of comparable maturity, infrared and NMR spectroscopy show a progression in aromatic carbon content ('aromaticity'), from low values for liptinites (~0.4) to higher values for vitrinites and highest for inertinites (up to ~0.9).

Specific gravities increase in a similar sequence from about 1.15 to about 1.45. Although individual macerals may be separated from ground coals by hand picking, less laborious procedures make use of the difference in the specific gravities of the maceral groups: liptinite 1.20–1.25; vitrinite, 1.30–1.35; and inertinite 1.40–1.45. Float-sink methods employ liquids, with densities in the range 1.2–1.5 g mL^{-1}, which may be organic solvents (*e.g.* tetrachloromethane/toluene mixtures) or aqueous salt solutions (*e.g.* zinc chloride or bromide or caesium chloride). Centrifugation is used to make the separation more rapid, and continuous centrifugation methods have been developed.

A considerable break-through was the development by Dyrkacz and Horwitz [1982] of a density-gradient centrifugation based method, adopted from medical practice. The method helped to separate finely ground (~3 μm) particles of demineralized coals into narrow-density range fractions in an aqueous caesium chloride gradient. A non-ionic surfactant was added to disperse the particles, which sometimes caused interference. Nevertheless, density differences allowed the separation of finely divided coal particles into their respective macerals [*e.g.* see Pandolfo *et al.*, 1988]. This technique allows maceral concentrates to be separated on the 20 g scale [Dyrkacz *et al.*, 1992]. Fractions separated in this way generally contain at least 90% by volume of a single maceral species although the efficiency of separation varies from coal to coal and with particle size. Recent improvements to the separation method include cryogenic treatment – immersing the sample in liquid nitrogen and then rapidly thawing to fracture particles along maceral boundaries [Crelling & Bensley, 1995]. For samples of higher rank, say, above ~87% C, microscopic and other physical differences diminish and distinct macerals become more difficult to detect and separations more difficult to achieve. The reader will find a wealth of information relating to the formation, morphology and geochemistry of coal macerals in the classic "Stach's Textbook" [Stach *et al.*, 1982] and in Given [1984].

2.2.1 Coal petrography

Coal macerals have different colors and fluorescence intensities and can be identified under the microscope. Coal petrography is the study of the organic constituents of sedimentary rocks, originally, by optical microscopy of thin sections (transmitted light) or of polished blocks (reflected light) [Unsworth *et al.*, 1991]. Fluorescence microscopy is used to complement the more traditional approaches. The latter technique is based on the ability of organic matter to absorb UV light and then emit (fluoresce) at longer wavelengths in the visible spectrum. All three major maceral groups may be recognized by this technique. Vitrinites show up as light grey or grey and fluoresce weakly, while liptinites show up as dark grey and fluoresce more intensely. They also reflect light less intensely than vitrinites. Inertinites show up as light grey or white. They reflect light more strongly than vitrinites but fluoresce less intensely than other macerals. Figure 2.1

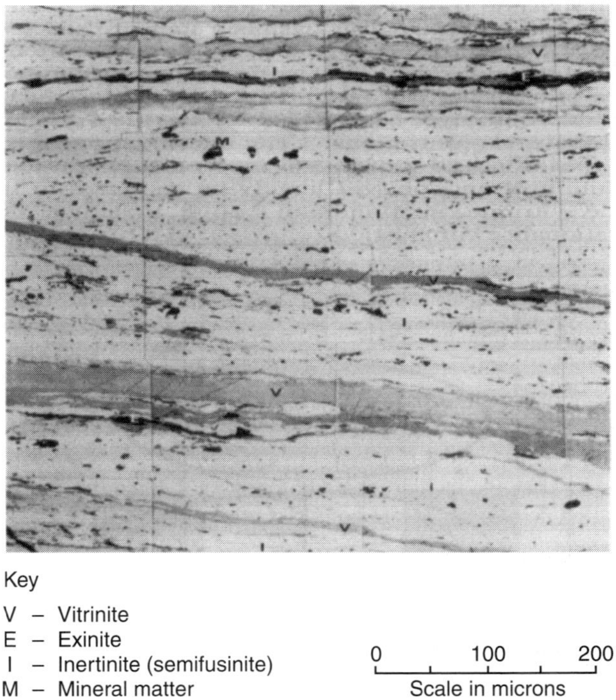

Figure 2.1 Maceral (V, vitrinite; E, exinite; and I, inertinite) and mineral matter (M) bands and particles in a bituminous coal. From Unsworth *et al.*, 1991. Copyright Elsevier; reproduced with permission.

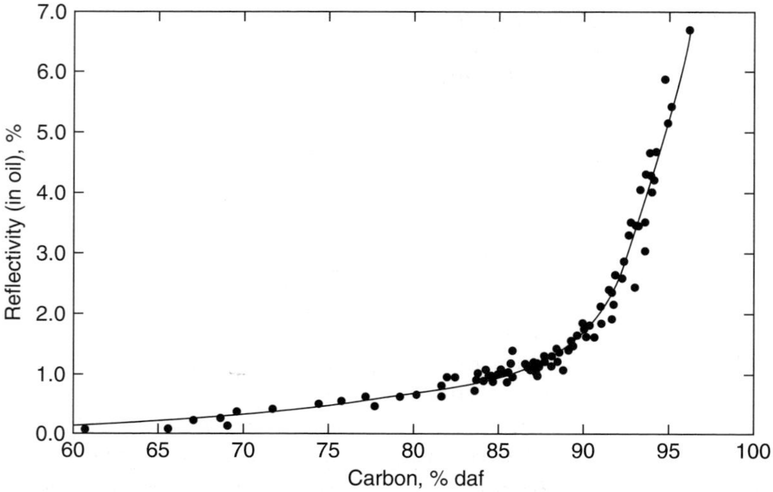

Figure 2.2 Dependence of vitrinite reflectance on coal rank (% carbon). From Berkowitz, 1997; reproduced with permission, copyright Academic Press.

shows a photograph of macerals and mineral matter bands and particles in a bituminous coal. The reflectance of individual coal macerals also increases with coal rank.

Figure 2.2 shows the relation between reflectance and the carbon contents of coals. A petrographic examination of a coal would normally involve first the optical microscopy of a polished surface with reflected white light to allow identification of vitrinite and inertinite group macerals, and by fluorescence to identify the liptinite group. This is usually followed by the measurement of the vitrinite reflectance, as the accepted indicator of coal rank. Automated procedures are available for the third major procedure, the determination of the volumetric abundance of the macerals [Hutton, 1995a].

2.2.2 *Coal macerals*

Not surprisingly, the maceral distribution in a coal is heterogeneous, with individual domains varying from 1 μm to 1 mm (Figure 2.1). Maceral compositions reflect the botanical origins of coals and the environmental conditions during their deposition. These are in turn, reflected in their morphologies and behavior during their thermochemical reactions [Given & Dyrkacz, 1988].

Table 2.4 presents a list of the more frequently occurring maceral types, alongside their plant precursors. Density gradient centrifugation has allowed good separations of the main maceral groups, as well as individual maceral types including cutinite, resinite, sporinite, fusinite and semifusinite [Crelling & Bensley, 1995]. Reeds from marshes may give rise to light brown coals with only small phyterals – recognizable fragments of plants – and much vitrodetrinite. On the other hand, forests produce dark brown coals containing coalified trees. Telinite (a vitrinitic maceral) from woody material is observed to predominant in this environment. Resinite-rich coals arise from plant material containing large amounts

Table 2.4 Classification of coal macerals[a]

Material group	*Maceral*	*Precursor*
Vitrinite[b] (humite in brown coals)	Telnite	Cell Walls
	Collinite	Wood, bark and cortical tissue
	Vitrodetrinite	Vitrinite fragments
Liptinite (Formerly exinite)	Alginite	Algal remains
	Cutinite	Cuticle
	Resinite	Resins, oils, fats and waxes
	Sporinite	Spores, pollen
	Suberinite	Walls of cork tissue
	Bituminite	Altered algal and humic materials
	Exsudatinite	Secondary 'resinite'
	Fluorinite	Fluorescing secondary 'resinite'
	Liptodetrinite	Fragments of other 'liptinite'
Inertinite	Fusinite	Well-preserved 'carbonized' woody tissue
	Semi-fusinite	Cellular 'carbonized' woody tissue
	Macrinite	Unspecified detrital matter (10–100 μm)
	Micrinite	Unspecified detrital matter (<10 μm)
	Sclerotinite	Fungal tissue and spores
	Inertodetrinite	Fragments less than one cell

[a] After Hutton, 1995a.
[b] Macerals of coals originating from the Gondwanaland supercontinent are classified differently

of resins, oils and waxes, while sporinite-rich coals are derived from spores and pollen. For any given rank of coal, the elemental composition of maceral groups would be expected to change as follows:

Carbon content:	inertinite > liptinite ≈ vitrinite
Hydrogen content:	liptinite > vitrinite > inertinite
Oxygen content:	vitrinite > inertinite > liptinite
Sulfur content:	liptinite > vitrinite > inertinite

H/C ratios increase in the order inertinite < vitrinite < liptinite. The aromaticity of maceral groups, f_a, represents the fraction of total carbon that is aromatic. This is usually measured by solid-state ^{13}C nuclear magnetic resonance and follows the reverse sequence as H/C ratios, increasing from liptinites to vitrinites and to inertinites [Botto, 1987]. Information concerning values of f_a within a maceral group is scarcer, but among liptinites, resinites and alginites have much lower aromaticities than sporinites. Within the inertinite group, fusinite has a higher f_a value than macrinite. While maceral aromaticity remains the most commonly reported structural parameter, further peak discrimination suggests (Kasueschke *et al.*, 1989) that the fraction of non-protonated aromatic carbon increases in the order liptinite < vitrinite < inertinite, and the fraction of protonated aromatic carbon remains approximately constant; aliphatic CH and CH_2 carbon decrease in the order liptinite > vitrinite > inertinite, but CH_3 in aliphatic structures is unchanged between macerals. FTIR data for maceral concentrates from French coals have been interpreted (Unsworth *et al.*, 1991) to make detailed assessments of the relative importance of a wide variety of structural elements in the three maceral groups. Consistent with trends in H/C ratio and aromaticity for the principal maceral groups, volatile matter release during pyrolysis normally decreases in the sequence liptinite > vitrinite > inertinite.

Southern Hemisphere Coals: In the Southern Hemisphere, large masses of coals are thought to have common depositional origins in the 'lost' continent of Gondwana. This was the immense landmass, once constituted from parts of southern Africa, India, eastern Australia, Madagascar, South America and Antarctica. The original plant material is thought to have been deposited in the Permian (275–220 million years before present) from remains of stunted broad leafed flora of the *Glossopteridae* group and some broad-leafed plants similar to those found today in sub-arctic zones. The greater thickness of coal seams than normally found in the Northern Hemisphere is ascribed to slower subsidence. Deposition of these coals is held to have taken place in sub-arctic conditions, whereas much of the Northern Hemisphere Carboniferous coals are thought to have been deposited in warmer climates, during ages when the sites were nearer the Equator [Stach *et al.*, 1982; van Krevelen, 1993]. The petrographic compositions of Gondwana coals vary more widely than Carboniferous coals of the Northern Hemisphere. Vitrinite contents rarely exceed 80% and occurrences of less than 50% have been encountered. Contents of liptinite group macerals are seldom greater than several percent, whilst inertinite concentrations can be unusually high. By contrast, Northern Hemisphere Carboniferous coals usually contain more than 70% vitrinites and less than 20% inertinites and 5–10% liptinites are not unusual.

The concentrations of semi-fusinites in Gondwana coals have been found to change in parallel with two other inertinite group macerals (macrinite and inertodetrinite). This suggests a different route for their formation than forest fires, as has been proposed for the fusinites and semi-fusinites of Northern Hemisphere coals. Instead, the high inertinite contents of Gondwana coals have been explained in terms of relatively dry conditions during coalification, with greater extents of peat oxidation. These semi-fusinites are thought to have formed via the alteration of vitrinites by bacterial and fungal action under mildly oxidizing conditions, rather than charring during vigorous oxidation [van Krevelen, 1993]. Interestingly, some of these 'semi-fusinites' are reported to be more reactive during coking and liquefaction, compared with macerals of similar morphology (i.e. semi-fusinites) in Northern Hemisphere coals. We will return to a discussion of the technological properties of Southern Hemisphere coals in Chapter 3 (Section 3.8).

2.2.3 Coal rank

Coalification of precursors is brought about by the influence of pressure, temperature and time during burial. The rank of a coal indicates its degree of maturity, i.e. the extent to which metamorphic transformations have taken place. Increasing depth of burial and volcanic or folding disturbances within the strata bring about increasing temperature and pressure leading to accelerated maturation of the deposits, and to increasing rank.

A number of classification systems for the determination of coal rank were originally devised on the basis of technologically important parameters relevant to the utilization of coals: volatile matter (or 'fixed carbon') content; calorific value; and behavior during pyrolysis (coke type, swelling or agglomeration properties). The "National Coal Board" (United Kingdom) classification system is based on volatile matter content and coke type and specifies numerical codes for coal rank. Its usefulness is restricted, however, by its applicability to UK Carboniferous period coals only. Other, more widely applicable, coal rank classification systems are the ASTM (USA) and the United Nations/European Economic Community (UN/ECE) systems, both based on volatile matter (or 'fixed carbon') and calorific value. The latter system also specifies numerical codes. A corresponding system, which extends the UN/ECE system, has also been proposed for Southern Hemisphere coals.

Some of these properties could be determined from the elemental composition, but this, and the other properties above are average properties, all depending not only on coal rank, but also on maceral composition. It is necessary, therefore, to determine rank from a single rank-related property that is measured for a simple maceral group in the coal. Optical reflectance is related to the coal carbon content (see Figure 2.2), and hence to the coal aromaticity (see Section 2.2.4). The percent random reflectance (R_0) of the vitrinite domains in a polished coal sample, determined by an automated procedure is, therefore, a suitable estimate of coal rank. A petrographic system based on vitrinite reflectance appropriate to vitrinite rich coals is shown in Table 2.5, along with the equivalent classes in the ASTM and international (UN/ECE) systems.

Table 2.5 A petrographic system for describing coal rank[a]

Coal class	*% vitrinite reflectance*	*% carbon content of vitrinite*[b]	*Equivalent classes*	
			ASTM	*UN/ECE*
Lignite	≤0.40	<75	Lignite A Lignite B	12–15
Sub-bituminous	0.41–0.50	75–80	Sub-bituminous A, B, C	10–11
Low-rank bituminous	0.51–1.00	80–85	High-volatile Bituminous A, B, C	6–9
Medium-rank bituminous	1.01–1.50	85–89	Medium-volatile Bituminous	4–5
High-rank bituminous	1.51–2.00	89–91	Low-volatile Bituminous	3
Semi-anthracite	2.01–2.50	91–93	Semi-anthracite	2
Anthracite	>2.50	>93	Anthracite	0–1

[a] After Unsworth *et al.*, 1991
[b] Approximate. Dry, mineral matter free basis

2.2.4 Variation of coal properties with rank

Many of the coal chemical properties change with rank. The trend of diminishing oxygen contents with increasing maturity is clear from Table 2.1. Another trend, which has one important bearing on the variation of coal reactivity with rank is the increase in aromaticity ($f_a = C_{ar}/C$), the fraction of the total carbon in aromatic rings [Miknis, 1995]. Quantitative NMR of carbonaceous solids is made difficult by the low abundance (1.1%) of the magnetic ^{13}C isotope, and the consequent necessity of transferring magnetization from the abundant ^1H nucleus. The coal sample is spun at the 'magic' angle to the magnetic field to eliminate dipolar band broadening, and dipolar dephasing is employed to improve spectral resolution. Values of f_a determined in this way clearly increase with rank (Figure 2.3) but there is considerable scatter in the data. The reliability of f_a values determined by NMR has been the subject of a 'debate in print' [Snape *et al.*, 1989]. It was concluded that measured aromaticities may be underestimating the aromatic content by up to 15%, due to systematic discrimination against aromatic carbon signal. The best 'recipe' for reliable f_a values was to work at low fields with low spinning speeds to avoid problems with side-bands. Radical quenching with *e.g.* SmI_2 and single pulse excitation work were found to lead to more promising results [Love *et al.*, 1993].

Carbon functionality in coals can also be determined by Fourier-transform infra red (FTIR) spectroscopy [Fredericks, 1989], although spectral de-convolution methods are generally necessary. Absorptions attributed to aromatic C–H bending are particularly useful as indicators of changes in aromatic substitution patterns with rank. FTIR has also been used to estimate CH_2 in long aliphatic chains, and heteroatom functionality such as hydroxyl, ether and carbonyl groups.

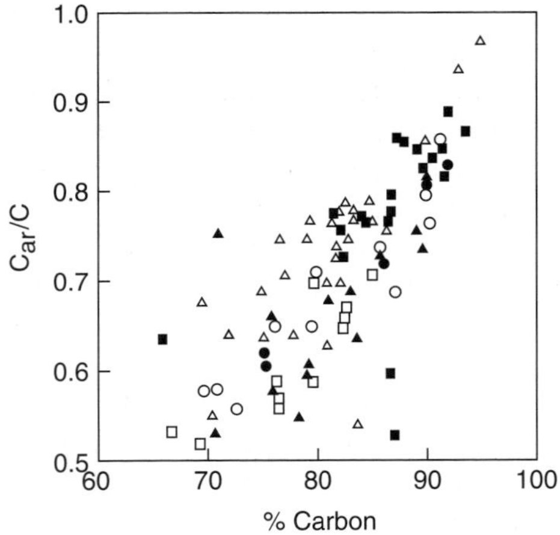

Figure 2.3 Variation of aromatic carbon with coal rank (% carbon). From Sfihi *et al.*, 1986. Copyright Elsevier, reproduced with permission.

2.3 The Chemical Composition of Fossil Hydrocarbons

2.3.1 Petroleum (crude oils)

Crude oils are complex mixtures containing paraffinic, naphthenic and aromatic compounds (Table 2.2). Lighter (smaller mass) material in crude oils may be distilled and otherwise refined to make transport fuels, heating oils, waxes and other refinery products such as liquefied petroleum gas (LPG). The less well characterized, higher molecular mass components designated as resins and asphaltenes make up most of the residue from distillation, which needs to be upgraded. The heaviest materials are often sent to delayed cokers to recover some hydrogen and light components before discarding as petroleum coke.

In Chapter 7, we will discuss the various analytical methods used for analyzing the smaller molecular mass materials. Normal-alkanes, iso-alkanes, and cyclo-alkanes and the mono-, di-, and poly-aromatics may be separated and analyzed by chromatographic means, with high-performance liquid chromatography and the various high-resolution gas chromatographic (GC) methods. Pre-eminent among these, GC with mass spectrometric detection is especially valuable in affording separation and identification. The presence of numerous homologous series makes comprehensive two-dimensional GC (GC $\times$ GC) especially valuable, since these form recognizable patterns in the 2D chromatogram. The latter method promises important advances in applications to lighter petroleum fractions.

The non-distillable fractions of oil cannot, of course, be separated or analyzed by GC, and have been studied by spectroscopic methods, including NMR, FT-ir. Much attention has focused on the so-called asphaltenes, and there is an extensive, if sometimes contradictory literature on the chemical nature of these materials (cf. Chapter 8). Based on such studies, a range of 'average' structures has been proposed [for a summary, see Bartle, 1988] in which large molecules are composed of alkyl and hetero-atom substituted aromatic and naphthenic structures linked together by chemical bonds as well as by hydrogen-bonding and aromatic pi-pi interactions. Generalizations concerning these structures are rendered difficult, since asphaltenes are an operationally defined *solubility class*: soluble in an aromatic hydrocarbon solvent, originally benzene and now generally toluene, but insoluble in an alkane solvent, usually *n*-pentane or heptane. This method of fractionation is not structure specific. The compounds with these properties have no particular chemical structure. Their solubility properties are conferred by a combination of: 1) the presence of polar groups, which take part in hydrogen bonding in alkanes, but not in aromatic solvents; 2) molecular mass (itself extremely difficult to determine for high MW fossil hydrocarbon materials – see Chapter 8; and 3) the degree of condensation of aromatic fused-ring structures, leading to pi-pi interactions and poor solubility in alkanes [Snape & Bartle, 1984].

The approach proposed by Snape & Bartle [1984] defines a wide range of asphaltenes from diverse sources, by classifying them along three dimensions. It proposes a specific volume in a three-dimensional graph (Figure 2.4) of percent acidic hydroxyl, to represent polar groups, number average molecular mass (M_n) and the fraction of carbon atoms that are in internal bridgehead aromatic positions (C_{INT}), to represent the degree of aromatic condensation. Alternatively, a solubility parameter

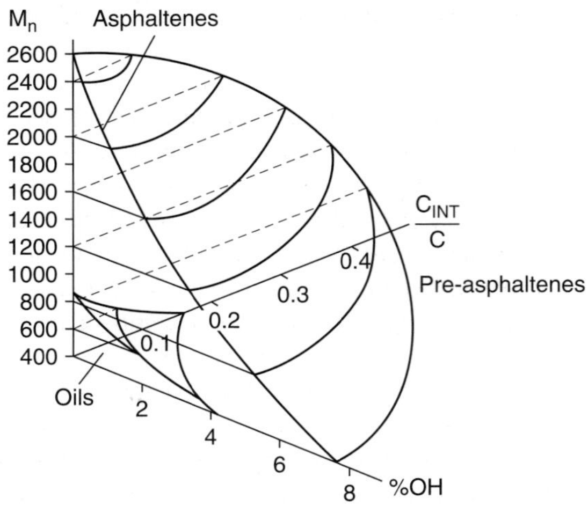

Figure 2.4 Three-dimensional graph defining solubility in pentane and benzene. Asphaltenes are defined by the volume between the boundary surfaces. From Snape & Bartle, 1984. Copyright Elsevier, reproduced with permission.

combining these factors could be used to define asphaltenes; for oil-derived asphaltenes, the dominant term is the number average molecular mass, M_n (see Chapter 8 for details).

Further (operational) solubility classes for the higher MM constituents of oil include 'pre-asphaltenes', which are insoluble in aromatic hydrocarbon solvents, but soluble in tetrahydrofuran. Asphaltenes have also been separated on the basis of solubility in carbon disulfide into 'carbenes' (soluble) and 'carboids' (insoluble) [Berkowitz, 1997].

2.3.2 *Bituminous materials*

Naturally occurring bituminous substances, such as the increasingly economically important Canadian tar sands [Berkowitz, 1997], have a highly variable composition related to that of the oil, from which they are thought to have originated by evaporation, microbiological alteration, and water washing during migration. The chief differences brought about by these influences are the very substantial depletion of lower MM alkanes ($<C_{15}$) and aromatics and the resulting increases in the proportions of asphaltenes and other high MM material. High concentrations of oxygen, nitrogen and sulfur may also be present.

2.3.3 *Oil shales*

Oil shales contain minor proportions of solvent-extractable bitumen containing geochemical markers (see Section 2.3.6) and other lower molecular mass compounds similar to those present in the heavier distillate fractions of oil and bituminous materials (Section 2.3.2), along with some macromolecular constituents [Tissot &

Vandenbroucke, 1983]. The principal organic constituent of oil shale is, however, the non-extractable kerogen. All of these are dispersed in an inorganic matrix of mineral matter precipitated in the sedimentary basin and introduced from outside, and mainly comprised of carbonates, silicates and sulfides. Traditional methods of removing these are by treatment with, respectively, dilute acid, hydrofluoric acid, and oxidation/reduction; aqueous BF_3 from HF/HBO_3 is a more modern and effective demineralization reagent which changes the kerogen little [Robl & Taulbee, 1995].

The chemical nature of the kerogen has been inferred from a combination of: a) extrapolation from the nature of the products of pyrolysis, particularly the retorted shale oil and also products identified during pyrolysis GC-MS; b) identification of the products of chemical degradation *e.g.* oxidation; and c) more direct spectroscopic analysis by solid state NMR and by IR methods. Structural models of oil-shale kerogen based on information derived by the above methods suggest predominantly aliphatic structures (long alkyl chains, some branched) joining naphthenic and (small) aromatic ring systems substituted with small alkyl and heteroatom-containing groups. Marked variations occur between the distributions of these constituent structures between different oil shale kerogens [Siskin *et al.*, 1995].

2.3.4 *Coal*

The chemistry and geochemistry of coals has been largely studied by methods similar to those mentioned in relation to petroleum fractions and shales. In this connection, much attention has been paid to the structures and molecular masses of solvent extracts. At room temperature, most coals are soluble to a limited extent in traditional solvents (Table 2.6). Recent work has shown extractions in 1-methyl-2-pyrrolidinone (NMP)

Table 2.6 Extraction yields from the Argonne premium coal sample program (APCSP)[a]. [Vorres, 1990]

Coal	*% C daf*	*% Pyridine extraction*	*% of coal extract amenable to study by GC*	M_n[b]	M_w[b]
Beulah-Zap	72.9	3.1	21.9	500	1360
Wyodak	75.0	7.4	13.1	460	1480
Illinois No 6	77.7	27.5	6.4	640	1680
Blind Canyon	80.7	24.4	19.3	ND	ND
Lewiston-Stockton	82.6	15.3	8.7	ND	ND
Pittsburgh No 8	83.2	29.8	11.0	670	2040
Upper Freeport	85.5	6.9	24.2	500	1300
Pocahontas	91.1	0.8	41.1	330	700

[a] From Chang *et al.*, 1992
[b] THF soluble fraction, determined by size exclusion chromatography
ND not determined

removing as much as 60% of some coals to be dissolved in solution at room temperature [Iino *et al.*, 1988]. Following heating to 300°C in NMP with lithium halides, up to 92% of a low rank coal could be extracted in the mixed solvent NMP-CS_2 [Li *et al.*, 2003].

When coals are extracted in a mixture of CS_2 and NMP and the extracts characterized by size exclusion chromatography using NMP as eluent, the extracts tend to show bimodal distributions. Evidence presented in Chapter 8 of this book suggests that the early eluting material may be of large molecular size, possibly showing three-dimensional structures. Such structures have not previously been considered in coal or petroleum derivatives. Similar excluded material has been observed in extracts of biomass, biomass pyrolysis tars, petroleum vacuum residues and asphaltenes.

Application of GC and GC-MS to the lower molecular mass constituents, comprising a minor fraction (see Table 2.6) of coal extracts has shown up the presence of normal and branched-chain alkanes alongside mainly alkyl substituted polycyclic aromatic hydrocarbons and their sulfur-, nitrogen- and oxygen- containing isosteres. Many of these compounds are related to the geochemical markers discussed in Section 2.3.6.

As in the case of oils and other fossil hydrocarbons, the higher molecular mass extractable coal constituents can be separated into solubility classes, which may then be studied by a battery of spectroscopic techniques. Detailed assignments of signals in ^1H and ^{13}C spectra of coal-extract fractions have allowed "average" structures to be deduced. However, these provide little information on the distribution of structures about the average. UV-fluorescence spectroscopy provides partial evidence that the more intractable fractions of coal liquids, biomass tars and petroleum residues contain increasingly more complex aromatic structures as the polarity of the solvents able to dissolve the fraction increases. However, the technique has its limitations. We will discuss in Chapter 8, how materials with molecular masses above about 3,000 u cannot be detected by UV-fluorescence spectroscopy, probably due to energy loss through intramolecular energy transfer.

Coal-extract asphaltenes generally present lower molecular masses than those derived from heavy petroleum derived fractions and shale oil. They normally contain more condensed-aromatic structures and more polar substituents (see Section 2.3.1) as well as show a greater proportion of material excluded in size exclusion chromatography (Chapter 8). Statistical structural analysis suggests the presence of linked small (1–4 ring) aromatic clusters in asphaltenes, with a variety of alkyl substituents [Bartle, 1988]. The actual presence of linked 2–4 ring aromatic structures has been confirmed by electrochemical analysis [Tytko *et al.*, 1987]. Again, the very high-MM constituents are less-well characterized, since they are insoluble in the more usual NMR solvents. Recent developments in the characterization of the high molecular weight fractions of coal liquids have been based on solubility in 1-methyl-2-pyrrolidinone (NMP) and will be outlined in Chapter 8.

Cross-linked macromolecular material comprises the major organic constituent of most coals, and has been investigated by a wide variety of methods, including oxidation, pyrolytic methods coupled GC and GC-MS analysis of products. Both NMR and infrared spectroscopy show the presence of small alkyl and naphthenic groups, in agreement with ruthenium VIII oxidation [Tse & Stock, 1983]. Products of oxidation by sodium dichromate [Hayatsu *et al.*, 1975] are 1–4 ring aromatic and heteroaromatic carboxylic acids (Figure 2.5), in agreement with findings from NMR. A comparison of products from different coals suggests greater contributions from larger ring systems

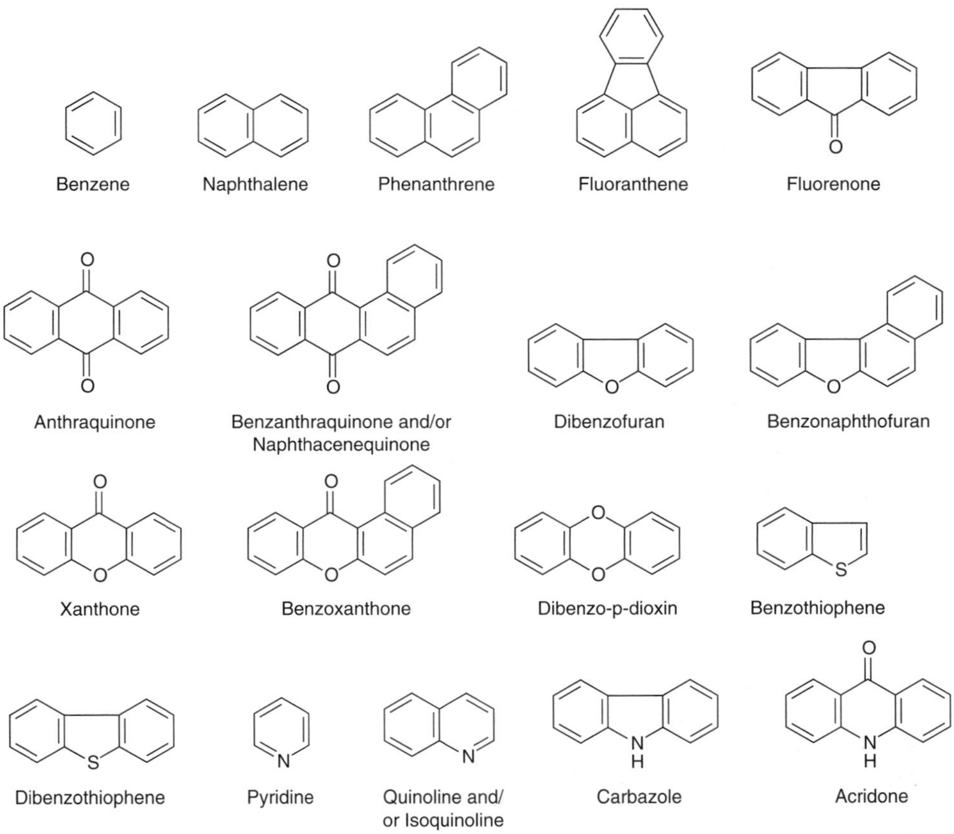

Figure 2.5 Aromatic structures produced by oxidation of coal with sodium dichromate. From Hayatsu *et al.*, 1975, reproduced with permission.

with increasing coal rank, consistent with solid-state NMR-derived aromaticities (section 2.2.4) and with findings from Py-GC-MS.

It is significant that these are just the aromatic clusters both identified directly in low-MM fractions of coal extracts and also indicated by statistical structural analysis and electrochemical analysis by differential-pulse voltammetry (Figure 2.6) as being linked together in higher-MM fractions. It may be concluded that the predominant organic matter present in coal is comprised of molecules of varying MM, but with similar basic structural types: alkyl and naphthenic-substituted small aromatic clusters. In these analyses, much of the larger molecular mass material remains undetected, however.

Many aspects of the reactivity of coals in thermal processes are thought to arise from the presence of hydroaromatic structures in the coal. As long ago as 1963, Ladner & Stacey [1963] inferred from broad-line ^{1}H NMR spectra the presence of hydroaromatics in exinites and vitrinites; however, their direct determination by modern ^{13}C NMR has proved difficult, although changes in NMR-determined average structural parameters of pyrolysis chars have been cited [Fletcher *et al.*, 1990] as evidence of the presence of

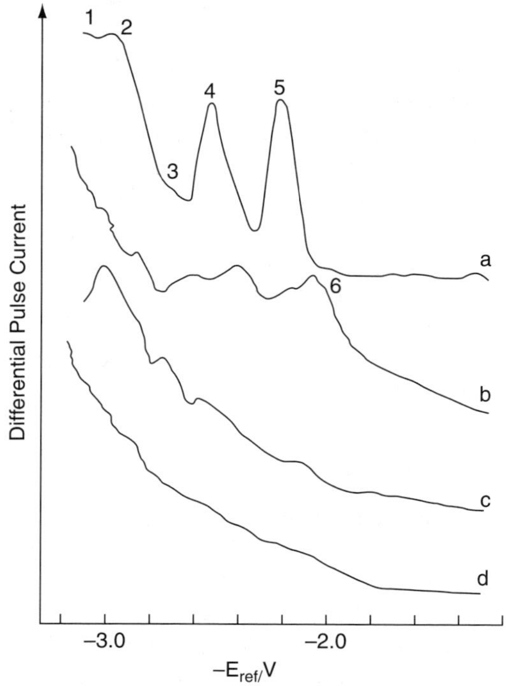

Figure 2.6 Differential-pulse voltammograms of asphaltenes from: (a), a hydrogen-donor solvent (HDS) extract of a UK bituminous coal; (b), a hydrocracking residue from an HDS extract of the same coal; (c), a supercritical-gas (SCG) extract of a UK bituminous coal; and (d), an SCG extract of a US lignite. Peaks assigned to the following structures: 1, two-ring aromatics (naphthalene, biphenyl etc.); 2, phenanthrene; 3, chrysene; 4, pyrene; 5, fluoranthene; and 6, more condensed aromatics. From Tytko *et al.*, 1987. Copyright Elsevier, reproduced with permission.

hydroaromatics in the original coals. The results of hydrogen-transfer reactions with *e.g.* 9-fluorenone (Choi & Stock, 1983) also show that such structures are responsible, at least in part, for the differing reactivities of coal macerals: exinite (especially sporinite and cutinite) > vitrinite > inertinite.

The heteroatom-containing functional groups in coal play a vital role in its structure and reactions. Oxygen, mainly present as phenolic hydroxyl and, especially in low-rank coals as carboxyl groups has been extensively characterized by wet chemistry and by IR spectroscopy. Ether groups, especially in furan rings in benzofuran and dibenzofuran structures, are significant contributors to oxidation products, and are also present in pyrolysis and liquefaction products. Nitrogen is present in (mainly pyridine and its benzologues) bases, which may be extracted by acids, but secondary, pyrrole, nitrogen is also present as shown by the indoles and carbazoles identified in pyrolysis products. X-ray photoelectron spectroscopy (XPS) allows ready and quantitative differentiation of pyridine and pyrrole nitrogen in solid coals and a clear rank dependence is evident in their relative concentrations [Wallace *et al.*, 1989].

A variety of techniques has been used to speciate sulfur in coal – vital information given the importance of this element in emissions and processing chemistry [Markuszewski & Wheelock, 1990; IEA Coal Research, 1989]. As well as pyrite, coal also contains sulfate and elemental sulfur – probably oxidation products of pyrite – but significant amounts of organic sulfur [Snape *et al.*, 1995]. The latter has been studied by XPS and by X-ray absorption near-edge spectroscopy (XANES); both techniques suggest that coals contain both aromatic bonded (*e.g.* in thiophenes and aryl sulfides) and aliphatic sulfides. Benzo-, dibenzo-, and naphthothiophenes are prominent in coal oxidation and pyrolysis products. Chemical analysis methods have also been applied to coals to determine sulfur functionality, in particular temperature-programed reduction [Mitchell *et al.*, 1994] and oxidation. These studies confirm that thiophene rings in 1–3 ring structures predominate, but that aliphatic sulfides are significantly present, especially in low-rank coals.

2.3.5 Oil and gas from coal beds?

Temperature and time are principal parameters in determining the rates and extents of maturation of coal and oil bearing deposits. "Each stage of petroleum maturation can be matched with a particular rank stage of coal". Thus "the importance of coal petrology in petroleum and natural-gas prospecting lies in the fact that the degree of diagenesis of both the source and reservoir rocks can be determined relatively quickly and accurately by reflectivity measurements on vitrinitic inclusions . . ." [Stach, 1982]. The occurrence of economically viable gas deposits originating from coal beds appears to be generally recognized (*e.g.* Slochteren, Netherlands) [Stach, 1982]. There are other North Sea *gas* fields associated with coal bearing strata. However, the extent to which bituminization/gelification may lead to the expulsion of *oil* from coal beds is a matter of debate. Introducing a collection of articles entitled "Coal and Coal-bearing Strata as Oil-prone Source Rocks?" Scott & Fleet [1994], have maintained that high-wax, low-sulfur oils may be "commonly" associated with coals and other terrigenous kerogens, although contributions from microbial biomass and non-marine aquatic matter cannot be excluded. While the direct recognition of *alginite* in coals is rare, it has been reported, mixed with other macerals in 'boghead' coals, and "consists largely of open-chain aliphatic structures, as would be predicted from its presumed origin" [Given, 1984]. Much of the fossilized alginites are thought to have been assimilated in vitrinites, particularly into desmocollinites, identifiable through the straight chain C_{17} component in pyrolysis tars [Stach, 1983; Given, 1984].

Resinites are chemically labile and may have given rise to significant oil generation in some coal bearing sequences. Pristane/phytane ratios > 4 would be taken to confirm the terrigenous origins of oil deposits. Both markers are thought to be decomposition products of chlorophyll. Deposits "generally acknowledged" to have given rise to significant oil accumulations (>50 m bbl recoverable) are found to be low latitude Tertiary deposits or, Jurassic-Palaeogene sequences in the Australian region. However, the possibility of the destruction by geological processes of oil from earlier coal sequences has not been ruled out. Oil-prone coal deposits are thought to be hydrogen-rich (H/C ratios > 0.9) with liptinite contents above 15%, although the recognition of oil-prone coals and associated mudrocks was said to be 'problematical'. "What controls the expulsion of petroleum in the liquid phase from coal-bearing sequences is probably the critical factor." Nevertheless ". . . oil-prone coal sequences are recognised as key oil source rocks in at least Southeast Asia, Australia and New Zealand" [Scott & Fleet, 1994].

It seems nevertheless difficult to find conclusive evidence. Biomarkers readily traceable to land based plant material may always have migrated to marine or lacustrine environments prior to assimilation into the deposits. Changes in isotope ratios provide another line of research, although the level of uncertainty in interpretation is said to be high. There appears to be doubt regarding the consistency of pre- and post-Miocene $\delta^{13}C$ values. At the level of oil exploration, "understanding why coal-bearing sequences are oil prone can unlock a predictive capability for petroleum exploration and so help to reduce exploration risk" [Scott & Fleet, 1994].

2.3.6 Fossil hydrocarbon biomarkers

Alkanes solvent-extracted from fossil hydrocarbons can provide much information concerning their source and original environments, and on the processes to which they were subjected during maturation and biodegradation. Such biomarkers, also referred to as molecular fossils, are molecules derived from the original organic compounds and subject to only minor alteration, or whose basic carbon skeleton remains intact. Thus, analyses of mild but high-yield (supercritical gas) extracts of a UK coal [Bartle *et al.*, 1979] suggested that the coal contained as much as 500 p.p.m. of the isoprenoidal hydrocarbons derived from the phytyl side chain of chlorophyll, accounting for more than 10% of the original. The topic has been reviewed [Philp, 1994]. The rapid formation of mature biomarker molecules in hydrothermal generation of petroleum has been described [Simoneit, 2003].

Murchison [1993] reviewed studies of biomarker molecules, arising from the development of high-resolution capillary gas chromatography columns with mass spectrometric detection. Molecular types constituting biomarkers include isoprenoid alkanes such as pristane and phytane, sesquiterpenoids, diterpenoids, tricyclic and tetracyclic terpenoids, pentacyclic terpenoids (of which the hopane structure is a well-known example giving the characteristic fragment ion at m/z 191 in mass spectra), and steroids (for which the fragment ion at m/z 217 is characteristic) [Philp, 1994]. Other common pentacyclic terpenoids not based on the hopane skeleton include oleananes, lupanes and oleanenes. The relative concentrations of different isomers of these structures are considered to reflect the maturity of the deposited material. These biomarker molecules have been detected in coals [Curry *et al.*, 1994; Bagge & Keeley, 1994]. The relevance of the biomarkers is that they are characteristic of material deposited from terrestrial or marine organic precursors and in some cases, can provide evidence, for instance if oleananes are present, of deposition from angiosperms (flowering plants) that only appeared in the Cretaceous period. Hopanes, steranes, pristane and phytane, as well as other biomarkers, have been found in very old coal seams of the Devonian period [Fowler *et al.*, 1991; Sheng *et al.*, 1992]. The analysis method based on GC-MS has led others to suggest the likely presence of a wealth of biomarker molecules beyond the range of GC-MS [Hedges *et al.*, 2000; Poirier *et al.*, 2000].

Developments in the study of nitrogen-containing macromolecules in the bio- and geo-sphere and their methods of decay [Stankiewicz & van Bergen, 1998] have provided indications of the ways in which nitrogen in these molecules may be preserved into the geosphere. Methods such as ^{13}C and ^{15}N NMR, and pyrolysis-GC-MS have been applied to biopolymer materials such as chitin and proteins in Quaternary and Tertiary deposits. Studies of humic and fulvic acids from Leonardite, a low rank lignitic coal, [IHSS, 2001] have shown the presence of small quantities of amino acids (neutral hydrophilic, neutral hydrophobic, acidic, basic and S-containing) and carbohydrates (glucose, galactose, mannose, rhamnose, arabinose). Porphyrins have been found in coal extracts associated with Ga and Fe, as described in Chapter 7. Although these studies of nitrogen-containing macromolecules may not yet be able to identify biomarker molecules within the large-mass molecules of coals and petroleum asphaltenes, the continual development of analytical methods might lead to such a conclusion. It seems advisable to watch developments in these fields.

References

Bagge, M.A. and Keeley, M.L. (1994) The oil potential of Mid-Jurassic coals in Northern Egypt. In Scott A.C. and Fleet A.J. (eds) *Coal and Coal-bearing strata as Oil-prone Source Rocks*? Geological Society Special Publication No 77, pp. 183–200.

Bartle, K.D., Jones, D.W., Pakdel, H., Snape, C.E., Calimli, A., Olcay, A. and Tugrul, T. (1979) *Nature*, 277, 287.

Bartle, K.D. (1988). In Yurum Y. (ed) *New Trends In Coal Science*. p.169. NATO ASI Series C. Vol. 244. Kluwer, Dordrecht.

Berkowitz, N. (1997). *Fossil Hydrocarbons: Chemistry and Technology*. Academic, San Diego.

Botto, R.E. (1987). *Energy and Fuels*, 1, 173.

Cai, H-Y., Megaritis, A., Messenbock, R., Dix, M., Dugwell, D.R. and Kandiyoti, R. (1998) *Fuel*, 77, 1273–1282

Chang, H.-C., Bartle, K.D., Markides, K.E. and Lee, M.L. (1992). In Meuzelaar L.C. ed, *Advances In Coal Spectroscopy*. p.141. Plenum, New York.

Choi, C.-Y. and Stock, L.M. (1984) in R.E. Winans and J.C.Crelling eds, *Chemistry and Characterization of Coal Macerals* p.157, ACS Symposium Series No. 252. American Chemical Society, Washington, D.C.

Crelling, J.C. and Bensley, D.F. (1995) in J.A.Pajares and J.M.D.Tascon eds, *Coal Science*, p. 235, Elsevier, Amsterdam.

Curry, D.J., Emmett, J.K. and Hunt, J.W. (1994) Geochemistry of aliphatic-rich coals in the Cooper Basin, Australia and Taranaki Basin, New Zealand: implications for the occurrence of potentially oil-generative coals. In Scott A.C. and Fleet A.J. eds, *Coal and Coal-bearing Strata as Oil-prone Source Rocks*? Geological Society Special Publication No 77, pp. 149–182.

Dyrkacz, G.R. and Horwitz, P. (1982). *Fuel*, 61, 3.

Dyrkacz, G.R., Bloomquist, C.A.A. and Ruscic, L. (1992). *Energy and Fuels*, 6, 357.

Fletcher, T.H., Solum, M.S., Grant, D.M., Critchfield, S., and Pugmire, R.J. (1990). Proc. 23rd Int. Symp. on Combustion, p.1231. The Combustion Institute, Pittsburgh PA.

Fowler, M.G., Goodzari, F., Gentzis, T. and Brooks, P.W. (1991) *Org. Geochem.* 17, 681–694.

Francis W. (1961) 'Coal, its formation and composition' 2nd Ed., Edward Arnold, London.

Fredericks, P.M. (1989). In Kershaw J.R. (ed) *Spectroscopic Analysis of Coal Liquids*. P. 129. Elsevier, Amsterdam.

Given, P. H. (1984) *Coal Science* (ed. Gorbaty, M. L., Larsen, J. W., I. Wender, I.), III, Academic Press, NY.

Given, P.H. (1988). In Yurum Y, (ed) *New Trends in Coal Science*. p. 1. NATO ASI Series C. Vol. 244. Kluwer, Dordrecht.

Given, P.H. and Dyrkacz, G.R. (1988). In Yurum Y. (ed) *New Trends in Coal Science*. p. 53. NATO ASI Series C. Vol. 244. Kluwer, Dordrecht.

Hayatsu, R., Scott, R.G., Moore, L.P. and Studier, M.H. (1975). *Nature*, 261, 77.

Hedges, J.I., Eglinton, G., Hatcher, P., Kirchman, D.L., Arnosti, C., Derenne, S., Evershed, R.P., Kögel, Knaber, I., de Leeuw, J.W., Littke, R., Michaelis, W. and Rullkötter, J. (2000) *Organic Geochemistry, 31*, 945.

Hutton, A.C. (1995a). In Snape C.E. (ed) *Composition, Geochemistry and Conversion of Oil Shales*. p.125. NATO ASI Series C Vol. 455. Kluwer, Dordrecht.

Hutton, A.C. (1995b). In Snape C.E. (ed) *Composition, Geochemistry and Conversion of Oil Shales*. p.17. NATO ASI Series C Vol. 455. Kluwer, Dordrecht.

IEA Coal Research, Reviews in Coal Science (1989) *The Problems of Sulfur*. Butterworths, London.

IHSS [2001], International Humic Substance Society, Data analysis sheets for humic and fulvic acid standards and reference materials, Department of Soil, Water and Climate, Univ. of Minnesota, St. Paul, MN, USA.

Iino, M., Takanohashi, T., Ohsuga, H. and Toda, K. (1988) *Fuel* 67, 1639.

Kasueschke, I., Riepe, W. and Gerhards, R. (1989) *Erdoel, Kohle, Erdgas, Petrochemie*, 42, 209.

Kershaw, J.R. (1989). In Kershaw J.R. (ed) *Spectroscopic Analysis of Coal Liquids*. p. 1. Elsevier, Amsterdam.

Ladner, W.R. and Stacey, A.E. (1963) *Fuel*, 42, 75.

Li, C., Takanohashi, T., Saito, I. and Iino, M. (2003) *Energy & Fuels* 17, 762–767.

Love, G.D., Law, R.V. and Snape, C.E. (1993). *Energy and Fuels*, 7, 639.

Lowry, H. H. (1963) Editor, *Chemistry of Coal Utilization Supplementary Volume*, Wiley, NY.

Markuszewski, R. and Wheelock, T.D. (eds) (1990) 'Processing and Utilisation of High-Sulfur Coals III' Coal Science and Technology Series vol.16, Elsevier, Amsterdam.

Miknis, F.P. (1995). In Snape C.E. (ed) *Composition, Geochemistry and Conversion of Oil Shales*. p. 69. NATO ASI Series C Vol. 455. Kluwer, Dordrecht.

Mitchell, S.C., Snape, C.E., Garcia, R., Ismail, K. and Bartle, K.D. (1994). *Fuel*, 73, 1159.

Mukhopadhyay, P.K.; (1994) Vitrinite reflectance as Maturity Parameter; Chapter 1 in *Vitrinite reflectance as a maturity parameter*; Mukhopadhyay P.K.; Dow W.G. Eds; ACS Symposium Series 570, American Chemical Society.

Murchison DG. (1993) in Geological Society Special Publication No. 32, *Coal and coal-bearing strata: recent advances*. Scott A.C. ed, The Geological Society, London, pages 257–302.

Pandolfo, A. G., Johns, R. B., Dyrkacz, G. R., Buchanan, A. S. (1988) *Energy and Fuels*, 2, 657–662.

Parks, B. C. (1963) *Chemistry of Coal Utilization Supplementary Volume* (ed. Lowry, H. H.), Wiley, NY, 4.

Philp, R.P. (1994) Geochemical characteristics of oils derived predominantly from terrigenous source materials. In Scott AC and Fleet AJ (eds) Coal and coal-bearing strata as oil-prone source rocks? Geological Society Special Publication No 77, pp. 71–91.

Poirier, N., Derenne, S., Rouzaud, J-N., Largeau, C., Balesdent, J. and Maquet, J. (2000) *Organic Geochemistry, 31*, 813.

Robl, T.L. and Taulbee, D.N. (1995) in C.E.Snape ed 'Composition, Geochemistry and Conversion of Oil Shales', p. 35.NATO ASI Series C Vol. 455. Kluwer, Dordrecht, The Netherlands.

Scott AC and Fleet AJ. (1994) In 'Coal and Coal-bearing Strata as Oil-prone Source Rocks?' Eds Scott AC and Fleet AJ. Geological Society Special Publication No. 77, Geological Society, London, page 1.

Sfihi, H., Quinton, M.F., Legard, M.F., Pregermain, S., Carson, D. and Chiche, P. (1986). *Fuel*, 65, 1007.

Sheng, G., Simoneit, B.R.T., Leif, R.N., Chen, X. and Fu, J. (1992) Fuel 71, 523–532.

Simoneit, B.R.T. (2003); Chapter 1 in *Natural and laboratory-simulated thermal geochemical processes*. R. Ikan ed, Kluwer Academic Publishing, Dordrecht, pages 1–30.

Siskin, M., Scouten, C.G., Rose, K.D., Aczel, T., Colgrove, S.G. and Pabst, R.E. (1995) in C.E.Snape ed 'Composition, Geochemistry and Conversion of Oil Shales', p. 35. NATO ASI Series C Vol. 455. Kluwer, Dordrecht, The Netherlands.

Snape C.E. and Bartle K.D. (1984). *Fuel*, 63, 883.

Snape, C.E., Axelson, D.E., Botto, R.E., Delpeuch, J.J., Tekely, P., Gerstein, B.C., Pruski, M., Maciel, G.E. and Wilson, M.A. (1989). *Fuel*, 68, 547.

Snape, C.E., Ismail, K., Mitchell, S.C. and Bartle, K.D. (1995). In Snape C.E. (ed) *Composition, Geochemistry and Conversion of Oil Shales*. p.125. NATO ASI Series C Vol. 455. Kluwer, Dordrecht.

Speight, J. G. (1989) *The Chemistry and Technology of Petroleum (2nd* Ed.) Marcel Dekker, New York.

Speight, J. G. (1994) *The Chemistry and Technology of Coal (2nd* Ed.), Marcel Dekker, New York.

Stach, E., Mackowsky, M. T., Teichmüller, M., Taylor, G. H., Chandra, D., Teichmüller, R. (1982) *Stach's Textbook of Coal Petrology*, Gebrüder Borntraeger, Berlin, Stuttgart.

Stadnichenko, T. (1934) *Econ. Geol.*, 29, 511.

Tissot, B.P. and Vandenbroucke, M. (1983). In Miknis F.P. and McKay J.F.M. (eds) *Geochemistry and Chemistry of Oil Shales*. p.1. ACS Symposium Series No. 230. American Chemical Society, Washington D.C.

Stankiewicz, B.A. and van Bergen, P.F. (1998) *Nitrogen-containing Macromolecules in the Bio- and Geo-sphere*, ACS Symposium Series 707, American Chemical Society, Washington, DC, USA.

Tse, K.T. and Stock, L.M. (1983). *Fuel*, 62, 974.

Tytko, A., Bartle, K.D., Taylor, N., Amaechina, I.O. and Pomfret, A. (1987). *Fuel*, 66, 1060.

Unsworth, J.F., Barrett, D.J. and Roberts, P.T. (1991). *Coal Quality and Combustion Performance*. Elsevier, Amsterdam.

van Krevelen, D. W. (1993) *Coal (3rd* Edition), Elsevier.

Vorres, K. S. (1990) *Energy & Fuels*, 4, 420.

Wallace, S., Bartle, K.D. and Perry, D.L. (1989) *Fuel*, 68, 1450.

CHAPTER 3

Pyrolysis: Thermal Breakdown of Solid Fuels in a Gaseous Environment

The initial pyrolytic decomposition stages of most solid fuel utilization routes involve chemical and physical processes that are broadly similar. The subsequent course of thermal breakdown depends on the particular reaction conditions and reactor design. Its evolution profoundly affects the outcome of subsequent process stages and chemical transformations.

This is the first of several chapters examining the thermochemical behavior of solid fuels and methods used for their characterization. The principal challenge faced in developing these methods is the establishment of reaction conditions accurately simulating those in selected locations within pilot or plant scale equipment. At bench scale, the two facets of this problem (i.e. fuel behavior and its measurement) are inextricably linked. This is because most products released during the thermal reactions of chemically complex solid fuels are reactive. As a result, the outcomes of experiments are often affected by the method of measurement. It is therefore essential to establish elements of design that distinguish clearly between the properties of the fuels and effects arising from the configuration of apparatus used in the measurements.

Bench-scale tests can be rapid and relatively inexpensive. With careful attention to detail, they may help in the design and development of larger scale equipment, as well as assist in trouble-shooting, in support of equipment in actual operation. In this Chapter, we will examine elements of reactor design relevant to the pyrolysis of solid fuels. The samples handled will range from forestry and other organic wastes to bituminous coals. Using some of these tools, we will also examine the pyrolytic behavior of coal macerals and several Southern Hemisphere coals. We will address the particular difficulties inherent in designing pyrolysis experiments and survey how the configurations of specific bench-scale pyrolysis reactors affect the results of measurements. Eventually, we will attempt the conceptual integration of sample characterization, reactor design and the evaluation of resulting product distributions. We will also pay close attention to reactor design with respect to tar recovery. A discussion of techniques for the chemical characterization of tars recovered from these experiments will be presented in Chapters 7 and 8.

The final sections of this Chapter will describe several examples, where bench-top experiments were used to develop a better understanding of the behavior of solid fuels in pilot or plant scale operation. We will describe how a study of the thermochemical reactions of coal macerals was extended to Southern Hemisphere coals, to evaluate their properties during pf-combustion. One reactor developed within this framework was used for characterizing kerogens, allowing a measure of accuracy not available in other techniques used for determining tar/oil and total volatile yields. The experiments helped distinguish between geologically younger and older kerogen samples. They also generated tar/oil samples for further characterization, linking kerogen molecular structures with parameters relevant to oil exploration, as well as providing a more fundamental understanding of the maturation processes. Several types of bench-scale reactors were compared, to select the best method for generating data on biomass behavior, to assist in the design and development of larger reactors. A method will be given in outline, developed for deriving the kinetics of tar cracking in the freeboards of fluidized-beds, using measurements in a reactor with sample sizes no greater than ~4 g. We will explore why synergistic effects are observed, in contrast to coals, during the pyrolysis of wood, between the polymeric species that make up this composite material. In extensions of this work in Chapter 4, we will show how the *absence* of synergistic effects during the co-pyrolysis and co-gasification of mixtures of coal and biomass can be established with two distinct bench-scale reactors, each using 50 mg samples. We will also describe an investigation of the (very low) reactivities of chars from the British Coal (Air Blown Gasification Cycle) gasifier, with a high-pressure wire-mesh reactor using 5 mg samples.

3.1 Introduction to the Design of Pyrolysis Experiments

Consider first the simple homogeneous gas phase reaction sequence, A → B → C. In this example, the final product composition depends, among others, on several parameters related to reactor design and operation. For example, short-to-medium residence times are likely to favor larger final concentrations of the intermediate "B". Longer residence times would favor higher concentrations of the product, C. If the energies of activation of the two reactions were significantly different, temperature changes during the reaction would translate into intervals where the relative rates of the two reactions would differ, possibly quite substantially. In this simple example, as in countless others, eventual product compositions depend primarily on the reactivity of the intermediate product.

When considering the pyrolysis of complex solids such as wood or coal, many *parallel* sets of *sequential* reactions need to be visualized. Each reaction would produce a reactive solid plus liquid and gaseous intermediates, which would interact with each other at widely differing reaction rates. The secondary reactions would produce new intermediates and final products. In the case of such complex reaction schemes, final product distributions depend critically on the time-temperature-pressure distribution – and history – of the reaction mixture. Amounts of tars and other volatiles recovered during a pyrolysis experiment are therefore sensitive to the method of heating, the patterns of flow as well as the design of the reaction zone and to the configuration (shape, particle size) of the sample.

In practice, the variety of designs on offer for coal/biomass pyrolysis experiments is testimony that generally accepted schemes, which offer unambiguous results, have proved elusive. The underlying complication in most cases turns out to be the reactivity of intermediate products. In particular, when coal particles are stacked together in the form of a fixed-bed, the outcome of the experiment is affected by reactions between evolving volatiles and heated solid particle surfaces. Evolving tar vapors are likely to deposit on these pyrolyzing solid surfaces, re-polymerize to a char or partially crack to release lighter volatiles. Carefully designed early experiments by Griffiths and Mainhood [1967] have shown that tar molecules move through heated fixed beds of solids in the manner of molecules moving through a chromatographic column. They sequentially adsorb onto, and desorb from successive heated particles. If temperatures are sufficiently high, the volatiles react with bed solids, producing more char, lighter tars and more gas. In fixed bed reactor experiments, measured volatile releases may differ by as much as 6–8%, or more, depending on the extent of solids-volatiles contact [Gonenc *et al.*, 1990]. The problem is complicated by wide differences in reactivity between tars of different origins, as determined by the chemical makeup of the original fuels. The more oxygenated pyrolysis tars evolved from pure cellulose and wood (*e.g.* silver birch) samples are thermally more sensitive and break down into gaseous products at lower temperatures, compared to lignite or bituminous coal tars [Stiles & Kandiyoti, 1989].

Most of the following discussion will focus on methods that aim to characterize the underlying behavior of the pyrolyzing material itself, with as little reference to sample or reactor configuration as possible. More formally stated, determining the fundamental pyrolytic behavior of a solid fuel requires the strict de-coupling of the actual observations from effects due to the design of the particular experiment and of the shape and dimensions of the sample. This is not different from requiring the result of *any* measurement to be entirely independent of the method of measurement. In actual practice, we will have to make do with results *as independent as possible* from the method of measurement. Nevertheless, the importance of striving to *minimize* the effect of sample and reactor configuration on the results of pyrolysis experiments cannot be overstated.

One practical corollary of this rather rigid dictum is the need to assess the behavior of sample particles – as much as possible – in isolation from one another. It is also important to subdivide the sample particles as finely as practicable, since *intra*-particle reactions of tar precursors do have an effect on the amount of tar and other volatiles released from individual particles. As will be discussed in detail below, wire-mesh instruments [e.g. Howard, 1981; Gonenc *et al.*, 1990; Cai *et al.*, 1998] and drop-tube (entrained-flow) reactors [e.g. Hindmarsh *et al.*, 1995] approach these stringent criteria more closely than other types of reactors, commonly used for pyrolysis experiments.

Within this framework, thermogravimetric (TG) balances do not readily lend themselves to accurate determinations of the pyrolytic behavior of solid fuel particles. With relatively good repeatability and ease of time-temperature programing, TG-balances offer a tempting combination of instrumental characteristics for undertaking pyrolysis experiments. However, they fail to conform to some of the requirements we have introduced above.

One major problem with these instruments is the shape of the sample "pile". The sample-holder in a TG-balance is usually a small pan on which sample particles are stacked. The particles are very far from being isolated from one another. When this assembly is heated, there is close contact between sample particles and evolving tars and volatiles. Furthermore, there is usually no provision for forcing a flow of gas through the bed to help evolving volatiles out of the heated zone. Due to secondary reactions of the volatiles, the product distribution inevitably reflects the size of the pile and the manner of stacking of the particles. What is determined during such an experiment is the pyrolytic behavior of the particular pile of sample particles. Furthermore, the dimensions of 'a small pile' of particles are likely to differ between one TG-balance and another (pan size, sample size, depth of pile), as well as changing from experiment to experiment. There is a danger that results remain specific to the geometry of the individual TG-instrument and to the particular sample configuration. How measured values or deduced parameters can be generalized to the behavior of the fuel then becomes a matter of conjecture [Kandiyoti, 2002]. In particular, kinetic parameters deduced from experiments where product distributions are specific to the sample configuration, must be open to question. There are analogous problems concerning the validity of results from combined TG-FT-ir and TG-MS instruments. In addition to the uncertainties already outlined, the lengths (and temperatures) of pathways that tars and volatiles travel through, on the way from the TG-furnace to the instrument create additional problems. Transmitting tar vapors over any distance without loss through condensation and/or cracking is difficult to achieve. There are questions that need to be answered regarding how results could be affected by simultaneous (i) secondary reactions of volatiles and (ii) selective tar condensation in successive sectors of such equipment [Herod & Kandiyoti, 1994].

Another limitation of TG-balances concerns the ranges of available heating rates. Often thermogravimetric systems are rated to work at up to 100 or even 200°C s^{-1}. However, furnaces of TG-balances are delicate and do not usually last long, if operated repeatedly at rates much above 30–40°C s^{-1}. In reality, heating rates relevant to the pyrolysis of solid fuels during pf-combustion or fluidized-bed gasification are far higher. They *begin* at around 1,000°C s^{-1}. There is therefore a gap between the capabilities of these instruments and heating rates required for studying aspects of many industrial applications. In any case, rapidly heating a small pile of coal in a TG-balance would give rise to additional difficulties. Outer particles of the pile might indeed experience the applied heating rate. However, the propagation of the temperature front within the pile would be slower and dependent on the bulk stacking properties and the thermal conductivity of the sample matrix.

In the past, the contribution of thermogravimetric balances to our understanding of the pyrolytic behavior of solid fuels has been path breaking. TG-systems were even used to demonstrate the effect of changes in heating rate on pyrolysis product distributions, although, admittedly this was done over a lower and rather limited range of heating rates [Howard, 1963]. It is now clear, however, that the course of pyrolytic reactions is altered by sample configuration and the presence of the TG-balance pan itself. It is still possible to put these admirable instruments to some very legitimate and important uses, including the determination of "relative combustion reactivities" of chars [*e.g.* cf. Cai & Kandiyoti, 1995] and boiling point distributions of heavy hydrocarbon mixtures [Zhang *et al.*, 1996]. In the case of pyrolysis experiments, however, product distributions are demonstrably affected by the TG-balance configuration. The continued use of these systems to analyse the fundamentals of *pyrolytic* reactions and, in particular, to evaluate pyrolysis reaction kinetics cannot, in our opinion, be considered a sound procedure.

In this context, it is as well to mention that the level of detail we have just outlined is not always necessary. It always makes sense to design laboratory scale pyrolytic tests with due regard to the purpose of the measurement. If, for example, the technological requirement is to understand the behavior of large lumps of coal, as in chain grate industrial boilers or in coke ovens, it might well be reasonable to conduct experiments with lumps of coal.

A brief word must also be said about standard volatile matter measurements, undertaken as part of "proximate" analyses. The tests prescribed by agencies such as the ASTM, the British Standards Institute and others do not differ greatly. Briefly, the procedure involves placing about 1 g of coal in a crucible covered with a lid, followed by heating in a furnace at 900 ± 5°C for about 7 minutes. The sample weight loss determined during this experiment allows calculating the proportion of volatile matter released. This procedure unabashedly suffers from many of the secondary effects described above, in relation to pyrolysis in fixed beds. The presence of the lid adds to complications of volatile escape. The measured "VM" values thus inevitably differ from results obtained using more realistic experiments in more sophisticated equipment, some of which will be described below. However, the method is simple to use under changing and possibly difficult field conditions. Results that are internally consistent to within perhaps several percent (often less) may be obtained, without recourse to extensive operator training. More important, it is a well-established test practically the world over, with databases accumulated over decades. Power station operators consider its results meaningful, in particular when *comparing* different feedstocks. The proximate analysis "VM" test is therefore likely to retain its pre-eminence in these applications for the foreseeable future. Procedures designed to adapt TG-balances for proximate analysis also exist [*e.g.* cf. Gaur & Reed, 1998]. Differences with the crucible-based method are small and the test is rapid and useful, so long as the limitations of the determination are well understood.

3.2 Product Distributions From Pyrolysis Experiments: General Trends

We have already seen that product distributions from pyrolyzing solid fuels depend on the combined effects of numerous experimental parameters. Considering these effects in isolation requires a level of simplification, but can serve as a convenient device for outlining an initially straightforward qualitative picture. In the next few paragraphs, we will review some basic trends before describing several types of pyrolysis experiments and their results in detail.

3.2.1 Effect of temperature

When a low-to-middle rank bituminous coal sample is heated at several degrees per minute in an inert atmosphere, initially, water vapor, hydrogen and light hydrocarbon gases evolve, together with some H_2S. Depending on the sample, the onset of covalent bond scission is expected somewhere between 310 and 350°C, when greater volumes of hydrocarbons and other gases begin to evolve. Between 350–400°C, sample weight loss is observed to rise rapidly. Within the solid matrix, tar precursors begin to crack and some of the lighter products evaporate. Tar evolution often continues until about 525–550°C [*e.g.* cf. Taupitz, 1977]. At the slower heating rates (say, at or below 5–10°C s^{-1}), the weight loss curve would be expected to begin leveling off at around 550°C. Above 700–800°C, the char residue progressively carbonizes, expelling diminishing amounts of CH_4, CO and hydrogen, up to perhaps 1,800°C [Kobayashi *et al.*, 1977] and beyond.

Lignocellulosic materials such as wood and paper usually contain up to 35–45% oxygen in chemical structures that appear thermally more labile than those in bituminous coals. These materials begin to decompose at lower temperatures (~300°C) than geologically more mature samples [*e.g.* cf. Shafizadeh, 1968]. Most volatile evolution is usually completed by about 450–500°C. Depending on heating rate and particle size, the proportion of volatile matter released from a lignocellulosic sample may range from 65 to nearly 99% of the original mass [Zaror *et al.*, 1985; Fraga *et al.*, 1991]. Flash-heating finely divided (~100 μm or less) wood particles either in inert or in reactive – gasification – atmospheres gives nearly similar volatile yields, leaving practically no char behind above 550–600°C [Pindoria *et al.*, 1998]. Larger particles of the same wood may give higher char yields, between 10–30% of the original sample mass, depending on whether tar vapors are swept away or left to linger in-situ.

3.2.2 *Effect of heating rate*

When samples are heated rapidly, the speed of the temperature rise overtakes the sequence of pyrolytic events observed during the stages of slow pyrolysis outlined above. Rapid heating does not allow completion of each stage *at or near the same temperature* (as in slow heating) before moving on. At rates above 100–200°C s^{-1}, therefore, the sequence of pyrolytic events is shifted up the temperature scale and telescoped into a shorter time interval. For example, at 1,000°C s^{-1}, the temperature interval for tar release is found between 600 and 700°C, whereas when coal samples are heated more slowly, say at ~1°C s^{-1}, tar release reaches completion between 550 and 600°C.

Greater heating rates have also been observed to boost volatile yields by as much as 6–8%, depending on the sample [Gibbins-Matham and Kandiyoti, 1988]. In coal pyrolysis, the increase in volatiles often tends to match the accompanying increase in tar yields. This is explained in terms of the greater survival of tars during faster heating. The rapid build-up of internal pressure in volatile bubbles is thought to force the faster ejection of tar precursors, thereby reducing the probability of retrogressive repolymerization reactions [Gray, 1988]. This mechanism is consistent with tars from fast-pyrolysis experiments showing wider ranges of molecular masses. We will also discuss evidence suggesting that locally available (i.e. sample derived) hydrogen may be incorporated into the pyrolyzing mass more effectively during rapid heating. It is likely that hydrogen released – or donated – during early stages of the pyrolysis process serves to quench and stabilize some of the more reactive free radicals associated with tar precursors. With increasing heating rates, pyrolytic events are triggered during progressively shorter time intervals, facilitating the overlapping of hydrogen release and covalent bond scission in a manner likely to assist the partial blocking of repolymerization reactions, in favor of tar survival. The resulting enhanced plasticity leads to cenosphere formation during pf-combustion and plasma heating, where peak temperatures are thought to reach several thousand degrees.

Heating rate and reactor shape: The combined effects of heating rate and reactor shape on the "temperature – tar yield" relationship may be quite complicated. When sample particles are introduced into an already heated fluidized-bed reactor, particle heating rates are high, forcing the release of correspondingly high proportions of tar vapors. However, provided the temperature is sufficiently high, the fluidized-bed configuration allows sufficient residence time for extensive secondary reactions to take place before tar vapors can sequentially exit from the bed and from the freeboard of the reactor. When separate experiments are performed at a succession of increasing temperatures in fluidized-beds, tar yields initially increase. With increasing temperature, tar cracking reactions begin to affect tar yields. When the increase in tar production is matched by tar cracking reactions, the tar-yield curve traces a maximum and declines with further increases in temperature (*e.g.* see Figure 3.6b). For bituminous coals, the tar-yield maximum has been observed between 550 and 600°C. For thermally more sensitive materials such as cellulose or softwood (*e.g.* silver birch), the analogous tar-yield maximum is observed at lower temperatures, between 400 and 425°C [Stiles & Kandiyoti, 1989].

In fixed bed reactors, the temperature-tar yield relationship again depends on a combination of several factors. We have seen how, in this type of reactor, tars released by coal particles may be destroyed by contact with other coal particles, through cracking and char forming reactions. Due to their thermal inertia, fixed bed reactors cannot be heated at rates much faster than about 10°C s^{-1} [O'Brien, 1986; also see below]. This type of reactor is normally operated by ramping the temperature from ambient. When the reactor is reasonably short, tars that survive secondary processes may exit from the reactor, well before the intended peak experimental temperature is reached. In other words, some tars may exit from the reactor without experiencing temperatures very much higher than those at which they were released, since the heating rate is relatively slow. The overall effect is of tar yields initially rising with temperature and flattening out between 500–600°C, irrespective of the final temperature reached by the reactor (see Figure 3.6a). We will see below, however, that when using similar samples, peak tar yields reached in fixed beds are usually lower than yields that can be attained in fluidized-bed reactors.

3.2.3 Effect of pressure

When pyrolysis experiments are performed under reduced pressure, both volatile and tar release tends to increase compared to operation at atmospheric pressure. Furthermore, the increase in *tar* yields due to operation under "vacuum" are often, somewhat counter-intuitively, a little larger than corresponding increases in the total volatiles. Depending on the nature of the sample (*e.g.* see Table 3.8) the incremental increase in tar yield could be as much as 5%. These observations suggest that reduced external pressure allows tar precursors to exit from parent coal particles more rapidly. The effect is not without its parallel with fast heating, where the rapid build-up of internal pressure tends to force tar precursors to exit from parent coal particles more rapidly. The data suggest that when the initial external pressure is reduced from atmospheric pressure to about 1×10^{-9} bar, intra-particle tar loss through re-condensation (re-polymerization) reactions of some tar precursors is reduced. Conversely, as the pressure is raised toward atmospheric pressure, some tar precursor material is lost through re-polymerization processes giving off gas and char and perhaps some lighter tar.

Experiments conducted at reduced pressures thus serve as a pointer regarding the level of *intra*particle reactions and loss of tar product (and tar precursors), when operating at atmospheric and higher pressures. Increasing the *external* pressure thus acts against internal forces, which tend to drive volatiles out of coal particles. Higher *external* pressures tend to slow down (i) the flow and diffusion of volatiles toward the external surfaces of particles, and, (ii) the diffusion from external particle surfaces to the surrounding bulk gas.

When the external pressure of inert gas is raised above atmospheric pressure, volatile and tar yields initially tend to diminish rapidly, up to about 5 bars. With increasing pressure, this trend slows down and appears to level off above 40 bars. Compared with atmospheric pressure results, the overall decline in total volatiles may be as much as ~10–12%. The effect was first reported and explained by Howard and co-workers [cf. Howard, 1981], in terms of the partial suppression of volatile release by the physical effect of increasing external pressure.

3.2.4 *Effect of particle size*

Volatile yields tend to diminish with increasing particle size, again providing indications of the extent of intra-particle volatile loss during pyrolysis. However, the effect is difficult to evaluate quantitatively at higher heating rates, as the propagation of the temperature front toward the center of a large particle is limited by the thermal conductivity of the intervening mass [Suuberg, 1977]. The high rates of heating imposed at the boundary would not be "seen" by the mass of sample inside large particles. Instead, the temperature front would advance at a rate modulated by the thermal conductivity of the mass of sample.

3.3 On the Design of Bench-Scale Pyrolysis Reactors: Wire-Mesh Reactors

The discussion in Section 3.1 sets us a nearly impossible task. The ideal pyrolysis reactor is required to heat all sample particles uniformly and at a precisely defined rate. Evolving volatiles must not contact *any* heated surfaces after being released from individually reacting, infinitesimally small particles. These volatiles must then be *instantaneously* quenched and quantitatively recovered. By our own exacting definitions, we appear to have stumbled on requirements that are difficult to work into the design of a real experiment.

Historically, numerous reactor configurations have been devised for determining the pyrolytic behavior of solid fuels. A useful review of pre-1963 literature on coal pyrolysis [Howard, 1963] describes "laboratory assay methods" for estimating coking properties, alongside experiments for examining the thermal reactions and behavior of coals. The "short-path vacuum still" developed by the same author and his team is an early attempt to apply uniform heating to sample particles and to suppress the secondary reactions of evolving volatiles [Sun *et al.*, 1958]. The use of relatively low pressures (~10^{-6} bar) tended to reduce sample plasticity and was deemed to help remove volatiles. "Concentric square-bottomed grooves" were machined into an electrically heated aluminum plate and only half-filled with coal particles. The grooves were intended to reduce heat transfer problems, "accentuated in vacuum systems". The apparatus had a top temperature of 550°C. It was equipped with a water-cooled plate, placed directly above the sample holder, to condense tars evolving from pyrolyzing sample particles. Interestingly, the heating rate did not receive much attention: "Approximately 90 minutes was required to reach the desired operating temperature . . ." With plenty of hindsight, the stacking of particles in this device may be considered a little too dense and the escape path for the volatiles a little ill defined. Nonetheless, many of the concepts used in the construction of this apparatus are essential to the design of pyrolysis reactors intended to resolve problems associated with sample geometry, heat transfer and volatile removal. The work was based on the recognition of uniform heating as a critical factor and that rapid cooling of tars might help reduce secondary reactions. Not unlike so many others who followed in their tracks, these experimenters could not help being drawn to the black arts of molecular mass estimation and speculating on structural aspects of pyrolysis tars. We will briefly review some of this early work in Chapters 7 and 8.

In the next few sections, we will describe several types of bench-scale reactors most frequently used in determining the pyrolytic behavior of solid fuels. These include fixed and fluidized-bed reactors, entrained-flow ('drop tube') reactors and a versatile wire-mesh ("heated-grid") instrument, developed expressly for pyrolysis and gasification experiments. The differences between results from these different reactor configurations show up the peculiarities of each particular design. We will compare results between these experimental systems, using similar coal samples. When matched against our 'wish-list' of idealized pyrolysis parameters, each reactor type turns out to suffer from some inherent shortcomings, which limit the ranges of conditions where valid experiments may be carried out. Our discussion will first focus on experimental design, operation and results under reduced ('vacuum') and atmospheric pressures. The ways in which these designs can be adapted for *high-pressure* operation will be described in Chapter 4, where we will examine results from high-pressure pyrolysis, hydropyrolysis and gasification experiments. Much of our discussion will focus on short residence time, fast heating experiments, simulating fuel behavior in fluidized-bed combustors and gasifiers, pf-combustors and entrained flow gasifiers.

3.3.1 Wire-mesh reactors

The basic design concept of these instruments is straightforward. Milligram quantities of sample particles are placed between two layers of folded wire-mesh. This assembly is weighed and stretched between two electrodes. Fine wire thermocouples are attached. A controlled current is then passed through the mesh, which also serves as a resistance heater. After the sample has been exposed to a pre-programed time-temperature profile, the weight change of the assembly is determined. Depending on the purpose of the experiment, volatiles and/or tars may be recovered and characterized. This reactor configuration allows experiments to be carried out using wide ranges of heating rates (1–20,000°C s^{-1}), temperatures (to 2,000°C) and pressures (to 160 bar). As will be discussed in detail, the design is intended to minimize the effect of sample and reactor geometry on the outcome of the experiments.

At this stage, a note of caution seems in order. Despite their manifest advantages, wire-mesh reactors should not be considered as instruments of first resort. Considerable investment is required in electronic hardware and software design, in purchasing sensitive balances and in substantial operator training. Another major drawback is the small (several milligram) size of tar, char and gas samples generated during individual experiments. By contrast, a fixed-bed pyrolysis reactor is quickly constructed and may be useful, even though results are never easy to interpret and the range of possible heating rates is limited. As will be discussed, fluidized-bed and drop-tube ('entrained-flow') reactors provide partial solutions to some of these problems, while introducing several difficulties of their own.

Let us begin by outlining the major stages in the evolution of this remarkably versatile instrument.

3.3.2 *Evolution of the wire-mesh ('heated-grid') reactor design*

The basic design: The first apparatus of its kind encountered in the literature was constructed by Loison and Chauvin [1964], working at the French coal research organization, CERCHAR. The original paper begins by mentioning that, initially, the authors built a vertical furnace pyrolyzer, what in our day would have been called a drop-tube reactor. However, the design does not seem to have found favor with these very creative researchers, because they found it difficult to quench and recover all the chars. These are indeed problems that have persisted and continue to worry drop-tube reactor operators of our day. The authors then go on to describe their original wire-mesh reactor.

A coal-water paste was pressed onto a single layer of metallic mesh, held between two electrodes. It appears sample particles were held within the "holes" of the mesh. One of the electrodes was spring-loaded, to take up the thermal expansion of the mesh during heatup. This feature prevents the buckling that would have occurred due to thermal expansion, had the mesh instead been held between two rigid electrodes. Heating was achieved by passing a single electrical pulse from a variable voltage transformer: "a thyratron time switch enabled current to be passed for times varying between 10 milliseconds and 1 second, in steps of 10 milliseconds." Temperatures up to 1,100°C were monitored with a Pt/Pt-Rh thermocouple, placed at the center of the sample holder. Experiments were restricted to the heatup ramp with an average heating rate of 1,500°C s^{-1}. Jüntgen and van Heek [1968] constructed and operated a similar wire-mesh reactor, working under vacuum and connected to a mass-spectrometer. They reported qualitative data on the release of light volatiles during the fast pyrolysis of coal. Relatively few results appear to have been published from this promising experiment.

The wire-mesh reactor configuration is best known through the work of Howard and co-workers [Howard & Anthony, 1976; Howard *et al.*, 1975, 1976; Anthony *et al.*, 1974, 1975; Suuberg *et al.*, 1978a, 1978b; Suuberg *et al.*, 1980] and Suuberg and co-workers [Suuberg & Unger, 1981; Unger & Suuberg, 1983, 1984; Suuberg *et al.*, 1985]. Coal pyrolysis and hydropyrolysis literature up to 1979 has been exhaustively reviewed by Howard [1981]. These researchers placed a coal sample of ~10–15 mg (5–10 mg in the early work by Anthony) between two layers of a folded mesh, fixed between "heavy" rigid electrodes, to absorb the resistive heat. This early version built at the Massachusetts Institute of Technology used direct current from a battery. Heating rates available to this instrument ranged between ~270–10,000°C s^{-1} (650–10,000°C s^{-1} in early work by Anthony). Temperatures reached by the mesh were determined using a thermocouple placed between the two layers of mesh, without physical contact. There are indications that a temperature lag may have existed between the thermocouple and the mesh due to this lack of contact. The instrument was operated over a range of pressures from ~10^{-3} to 70 bars. As in the case of Loison and Chauvin [1964], tars were collected by washing internal surfaces with solvent. In later work, these researchers characterized tars by size exclusion chromatography, using tetrahydrofuran as eluent [Fong *et al.*, 1986a, 1986b; Unger & Suuberg, 1984; Suuberg *et al.*, 1985].

Another wire-mesh instrument constructed at Bergbau-Forschung used a 10 kHz heating current and analogue feedback control [Arendt, 1980; Arendt & van Heek, 1981]. The system operated at heating rates above 210°C s^{-1}. Pyrolysis and hydropyrolysis experiments were carried out at pressures up to 100 bars. The tar yield determination relied on an indirect calculation. In constructing the cell, polymeric materials were used to make the electrode holders. The thermal sensitivity (danger of melting) of these components limited experiments to about 2 seconds holding time at peak temperatures between 700 and 1,100°C, the top design temperature. Instruments commercially acquired from Bergbau Forschung (later called DMT) by the British Gas Research Station in Solihull (UK), by the Coal Research Establishment at Stoke Orchard (UK) and by ABO Academy University (Finland) suffered from similar limitations.

The amount of sample used by different laboratories has varied over the years between 5–7 milligrams at Imperial College (see below) and up to 35 milligrams in some of the experiments conducted at Bergbau Forschung. Similarly, the rectangular mesh size has varied between approximately 15 × 60 mm at Bergbau Forschung and 30 × 90 mm at Imperial College. Sample sizes and mesh dimensions used in most other laboratories have been intermediate between these extremes.

3.3.3 Expanding the heating rate range & improving tar recovery

In this type of apparatus, a large current pulse is all that is required to achieve a reasonably well defined time-temperature ramp during rapid heating, at or above 200°C s^{-1}. All the early wire-mesh instruments were comfortably able to run at high heating rates. Arranging for a steady holding period at peak temperature may require some manipulation, but the technology required to do this could still be relatively simple. However, heating the samples at reproducible and reasonably linear temperature-time ramps over the lower heating rate ranges (~1–100°C s^{-1}) requires adequate online feedback control capabilities. In the late 1970s and early 1980s, the hardware components to facilitate achieving low heating rates were not commercially available.

In early rapid heating experiments, Howard and co-workers observed volatile matter and tar yields significantly greater than those from slow heating experiments, performed in standard volatile matter test crucibles or by thermogravimetric balances. However, they were unable to distinguish experimentally between the enhancement of tar and volatile yields due to fast heating from the effect of fine sample dispersion in the wire-mesh reactor. The enhanced yields observed in wire-mesh reactors operated at high heating rates were, thus, thought to result from fine sample dispersion rather being due to high heating rates [Howard, 1981].

In the early 1980s, researchers at Bergbau Forschung attempted to close this heating rate gap. They supplemented results from fast heating experiments in a wire-mesh reactor with weight loss data from a TG-balance operated at slow heating rates [Wanzl, 1988]. Once again, effects due to slow heating could not be resolved from effects due to particle stacking in the TGA. Lower heating rates in wire-mesh reactors first became available in laboratories with facilities of their own for designing and constructing data acquisition and temperature programing systems.

Expanding the heating-rate range: The first wire-mesh instruments capable of operating at slow as well as fast heating rates have been described by Hamilton [Hamilton *et al.*, 1979; Hamilton, 1979] at CSIRO in Sydney (Australia) and by Williams and co-workers at the University of Leeds (UK) [Desypris *et al.*, 1982]. Relatively little data was published from the latter instrument, which appears to have produced data with surprisingly wide bands of scatter, substantially swamping effects due to changes in the heating rate. Hamilton's wire-mesh reactor was also used for only a short period of time. This apparatus was capable of achieving heating rates between 10^{-1} and 10^{4}°C s^{-1}. The purpose built power supply was innovative for its time. It provided a current interrupted for ten milliseconds in every hundred, to enable reading the temperature via a thermocouple. Before being abandoned, apparently for reasons unrelated to its technical performance, the system was used for preparing coal and coal maceral chars, in order to examine changes in morphology as a function of heating rate and peak temperature. The chars were examined by scanning electron microscopy and valuable matrices of photomicrographs were published. Overall, the work showed that increasing heating rates enhanced the plasticity of samples of *all* ranks. Volatile and tar yield measurements were not attempted.

Problems of quantitative tar recovery: Accurate measurements of tars evolved during the pyrolysis of solid fuels may be useful in several ways. They provide maximum (limiting) tar production data for process schemes seeking to maximize or minimize liquid product yields from pyrolysis and hydropyrolysis processes. For example, in biomass gasification, the presence of tars in the product gas is a nuisance that must be suppressed, while in pulverized fuel combustion, evolved tars carry a substantial part of the calorific value assigned to individual coal particles. Tar yields also provide valuable information regarding mechanisms of thermal breakdown: the evolution and subsequent reactions of tars must therefore be tracked carefully.

In wire-mesh reactors, tars released by the sample may be seen to rise in the shape of a small cloud. In the absence of a facility for sweeping volatiles away from the heated zone, the cloud of tars slowly circulates. Some tar tends to settle down on the mesh. This affects the measurements [Howard, 1981]. Initially, tar recovery in wire-mesh reactors was achieved by a combination of filtering the gas from the chamber and washing chamber walls with some such solvent as dichloromethane, chloroform or methanol. In Canada, Stangeby and Sears [1978, 1981a, 1981b] developed an atmospheric pressure wire-mesh instrument with a *lateral* gas sweep flowing at 3 cm s^{-1}, to remove volatiles away from the reaction zone. The instrument was battery powered with operation confined to relatively high heating rates, between 250 and 6,000°C s^{-1}. "Heating rate was found to have little effect on total weight loss of the coal, but a dramatic effect on the actual composition of the products. High heating rates substantially increased the yield of light hydrocarbons." [Stangeby & Sears, 1981a] With hindsight, the results suggest substantial extents of secondary cracking of tar vapors as they were swept laterally *across* the surface of the rapidly heated mesh. It is likely that radiation from the mesh at higher temperatures had a significant effect on the composition of the recovered products.

At about the same time, Niksa *et al.* [1982a, 1982b, 1984] used a DC "operational power supply" equipped to deliver two independent "cycles", a constant current during heatup, followed by adjustable constant voltage to maintain an isothermal reaction temperature. A preheated gas stream was allowed to sweep *parallel to* ("across") the mesh, to carry volatiles toward a set of filters. The instrument described in 1982 [Niksa *et al.*, 1982a] was designed for heating at up to 1,000°C at rates up to 10^4°C s^{-1} and was equipped with a liquid nitrogen spray for quenching the system. One of the electrodes was spring-loaded to absorb the thermal expansion of the mesh. 50 micrometer diameter chromel-alumel thermocouple wires were spot-welded "to the outside" of the wire-mesh support. While the spot-welds would be expected to cause local temperature distortions, little temperature variation was reported from experiments conducted using two pairs of thermocouples. Avoiding interference in temperature measurement from the power circuit was achieved through the high impedance of the storage oscilloscope and floating the oscilloscope and power supply in common.

Initial experiments were conducted between 13 Pa and 0.22 MPa (vacuum to 2.2 bara) pressure although the system was placed in a 13.5 MPa pressure vessel. Niksa *et.al* [1982b] reported results from variable heating rate experiments in the 10^2–10^4°C s^{-1} range, under reduced pressures. At "the highest heating rate the uncertainty involves whether or not the actual rate was 10^4 K s^{-1}". The authors state that short residence times at peak temperature may not have allowed experiments to reach completion, particularly at the lower temperatures. Nevertheless, differences of 8–10% in weight loss were reported between heating at 10^2 and 10^4°C s^{-1}, although this may have partly resulted from incomplete pyrolysis, since "experiments at 10^2 were terminated after heatup." The initial set of papers does not mention tar yield measurements. The instrument was subsequently used for tar yield determinations during hydropyrolysis experiments, with up to 25 bars of hydrogen [Bautista *et al.*, 1986].

Freihault and co-workers also developed an atmospheric pressure wire-mesh instrument. Initial publications [Freihaut *et al.*, 1982; Freihaut & Seery, 1983] suggest a vertical pyrolysis chamber construction. One interesting feature of this apparatus was the direct connection of the chamber to an infrared cell, for analyzing light volatiles. Tars were defined as material condensed on reactor cell walls, liners and the glass wool filter placed between the chamber and the IR cell. The pyrolysis chamber appears to have been initially "slightly pressurized" to remove volatiles from the vicinity of the wire-mesh. Heating rates ranged from less than 1°C s^{-1} to 1,000°C s^{-1}. Photomicrographs in a later publication show that thermocouples have been spot-welded to the mesh, creating a metal bead of about 180 μm across [Freihaut & Proscia, 1989]. Positioning such a large mass of thermocouple metal onto the mesh would be expected to distort the temperature distribution as well as create a probably unacceptable time lag in the temperature measurement. In an entirely different context, the present author has found it perfectly possible – although not easy – to weld 25 μm wires across each other with no apparent distortion in the wire [Kandiyoti, 1969; Kandiyoti *et al.*, 1972]. Tar yields reported in the published work of Freihaut and co-workers showed evidence of substantial decline at higher experimental temperatures. This suggests that secondary cracking of volatiles could not be avoided, probably due to the absence of flow induced movement away from the vicinity of the heated mesh.

Installing a vertical sweep flow: Two wire-mesh reactors were constructed at Imperial College, the first for vacuum and atmospheric pressure operation and the second for high-pressure work up to 160 bar [Gibbins, 1988; Gibbins-Matham & Kandiyoti, 1988; Gibbins & Kandiyoti, 1989a; Gibbins & Kandiyoti, 1989b]. Both reactors were designed with a carrier gas stream flowing normal to the plane of the wire-mesh, to sweep volatiles away from the reaction zone. Initially, the heating rate range could be changed between 0.1°C s^{-1} and 1,000°C s^{-1}, using chromel-alumel thermocouples and stainless-steel mesh. Details of the cell and purpose built electronic instrumentation have been described in the original publications [Gibbins, 1988; Gibbins-Matham *et al.*, 1989c]. Cai *et al.* [1996] subsequently extended the heating rate range of these instruments up to 10,000°C s^{-1} and the temperature range to 1600°C, by using Pt-Pt/Rh thermocouples and molybdenum mesh. Recent work has extended the operating range of both the atmospheric pressure and the high-pressure instruments to 2,000°C, using tungsten-rhenium thermocouples. As outlined in Chapter 4, the evolution of the high-pressure wire-mesh reactor has successively allowed mimicking of conditions in hydropyrolysis (850°C, 80 bara), steam gasification (1,100°C and 30 bar) and eventually entrained flow, steam-oxygen gasification reactors (2,000°C at 30 bar) [Peralta *et al.*, 2002; Peralta *et al.*, 2004; Peralta *et al.*, 2005].

Figure 3.1a shows the atmospheric pressure reactor equipped with an early version of the tar trap. Unlike previous designs, a brass plate with a central hole was installed underneath the mesh. This allowed the sweep gas to be directed vertically upwards, *through* a 3 cm diameter circular area of mesh, where 106–152 μm sample particles (~5 mg) were evenly distributed. The flow served to sweep volatiles released by sample particles away from the reaction zone and into the cold-trap.

In order to prevent overheating, the two electrodes and the connecting tubes (Item 8 in Figure 3.1) were cooled with circulating water. This was particularly useful during slow heating experiments, when the electrodes were capable of absorbing large amounts of heat. One of the electrodes was spring-loaded for taking up the thermal expansion of the mesh, which was heated by a low voltage (12–24 V), alternating current.

The initial tar tap design in Figure 3.1a consisted of an off-take tube, placed vertically above the wire-mesh sample holder and closed-off near the upper end with a Pyrex

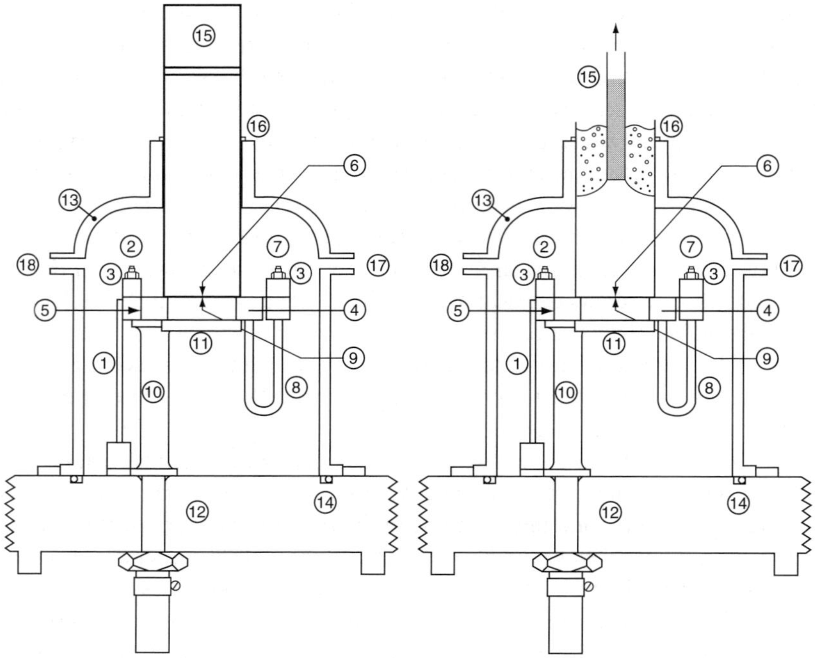

Figure 3.1 The atmospheric pressure wire-mesh reactor with the early (a) and present (b) tar trap designs. Legend: [1] Copper Current Carrier; [2] Live Electrode; [3] Brass Clamping Bar; [4] Sample Holder Support Plate; [5] Mica Strip; [6] Wire-mesh Sample Holder; [7] Electrode; [8] Stainless Steel Tubes; [9] Mica Layer; [10] Brass Pillars; [11] Sintered Pyrex Glass Disk; [12] Base Plate; [13] Pyrex Bell; [14] O-ring Seal; [15] Off-take Column; [16] O-ring; [17] Carrier Gas Entry Port; [18] Connection for Vacuum Pump (Fig 1a reproduced with permission: Fuel *68*, (1989), 895; Copyright 1989 Elsevier).

sinter. Liquid nitrogen poured above the sinter served to chill the flowing stream of gas. The sweep gas (He) and lighter products (CO, H_2 etc) exited without significant increase in reactor pressure, while heavier volatiles condensed on or near the cooled sinter. The determination of weight uptake by the traps proved to be a reliable method for quantifying tar yields. Problems emerged, however, when the quantitative *recovery* of the tars was required for structural characterization work. Small amounts of tar were found to adhere to the sinter, even after careful solvent washing. Figure 3.1b shows the reactor assembly, together with the "Mark II" trap design that was eventually adopted for vacuum and atmospheric pressure work. The new traps were made of Pyrex glass and could be operated at up to 1,000°C for short periods. Quartz was used for higher temperature experiments. Evolving volatiles were continuously swept into the central "chimney" packed with fine strips of stainless steel wire-mesh. The traps were dried at 50°C and weighed before and after an experiment, to quantify tar deposition. Tar yields from the new traps were found to be indistinguishable from results obtained with the trap in Figure 3.1a, but the tars did not adhere to metal mesh in the same way and could be recovered quantitatively [Li *et al*., 1993a, 1993b].

The use of a stream of gas through the sample holder, directing evolved volatiles into the liquid-nitrogen cooled traps (Figure 3.1), has enabled the determination of tar yields more reproducibly than could have otherwise been achieved. As described above, previous work in wire-mesh cells with stagnant atmospheres had relied on washing internal surfaces for tar recovery and accepted an unspecified level of error, due to recirculating tar vapors settling onto the heated mesh. Where a lateral flow was made available, traps placed at the exit served to capture some of the tar. However, it seems difficult to imagine that tars sweeping laterally over a mesh heated at anywhere up to 6,000°C s^{-1} would not be cracked or otherwise affected by radiation from the heated mesh. Indeed the results of Stangeby and Sears (1981a) suggest the presence of these effects, particularly when operating at the higher heating rates.

In the Imperial College reactors, two pairs of thermocouples were used to monitor lateral temperature variations in the sample holding part of the mesh. The average of outputs from the two thermocouples was used as the feedback signal for on-line control. Thermocouple pairs were spot-welded to each other and pulled tight for contact with the mesh but were not *welded* to the mesh. Thermocouple readings were taken every 20 milliseconds during 3 millisecond interruptions in the power supply, triggered by a purpose built pulse generator. Achieving good linear temperature-time ramps during fast heating was shown to be feasible even with relatively simple power supplies. Figure 3.2 shows that adequate linear temperature-time ramps could also be achieved during slow heating, at 1°C s^{-1}. The wider heating-rate ranges available to the Imperial College instruments (0.1 and 1,000°C s^{-1}) combined with the improved tar yield measurement method has enabled demonstrating unambiguously that increasing heating rates have a definite effect on pyrolysis tar *and* total volatile yields [Gibbins-Matham & Kandiyoti, 1988]. These data will be presented below.

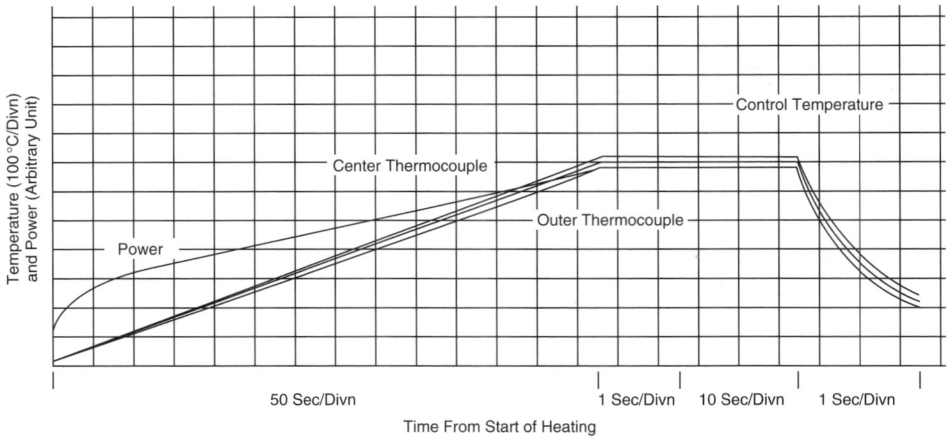

Figure 3.2 Time-temperature ramp: heating at 1°C s^{-1} (Reproduced with permission: Fuel *68*, (1989), 895; Copyright 1989 Elsevier). Comparison of two thermocouple outputs shows the lateral variation in temperature within the sample holding part of the mesh. The averaged output (in the middle) was used as the feedback signal for on-line control.

3.3.4 Product trends from pyrolysis in atmospheric pressure wire-mesh reactors

Effect of temperature: Figure 3.3a presents data from the atmospheric pressure wire-mesh reactor, showing tar and total volatile yields from a sample of Linby (UK) coal heated at 1°C s^{-1} with 30 s holding at peak temperature. In these experiments, the total volatile yield was observed to increase rapidly up to about 600–700°C, while the tar yield flattened out just above 500°C. At higher heating rates, tar yields level off at somewhat higher temperatures (Figure 3.3b). Above about 700°C, tar yields do not usually increase by very much. Total volatile yields from coal samples tend to increase, albeit slowly, as the sample temperature is raised. Kobayashi *et al.* [1977] have presented evidence of weight loss at up to 1,800°C. In most chars, there would still be a percent or two of volatiles to squeeze out by keeping the sample at these higher temperatures for longer times [Howard, 1981].

We have already seen that fluidized-bed reactors allow tar loss through secondary reactions in the bed itself as well as in the reactor freeboard. For most coals, the tar yield measured in a fluidized-bed reactor will go through a maximum near 550–600°C. In wire mesh-reactors, tar yields would be expected to hold stable and not to decline with increasing temperature, above about 700–800°C, provided that *no* tar product is destroyed *after* its release from parent coal particles.

Effect of heating rate: When the heating rate is increased from several degrees per second to several *hundred* or several *thousand* degrees per second, the temperature rise runs ahead of the sequence of pyrolytic events outlined in the first paragraph of Section 3.2.1. During rapid heating, chemical reactions have no time to reach completion within narrow temperature bands – as in slow heating. On the other hand, the quicker moving temperature ramp during faster heating tends to accelerate chemical processes. The

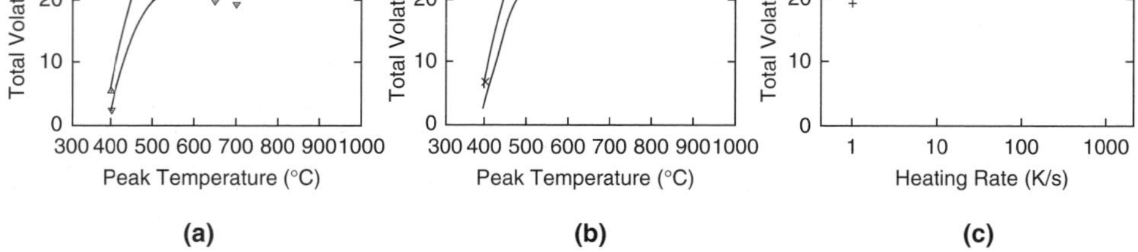

Figure 3.3 Effect of peak temperature, and heating rate on tar and total volatile yields. (a) Tar and total volatile yields at 1°C s^{-1}. (b) Tar and total volatile yields at 1,000°C s^{-1}. (c) Effect of heating rate on tar and total volatile yields for a peak temperature of 700°C. Linby coal. 30 s holding at peak temperature; sweep gas, helium at 1.2 bar flowing at 0.1–0.2 m s^{-1}. Particle size range: 106–152 μm. (Reproduced with permission: Fuel 1989, *68*, 895; Copyright 1989 Elsevier).

sequence of events is telescoped into a narrower time band and shifted up the temperature scale. Furthermore, changes are observed in product distributions with increasing heating rate, suggesting that the *relative* rates of competing reactions are being altered, arguably due to differences in their activation energies.

Figure 3.3c shows that when Linby coal is heated at 1,000°C s^{-1}, tar and total volatile yields increase by about 6%, compared to heating at 1°C s^{-1}. It was noticed during subsequent campaigns, that the choice of Linby coal for these initial experiments was somewhat fortunate. The response of this coal to changes in heating rates was somewhat greater than many other samples tested later. The effect is observable, however, for many low-to-middle rank coals [*e.g.* see Li. *et al.*, 1994a]. As will be discussed below, the vitrinite component of coals seems more sensitive to changes in heating rates. Occasionally, however, liptinite or inertinite group concentrates also show a measure of sensitivity to changes in heating rate.

Despite the experimental scatter, the data in Figure 3.3c indicate that between 1 and 1,000°C s^{-1}, the difference between tar and total volatile yields remained approximately constant. Broadly similar behavior has been observed for a number of other samples (see Table 3.6). It thus appears that the greater volatile evolution observed at higher heating rates consists mostly of additional tar release. One likely explanation is that greater tar survival should be possible during rapid heating, due to the faster expulsion of tars and tar precursors. Slow heating would allow more time for the repolymerization of tar precursors. One further mechanism is likely to have contributed to this effect. Compared to slow heating (~1–100°C s^{-1}), rapid heating has been observed to give rise to greater amounts of solvent extractable material in sample particles heated to temperatures in the 400–550°C range [Fong *et al.*, 1986a, 1986b; Fukuda, 2002; Fukuda *et al.*, 2004]. It seems reasonable to expect that more tar would evaporate from the greater amount of the liquid-like "extractables" present in rapidly heated coal particles. Clearly, we still need an explanation for reasons why greater amounts of extractables are actually formed in some coals during rapid heating. The work of Howard and co-workers in the mid-1980s and more recent work at Imperial College go some way toward providing a likely mechanism for "additional" extractable formation and will be reviewed in Chapter 6.

Effect of particle size: Volatile yields tend to diminish with increasing particle size [Suuberg, 1977]. Two effects need to be considered. The first is the extent of intraparticle reactions taking place and how they may increasingly block the evolution of volatiles, with increasing particle diameter. Intuitively, one expects the initial sharp *drop* in total volatiles to flatten out for much larger particle sizes. This is because, tar molecules that survive and continue to diffuse through the heated mass as well as tars formed during secondary reactions are likely to be lighter and chemically more stable. However, at high heating rates, the phenomenon is difficult to evaluate. The heating rate of the coal mass *inside* large particles would be modulated by the rate of heat transmission – governed by the thermal conductivity of the coal mass itself. The internal transmission rate of the high temperature front is likely to be far lower than the rate imposed at the particle periphery. There appears to be no straightforward way of heating large particles uniformly. We are thus unable to determine extents of intraparticle secondary reactions in isolation from heating rate effects.

The interrelationship between particle size and heating rate is nevertheless interesting. It seems possible to get a handle on these reactions from vacuum pyrolysis experiments (see below). We will also see that the effect of heating rate on tar yields is different for somewhat smaller sized particles than those ordinarily used in wire-mesh instruments [Griffin *et al.*, 1993; Howard *et al.*, 1994.]. For particles in the 106–152 μm size range, Cai found little change in tar yields with increasing heating rate, above 1,000°C s^{-1}. However, volatile release continued to increase with heating rate, in relatively small increments of perhaps several percent of the sample mass [Cai *et al.*, 1996].

In a parallel development, Howard and co-workers [Griffin *et al.*, 1993; Howard *et al.*, 1994] constructed a new wire-mesh reactor, where volatiles were drawn away from the heated-mesh by suction tubing via glass funnels, and subsequently quantified. This reactor was powered by an alternating current, avoiding problems of interference between the heating current and temperature readings. Data was reported for atmospheric pressure pyrolysis experiments in helium, with heating rates ranging from 10 to 20,000°C s^{-1}. Howard and co-workers also found that tar yields remained constant above 1,000°C s^{-1} for particles in the 106–125 μm diameter range. However, for 63–75 μm coal particles, they found small increases in tar yields when heating rates were raised above 1000°C s^{-1}. The result is important in underlining the interrelationships between tar yields, heating rates and the effects of *intra*particle recombination reactions of tars. It is also consistent with vacuum pyrolysis experiments showing approximately 5% more tar yield, i.e. 5% *intra*particle tar loss *at atmospheric pressure* for particles in the 106–152 μm size range.

As an aside, it is relatively difficult to reduce the sizes of sample particles used in wire-mesh reactors to much below the mentioned ranges (63–75 μm, 106–152 μm). Commonly used stainless meshes have 64 μm holes and the presently available molybdenum meshes, about 100 μm. Smaller mesh sizes weaved in stainless steel may be used although smaller-weave molybdenum wire-mesh does not seem to be commercially available. Clearly, particles must not be so small that they would drop through the mesh. In any case, difficulties due to static electricity do not allow easy handling of sample particles much below 60 μm. However, the size of thermocouple wires can also be a problem. The common stainless steel meshes with 64 μm holes can accommodate thermocouple wires of 50 μm diameter. When smaller stainless steel mesh sizes are used, the next available standard size of thermocouple wire is 25 μm in diameter and quite difficult to handle on a routine basis, even by experienced operators. Moreover, the use of mesh with much smaller aperture sizes would be impractical, because of the reduced permeability for the carrier gas and coal volatiles through the mesh. The possibility of tar cracking on the mesh has been considered and found to be negligible for the reported combinations of particle/mesh sizes [Gibbins-Matham & Kandiyoti, 1988]. Let us review the detailed configurations of, and product trends from, several other types of basic pyrolysis reactors.

3.4 The Design of Bench-Scale Fixed-Bed ("Hot-Rod") Pyrolysis Reactors

In a fixed bed pyrolysis reactor, sample particles are stacked to the desired bed-depth. Heat usually diffuses inward from the reactor walls. Released volatiles expand, raising the local pressure in small increments. The resulting pressure gradient assists volatiles in slowly exiting from the bed and from the reactor [*e.g.* cf. Berk, 1978]. Passing an inert gas stream through the fixed-bed would help reduce volatile residence times in the reaction zone. A constant carrier gas velocity is required to maintain reasonably steady volatile residence times inside the reactor. Comparable residence times would then allow comparison of data from reactors of different sizes but of similar shape [Dryden & Sparham, 1963]. The reactor body can also be employed as resistance heater, if electrodes are clamped at both ends of a tubular reactor. The original "hot-rod" reactor configuration [Hiteshue *et al.*, 1957] is completed when the tubular reactor body is made of an alloy able to withstand high pressures.

Despite inherent problems associated with secondary reactions between stacked particles and evolving volatiles, the "hot-rod" reactor configuration has proved useful. It is relatively easy to construct and to operate. It was originally conceived and constructed at the U.S. Bureau of Mines laboratories for examining the hydropyrolysis of coals. It was used at the Coal Research Establishment (British Coal) for investigating the production of benzene, toluene and xylenes ("BTX") during hydropyrolysis.

'Hot-rod' reactors are best known from the work of Hiteshue and co-workers at the U.S, Bureau of Mines [1957, 1960, 1962a, 1962b] who conducted coal hydropyrolysis experiments at up to 400 bara and nearly 900°C. Sample beds of 25–40 cm length were mounted in small-bore stainless steel tubing. Reactors with different aspect ratios (length/diameter) were tested by Graff *et al.* [1976] and Kershaw & Barras [1979]. The version constructed and operated by Ladner and co-workers [Finn *et al.*, 1980; Fynes *et al.*, 1984] held about 10 gm of coal in 75 cm or longer reactor tubes with 8 mm internal diameters.

At Imperial College, a smaller (6 mm i.d., 20 cm long) reactor was constructed, originally to produce larger amounts of tar during hydropyrolysis experiments than was possible with the wire-mesh instrument. Between 0.5 and 1 gm of coal sample were used during initial experiments, where the effect of carrier gas flow rate and the heating rate were investigated. A comparison was made with results from the larger "hot-rod" reactor at British Coal [O'Brien, 1986; Bolton *et al.*, 1987]. Efforts to examine the precise role of the bed height led to the use of much shallower (~4 mm) fixed-beds, corresponding to about 50 mg of sample (Figure 3.4a). The use of smaller sample sizes has served to partially suppress secondary reactions within the bed as well as in the reactor freeboard [Gonenc *et al.*, 1990]. The reduced bed-height also improved axial temperature uniformity. This reactor was operated at heating rates between 10°C min^{-1} and 10°C s^{-1}. Faster heating rates were calculated to lead to unacceptably steep radial temperature gradients in the 6 mm diameter sample bed [O'Brien, 1986].

As outlined above, the "hot-rod" configuration requires the reactor body to act as a resistance heater as well as the pressure vessel. The tube material must therefore be able to withstand the internal stresses due to high-pressure operation at reaction temperatures – depending on the particular experiment – up to perhaps 1,000°C. This imposes rather stringent requirements on the tubing material. The original U.S. Bureau of Mines reactors consisted of small-bore, thick walled stainless steel tubing, which were discarded after each or several experiments. At British Coal, standard practice was to *make* the reactor bodies of stainless steel and discard them after several runs. This appears to have been the compromise arrived at, once the decision had been taken

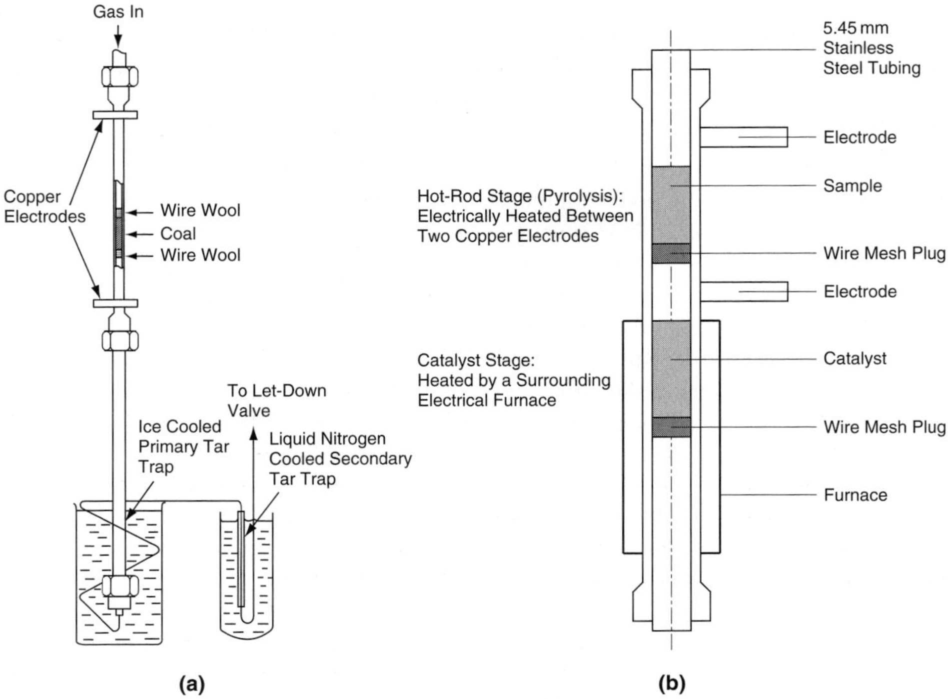

Figure 3.4 The 'hot-rod' reactor configuration. (a) The common single bed reactor. (b) Two fixed beds in tandem; the lower bed is heated by a separate furnace and packed with catalyst for hydrotreating volatiles released by the sample pyrolyzing in the upper bed. [Reproduced with permission: (a) Fuel 1987, *66*, 1414; Copyright 1987 Elsevier; (b) Fuel, 1998, *77*, 1715; Copyright 1998 Elsevier.]

to use 8 mm bore reactors that were long (>75 cm) and required drilling from both ends. Reactors that are more durable could have been made with specialized alloys. Although the Nimonic series constitute an extreme case, many of these alloys are more difficult to machine than stainless steel. However, considerable machining time and expense goes into making each such reactor.

More durable bodies have been made for the shorter (20 cm) reactors used at Imperial College, of various specialized alloys. Nimonic 80 and Nimonic 105 alloys (Henry Wiggin Alloys) were initially used for experiments carried out at 850°C and pressures up to 100 bar. However, these alloys require heat treatment to soften the material before machining and, afterward, to harden the reactor following the machining stage. Incolloy 800 HT is far easier to machine and was used to make reactor bodies for CO_2 and steam gasification experiments up to 1,000°C and 40 bar. Above 700°C, the thermal expansion of the reactor body must be taken into account. The use of rigid electrodes may cause reactor tubes to distort. In the present design, power is supplied to one of the electrodes through woven copper-cables and both electrodes are water cooled to avoid large changes in resistivity [Pindoria *et al.*, 1998a; Collot *et al.*, 1999].

In another application of the 'hot-rod' reactor configuration, trace element releases from various solid fuels were measured during the co-gasification and co-combustion of coal and biomass. In order to block contamination from the metal walls, a larger (13.8 mm) internal diameter reactor, constructed of Incolloy 800 HT, was lined with a quartz sleeve and operated at up to 40 bar and 1,000°C [Collot *et al.*, 1998]. The same reactor was also used for investigating whether the co-pyrolysis and co-gasification of coal and biomass produced significant synergistic effects.

Relatively few two-stage experiments using the "hot-rod" reactor configuration have been described in the literature. Bolton *et al.* [1988] attached a second bed packed with hydrous titanium oxides for catalytic cracking of hydropyrolysis tars. The runs were mostly carried out at 150 bar, with the hydropyrolysis section being ramped to 500°C while the catalyst stage was kept at a maximum temperature of 400°C. The tar yield from the first stage was about 25% of the original coal mass. The authors reported conversion of the tars into "into colourless liquids low in heteroatoms" with about 40 percent boiling below 140°C. Figure 3.4b shows the two-stage fixed-bed reactor constructed at Imperial College. In the upper (shortened 'hot-rod') section, tar/oil vapors were produced by mild hydropyrolysis (H_2-pressure up to 40 bar). Gas flowed through the fixed-bed and swept evolving volatiles into the second, catalyst packed stage, positioned below the sample and heated independently by a small furnace [Pindoria *et al.*, 1998b].

In the next section, we will introduce the third major reactor type, a bench-scale fluidized bed reactor, successfully used in coal pyrolysis experiments. In Section 3.6, we will compare product trends from atmospheric pressure coal pyrolysis experiments in the "hot-rod" reactor with results from a wire-mesh apparatus. Results from high-pressure experiments in the "hot-rod" reactor will be presented in Chapter 4.

3.5 Bench Scale Fluidized-Bed & Entrained Flow Pyrolysis Reactors (Atmospheric Pressure)

3.5.1 *Bench scale fluidized-bed pyrolysis reactors*

Experimental fluidized-bed pyrolysis reactors normally operate in flash pyrolysis mode. The sample consisting of coal or biomass particles is injected into an already heated bed of inert solids. The fluidizing gas sweeps evolving volatiles out of the reaction zone.

Figure 3.5a shows a very useful bench-scale fluidized-bed design, developed by Tyler [1979; 1980] at CSIRO in Sydney. The reactor (~50 mm i.d.; ~350 mm high) was made of quartz. Several grams of fuel particles were dropped at the rate of 1 gm min^{-1} through the inner of two concentric tubes, into the heated bed of sand particles. The outer annulus carried cold nitrogen, to keep sample particles from beginning to react before they reached the fluidized-bed. Tyler used superficial gas velocities of about 5 times minimum fluidization and operated this reactor to temperatures up to 900°C. The char residue was weighed together with the bed materials at the end of an experiment, in order to determine total weight loss. In the original design, tar product was captured and recovered using an extraction thimble to filter product gases in a liquid nitrogen cooled chamber.

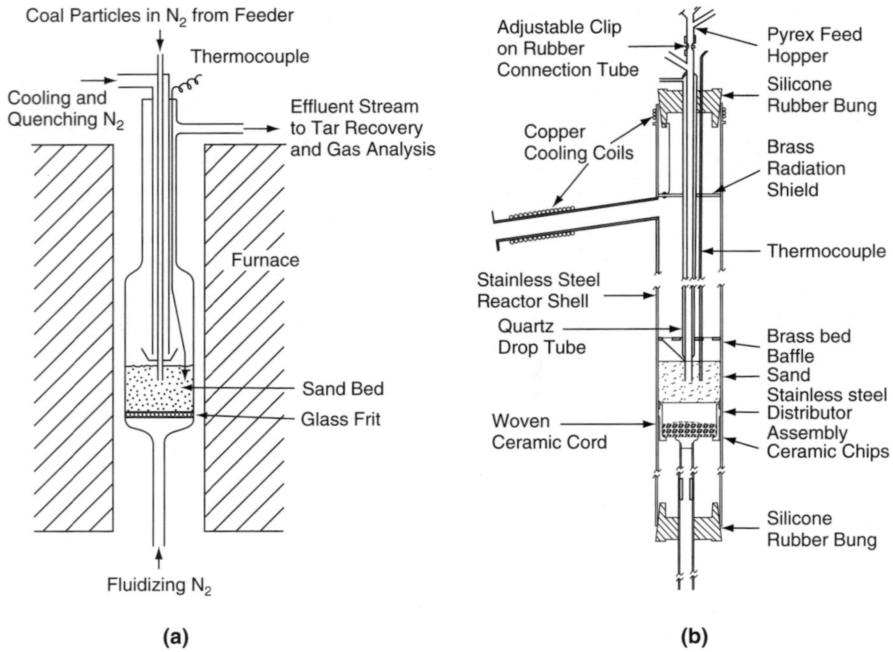

Figure 3.5 (**a**) Fluidized-bed pyrolyzer in quartz by Tyler [1979, 1980]. (**b**) Fluidized-bed pyrolyzer with movable support plate, where volatile residence times in the freeboard may be varied without altering fluidizing conditions [(a) Reproduced with permission: Fuel, 1979, *58*, 680; Copyright 1979 Elsevier. (b) Reproduced with permission: Fuel, 1989, *68*, 275; Copyright 1989 Elsevier].

Figure 3.5b shows the design of a larger (76 mm i.d.; 1000 mm high) fluidized-bed pyrolysis reactor made of stainless steel. The design was inspired by Tyler's concept, but incorporated several additional features. The fluidized-bed was mounted on a vertically mobile support-plate, enabling the residence times of evolved volatiles in the reactor freeboard to be changed without altering fluidizing conditions. Freeboard residence times could be changed between 0.8 and ~4 s, normally using 3–3.7 times the incipient fluidizing velocity [Stiles & Kandiyoti, 1989]. In this reactor, the variability of the free-board height allowed collecting data to calculate kinetic constants for tar cracking reactions [Stiles, 1986].

During trial runs, it was observed that tar and char yields were distorted by solid particle carryover into the quench zone. The effect was particularly severe during operation at shorter freeboard residence times and during runs with lower density substrates, such as cellulose and wood dust. A wire screen (denoted as "brass bed baffle" in Figure 3.5b) was placed above the fluidized-bed and served to block the escape of solid particles from the bed. Careful cold-trap design pays off in this reactor configuration, to stop fine tars from escaping with the lighter hydrocarbons into the gas product. Product trends from this reactor will be presented and compared with those from a wire-mesh reactor in Section 3.6.

3.5.2 *Bench scale drop-tube ('entrained-flow') pyrolysis reactors*

Drop-tube reactors consist of pre-heated reactor tubes through which sample particles are "dropped" in batch or continuous mode. Entraining (i.e. co-current) gas velocities may be varied over considerable ranges. However, the flow is usually kept laminar. Due to radiation from the walls, heating rates are estimated to be high (>10^4°C s^{-1}), but cannot be measured directly. Pre-heating the carrier gas stream forced into the reactor can further enhance the heating rate. The drop-tube configuration lends itself to pyrolysis, gasification and combustion experiments, where temperatures as high as 2200°C have been attained [Kimber & Gray, 1967]. It has been widely used to simulate coal pyrolysis and combustion under conditions thought to approach pulverized-fuel firing conditions [*e.g.* cf. Freihaut & Seary, 1981; Fletcher, 1989]. Limitations of space allow citing only a few applications. The system is also readily adaptable to high-pressure operation.

Drop-tube reactors are normally used in dilute particle injection mode, which allows monitoring the behavior of individual coal particles, without interference from neighboring particles. However, provision cannot be made to determine or control particle heating-rates. Data presented in Section 3.6 below indicate that particle residence times are not as uniform as one would have initially expected, leading to differences in extents of pyrolysis. The apparatus is known to give variable mass closures and sample weight loss is nearly always calculated indirectly, by using the 'ash-tracer' method. This latter method involves back calculation from the increase in the proportion of mineral matter within an aliquot char sample.

One clear disadvantage of this type of apparatus is the uncertain fate of tars evolved from pyrolyzing particles in free or forced fall. Within the drop-tube, the environment is one where contact of tar vapors with falling coal particles and reactor walls is probable over a sizeable length of the reactor. Furthermore, at temperatures above 550–600°C, tar residence times longer than several hundred milliseconds in the heated zone would lead to substantial extents of cracking [Stiles & Kandiyoti, 1989]. Thus amounts and structures of tars from drop-tube ('entrained flow') reactors are imperfectly related to tars originally released from the sample coal particles. The structural characterization of tars produced in drop-tube reactors is therefore, bound to lead to information of questionable value. On the benefit side, drop-tube reactors do not need sophisticated instrumentation and are relatively easy to construct, provided the central ceramic tube is able to withstand the intended temperatures. Despite reservations, they have been used widely in industrial research, to collect data for modeling pulverized-fuel combustion.

3.6 Comparison of Results From Bench-Scale Pyrolysis Reactors

We have already discussed the general requirement that results from *any* valid measurement needs to be independent of the method of the measurement. In translating this principle to the design of pyrolysis experiments, the discussion of Section 3.3 has already suggested that we need to trim our sails a little. The aim must nonetheless be to measure product distributions in a manner that is *as free as possible* from effects due to reactor design and sample configuration. In this section, we will compare results from several techniques, trying to understand how closely data from each method reflects actual sample properties. We will observe that different reactor designs are able to produce reliable data under particular ranges of reaction conditions.

3.6.1 Results from wire-mesh, fluidized-bed and 'hot-rod' reactors

Table 3.1 presents characteristic parameters of the reactors used at Imperial College for the "*Three-Rig*" comparison. Since the publication of the original study [Gonenc *et al.*, 1990], improvements have been made to reduce the scatter in the data and increase the range of experimental conditions. The original study was conducted at atmospheric pressure, between 400–800°C, using a common sample of Linby (UK) coal (moisture, a.r.: 2.1%; ash, d.b.: 5.3%; volatile matter, d.a.f.: 40.1%. C: 81.5, H: 5.2, N: 1.8, O: 10.6% w/w daf; S: 1.5% d.b.; particle size range: 106–150 μm). We will first compare results from the 'hot-rod' reactor with slow-heating rate data from the wire-mesh reactor, followed by a comparison of fluidized-bed data with high heating-rate data from the wire-mesh reactor.

Table 3.1 Summary of experimental parameters for the three reactor systems used at Imperial College by Gonenc *et al.* [1990]. (Reproduced with permission: Fuel 1990, *69*, 383; Copyright 1990 Elsevier)

	Fluidized-Bed Reactor	*Wire-Mesh Reactor*	*Hot-Rod Reactor*
Sample Size	1–15 g	5–15 mg	50–1000 mg
Temperature Range	To 900°C	To 1,200°C	To 900°C
Pressure Range	Atmospheric	Vacuum to 160 bar	Atmospheric to 150 bar
Heating Rate	Flash heating. Rate function of temp; can also operate as slow heating reactor	Very slow to 5,000°C s^{-1}	Very slow to 10°C s^{-1}
Sweep Gas Flow Rate	3–5 × minimum fluidization	Very slow to 0.3 m s^{-1} Very slow at pressure	Very slow to 10 m s^{-1}
Secondary Reactions	Potentially intense in the bed and freeboard; better quantified than "hot-rod"	Minimal but non-zero	May be minimized at very high flow rate
Accuracy/Repeatability	Tar: ±2–3% Char: ±5%	Tar: ±2–3% Char: ±1–2%	Tar: ±2–3% Char: ±3–4%

3.6.2 *Comparing results from the 'hot-rod' and wire-mesh reactors*

Figure 3.6a presents tar and total volatile yields from the 'hot-rod' reactor experiments at temperatures up to 800°C. These experiments were performed by heating in helium at 5 °C s^{-1}, with 200 s holding at peak temperature and 300 mg sample. The data is superposed on wire-mesh tar and total volatile yields (shown as solid lines) at 1 and 1000°C s^{-1} over the same temperature range. A carrier gas superficial velocity of 0.1 m s^{-1} was used to sweep through the bed of sample, during both sets of experiments.

Figure 3.6a shows that both tar and total volatile yields recovered from the "hot-rod" reactor (operated at 5°C s^{-1}) were measurably lower than those obtained in the wire-mesh reactor (solid lines B and D), operated at the lower heating rate of 1°C s^{-1}. The differences between the two sets of data reflect product loss through contact between evolving volatiles and the fixed bed of pyrolyzing coal in the "hot-rod" reactor. An additional factor contributing to product loss arises from an increase in pressure drop across the

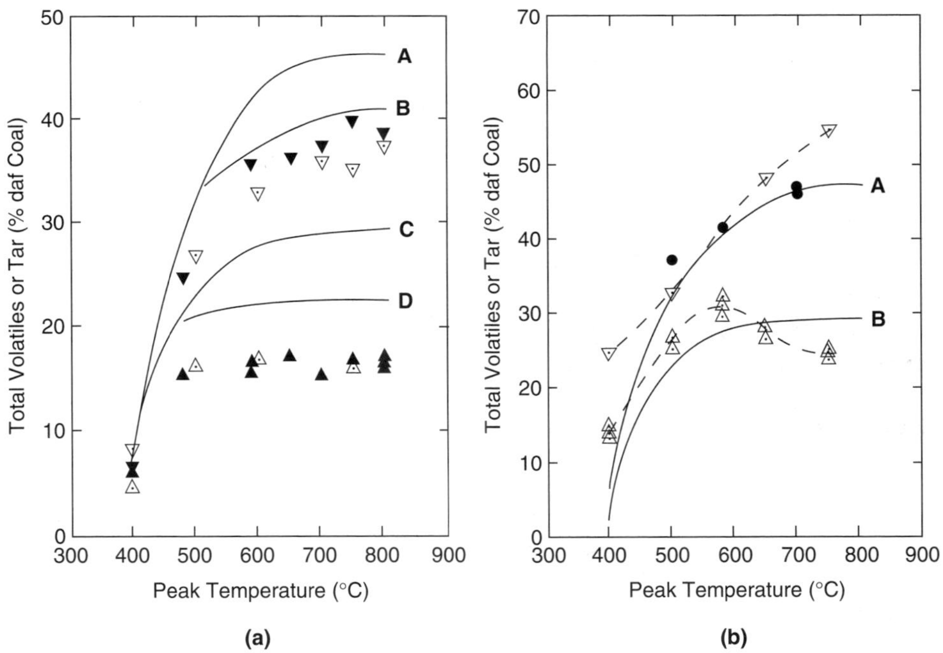

Figure 3.6 (**a**) Comparison of results from the wire-mesh (solid lines) and "hot-rod" reactors. (▽,▼): Total volatile yield from the "hot-rod" reactor with helium (▽) and H_2 (▲) as carrier gas, respectively. (△,▲): Tar yields from the "hot-rod" reactor with helium (△) and H_2 (▲) as carrier gas, respectively. Solid lines show data from the wire-mesh reactor. (**A**:) Total volatiles; 1,000°C s^{-1}. (**B**:) Total volatiles; 1°C s^{-1}. (**C**:) Tar yield; 1,000°C s^{-1}. (**D**:) Tar yield: 1°C s^{-1}. (Reproduced with permission: Fuel 1990, *69*, 383; Copyright 1990 Elsevier).
(**b**) Comparison of results from the wire-mesh and fluidized-bed reactors. (▽): Total volatile yield from the fluidized-bed reactor. (△): Tar yield from the fluidized-bed reactor. Solid lines show data from the wire-mesh reactor (heating at 1,000°C s^{-1} with 30 s holding). **A**: Total volatiles. **B**: Tar yield. (●): Total volatiles from wire-mesh reactor during heating at 1,000°C s^{-1} with 1,000 s holding at peak temperature. (Reproduced with permission: Fuel 1990, *69*, 383; Copyright 1990 Elsevier).

sample bed by several bars around 500–550°C. This appears due to sample softening, which probably leads to an ill defined gas flow pattern that would tend to exacerbate tar loss within the fixed bed [Gonenc *et al.*, 1990]. The solid triangles in Figure 3.6a represent tar and total volatile yields from operation with hydrogen at 1 bar. Tar yields were unchanged while total volatiles increased by a little more than experimental scatter. At these lower pressures, the reactivity of hydrogen appears to be quite low.

We would normally expect that extended tar residence times (order of seconds) in the heated zone would cause some of the tar product to be destroyed, with increasing reactor temperatures. In Figure 3.6b, the tar yield curve from the fluidized-bed (△) shows just such a maximum. However, in Figure 3.6a, tar yields from the 'hot-rod' reactor leveled off at ~500°C. The expected decline of tar yields at higher temperatures was clearly absent, suggesting that tars released at up to about 500°C exited from the reactor intact, *before* the reactor could reach much higher temperatures – due to the slow rate of heating. This result, obtained using a sample bed depth of ~18 mm, contrasts favorably with the drop in tar yield at high temperatures observed in the much deeper bed of the USBM reactor [Hiteshue *et al.*, 1962a] and tends to confirm the approach regarding secondary reactions of tars in fixed-beds of coal outlined above.

Effect of gas sweep velocity and bed depth in the 'hot-rod' reactor: We have seen how gas-solid interactions in the fixed-bed can lead to loss of tar and volatile products. Table 3.2 compares the effect of bed depth (extent of contact with solids) and carrier gas velocity (gas residence time) on tar yields. It was not possible to operate at sample sizes below 50 mg, as smaller sample weights gave unacceptable scatter in the results due to weighing errors. Similarly, velocities greater than 9.5 m s^{-1} could not be attempted. Greater volumes of carrier gas could not be preheated adequately and caused cooling of the sample bed. Increasing pressure drop problems also imposed an upper limit to the sweep gas velocity range possible with the present design. Hydrogen was used as sweep gas, to minimize the increases in pressure drop across the bed encountered during the coal softening and tar evolution stages. The effect was severe at the higher flow rates. Experiments summarized in Table 3.2 were carried out by heating at 5°C s^{-1} to 590°C.

Table 3.2 Comparison of the effects of bed height and sweep gas flow rate on tar yields from the "hot-rod" reactor. (Reproduced with permission: Fuel 1990, *69*, 383; Copyright 1990 Elsevier)

	1	*2*	*3*	*4*
Sample Size (mg)	300	300	50	50
Approximate Bed Depth (mm)	20	20	4	4
Superficial Velocity (m/s)	0.1	9.5	0.1	9.5
Approximate Gas Residence Time in the Coal Bed (s)	0.2	0.002	0.03	0.0003
Tar Yields (%w/w daf coal)	16.0*	18.0	18.6*	21.9**
% Change in tar yield over the 'Base Case' (%w/w daf coal)	Base Case	+2.0	+2.6	+5.5

* Average of two runs; ** average of three runs

In Table 3.2, increasing the gas velocity appears to only partially counteract the "bed-depth-effect" by removing some of the tars more rapidly from the reaction zone [Gonenc *et al.*, 1988]. There is no evidence to suggest that any additional tar is removed from the pyrolyzing coal by the substantial increase in gas velocity, above and beyond what can be collected from a virtual monolayer of sample in the wire-mesh reactor using a mere 0.1 m s^{-1} gas sweep. A similar exercise involving the wire-mesh reactor showed no sensitivity to flow rate above and beyond what was needed to remove volatiles away from the reaction zone. Table 3.2 also shows that for the shorter bed and the higher flow rate, results *tend* toward the 24% tar yield expected in the wire mesh reactor, when heating at 5°C s^{-1}.

Tar travel through a fixed bed of pyrolyzing coal: Comparing data in Columns 2 and 3 with Column 1 in Table 3.2, it appears a *hundred*-fold increase in the superficial gas velocity causes an increase in tar yield comparable to reducing the bed depth by a factor of *five*. In an important early study, Griffiths and Mainhood [1967] passed tars and model compounds through a bed of active carbon at 500°C. They reported that the "dependence of retention time on boiling point is similar to that found in gas-liquid chromatography" and that the active carbon acts as a cracking catalyst for the lower boiling constituents of the tar, but promotes charring of the higher boiling constituents. It would appear that the speed at which tar travels along the fixed bed of coal is less than directly related to the sweep gas velocity. Table 3.2 confirms that reducing the "chromatographic" travel time (i.e. reducing the bed depth) affects tar survival more directly than the superficial velocity of the sweep gas, which appears to help the motion of tar molecules along the length of the bed only to a limited extent. Axial tar transport may thus be viewed as taking place in forced convection mode, interrupted by a sequence of "sticky", reactive collisions with pyrolyzing coal particles. The re-examination of data from the original "hot-rod" reactor [Hiteshue *et al.*, 1962a] supports this proposition. A more detailed comparison of results from the wire-mesh and "hot-rod" reactors may be found in Gonenc *et al.* [1990].

3.6.3 Comparing results from the fluidized-bed and wire-mesh reactors

Figure 3.6b compares tar and total volatile yields from the wire-mesh (solid lines) and fluidized-bed reactors. The latter was operated with a freeboard height corresponding to 0.8-second volatiles residence time. This was the shortest freeboard residence time feasible with this fluidized-bed and was intended to minimize tar cracking in the freeboard [Stiles & Kandiyoti, 1989]. Several points have emerged from this comparison, showing up some shortcomings in *both* designs.

First, the wire-mesh reactor gave less tar and volatiles at 400°C compared to the fluidized-bed. During this set of runs, the time-temperature ramp in the wire-mesh reactor was followed by 30 seconds holding at peak temperature. At the relatively low temperature of 400°C, 30 seconds appears to have been too short for pyrolysis reactions to run to completion. By contrast, in the fluidized-bed reactor, chars remain in the heated bed for several minutes after the furnace heating current is turned off. This is done for the solids to cool sufficiently to be removed and apparently long enough for some more volatiles to evolve from the sample. In addition to heat transfer being more efficient due to the vigorous circulation of particles in the bed, the sample residence time at or near peak temperature in the fluidized-bed thus turns out to be somewhat longer. In

Table 3.3 Linby vitrinite concentrate total volatile and tar yields (% w/w daf basis) as a function of holding time at 400°C in the wire-mesh reactor. Heating rate: 1000°C s^{-1}. (Reproduced with permission: Fuel 1993, 72, 3; Copyright 1993 Elsevier)

Holding Time, s	30	100	200	250
Tar Yields; % daf	6.7	13.1	12.3	14.3
Total Volatiles; % daf	10.1	20.8	19.9	22.1

Figure 3.6b, the points marked as (●) represent data for much longer (1,000 s) hold times in the wire-mesh reactor. The total volatile yield at 500°C increased sharply, but results at 600 and 700°C remained unchanged. Clearly, 30 seconds at 400–500°C in the wire-mesh reactor was not sufficient for pyrolysis reactions to reach completion.

Subsequent work by Li *et al.* [1993a] presented in Table 3.3 shows this effect more clearly. At 400°C, the main changes for Linby coal occurred between 30 and 100 seconds. Recent work by Fukuda showed variations of 30 to 400 seconds between different coals, regarding the length of time required in the wire-mesh reactor for pyrolysis reactions to reach completion at temperatures below 500°C [Fukuda, 2002; Fukuda *et al.*, 2004]. The effect appears dependent on structural features of particular coals. By contrast, the analogous length of time needed at 700°C for completion of volatile release – for *all* coals so tested – was less than 1-second.

In the wire-mesh reactor, the longer times required for reactions to reach completion at low temperatures are probably due to overlapping effects of slow reaction kinetics, compounded by poor heat transfer at these low temperatures. Above 500°C, where the mesh can be observed to glow with the naked eye, radiative heat transfer would play an increasingly important role. It is also likely that thermal contact between mesh and sample improves if/when coals soften, usually near and above 500°C during fast heating experiments. In any case, some care seems necessary in the evaluation of results obtained at temperatures between 400–500°C, due to the sensitivity of the yields to small variations in the final experimental temperature attained. The difference between yields from the two reactors may also have been somewhat exaggerated by the larger estimated error inherent in fluidized-bed char determinations (Table 3.1). The latter measurement is made by tipping out the contents of the fluidized-bed and comparing sand bed weight plus coal before the pyrolysis experiment with the weight of total solids afterward. With the best of operator care, this turns out to be a difficult measurement to perform reproducibly.

In Figure 3.6b, the tar yields determined using the fluidized-bed reactor might be observed to trace a maximum with increasing temperature and to diminish above about 580–590°C. This is the other major difference between the two reactors. Maxima in "tar vs. temperature" data are a characteristic of pyrolysis reactors, where significant tar destruction occurs within the reactor through secondary charring and/or cracking reactions. Figure 3.6 indicates these reactions are more intense at higher temperatures. As previously pointed out by Tyler [1979; 1980], the simultaneous monotonic increase of the total volatile curve from fluidized-bed reactors suggests that a significant proportion of tar was "lost" by cracking to gaseous products. For biomass samples and low rank coals, the position of the maximum shifted to lower temperatures, depending on the thermal stability of evolving tars. For cellulose and silver-birch (soft-wood) tars, the maximum occurred between 400–425°C, while for a lignite, it was observed nearer 530 °C [Stiles & Kandiyoti, 1989].

The ability to change freeboard residence times of volatiles enables the calculation of tar cracking kinetics. It also allows estimating the amount of tar *destroyed* in the bed itself, as a function of reactor temperature. The procedure has been described by Gonenc *et al.* [1990]. For the present experiments, ~31% of the original sample mass was estimated to have *entered* the freeboard at 580°C as tar, while at 750°C, the figure was down to 26%. By contrast, relative freedom of tars from secondary reactions in the wire-mesh reactor enables tar yields to remain constant within experimental scatter, once the temperature for maximum tar production has been reached. However, in the case of thermally more sensitive cellulose tars, or very high heating rates [Peters *et al.*, 1980; Griffin *et al.*, 1993], some tar loss has been observed in wire-mesh reactors, showing a shallow tar-temperature maximum. In wire-mesh reactors developed at Imperial College, the flow of gas through the sample holder has served to minimize this effect. However, when peak temperatures above 1,000°C are used, some tar loss has been observed – particularly for long hold-times, when any tar deposited on the ceiling of the trap (near the entrance to the off-take tube; see Figure 3.1b) "sees" radiation from the glowing mesh. The effect can be minimized by shortening the hold time at high peak temperatures during tar yield determinations.

3.6.4 Comparing results from a "drop-tube" (entrained flow) and a wire-mesh reactor

These two reaction systems have all but occupied center-stage in most combustion related research during the past two decades. Data from both techniques have been used in mathematical simulations of pf-combustion. Inexplicably, there has been little effort to compare results between the two techniques.

Hindmarsh *et al.*, [1995] have compared results from these two types of reactors at atmospheric pressure, using a common set of coal samples. Table 3.4 presents pyrolysis data from the two reactors using similar, but not quite identical, experimental conditions. Two different particle size ranges were used in the two reactors: a 38–75 μm fraction in the entrained flow reactor (EFR) and the 106–152 μm fraction in the wire-mesh reactor. Furthermore, the 'drop-tube' reactor was operated with nitrogen rather than helium and at 50°C higher than the peak temperature used in the wire-mesh apparatus. Nominal residence times were on the order of 1 s and sample weight loss in the EFR was calculated by the ash tracer method.

The results in Table 3.4 show systematically lower volatile yields from the EFR compared to the wire-mesh reactor. The differences were large for the three softening coals: Linby and the two samples of Illinois No.6. Columns 3 and 6 in Table 3.4 provide a direct indication of the level of residual volatile matter remaining in the chars. The 3–9% range of values recorded for the EFR chars suggests that pyrolytic reactions have not run to completion. These data may be interpreted as indications that the 1 s *nominal* residence time of particles in the EFR was not sufficient for heatup to 1,000°C to be

Table 3.4 Comparison of pyrolysis data from an EFR and a wire-mesh reactor, using common samples. (Reproduced with permission: Fuel 1995, 74, 1185; Copyright 1995 Elsevier)

	Entrained-Flow (drop-tube) Reactor Pyrolysis Results (wt % daf) at 1,000°C under nitrogen Estimated residence time: 1 s		*Wire-Mesh Reactor Pyrolysis Results (wt % daf) at 950°C 5000°C s^{-1} under helium 2 s holding at peak temperature*		
	Total Volatiles	*Residual volatile matter in char*	*Total Volatiles*	*Tar*	*Residual volatile matter in char*
Taff Merthyr	16.0	3.6	17.2	9.5	1.7
Emil Mayrisch	20.0	3.7	21.2	11.6	1.9
Linby	39.1	8.2	49.6	29.9	2.3
Illinois No. 6 (SBN**)	46.8	9.3	53.6 (53.4)*	28.9 (29.2)*	3.7 (3.8)*
Illinois No. 6 (APCS**)	48.0	4.9	59.4	n.d.	2.4

* Values in parentheses are for 5 seconds hold; char VM at zero hold time, 4.0 wt% (daf)
** SBN: Steinkohlebank, Netherlands; APCS: Argonne Premium Coal Sample program [Vorres, 1991]

completed, at least for some proportion of the particles. It is also likely that a small distribution of residence times developed around the 1-second nominal time. The observed problem would probably be overcome by designing a longer "drop-tube". When working at 1,000°C in the wire-mesh reactor, sample weight loss was found to be largely completed during heatup. Table 3.4 shows a marginal difference in total volatiles in the wire-mesh reactor, when longer hold-times were used for "Illinois No. 6 (SBN)".

As already indicated, different carrier gases were used in the two reactors. Several further experiments indicate that the properties of the gases used in the two reactors may account for some of the differences observed between the two sets of results shown in Table 3.4. Table 3.5 shows that operation with nitrogen *in the wire-mesh reactor* has the

Table 3.5 Pyrolysis of Illinois No. 6 (SBN) Coal; 1,000°C in different atmospheres. EFR: entrained flow reactor; WMR: wire-mesh reactor. (Reproduced with permission: Fuel 1995, 74, 1185; Copyright 1995 Elsevier)

Reactor	*Atmosphere*	*Total Volatiles (wt% daf)*	*Tar (wt% daf)*
EFR	N_2	46.8	n.d.
WMR	He	51.4	27.2
	N_2	48.2	n.d.
	N_2 (O_2 free)	48.1	n.d.

effect of reducing the weight loss by about 3%. This is likely to be due to the lower thermal conductivity of nitrogen compared to helium. It would appear that somewhat longer residence times in the 'drop-tube' reactor coupled with the common use of helium would go some way toward aligning results more closely between the two reactors. The relative combustion reactivities of the chars determined by thermogravimetric analysis seemed in reasonably close agreement [Hindmarsh *et al.*, 1995]. However, the far greater scatter in data from entrained flow reactors appears endemic to the design and is probably associated with difficulties relating to complete char recovery.

3.6.5 Overview of results from different pyrolysis reactors

We have found that the level of contact between heated solids and evolved volatiles in fixed bed reactors have measurable effects on results from pyrolysis experiments. Compared to yields from the wire-mesh reactor, the fixed-bed ("hot-rod") experiments gave less tar and volatiles, when working at similar temperatures, pressures and heating rates. Moreover, tar losses in the 'hot-rod' reactor were found to be more sensitive to sample bed depth than to carrier gas superficial velocity. This is understandable in terms of collisions between larger and possibly more reactive molecules with heated solids having entirely different outcomes than contact with light (carrier) gas molecules.

By contrast, greater consistency between tar and total volatile yields from the fluidized-bed and the wire-mesh reactor was observed, when the latter was operated at high heating rates. At lower temperatures, divergences observed between the two reactors could be accounted for by the relative shortness of the holding time in the wire-mesh reactor. At higher temperatures, tar survival in the fluidized-bed deteriorated, due to secondary tar destruction reactions within the fluidized-bed itself and in the reactor freeboard. Tar cracking reactions within the bed itself were found to be more significant above about 650°C. By contrast, tar yields in the wire-mesh reactor reached a steady level between 600–700°C and mostly remained constant at higher temperatures.

Compared to the entrained flow ("drop-tube") reactor, the wire-mesh reactor appears as the more reliable instrument for measuring pyrolysis yields. Some of the problematic aspects of the EFR data may be improved by lengthening reactor tubes. However, avoiding – or narrowing – the distribution of residence times for exiting sample particles does not seem quite as straightforward. This in itself may not be an important shortcoming for pyrolysis experiments, but would introduce an extraneous factor when working with reactive gases such as steam or CO_2. The difficulties encountered with char recovery and mass closures appear to be inherent to the configuration of the apparatus. Furthermore, EFRs were never really intended for accurate measurements of the amounts of tar released from sample particles. The yields would have declined and the structures of surviving tars would be expected to have thermally degraded. We would expect tars from EFRs to differ substantially from primary tars.

We have observed, so far, that wire-mesh reactors are capable of operating over wider ranges of reaction conditions than other reactor configurations. They require small amounts of sample and enable the recovery of products for further characterization, in a state relatively uncontaminated by extensive secondary reactions. These advantages make the wire-mesh reactor the instrument of choice in characterizing samples of special interest, or scarcity, such as macerals and kerogens. However, two distinct *dis*advantages of wire-mesh reactors compared to fluidized-bed and drop-tube (entrained-flow) reactors must be kept in mind:

1. Fluidized-bed and entrained-flow reactors may be operated using a range of particle sizes, whereas the wire-mesh reactor is limited in the sizes of particles it can handle, due to the fixed sizes of holes in the mesh. In particular, the wire-mesh reactor configuration would prove clumsy in experiments aiming to determine the behavior of particles much larger than several hundred micrometers. At the other extreme, static electricity usually causes coal particles below about 60 μm to adhere to feeder and reactor walls, making it difficult to carry out *quantitative* experiments in *any* reactor configuration.
2. Fluidized-bed and entrained-flow reactors are normally designed to handle variable quantities of sample and are capable of producing chars in quantities required for, say, surface area determinations. By contrast, sample sizes in the wire-mesh reactor are of the order of ~5–6 mg and amounts of char recovered are normally even smaller.

The rest of this Chapter will focus on several studies carried out using some of the tools and concepts relating to pyrolysis experiments developed in the foregoing Sections.

3.7 Case Studies – Pyrolysis of Coal Macerals

In the first of these studies, the pyrolytic behavior of coal maceral concentrates will be examined. We have already touched upon the botanical origins and the organic geochemistry of these materials in Chapter 2. Technological interest in the diversity of the constituent components of coals arises from their differences in behavior under particular processing conditions. The thrust of the present study required accurate determination of tar yields and the quantitative recovery of captured tars for further examination. An atmospheric pressure wire-mesh reactor was used, due to its greater accuracy in tar capture and recovery. In the event, the requirements of the study provided the stimulus for developing the tar trap described in Figure 3.1b. The capability of the wire-mesh reactor to work at selected heating rates enabled disentangling effects due to this parameter from observations related to sample dispersion.

Experimenters working with coal macerals are faced with several peculiarities not otherwise encountered. As outlined in Chapter 2, coals are composite materials. The separation of macerals from the parent coals takes place in solvents of variable density and relies on relatively small density differences between finely ground particles. [*e.g.* see Pandolfo *et al.*, 1988 and references therein]. Most maceral separation methods therefore require fine subdivision of the sample – at the level of several μm or less. In the wire-mesh reactor, where sample particle sizes normally need to be greater than 100 μm (60–70 μm at its finest), this creates a problem. Pellets need to be re-constituted using these fine particles in a hydraulic press (*e.g.* for making KBr discs for infrared work). The pellets are then re-ground and sieved, to obtain particles in the requisite size range. Inertinites do not usually compact very well, adding to the difficulty by losing weight in the form of fine powder and contributing to experimental scatter. Furthermore, despite the fine grinding, maceral sample purities vary between different batches. Samples are rarely pure (i.e. 100%). It is important therefore to distinguish between actual macerals in coals and the maceral *concentrates* that most experimenters are constrained to work with. For laboratories other than those specialized in maceral separation, the choice of maceral concentrates to be examined is largely a matter of sample availability.

3.7.1 Pyrolysis of coal macerals

Two sets of experiments will be described. The first compares the sensitivity of macerals to changes in heating rate between 1 and 1,000°C s^{-1}. The second set of experiments compares product distributions from different macerals during atmospheric pressure and vacuum pyrolysis.

Effect of heating rate on product distributions from maceral concentrates: In a concise and informative (and surprisingly widely ignored) review, Taupitz [1977] suggested, "... only vitrinite is susceptible to the speed of heating ..." Most Northern Hemisphere coals are predominantly vitrinitic and thus should be expected to show some sensitivity to "the speed of heating". We have since quantified the levels of sensitivity of different coals to changes in heating rate and found that differences between some coals can be significant [Gibbins-Matham & Kandiyoti, 1988]. Clearly, the behavior of individual macerals under different heating regimes may provide clues regarding the variations in the sensitivities of "whole" coals to changes in heating rate.

The basic trends describing the evolution of tars and volatiles from maceral concentrates as a function of temperature are broadly similar to those of whole coals (*e.g.* Figure 3.3). Table 3.6 shows that, liptinites gave the highest and inertinites the lowest conversions to volatile matter. The vitrinite results were numerically close to those from the predominantly vitrinitic whole coal samples. All liptinites melted upon heating and most inertinites did not. Linby coal and its vitrinite concentrate are interesting because they melt only upon being heated rapidly (1,000°C s^{-1}) and *not* if heated slowly (1°C s^{-1}). Linby (a typical vitrinitic) coal also happens to give volatile and tar yields which showed the greatest sensitivity to changes in heating rate, among the samples tested. In Chapter 6, we will present evidence attempting to show *how* the transitional melting behavior of Linby coal is related to its sensitivity to changes in heating rates. We will see that this makes for strong coke from unexpected coals.

The data in Table 3.6 indicate that clear-cut generalizations are difficult to make. Above 500°C, many of the samples showed *some* sensitivity to changes in heating rate, with liptinite and inertinite concentrates broadly showing less sensitivity. As already observed, however, maceral samples are never quite pure. In evaluating results presented in Table 3.6, therefore, some suspicion remains that unintended vitrinitic inclusions may have contributed to the unexpected sensitivity to heating rates of liptinite and inertinite samples. The compositions of the samples may be gleaned from Table 3.9a [also see Li *et al*., 1993a, 1993b].

Working under 20 bar He, with relatively pure liptinites and inertinites from two different coals, Messenbock [1998] again found the sensitivity of these macerals to the

Table 3.6 Pyrolysis of coals and their maceral concentrates as a function of heating rate in atmospheric pressure helium. 30 s holding at 700°C. (Reproduced with permission: Fuel 1994, *73*, 851; Copyright 1993 Elsevier). Please refer to Table 3.9a and Li *et al.* [1993a, 1993b] for the elemental and petrographic compositions of the samples

	Carbon Content	*Heating Rate: 1000°C s⁻¹*		*Heating Rate: 1°C s⁻¹*	
		Tar Yields	*Total Volatiles*	*Tar Yields*	*Total Volatiles*
		Point of Ayr			
Whole Coal	85.2	26.1	42.4	20.7	33.6
Clarain	82.6	24.4	40.8	18.1	34.7
Durain	84.9	30.1	44.2	26.8	40.5
Vitrinite conc.	84.8	24.6	40.1	20.5	33.9
Liptinite conc. 1	85.7	47.1	62.0	43.4	56.8
Liptinite conc. 2	84.8	47.9	62.5		
Inertinite conc.	84.2	16.1	31.3	15.4	30.4
		Linby			
Whole Coal	82.3	30.7	46.6	24.2	40.2
Vitrinite Conc.	77.6	29.5	45.2	20.3	37.8
Liptinite Conc.	79.1	48.9	64.9	45.5	59.9
Inertinite Conc.	78.2	26.3	42.4	19.6	35.8
		Cortonwood			
Whole Coal	86.5	29.8	40.9	26.7	37.6
Vitrinite Conc.	85.9	26.5	42.1	24.7	39.2
Liptinite Conc.	85.1	53.7	70.7	54.9	65.5
Inertinite Conc.	85.7	22.5	35.8	21.0	33.4
		Freyming			
Whole Coal	82.3	28.4	44.2	20.3	36.7
Vitrinite Conc.	83.5	26.3	43.0	18.2	34.4
		Dinnington			
Vitrinite Conc.	81.4	21.2	34.2	14.3	31.0
Liptinite Conc.	84.3	48.8	63.1	47.3	58.9

heating rate to be limited (Table 3.7). Thus after much toil and a quarter of a century, we seem able to add to Taupitz's dictum about vitrinites that *some* liptinites and *some* inertinites may also be "susceptible to the speed of heating", albeit generally to a lesser extent than the corresponding vitrinites.

Pyrolysis of maceral concentrates under vacuum: Early work summarized by H.C. Howard [1963] and J.B. Howard [1981] indicates that tar and total volatile evolution during pyrolysis should be enhanced when the external pressure is reduced. The effect is related to faster mass transfer out of the particle, due to a steeper outward pressure

Table 3.7 Sensitivity of total volatile yields to changes in heating rate. Pyrolysis at 20 bar; 10 s holding at 1000°C. [Messenbock, 1998]

Sample	*Vol. yield; 10°C s^{-1} %, daf basis**	*Vol yield; 1000°C s^{-1} %, daf basis**	*Maceral Purity*
Liptinite No. 11	66.5	67.0	91.2
Repeat run	65.6	67.1	91.2
Inertinite No. 18	18.9	20.3	97.8
Repeat run	20.3	22.5	97.8

* daf: dry, ash free basis

gradient [see Suuberg, 1985 on mass transfer effects]. In the process, tar precursors that normally act as binder in softening coals are removed from the pyrolyzing mass. Residual chars from pyrolysis under "vacuum" are, therefore, less agglomerated. It is still possible, however, to depress volatile yields by hasty reactor design. Roy *et al.* [1985] reported not much more than ~36–37% total volatiles at 1,000°C, using a coal with a "proximate analysis" volatile matter content of 34%. The experiments were done in a 50 mm diameter tube, where 120 grams of sample particles had been placed in the form of a fixed bed.

Relatively little attention has been paid to the possible oxidative re-polymerization of freshly produced coal liquids. H.C. Howard's [1963] description of the 'short-path vacuum still' [Sun *et al.*, 1958] contains a fascinating note regarding the amber color of material initially condensed onto cold surfaces. This material gradually darkened upon exposure to air. The observation was made possible by the large size of the apparatus and of the relatively large amount of sample used – not possible in wire-mesh reactors. The low experimental temperature (up to 500°C) would perhaps explain the lightness of the oil. However, the darkening "upon exposure to air" certainly raises questions about whether our current methods of tar recovery lead to the polymerization of the tars, with attendant increases of molecular mass distributions and other changes in structural features. Size exclusion chromatograms and other properties of pyrolysis tars appear to change in a fashion that we think is consistent with changes in *reaction* conditions [Suuberg *et al.*, 1985; Li *et al.*, 1993a, 1993b]. Such trends suggest that the fundamental characteristics of the tars could not have been radically altered by current methods of tar handling. Nevertheless, the image of the amber colored coal tar is compelling and the subject well worth revisiting.

Wire-mesh reactors have proved particularly adaptable to work under vacuum. An early wire-mesh reactor [Jüntgen & van Heek, 1968] was in fact designed to operate at about 10^{-7} bar and the volatiles passed directly into a mass spectrometer. Howard and co-workers [Howard, 1981; Anthony *et al.*, 1975; Suuberg *et al.*, 1978a, 1978b], worked between 10^{-4} and about 100 bar. They found sample weight loss could be increased systematically by reducing the pressure, by about 7% above the weight loss at atmospheric pressure. The effect appears associated with competition between tar precursor transport out of the coal particle and intraparticle secondary char formation and cracking reactions.

Researchers at Princeton University have also observed greater weight loss under reduced pressures. Furthermore, the effect of increased heating rate on weight loss was observed to decrease when the pressure was increased from about 10^{-4} to 2 bar [Niksa *et al.*,1982a; 1982b; 1984]. Similar larger increases in yields at high heating rates when operating at reduced pressures were also reported by workers at Imperial College [Gibbins & Kandiyoti, 1988; 1989b; Gibbins *et al.*, 1989c].

Winans and co-workers have analyzed vacuum pyrolysis tars, either released from a probe attached to the inlet of a mass spectrometer, *or* prepared separately "in batch mode" [Winans *et al.*, 1986; Winans & Neal, 1990; Winans, 1991; Winans *et al.*, 1991; Hunt *et al.*, 1991]. No yields were reported from the samples prepared in batch mode nor would it have been possible to determine product distributions from the probe configuration. The attendant mass spectrometric work was capable of identifying small hydrocarbon molecules.

The recovery of tars during wire-mesh pyrolysis experiments under reduced pressure presents peculiar difficulties. One common difficulty is the recirculation of evolving tars in the vicinity of the heated mesh: "a certain amount of carbon deposition from cracked volatiles occurred on the stainless steel screen sample holder, particularly at lower pressures" [Howard, 1981]. Suuberg *et al.* [1978a, 1978b] recovered tars by solvent washing of inner surfaces and linings of the wire-mesh cell. In later work, Suuberg and co-workers [Unger & Suuberg, 1984; Suuberg *et al.*, 1985] using similar procedures, reported greater concentrations of large molecular mass (MM) components in tars produced under vacuum compared to atmospheric pressure, but suggested that their tar recovery procedure was "somewhat inefficient".

As already explained, tar capture in the Imperial College wire-mesh reactors was achieved by sweeping evolved volatiles away from the reaction zone. Clearly, an efficient externally imposed flow field cannot be set up under vacuum. In the absence of a carrier gas stream, there can be no preferential direction for volatile evolution. Volatiles could diffuse downward from the mesh, as well as upward. On the other hand, quantitative tar capture in wire-mesh reactors clearly depends on minimizing the recirculation of tar vapors within the cell and in particular, in the vicinity of the mesh. Another constraint is the necessity to minimize the cracking of tars deposited on cold surfaces by the heated, glowing wire-mesh, which radiates intensely at temperatures above 500°C.

In arriving at a tar-trap design suitable for operation under reduced pressures (described below), a compromise had to be arrived at between two extremes:

(i) Placing flat parallel cold-plates very close to the sample holder, as in the "short-path vacuum still" [Sun *et al.*, 1958], carried the attendant risk of inducing tar charring and cracking by radiation from the mesh.

(ii) At the other extreme, washing tars off the reactor walls would not allow suppressing tar recirculation, with attendant dangers of charring and cracking on the mesh itself.

Figure 3.7 presents a schematic diagram of the tar trap assembly designed for the recovery of tars in the wire-mesh reactor operating under reduced pressures [Li *et al.*, 1993b]. This design aims to condense evolved tars on exposed trap surfaces, to minimize tar recirculation and simultaneously reduce the exposure of condensed tars to radiation

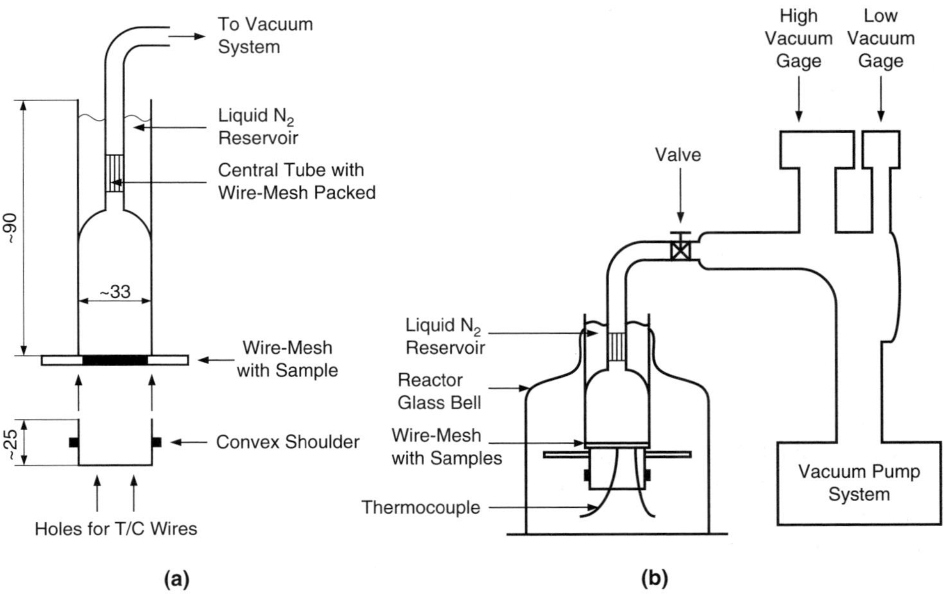

Figure 3.7 Schematic diagram of tar traps designed for use in vacuum pyrolysis experiments. (a) Off take tube above the mesh and 'shallow cup' underneath. (b) Sketch of assembled system showing the positioning of the cell and the vacuum pump. All dimensions are given in millimeters. (Reproduced with permission: *Fuel* 1993, 72, 1459; Copyright 1993 Elsevier).

from the mesh. The top part of the trap configuration in Figure 3.7 was similar to the atmospheric pressure trap (Figure 3.1b, above). The 'chimney' was packed with wire-mesh, cooled with liquid-N_2 and connected directly to the vacuum pump. The bottom trap was a shallow cup mounted directly underneath, intended to capture tars released below the mesh and to block tar recirculation. Tiny holes in the cup were used for threading through thermocouple wires and for equilibrating the pressure. In order to limit losses due to cracking by radiation, holding times at peak temperature were reduced from 30 to ~5 s at 700°C. Previous experiments at atmospheric pressure have shown volatile release at 700°C to be virtually completed in about 1 s [see Section 3.3.1 above]. Following initial trials with the new set of traps, it was concluded that this design worked well and would yield the largest tar yield identifiable in the present reactor configuration.

Table 3.8 shows that, with the exception of the inertinite concentrates, all samples gave enhanced yields when the pressure was reduced from the atmospheric. Under vacuum, as well as at atmospheric pressure, total weight loss and tar yields were greatest for liptinites and smallest for inertinites:

liptinites > vitrinites > inertinites.

Comparing yields from vacuum and atmospheric pressure pyrolysis experiments, Table 3.8 shows that observed increases in *tar* yields under vacuum were measurably greater than corresponding increases in *total volatile* yields. About 5% more tar was recorded,

Table 3.8 Pyrolysis tar and total volatile yields from Linby, Point of Ayr and Freyming *whole*-coals and maceral concentrates at atmospheric pressure and under vacuum. Heating at 1000°C s^{-1} to 700°C with 30 s holding at atmospheric pressure and 5 s holding under vacuum. These values represent the average of between 2 and 7 determinations. The data were presented on a w/w% daf basis. Corresponding petrographic compositions may be found in Table 3.9a and Li *et al.* [1993a, 1993b]. (Reproduced with permission: Fuel 1993, 72, 1459; Copyright 1993 Elsevier)

Sample	*Atmospheric Pressure*		*Vacuum*	
	Tar Yields	*Total Volatiles*	*Tar Yields*	*Total Volatiles*
		Linby		
Whole coal	30.7	46.6	37.9	49.7
Vitrinite concentrate	29.5	45.2	36.8	47.4
Liptinite concentrate	48.9	64.9	58.2	67.4
Inertinite concentrate	26.3	42.4	31.5	41.3
		Point of Ayr		
Whole coal	26.1	42.4	33.1	41.8
Vitrinite concentrate	24.6	40.1	31.3	42.8
Liptinite concentrate	47.1	62.0	62.7	72.6
Inertinite concentrate	16.1	31.3	18.7	29.5
		Freyming		
Whole coal	28.4	44.2	35.6	45.4
Vitrinite concentrate	26.3	43.0	30.7	43.5
Liptinite concentrate	42.7	55.1	47.5	57.3

compared to a corresponding increase in total volatile yields of ~1–2%. The result indicates that enhanced volatilization at reduced pressure primarily affects the devolatilization of tars. It appears to facilitate the escape of tar precursors, which at atmospheric pressure would have cracked and would have given rise to about 5% more gaseous product. It is likely that the effect is associated with both enhanced bubble transport within particles and enhanced evaporation due to a greater driving force at the external particle boundary.

3.7.2 *Probing for synergistic effects between maceral components of coals during pyrolysis*

Table 3.9a presents the maceral analyses of Linby and Point of Ayr coal derived samples used in the study. The maceral concentrate samples are not pure. However, combined with the yield data from Tables 6 and 8, it is possible to back-calculate the would-be tar and volatile yields corresponding to the "pure" macerals of these coals. The calculation is straightforward [Li *et al.*, 1991], simply requiring three equations and three unknowns. Table 3.9b presents these hypothetical yields calculated for the "pure" macerals of these two coals.

Table 3.9a Petrographic analyses of Linby and Point of Ayr maceral group concentrates. (Reproduced with permission: Fuel 1993, 72, 1459; Copyright 1993 Elsevier)

	Vitrinites % v/v, dmmf	*Liptinites % v/v, dmmf*	*Inertinites % v/v, dmmf*
Linby (whole) coal	73	15	12
Linby vitrinite concentrate	85	6	9
Linby liptinite concentrate	16	70	14
Linby inertinite concentrate	35	4	61
Point of Ayr (whole) coal	84	6	10
Point of Ayr vitrinite concentrate	91	5	4
Point of Ayr liptinite concentrate	30	61	9
Point of Ayr inertinite concentrate	17	3	80

Table 3.9b Calculated tar and total volatile pseudo-yields for the pyrolysis of Linby and Point of Ayr coal derived "pure" maceral groups at atmospheric pressure and under vacuum. Heating at 1000°C s^{-1} to 700°C with 30 s holding (atmospheric pressure) and 5 s (vacuum). (Reproduced with permission: Fuel 1993, 72, 1459; Copyright 1993 Elsevier)

Sample	*Atmospheric pressure*		*Vacuum*	
	Tar Yields	*Total Volatiles*	*Tar Yields*	*Total Volatiles*
	(w/w%, daf sample)			
Linby				
Vitrinites	28.2	44.0	35.6	46.4
Liptinites	59.1	75.1	69.7	78.5
Inertinites	23.1	39.3	26.7	35.9
Point of Ayr				
Vitrinites	22.5	38.8	29.0	40.7
Liptinites	64.2	78.4	86.5	95.4
Inertinites	12.9	27.9	14.0	24.6

The results largely reflect trends already observed in the data of Table 3.8. Compared to atmospheric pressure pyrolysis, greater yields were observed under vacuum in the case of whole coals, and of vitrinite and liptinite concentrates. However, inertinite concentrates gave smaller increases in tar yield on the application of vacuum, and the total volatile yields actually *decreased*. The latter result appears counter-intuitive; repeated experiments with the present samples produced similar results and did not help clarify reasons for this observation. In a previous study, similar observations on an inertinite concentrate derived from Treeton-Barnsley coal have been reported by Li *et al.* [1991a].

Table 3.10 Comparison of experimental and calculated tar and total volatile yields, for the pyrolysis of Linby and Point of Ayr *whole*-coals. Operation at atmospheric pressure and under vacuum. The calculated results are based on assuming additivity of yields from individual pure maceral groups. Heating at 1,000°C s^{-1} to 700°C with holding 30 s (atmospheric pressure) or 5 s (vacuum). (Reproduced with permission: Fuel 1993, 72, 1459; Copyright 1993 Elsevier)

Substrate	*Atmospheric pressure*		*Vacuum*	
	Tar yields	*Total volatiles*	*Tar yields*	*Total volatiles*
	(w/w%, daf sample)			
Linby whole coal				
Calculated	32.2	48.1	39.6	50.0
Experimental	30.7	46.6	37.9	49.7
Point of Ayr whole coal				
Calculated	24.0	40.1	30.9	42.3
Experimental	26.1	42.4	33.1	41.8

When combined with the maceral compositions of each coal (Table 3.9a), the tar and volatile yields reported in Table 3.9b allow the calculation of yields from the corresponding whole coals, as a weighted sum of pyrolysis yields from individual "pure" macerals. The calculation may be likened to "reconstituting" the pyrolysis yields of "whole" coals from those of individual maceral components present in each coal.

Table 3.10 compares *experimental* pyrolysis yields measured using ordinary Linby and Point of Ayr coal samples (from Table 3.8), with those of the *calculated* ("*reconstituted*") yields for the same samples given in Table 3.9b. For both vacuum and atmospheric pressure data, the level of agreement in Table 3.10 between calculated and experimental values was well within experimental repeatability. In view of errors inherent in the pyrolysis experiments themselves and the usually larger errors in petrographic analyses, the level of agreement is a little surprising. However, the results clearly indicate the absence of measurable, experimentally significant (i.e. large) synergistic effects between different maceral groups during the pyrolysis of coals.

Synergistic effects between dissimilar components of solid fuels: The absence of synergistic effects between coal macerals during pyrolysis is an interesting result.

During the pioneering days of *biomass* pyrolysis, it was standard practice to introduce little metallic boats laden with powdered sample into horizontal furnaces. It has been recognized, since those early days, that the linear polymer cellulose (Figure 3.8) decomposes to give a mixture, composed predominantly of the monomer, *laevoglucosan* (1,6-anhydro-β, D-glucopyranose) [Shafizadeh, 1968]. Shafizadeh & Fu [1973] reported laevoglucosan yields of about 50%. In the absence of thermal degradation, theoretically, a nearly 100 percent yield of laevoglucosan would have been expected. Later work by Stiles & Kandiyoti [1989] was carried out in a fluidized-bed reactor, where tars recovered at moderate temperatures would be less degraded by secondary reactions. In these experiments, nearly 85% of the original cellulose mass was recovered as laevoglucosan.

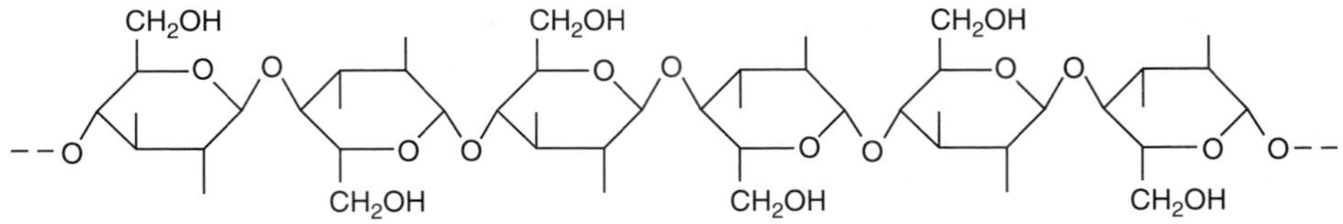

Figure 3.8 The structure of cellulose.

On the other hand, when *wood* samples, normally containing up to ~50 percent cellulose, were pyrolysed in Shafizadeh's inescapable metallic boats, only about 3% laevoglucosan could be recovered. All subsequent work has confirmed the low levels of occurrence of laevoglucosan in tars from composite lignocellulosic materials [*e.g.* Fraga, 1990], notwithstanding the much higher proportions of cellulose in the original sample. These observations unambiguously point to synergistic effects between the several polymeric components of pyrolyzing wood particles. In this respect, as in many others, lignocellulosic biomass behavior seems to differ from that of coals. In this particular case, we look to the different thermal stabilities of the component parts (and also of their tars) for a likely explanation.

Figure 3.6b shows the temperature of the maximum in the fluidized-bed tar yield curve for Linby coal at around ~580–590°C. For a lignite, the analogous maximum had occurred at ~530°C and between 400–425°C for a softwood and for cellulose itself [Stiles & Kandiyoti, 1989]. This reduction in the temperature of the tar-yield maximum signals the progressively earlier onset of extensive tar destruction reactions in the fluidized-bed and in the reactor freeboard. It provides an indication of the relative thermal stability of tars evolved from different feedstocks. We also know from previous work that the thermal decomposition temperatures of lignins are usually intermediate between those of cellulose and those of coals [e.g. Fraga *et al.* 1990]. There seems to be about ~100°C difference between the onset of decomposition temperatures of cellulose and lignin.

One likely explanation for synergistic effects observed in wood pyrolysis relies, therefore, on the difference between the decomposition temperatures of the component parts. Thus, when the cellulose in wood begins to decompose, the very fine intermeshing with lignin – and indeed with hemicelluloses – would provide a heated solid matrix in contact with thermally sensitive cellulose tars. This set of secondary reaction paths appears to decompose the laevoglucosan, which under other experimental conditions has been shown to constitute the predominant primary product in cellulose tar.

In attempting to test this hypothesis, mixtures of cellulose and lignin powders were pelletized before pyrolysis. However, the results could not reproduce the synergistic effects observed with natural wood [Fraga, 1990]. If the model outlined above does correspond to the actual sequence of events, it appears to require intermeshing at a more intimate (perhaps molecular) level, as opposed to the compaction of small globules (powdered particles) into 106–152 μm pellets. By contrast, the temperatures of decomposition of major macerals in coals do not appear to be as different and their occurrence within coal particles correspond more closely to the compacted powders – with which we conducted unsuccessful tests looking for synergistic effects between cellulose and lignin.

3.8 Case Studies: The Reactive Inertinites of Southern Hemisphere Coals

We have described in Chapter 2 how in the Southern Hemisphere, vast deposits of coal are thought to have common depositional origins in Gondwanaland, the paleo-continent that eventually fragmented into parts of southern Africa, India, eastern Australia, Madagascar, South America and Antarctica. For present purposes, the more intriguing aspect of these coals is the different maceral distribution patterns compared to Northern Hemisphere coals. Their vitrinite occurrences occasionally drop below 50 percent and never quite exceed 80 percent. Liptinites rarely occur above the several percent level. On the other hand, their high inertinite concentrations include material classed as "semi-fusinites", which are thought to have formed under cold, dry depositional environments with more than usual peat oxidation. They are more reactive than would have been expected from their classification by microscopic techniques.

Gondwanaland coals of the Permian age are mostly bituminous coals. The occurrence of anthracites is rare. Their generally thicker coal seams are cheaper to mine compared to Northern Hemisphere Carboniferous deposits. Furthermore, they have comparatively low sulfur contents, which makes them more marketable. On the other hand, operational problems relating to the slagging and fouling of these coals are different than those of their Northern Hemisphere counterparts. Much of the mineral matter in Gondwanaland coals is finely divided within the carbonaceous matter, which makes it difficult to reduce ash contents by usual colliery washing procedures. The high proportion of inertinites often found in Gondwanaland coals is also a source of commercial concern. Much discussion has focused on the exceptional reactivity of "semi-fusinites" in Gondwanaland coals during coking and liquefaction [Given, 1984]. Power generators are wary of lower levels of volatile release during the pyrolytic stage and what that implies in terms of flame stability in the near burner zone of pulverized-fuel combustors. Considerable ink has been spilt over the combustibility of these coals – and the implications of high inertinite contents on their marketability.

In some respects, the structural properties of Gondwanaland coals are similar to coals of the Carboniferous age found in the N. Hemisphere: (i) When their vitrinite reflectances are plotted against carbon content, the plots follow the same curves as other Paleozoic, Mesozoic and Tertiary coals [Chandra, 1965a; 1965b]. (ii) Correlations between H/C ratios of Australian coals and their pyrolysis yields appear to accommodate Northern Hemisphere coals [Tyler 1979; 1980]. (iii) Forms and properties of individual macerals are said to be indistinguishable from their Carboniferous counterparts. Physical and physico-chemical changes which accompany increases in rank appear to be analogous to those of N. Hemisphere coals [Stach *et al.*, 1983].

Regarding their *coking* behavior, inertinite rich Southern Hemisphere coals are considered to provide good value for money. Given [1984], citing Roberts [1982] and Diessel [1983], has summarized the matter as follows: "in predicting the strength and reactivity of cokes made from Northern Hemisphere Carboniferous coals by petrographic analysis, conventionally one third of the semi-fusinite is added to the total of "reactive macerals". Use of the same methods does not lead to useful predictions with Australian coals, which produce better cokes than would be expected from their performance in dilatometer or plastometer tests or their petrographic analyses". The data of Cudmore [1978] quoted by Durie [1980] indicated a greater dependence of *liquefaction* conversions on vitrinite reflectance (i.e. rank) than on maceral composition. This suggests that the greater inertinite contents were *not* reflected in lower extractabilities, as would have been usual for Northern Hemisphere coals.

These findings have allowed apparently quite legitimate claims that rank related effects might provide a better guide to the *coking* and *liquefaction* performance of Gondwanaland coals, compared to their petrographic compositions. However, analogous claims have occasionally been made, regarding the *combustion* reactivities of Australian coals and coal chars [Jones *et al.*, 1985; Thomas *et al.*, 1989; Phong-anant & Thomas, 1990]. The discussion has been reviewed by Cai *et al.* [1998]. Put simply, what needs to be resolved is whether it is warranted to extend observations from coking and liquefaction to the combustion performance of inertinite rich Gondwanaland coals?

In order to achieve a measure of clarity relating to combustion properties, it is necessary to compare both volatile yields *and* char reactivities of Northern and Southern Hemisphere coals directly. It also seemed useful to compare yields and char reactivities of vitrinites and inertinites from the *same* Southern Hemisphere coals. To this end, experiments were conducted using two sets of inertinite-graded South African coals and a set of maceral concentrates from a Northern Hemisphere coal (Point of Ayr; UK). The first set of inertinite-graded samples was prepared from Vryheid Coronation Colliery (VCC) coal of 87.5% elemental carbon content. Sub-samples were prepared containing 73.5, 59 and 43.5% vitrinites, the balance being largely inertinites. The second South African coal was a lower rank sample from the Durban Navigation Colliery (DNC) with 83.5% elemental carbon, similarly graded to give samples containing 86.5, 71 and 57% vitrinites, the balance once again being largely made up of inertinites. Pyrolysis experiments were carried out in the atmospheric pressure wire-mesh reactor, by heating the samples at 5,000°C s^{-1} to temperatures up to 1500°C [Cai *et al.*, 1998].

Results from these experiments were much as would be expected from *any* set of maceral concentrates. For samples of similar elemental carbon content, total volatile release was found to decrease in the order: liptinite > vitrinite > inertinite. Despite the wide variation in origins – elemental carbon content was found to be the dominant parameter. Tar and total volatile yields from the South African samples decreased with increasing inertinite concentration. The decrease in yields was in line with trends observed for the set of maceral concentrates from Point of Ayr and other Northern Hemisphere coals. The variation of pyrolysis yields with inertinite content was found to be less sensitive in the case of the samples from the higher rank (87% carbon) Vryheid Coronation Colliery coal. This reflected observations based on Northern hemisphere coals. As in the case of Linby and other coal samples discussed above, the data showed no evidence of synergistic effects between the vitrinites and inertinites during pyrolysis.

The relative combustion reactivities of *chars* from these experiments were also determined, using a standard thermogravimetric method at 500°C. In marked contrast to the order established for volatile release, the relative combustion reactivities of maceral *chars* prepared at 1,500°C could be ranked in the order:

inertinite > vitrinite > liptinite.

A direct comparison with maceral concentrates from the same coal was undertaken. For VCC, the higher rank coal, char reactivities were found to be essentially independent of original inertinite concentration. Once again, this result is consistent with diminishing differences generally observed between macerals of increasingly high rank Northern Hemisphere coals. However, the lower rank DNC *chars* clearly exhibited *increasing* reactivity with *increasing* inertinite concentration, as observed with the other samples.

Taken together, these results show that it is clearly possible for *chars* from low volatile coals and from high inertinite coals, to be relatively reactive, provided they are formed at sufficiently high temperatures, at or above 1,500°C. In this respect, Southern hemisphere coals may hold their own, when compared to other coals. As an aside, it is worth noting that the temperature at which the chars are formed is critical. For chars formed at 700°C, inertinite chars were found to be less reactive than the corresponding vitrinite chars [Cai & Kandiyoti, 1995].

However, when properties of Southern Hemisphere coals relevant to ignition and flame stability are compared with those of coals containing less inertinite, the usual problems of high-inertinite coals are inevitably encountered. This is because of the lower volatile release from inertinite rich Southern Hemisphere coals. Certainly, there appears to be no evidence to suggest that "reactive inertinites" might release any more volatile matter than any other inertinite of similar maturity [Cai *et al.*, 1998]. In any case, judging by current power station practice, the low volatile contents of South African coals is indeed seen as a disadvantage to be circumvented. It is normally rectified by judicious blending with higher volatile coals, just like one would have done with inertinite rich Northern Hemisphere coals.

3.9 Case Studies: The Pyrolysis of Kerogens

Kerogens are solid organic occlusions found in sedimentary rocks. Depending on the original composition of deposited material and the history of maturation, they may yield oil and/or gas that then migrates from the source rock toward reservoir areas. Kerogens are defined as the fraction of the organic deposit insoluble in common solvents such as dichloromethane. Type I kerogens have high atomic H/C and low O/C ratios and are classed as oil-prone. They are often composed of algal material (*e.g.* botryococcus algae). Type II kerogens are also considered as oil-prone, with intermediate atomic H/C and O/C ratios. They normally contain mixed macerals (including nonvascular phytoplanktons) and terrestrial liptinites (sporinites, cutinites and resinites). Type III kerogens with lower atomic H/C ratios and higher O/C ratios are considered as gas-prone. Typical components of these kerogens are vascular terrestrial and humic macerals. Maturity levels are critical in oil generation, which is expected to commence when kerogens have a maturity equivalent to a vitrinite reflectance of between 0.4 and 0.6% and is at its peak at about 1.3, passing to gas production by about 1.7% vitrinite reflectance [Madrali *et al.*, 1994].

One important element of kerogen characterization is the determination of product distributions during thermal breakdown. The standard Rock-Eval test [*e.g.* see Tissot & Welte, 1984; Tyson, 1995] is widely used in work related to oil exploration, in the field as well as in the laboratory. The method is empirical in nature and does not distinguish between tars and lighter combustible volatiles. Furthermore, since volatiles are combusted during the test, there is no recovery of tar sample for structural characterization. Although the Rock-Eval test appears to be designed perfectly sensibly with regard to field conditions, this appears as an area where some care in designing pyrolysis experiments may be of assistance to the organic geochemist.

Pyrolysis tar/oil and volatile yields from the set of samples in Table 3.11 were determined with the atmospheric pressure wire-mesh reactor described in Figure 3.1b. Two of the samples were Type I kerogens, two of Type II and two of Type III. All six samples were of low maturity, with vitrinite reflectances below 0.5% [Madrali *et al.*, 1994]. Within each pair of samples, the geologically younger kerogen showed higher overall aliphatic and hydroaromatic content by FT-ir and higher tar/oil and total volatile yields in pyrolysis. Similarly, within each sample pair, FT-ir showed the geologically younger kerogens *and* their tars to contain greater concentrations of O-bearing groups. The tar and total volatile yields correlated well with the Rock-Eval derived Hydrocarbon Index [Rahman *et al.*, 2000] and with the FT-ir spectra of the original kerogens. The volatile-yield trends did not, however, correspond closely with results expected from the elemental analyses, as understood in terms of the van Krevelen diagram. Elemental analyses on their own, turn out to be blunt instruments in evaluating the oil potential of individual kerogen samples.

As explained earlier, wire-mesh pyrolysis reactors allow a measure of accuracy not available to many other techniques, for determining tar/oil and total volatile yields. The pyrolysis experiments outlined above have been found to distinguish between geologically younger and older kerogen samples. A wider range of samples need to be tested, to explore the extent to which the agreement found between wire-mesh tar yields for this set of samples and the Hydrocarbon Index may be considered as more generally valid. Tar characterization, combining standard spectroscopic methods with size exclusion chromatography (see Chapters 7 and 8) should prove rewarding in linking kerogen molecular structures with parameters relevant to oil exploration, as well as providing a more fundamental understanding of the maturation processes.

3.10 Case Studies: The Pyrolysis of Biomass Materials

In principle, most of what has already been said about aspects of pyrolysis in this Chapter applies to the thermochemical processing of biomass and waste as fuels. In some critical respects, however, the behavior of wood or, more generally, lignocellulosic waste, differs from that of coals. Volatiles evolved during the pyrolysis of such materials often make up a much larger proportion of the product slate. Above 600°C, Fraga *et al.* [1991] have reported total volatile yields above 95 percent during fast heating (1,000°C s^{-1}) experiments in a wire-mesh reactor. "Slow" heating rate experiments (1°C s^{-1}) in the same apparatus produced about 85 percent total volatiles, which was still quite high when compared with coals.

As in the case of coal pyrolysis, the amount and reactivity of residual biomass chars are largely dependent on conditions during the pyrolytic process. The heating rate, residence time at high temperatures and the level of tar re-condensation on solid surfaces are significant parameters in determining eventual extents of char gasification and reactivity

Table 3.11 Wire-mesh pyrolysis yields as wt%, on dry basis. Pyrolysis conditions: heating rate 1000 K/s; final temperature 700 °C; hold time 30 s; sweep velocity 0.1 m/s. (Reproduced with permission: Fuel 1994, *73*, 1829; Copyright 1994, Elsevier)

Sample	*Geographic origin*	*Sedimentary basin*	*Kerogen Type/Deposition environment*	*Geological age (Ma)*	*Tar Yield %*	*Total Volatile Yield %*
A	Philpstoun, Scotland	Midland Valley	Type I/Lacustrine	Lower Carboniferous 330 $\pm$ 5	48.1	73.5
B	Queensland Australia	Lowmead	Type I/Lacustrine	Eocene 50 $\pm$ 5	61.8	82.1
C	Salt Range, Pakistan	Kohat-Potwar	Type II/Marine	Lower Triassic 240 $\pm$ 5	41.5	64.8
D	Lurestan, Iran	Zagros	Type II/Marine	Palaeocene 60 $\pm$ 5	48.6	73.0
E	East Midlands, England	East Midland	Type III/Deltaic	Upper Carboniferous 305 $\pm$ 5	17.7	36.7
F	Isle of Wight, UK	Wessex	Type III/Deltaic	Lower Crateceous 125 $\pm$ 5	18.4	42.9

[Messenbock *et al.*, 1999]. Char yields from lignocellulosic biomass are even more sensitive to these conditions. In the extreme example of pure cellulose, almost no solids could be recovered when volatiles were deliberately and effectively removed, during pyrolysis in the wire-mesh reactor [Fraga, 1990]. By contrast, up to 26% char could be obtained from a similar sample of pure cellulose, when all forced gas circulation was eliminated [Zaror, *et al.*, 1985]. We will see from results presented here and in Chapter 4 on gasification that much less char residue remains behind during the pyrolytic step in the processing of many biomass materials and furthermore that these residues are usually quite reactive.

3.10.1 Pyrolysis of biomass in a wire mesh reactor

In the early 1980s, Scott & Piskorz [1982a; 1982b; 1984] published results from the pyrolysis of finely divided (<500 μm) aspen poplar wood in a bench scale fluidised-bed. The temperature range used in these experiments was 400–700°C. The basic design was similar to that of Tyler [1979; 1980]. The wood powder was introduced into an already heated bed and heating rates were estimated to be quite high. In these experiments, nearly 65% of the original fuel mass could be recovered as a tar/oil. When the temperature of the fluidized-bed was increased, more of the tar/oil could be cracked to gaseous products. As the temperature was raised, the highly oxygenated nature of these oils was made more apparent from increasing concentrations of carbon monoxide in the product mix. In subsequent work, the same authors scaled up the work to a pilot sized bed and attempted to find a use for the liquid product.

Table 3.12 presents tar and total volatile yields from the atmospheric pressure pyrolysis of silver birch wood, determined with the wire-mesh reactor. Results changed in predictable ways as a function of increasing heating rate and temperature. As in the case of experiments with most coals, larger tar/oil yields were obtained with the faster heating rates. However, *near total* conversion to gases and liquids was observed with fast heating to between 500 and 700°C. Experiments with pure cellulose gave total conversion from about 500°C onward [Fraga *et al.*, 1991].

Table 3.12 Tar and total volatile yields from the atmospheric pressure pyrolysis of silver birch wood, determined using the wire-mesh reactor. Helium was used as ambient gas in all experiments

	Total Volatiles % daf basis		*Tar/oil yield % daf basis*		*Gas (by difference) % daf basis*	
Temperature °C	*Heating rate*					
	1°C/s	*1000°C/s*	*1°C/s*	*1000°C/s*	*1°C/s*	*1000°C/s*
400	77	89	43	56	33	33
500	89	96	49	58	40	39
700	93	99	54	57	39	43
900	93	99	52	57	41	43

On the face of it, this result suggests that there is not much need to *gasify* biomass materials with reactive gases such as air or steam, in order to convert these solid fuels into volatile products, provided small particle sizes are used and high heating rates are maintained. Clearly, some air may be introduced to run the reactor in autothermal mode. As outlined above, Scott and Piskorz could achieve this in their bench scale and pilot plant reactors. The system could have been scaled up to plant size without much difficulty. The question remains, however, about what to do with the tars. Biomass pyrolysis liquids are normally quite corrosive, unstable during storage, readily forming gums and separating into an aqueous and an organic phase. They turn out to be difficult to process to convert into more usable materials. We will briefly return to this problem below.

3.10.2 Comparing the pyrolysis and gasification of biomass in a "hot-rod" reactor

In order to resolve some of these questions, a direct comparison has been made between conversions from the pyrolysis of biomass in an inert gas, with conversions from gasification in H_2, CO_2 and steam-helium mixtures, at pressures up to 20 bars [Pindoria *et al.*, 1998]. Even with experiments carried out in a fixed-bed "hot-rod" reactor, working with a peak temperature of 850°C in helium, samples of Eucalyptus (*Eucalyptus globulus*) sawdust gave conversions in the 85–88 percent range. Clearly, there was relatively little scope for improving on these pyrolysis yields.

The greatest difference between treatment of eucalyptus wood in an inert gas and a reactive gas was 8.5%, observed when comparing weight loss in helium and in hydrogen at 20 bars. It may be observed, furthermore, that these results were obtained at a relatively slow heating rate (10°C s^{-1}), while using a shallow (~4 mm) fixed-bed sample configuration. Both slow heating and stacked particles are factors known to give rise to some tar repolymerization – tending to increase the *char* yield. At fast heating rates and with no impediment to volatile release by neighboring sample particles, sample weight loss from pyrolysis alone would be expected to register values in the mid-to-high 90%, leaving even less scope – or need – for reactive gases to make any impact at all.

On the other hand, operators of pilot and plant scale equipment would normally reject the idea of operating with sawdust – or feedstocks with similar particle size distributions, unless special provision had been made to suppress fuel loss through elutriation from the bed. The standard remedy is to increase particle size. This in turn tends to increase the amount of residual char. Larger particles cannot be pyrolyzed quickly, due to the slower transmission of heat within the particle. Howard and Anthony [1976] have calculated that when heating at 1,000°C s^{-1}, the maximum diameter of coal particles (which have higher thermal conductivities than wood particles) that could reasonably be expected to show uniform temperature profiles is about 100 μm. The resulting slower heating temperature front within larger particles would allow more char formation, which in turn would provide a barrier to the passage of tar vapor from further inside the particles.

If the resulting chars cannot gasify quickly because of large particle sizes, they may quickly lose reactivity at temperatures above 800–850°C. The effect is akin to the annealing of carbons and chars at high temperature, resulting in rapid loss of reactivity with increasing levels of exposure. We will present data in the next Chapter showing that chars lose nearly 70 percent of their reactivity within the first ten seconds at 1,000°C. The presence of steam, with high heat transfer coefficients between the bulk gas and sample particles – coupled to its reactivity, could be instrumental in improving the conversion of larger feedstock particles. In pilot and plant size gasifiers, the use of reactive gases would appear therefore to partially compensate for choices of fluidization parameters that lead to char formation. Cracking of evolved tars is another useful function for reactive gases such as steam. In the presence of steam above 850°C, very little tar remains intact in the freeboard [Cai *et al.*, 1996]. However, as we will see in the next section, the small residue of tar that survives at the higher temperatures is a highly aromatized, chemically stable and toxic mixture.

3.10.3 Biomass pyrolysis in a bench-scale fluidized-bed

The high conversions possible during biomass pyrolysis (and gasification) have implications for the modeling and design of fluidized-bed reactors. At bench scale (Figure 3.5), as well as at pilot and plant scale, these reactors are normally operated by injecting sample into a heated bed of inert solids. This gives rise to high heating rates. However, unlike the case of wire-mesh instruments, the heating rate cannot be controlled and may only be calculated by making (often suspect) assumptions about heat transfer rates to particles. The reactivity of lignocellulosic biomass translates into steady-state fuel inventories no larger than several percent of biomass. The rest is made up of inert bed solids that insure adequate fluidization.

As explained in presenting the design in Figure 3.5b, the vertical motion of the support plate allows residence times of volatiles in the reactor freeboard to be altered between 0.8 and ~4 s. Figure 3.9 presents tar yield data from this reactor, as a function of temperature and freeboard residence time. The same sample of silver birch wood was used here as in subsequent wire-mesh experiments by Fraga *et al.* [1991] where similar peak tar yields were observed (Table 3.12).

As freeboard residence times are increased, Figure 3.9 clearly shows that the maximum of the tar yield curve recedes from a little above 450°C to nearly 425°C. The trend is in line with what we can expect from intensified tar cracking reactions at longer residence times. However, for fixed freeboard heights, the temperature of the tar yield maximum provides direct evidence of the relative thermal stability of the tars, which also reflects the chemical makeup of the sample. The analogous maximum for low rank bituminous Linby coal was observed near 590–600°C and that of Çan lignite (Turkey) was intermediate at about 530°C.

The second important observation arising from these data is the rapidity with which tar is destroyed in the reactor freeboard, with increasing temperatures and residence times. Clearly, compared to coals, lignocellulosic biomasses produce more tar per unit amount of original fuel. One major and largely still unsolved problem of biomass gasification in updraft-fluidized-beds is the tar content of fuel gases. The problem is sufficiently severe for catalytic methods to have been attempted [Corella *et al.*, 2004]. Another aspect of the problem, however, is the constantly changing *composition* of the biomass tars.

The pioneering GC-MS analyses of Evans and Milne [1987a; 1987b] have shown that biomass pyrolysis tars contain very large numbers of chemical components, significant proportions of which are oxygenated compounds. All the work since has confirmed the

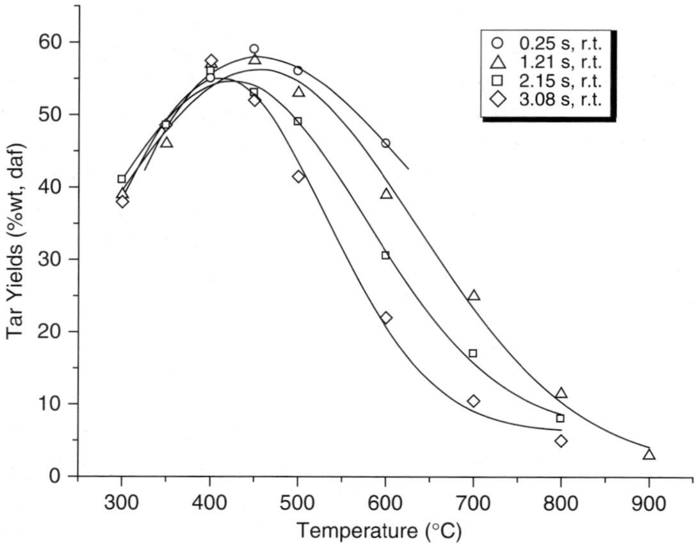

Figure 3.9 Silver Birch pyrolysis tar yields as a function of temperature and freeboard residence times. Operation at atmospheric pressure. (Reproduced with permission: Fuel *68*, (1989), 275; Copyright 1989 Elsevier).

soundness of the initial work [e.g. Guell *et al.*, 1993]. However, GC-MS is only able to analyze molecules that are able to pass through the chromatographic column. For most GC-columns, the upper limit for the passage of aliphatic compounds is about 500 u; for aromatics and polar molecules, this limit would be lower, at about 300 u. Recently, higher temperature columns have been able to push this limit to about 550–600 for aliphatics and nearly 400 u for aromatics. Meanwhile, characterization efforts by size exclusion chromatography and MALDI-mass spectrometry have since indicated that a large proportion of biomass tars shows up at molecular mass levels far above the range identifiable by GC-MS. Most of the tar is unable to pass through the GC-MS column. Applications of these techniques to pyrolysis tars from sugar cane bagasse [Pindoria *et al.*, 1999] and eucalyptus [Pindoria *et al.*, 1997] indicate molecular masses up to at least 6,000 u and perhaps a little beyond. We do not, therefore, have direct compound identification for more than 90 percent of the tars. *They* do *not* show up in GC-MS.

These developments do not invalidate findings by GC-MS but simply put it in context. Furthermore, findings from GC-MS may provide a general framework for visualizing the types of structures to be expected in the heavier tars – although direct extrapolations may not always be justified. Observations by GC-MS on silver birch tars, recovered during experiments described by Figure 3.9 are relevant in showing the shift in chemical speciation with increasing temperature. At the lower range of reactor temperatures (300–400°C) compounds identified in relative abundance included oxygenated species such as dimethoxypropene, tetrahydrofurylmethanol and of course, laevoglucosan. Phenols and ethoxybenzenes were observed at medium and higher temperatures, while polynuclear aromatic species, such as naphthalenes, acenaphthylenes, 9H-fluorene and pyrene occurred with greater frequency at the higher end of the temperature spectrum (900°C) [Stiles & Kandiyoti].

The scope of these GC-MS analyses was limited. However, the gradual increase in concentrations of polynuclear aromatic species identified in biomass tars exposed to higher temperatures signals potential difficulties for product gas cleanup. Recalling that above 700°C ring closure and dehydration reactions are intensified, this result might have been expected. However, polynuclear aromatic species are thermally far more stable than fresh oxygenated tars recovered during experiments at lower temperatures. If higher freeboard temperatures are used as the single blunt instrument for cracking biomass tars, it appears we are able to decompose most of these condensable materials, but would face the prospect of a reduced but much tougher residue. In that case, another cleaning step would be necessary to avoid harm to engines or gas turbines downstream, when the fuel gas is intended for power generation.

Several contradictory factors need to be reconciled before optimum conditions are found for tar destruction in biomass utilization, without excessive effort and expense. Thermal as well as economic efficiency considerations require that biomass gasifiers are operated at relatively low temperatures compared to coal gasifiers, where "tar in the fuel gas" is rarely found to be a problem. In steam-oxygen-blown coal gasification, temperatures well above 1800°C are reached and no tar survival could be expected. However, air blown gasifiers operate between 900–1050°C. Unpublished work at Imperial College found no extractables in bed solids recovered from the cyclones at the exit of the British Coal ABGC gasification pilot plant. Tar destruction appeared completed, presumably due to the presence of steam at around 950°C and probably also to a tall reactor freeboard. The solution to the tar-survival problem in the fuel gas from biomass gasifiers must ultimately lie in a combination of a reactive gas environment, perhaps at somewhat higher temperatures in the freeboard. A focused set of experiments in steam cracking of polynuclear aromatic species would much improve our overview of the problem.

References

Anthony, D. B., Howard, J. B., Hottel, H. C., Meissner, H. P. (1975) *15th Symp. (Int.) Combustion, Pittsburgh*, 1303

Anthony, D. B., Howard, J. B., Meissner, H. P., Hottel, H. C. (1974) *Rev. Sci. Instrum.*, 45, 992

Arendt, P (1980) *Ph.D. thesis*, University of Aachen, Germany

Arendt, P., van Heek, K.-H. (1981) *Fuel, 60*, 779

Bautista, J. R., Russel, W. B., Saville, D. A. (1986) *Ind. Eng. Chem. Fund.*, 25, 536

Berk, F. (1978) *MSc Thesis*, Bogazici University, Istanbul, Turkey

Bolton, C., Snape, C.E., O'Brien, R.J. and Kandiyoti, R., (1987) *Fuel*, 66, 1413

Bolton C., Snape C.E., Stephens H.P. (1988) *Fuel*, 68 161

Cai, H-Y., Kandiyoti, R. (1995) *Energy and Fuels*, 9, 956

Cai, H-Y., Guell, A. J., Chatzakis, I. N., Lim, J-Y., Dugwell, D. R., Kandiyoti, R (1996) *Fuel*, 75, 15

Cai, H-Y., Megaritis, A., Messenbock, R., Dix, M., Dugwell, D.R. and Kandiyoti, R. (1998) *Fuel*, 77, 1273

Chandra, D. (1965a) *Econ. Geol.*, 60, 1041

Chandra, D. Quart. (1965b) *J. Geol. Min. Metal. Soc. India*, 37, 37

Collot, A-G., Zhuo, Y., Dugwell, D. R., Kandiyoti, R. (1999) *Fuel*, 78, 667

Corella, J., Toledo, J.M. and Padilla, R., (2004) Ind. Eng. Chem. Res. 43, 2433

Cudmore, J. F. (1978) *Fuel Proc. Tech.*, 1, 227

Desypris, J., Murdoch, P., Williams, A. (1982) *Fuel, 61*, 807

Diessel, C. F. K., (1983) *Fuel*, 62, 883
Dryden, I. G. C., Sparham, G. A. (1963) *B.C.U.R.A. Monthly Bull.*, 7(1), 1
Durie, R. A., (1980) *Coal Liquefaction Fundamentals* (ed. D. D. Whitehurst), ACS Symp. Ser. No. 139, 53
Evans, R.J. and Milne, T.A., (1987a) *Energy and Fuels* 1, 123
Evans, R.J. and Milne, T.A., (1987b) *Energy and Fuels* 1, 311
Finn, M. J., Fynes, G., Ladner, W. R., Newman, J. O. H., (1980) *Fuel*, 59, 397
Fong, W. S., Khalil Y. F., Peters W. A., Howard J. B.(1986a) *Fuel*, 65, 195
Fong W. S., Peters W. A., Howard J. B. (1986b) *Fuel*, 65, 251
Fraga, A-R., (1990) Ph.D. Thesis, University of London
Fraga, A-R., Gaines, A. F., Kandiyoti, R. (1991) *Fuel*, 70, 803
Freihaut, J. D., Seery, D. J., Zabielski, M. F. (1982) *19th Symposium (Intl.) on Combustion*, the Combustion Institute, Pittsburgh, PA, 1159
Freihaut, J. D. and Seery, T. D. (1983) *Am. Chem. Soc. Div. Fuel Chem. Preprints*, 28(4), 265
Freihaut, J. D., Proscia, W. M. (1989) *Energy & Fuels*, 3, 625
Fynes, G., James, R.G., Ladner, W.R. and Newman, J.O.H., (1984) *Fuel*, 63, 897
Gaur, S., Reed, T. B. (1998) *Thermal data for natural and synthetic fuels*, Marcel Dekker, NY
Gibbins-Matham, J. R. (1988) *PhD Thesis*, University of London
Gibbins-Matham, J. R., Kandiyoti, R. (1988) *Energy & Fuels*, 2, 505
Gibbins, J. R., Kandiyoti, R. (1989a) *Energy & Fuels, 3*, 670
Gibbins, J. R., Kandiyoti, R. (1989b) *Fuel, 68*, 895
Gibbins-Matham, J. R., King, R. A.V., Wood, R. J., Kandiyoti, R. (1989c) *Review of Scientific Instruments, 60*, 1129
Given, P. H. (1984) *Coal Science* (ed. Gorbaty, M. L., Larsen, J. W., I. Wender, I.), Volume III, Academic Press, NY, p. 65
Gonenc, Z.S., Gibbins, J.R., Katheklakis, I.E. and Kandiyoti, R., (1990) *Fuel*, 69, 383
Gray V.R. (1988) *Fuel*, 67, 1298
Graff, R. A., Dobner, S., Squires, A. M., (1976) *Fuel*, 55, 109, 113
Griffin, T. P., Howard, J. B., Peters, W. A. (1993) *Energy & Fuels*, 7, 297
Griffiths, D. M. L., Mainhood, J. S. R (1967) *Fuel*, 46, 167
Guell, A. J., Kandiyoti, R. (1993) *Energy & Fuels*, 7, 943
Guell, A.J., Li, C.Z., Herod, A.A., Stokes, B.J., Hancock, P. and Kandiyoti, R., (1993) *Biomass and Bioenergy* 5, 155
Jüntgen, H., van Heek, K. H. (1968) *Fuel, 47*, 103
Jones, R. B., McCourt, C. B., King, K., (1985) *Fuel*, 64, 1460
Hamilton, L. H., Ayling, A. B., Shiboaka, M. (1979) *Fuel*, 58, 873
Hamilton, L.H. (1980) *Fuel*, 59, 112
Herod, A.A. and Kandiyoti, R., (1994) *Fuel*, 73, 470–471
Hindmarsh, C. J., Thomas, K. M., Wang, W., Cai, H-Y., Güell, A. J., Dugwell, D. R. and Kandiyoti, R., (1995) *Fuel*, 74, 1185–1190
Hiteshue, R. W., Anderson, R. B., Schlesinger, M. D. (1957) *Ind. Eng. Chem.*, 47, 2008
Hiteshue, R. W., Anderson, R. B., Friedman, S. (1960) *Ind. Eng. Chem.*, 52, 577
Hiteshue, R. W., Friedman, S., Madden, R., (1962a) *USBM Rep. Invest.*, 6027
Hiteshue, R.W., Friedman, S. and Madden, R., (1962b) USBM Rep. Invest. 6125
Howard, H. C. (1963) *Chemistry of Coal Utilization Supplementary Volume* (ed. Lowry, H. H.), Wiley, NY, 340–394
Howard, J. B., Anthony, D. B., Hottel, H. C. (1975) *15th Symposium (Intl.) on Combustion*, The Combustion Institute, Pittsburgh, PA, 103
Howard, J. B., Anthony, D. B., Hottel, H. C., Meissner, H. P. (1976) *Fuel*, 55, 121

Howard, J. B., Anthony, D.B. (1976) *AIChE Journal*, 22, 625
Howard, J. B. (1981) *Chemistry of Coal Utilization Second Supplementary Volume* (ed. Elliott, M. M.), Wiley, NY, Chapter 12
Howard J. B., Peters W. A., Derivakis G. S. (1994) *Energy & Fuels*, 8, 1024
Hunt, J. E., Lykke, K. R., Winans, R. E., (1991) *Prpnt. ACS DFC*, 36(3), 1325
Fong, W. S., Khalil Y. F., Peters W. A., Howard J. B.(1986a) Fuel, 65, 195
Fong W. S., Peters W. A., Howard J. B. (1986b) Fuel, 65, 251
Fukuda, K. (2002) *PhD Thesis*, University of London
Fukuda, K., Dugwell, D.R., Herod, A.A. and Kandiyoti, R., (2004) *Energy & Fuels*, 18, 1140–1148
Kandiyoti, R., (1969) PhD Thesis, University of London
Kandiyoti, R., McLaughlin, E. and Pittman, J.F.T., *J. Chem. Soc., Farad. Trans. I*, 68, 860–866.
Kandiyoti, R., (2002) *Fuel*, 81, 975
Kershaw, J. R., Barras, G. (1979) *Am. Chem. Soc. Div. Fuel Chem. Prepr.*, 24(3), 99
Kobayashi, H., Howard, J. B., Sarofim, A. F. (1977) *16 Symp. Combustion*, The Combustion Institute, Pittsburgh, 411
Li, C-Z., Gaines, A. F., Güell, A. J., Kandiyoti, R. (1991a) *Proc. Intl. Conf. Coal Sci.*, Newcastle-upon-Tyne, UK, 723
Li, C-Z., Bartle, K. D., Kandiyoti, R. (1993a) *Fuel*, 72, 3
Li, C-Z., Bartle, K. D., Kandiyoti, R. (1993b) *Fuel*, 72, 1459
Li, C-Z., Madrali, E. S., Wu, F., Xu, B., Cai, H-Y., Guell, A. J., Kandiyoti, R. (1994a) *Fuel*, 73, 851–865
Loison, R., Chauvin, R. (1964) *Chim. Ind. Paris, 91*, 269; Translation by University of Sheffield (DJB/WBD) May 1964; National Coal Board, Coal Research Establishment Library (Sept 1964)
Madrali, E. S., Rahman, M., Kinghorn, R. R. F., Wu, F., Herod, A. A. and Kandiyoti, R. (1994) *Fuel*, 73, 1829
Megaritis, A., Zhuo, Y., Messenbock, R., Dugwell, D. R., Kandiyoti, R. (1998a) *Energy & Fuels*, 12, 144
Messenbock, R.C. (1998) *PhD Thesis*, University of London
Messenbock, R. C., Dugwell, D. R., Kandiyoti, R. (1999) *Energy & Fuels*, 13, 122
Messenbock, R. C., Paterson, N., Dugwell, D. R., Kandiyoti, R. (2000) *Fuel*, 79, 109
Niksa, S. J., Russel, W. B., Saville, D. A. (1982a) *Fuel, 61*, 1207
Niksa, S. J., Russel, W. B., Saville, D. A. (1982b) *19th Symp.(Int.) Combustion*, The Combustion Institute, Pittsburgh PA, 1151
Niksa, S. J., Heyd, L. E., Russel, W. B., Saville, D. A. (1984) *20th* Symposium (Intl.) on Combustion, The Combustion Institute, 1445
O'Brien, J. R. (1986) *PhD Thesis*, University of London
Pandolfo, A. G., Johns, R. B., Dyrkacz, G. R., Buchanan, A. S. (1988) *Energy and Fuels*, 2, 657
Peralta, D., Paterson, N. and Kandiyoti, R., (2002) 3rd Periodic Report; UK DTI Project No. AEA C/07/00298/00/00
Peralta, D., Li Xiaoyu, Xu Shisen, Paterson, N., Dugwell, D. R., Kandiyoti, R., (2004) *Proc. 2004 Intl. Hi-Tech Symposium on Coal, Chemical Industry and Coal Conversion*, 30–31 Oct., Shanghai, China, pp. 136–142.
Peralta, D., Paterson, N., Dugwell, D. and Kandiyoti, R., *Energy and Fuels* 19 (2005) 532–537
Peters, W. A., Hajaligol, M., Howard, J. B., Longwell, J. (1980) *Specialist Workshop on Fast Pyrolysis of Biomass*, SERI/CP-622–1096, Solar Energy Research Institute, US Dept. of Energy, Golden CO, USA
Phong-anant, D., Thomas, C. G. (1990) *Fourth Australian Coal Science Conference*, Brisbane, Australia, 256

Pindoria, R.V., Lim, J-Y., Hawkes, J.E., Lazaro, M-J., Herod, A.A. and Kandiyoti, R., (1997) *Fuel* 76, 1013
Pindoria, R.V., Megaritis, A., Messenböck, R., Dugwell, D. R., Kandiyoti, R. (1998a) *Fuel*, 77, 1247
Pindoria, R.V., Megaritis, A., Herod, A.A., Kandiyoti, R. (1998b) *Fuel*, 77, 1715
Pindoria, R.V., Chatzakis, I.N., Lim, J-Y., Herod, A.A., Dugwell, A.A. and Kandiyoti, R., (1999) *Fuel* 78, 55
Rahman, M., Herod, A. A., Kandiyoti, R. (2000) *Fuel*, 79, 201
Roberts, O. C. (1982) *Published Report 82–8*, ISBN No.0 86722 224 X, Australian Coal Industry Research Laboratories Ltd
Roy, C., de Caumia, B., Kalkreuth, W. (1985) *Fuel*, 64, 1662
Scott S.C.; Piskorz J., (1982) *Canad. J. Chemical Eng*. 60, 666
Scott S.C.; Piskorz J., (1982) in "Flash pyrolysis of biomass" Chapter 23, Ann Arbor Science Pub., Ann Arbor Michigan
Scott S.C.; Piskorz J., (1984) *Canad. J. Chemical Eng*. 62, 404
Shafizadeh, F. (1968) *Advances in Carbohydrate Chemistry*, 23, 419
Stiles, H. N., Kandiyoti, R.(1989) *Fuel*, 68, 275
Sun, B., Ruof, C.H. and Howard, H.C., (1958*) Fuel*, 37, 299–308
Suuberg, E. M. (1977) *Sc.D. Thesis*, Massachusetts Institute of Technology, Cambridge, Massachusetts
Suuberg, E. M., Peters, W. A., Howard, J. B. (1978a) *Ind. Eng. Chem. PDD*, 17, 37
Suuberg, E. M., Peters, W. A., Howard, J. B. (1978b) *17th Symposium (Intl.) on Combustion*, The Combustion Institute, Pittsburgh, PA, 117
Suuberg, E. M., Peters, W. A., Howard, J. B. (1980) *Fuel*, 59, 405
Suuberg, E. M., Unger, P. E. (1981) *18th Symposium (Intl.) on Combustion*, The Combustion Institute, Pittsburgh, PA, 1203
Suuberg, E. M., Unger, P. E., Lilly, W. D. (1985) *Fuel*, 64, 956
Suuberg, E. M. (1985) *Chemistry of Coal Conversion* (ed. Schlosberg, R. H.), Plenum Press, NY, 67
Stangeby, P. C., Sears, P. L. (1978) *Final Report to Can. Dept. of Energy, Mines and Resources*
Stangeby, P. C., Sears, P. L. (1981a) *Fuel*, 60, 131
Stangeby, P. C., Sears, P. L. (1981b) *Fuel*, 60, 125
Stiles, H.N., (1986) PhD Thesis, University of London, UK
Stiles, H. N., Kandiyoti, R., (1989) *Fuel*, 68, 275–283
Taupitz K. C. (1977) *Hydrocarbon Processing*, 219–225
Thomas, C. G., Holcombe, D., Shibaoka, M., Young, B. C., Brunckhorst, L. F., Gawronski, E. (1989) *Proceedings of International Conference on Coal Science*, 257
Tissot, B.P., Welte, D. H., (1984) *Petroleum Formation and Occurrence*, Springer-Verlag, Berlin, NY, 509
Tyler, R. J. (1979) *Fuel*, 58, 680
Tyler, R. J. (1980) *Fuel*, 59, 218
Tyson, R.V. (1995) *Sedimentary Organic Matter London*, Chapman & Hall, Chapter 21
Unger, P. E., Suuberg, E. M. (1984) *Fuel*, 63, 606
Unger, P. E., Suuberg, E. M. (1983) *Am. Chem. Soc. Div. Fuel Chem. Preprints*., 28, 278
Wanzl, W. (1988) *Fuel Proc. Tech*., 20, 317–336
Winans, R. E., Scott, R. G., Neill, P. H., Dyrkacz, G. R., Hayatsu, R. (1986) *Fuel Proc. Tech*. 12, 77
Winans, R. E., Neill, P. H. (1990) *Geochem. Sulfur Fossil Fuels*, ACS Symp., SN 429, 249
Winans, R. E. (1991) *J. Anal. App. Pyrolysis*, 20, 1
Winans, R. E., McBeth, R. L., Hunt, J. E., Melnikov, P. E. (1991) *Proc. Intl. Conf. Coal Sci*., Newcastle, UK, 44
Zaror, C.A., Hutchings, I.S., Pyle D.L., Stiles, H.N. and Kandiyoti, R., (1985), *Fuel, 64*, 990–994
Zhang, S-F., Xu, B., Herod, A.A. and Kandiyoti, R., (1996) *Energy and Fuels, 10*, 733–742

CHAPTER 4

High-Pressure Reactor Design & Applications: Pyrolysis, Hydropyrolysis and Gasification

The reactor design methods described in these chapters attempt to bridge the gap between bench-scale fuel-characterization tests and conditions prevailing in thermochemical fuel conversion processes. One useful function of laboratory experiments is the ability to simulate process conditions within well-defined zones of pilot or plant scale equipment. As we turn to experiments at elevated pressures, we will retain many of the developments described in the context of atmospheric pressure pyrolysis. However, the new designs will incorporate features relating to pressure containment and to reactor safety. The present Chapter focuses on high-pressure experiments in inert atmospheres (pyrolysis), in hydrogen (hydropyrolysis and hydrogasification), as well as gasification in steam-air and steam-oxygen atmospheres. We will also examine the effects of high-pressure on the morphologies and reactivities of the solid products – the chars.

Clearly, gasification is a mature art and vast numbers of experiments have been done in the past, to investigate the pyrolysis and gasification of coals and biomass materials. Many coal gasification studies have made use of chars prepared and gasified in two separate experimental steps [*e.g*. Goring, *et al*., 1952; Sha *et al*., 1990; Guo & Zhang, 1986; Haga & Nishiyama, 1988; Nozaki *et al*., 1991; Ginter *et al*., 1993; Meijer *et al*., 1994; Yang & Watkinson, 1994]. Typically, coal samples are either heated in a furnace or pyrolyzed in an atmospheric pressure fluidized-bed reactor. Subsequent gasification reactivity measurements have often been based on thermogravimetric balances or differential thermal analyzers, instruments capable of heating at, at most, some hundreds of degrees per *minute* [Goring, *et al*., 1952; Sha *et al*., 1990; Muhlen, *et al*., 1985; Hurt *et al*., 1991; Shufen & Ruizheng, 1994; Bota & Abotsi, 1994]. This approach has worked reasonably well in dealing with moving-burden ("fixed-bed") gasifiers, where heating rates are slow during pyrolysis and char residence times are on the order of minutes or hours. Such reactors are still used frequently. In South Africa, where several phalanxes of fixed-bed gasifiers serve as the backbone of SASOL's synthetic fuel and chemicals production, they process nearly 50 million tons per annum.

However, most modern gasifier designs are based on short residence time fluidized-bed or entrained flow reactors, where heating rates are high and exposure to reaction conditions is short. Mostly, solids exit from the reactors in a matter of seconds or, at most, tens of seconds. The slurry-fed Texaco [Shorter & Marchese, 1988; Jahnke, 1995] and dry feed Shell gasifiers [Doering & Cremer, 1994; Koopman *et al.*, 1994] as well as some of the new designs being developed in China [Ren *et al.*, 2004] are oxygen-blown and operate at higher-temperatures (~1,400–2,000°C). These processes are based on entrained flow designs, where the solids inventory is entirely transitory.

An air blown pilot scale spouted bed gasifier operating at lower temperatures was developed by British Coal [Dawes *et al.*, 1993; Dawes *et al.*, 1995]. A similar spouted-fluidized-bed design [Higginbotham *et al.*, 1994] was used in the Piñon Pine Power Project [1998]. In these two reactor systems, residence times for a large fraction of the solid fuels were of the order of seconds. However, both the Piñon Pine and the British Coal gasifiers allowed longer residence times for part of the solids, resulting in the formation of some unexpectedly stable chars.

The requirements arising from the expanded use of fluidized and entrained flow designs have substantially changed the face of solid fuel characterization. The emerging designs of fuel characterization tests are driven by the necessity to match rapid processing requirements. Another critical factor is the recognition that the outcomes of these tests depend on reactor configuration and reaction conditions during prior thermal breakdown, as well as the original compositions of the fuels. We have already introduced many of these concepts in Chapter 3.

Meanwhile, much of the fuel characterization data in the coal gasification literature relies on bench scale determinations, acquired under conditions not representative of modern gasifiers. The two-step procedure mentioned above significantly affects the results of char gasification-reactivity experiments. There is ample evidence indicating that chars from the same coal can give different reactivities, depending on the way in which the initial pyrolysis step has been carried out [Sha *et al.*, 1990; Ginter *et al.*, 1993; Muhlen *et al.*, 1986; Silveston, 1991; Alvarez *et al.*, 1994]. In one case, *in situ* generated chars were reported to be up to six-times more reactive (in steam) than those of corresponding *ex situ* prepared chars [Peng *et al.*, 1995].

Coals do not react to thermal processing in quantitatively similar ways. The relative ordering of reactivities within particular sets of coals may be affected by the char preparation procedure. Overall, char reactivities are just as sensitive as product distributions to reaction conditions during the pyrolysis stage and, above all, on the time-of-exposure at peak process temperatures. Inevitably, the acquisition of realistic bench-scale conversion and reactivity data requires that fuel particles undergo sequences of conditions more akin to those of the actual processes they aim to simulate.

A quick word needs to be said about past work in this area. In sifting through the literature for relevant comparisons, we will look for experiments (i) designed to distinguish between fuel related and reactor related effects during the pyrolytic step, (ii) designed to determine residual char reactivities under conditions reflecting the actual processing environment with relevant residence times, and, (iii) designed to couple the pyrolytic and gasification steps together and treat them as a single integral process. The last requirement arises from observations, outlined above, that splitting the gasification process into separate pyrolysis and gasification experiments causes measurable changes in final char reactivities. Much previous work suffers from the lack of at least one these features and will need to be left aside. Excellent reviews of early work will be found in

Chapters 20 and 21 of the "Chemistry of Coal Utilisation" supplementary volume of Lowry [1963] and Chapters 23–26 of "Second Supplementary Volume" by Elliott [1981]. Since these authoritative reviews, vast amounts of new research have been carried out, leading to well over 1,000 listed publications. A relatively recent review of coal gasification technologies by Hotchkiss [2003] and the collection of IEA Clean Coal Centre [http://www.iea-coal.org.uk/site/ieaccc/home] reviews would provide a reasonable overall view of the field.

To the extent that gasifier modeling and design has relied on perceived kinetics during pyrolysis and gasification experiments, there has been a temptation in the literature to treat chars as if they were well-defined, identifiable compounds. Occasionally, even graphite has been used as a model substrate. We will begin this Chapter by describing experiments, which show that rapidly heated char particles lose nearly two-thirds of their combustion reactivity during the first 10 *seconds* of exposure *at* 1,000°C. In this way, we will attempt to put on a firm foundation the conditions that must be fulfilled if we wish to replicate actual process conditions in bench-scale fuel characterization tests. These are the necessity of working with high heating rates and short exposure/reaction times at peak process temperatures. We will then describe the development of several bench-scale fixed and fluidized-bed test reactors of more recent design, before returning to examine the implications of such tests. This will lead us to several case studies outlining how bench-scale experimental work can assist in the design and operation of pilot and plant scale equipment.

4.1 Rates of Char Deactivation and Implications for Reaction Design

4.1.1 *"Relative char combustion reactivity" measurements*

Several references have already been made to char reactivity measurements in Chapter 3. Clearly, the combustion reactivity of a fuel would depend on the actual combustion conditions as well as the properties of the fuel. However, there is a need to compare reactivities of chars and to establish some relative order of reactivity between them. In much of the work described below, the simplest possible test has been adopted. The measurement is based on the original work of Jenkins *et al.* [1973] and involves the combustion of 1–2 mg of powdered char in a thermogravimetric balance. A stream of nitrogen (~40 ml min^{-1}) is passed over the char sample during heatup at 25–50°C min^{-1} to the test temperature of 500°C. About 5 minutes are allowed at this temperature for thermal equilibration. The "isothermal" combustion of the char is then initiated by switching from pure N_2 to *air*, flowing at the same rate. The "relative char combustion reactivity" is calculated from the maximum weight loss rate during the combustion of the sample:

$$R_{max} = -(1/W_0)[dW(t)/dt]_{max}.$$

In this equation, t is defined as the time, W_0 the initial and W(t) the time dependent char weight. The temperature of 500°C was adopted as being sufficiently low to restrict the combustion processes to chemical control. Under these conditions, effects due to pore diffusion limitations and pore surface accessibility should be minimized. Clearly, combustion rates are still directly related to the porosity, as expressed in terms of total surface area. The latter is a key parameter in determining "relative combustion reactivities". To minimize *extra*-particle diffusion related effects, sample (pile) sizes were limited to around 1.5 mg. The orders of reactivity between chars established using this combustion test were found to be similar to trends found during analogous CO_2 gasification tests at 950°C [Cai, 1995].

4.1.2 Char deactivation at 1,000°C

The relative reactivity test just described was used to characterize a set of chars from the pyrolysis of Daw Mill (UK) coal in helium at atmospheric pressure. Figure 4.1 presents data for samples heated at rates between 1 and 10,000°C s^{-1} to 1,000°C, with holding at peak temperature for 0, 10 or 60 seconds. The relative combustion reactivities of chars from the 0-second holding runs were observed to increase nearly linearly with increasing heating rate. On the other hand, when held at 1,000°C for intervals as short as 10 seconds, the reactivity of the most rapidly heated chars (10,000°C s^{-1}) dropped by a factor of almost 3.5. The effect was reproducible and appears to indicate deactivation by annealing at the higher temperatures.

These data also indicate that slow heating to 1,000°C – even with 0-second holding at peak temperature – causes the chars to deactivate during heatup. In these cases, the

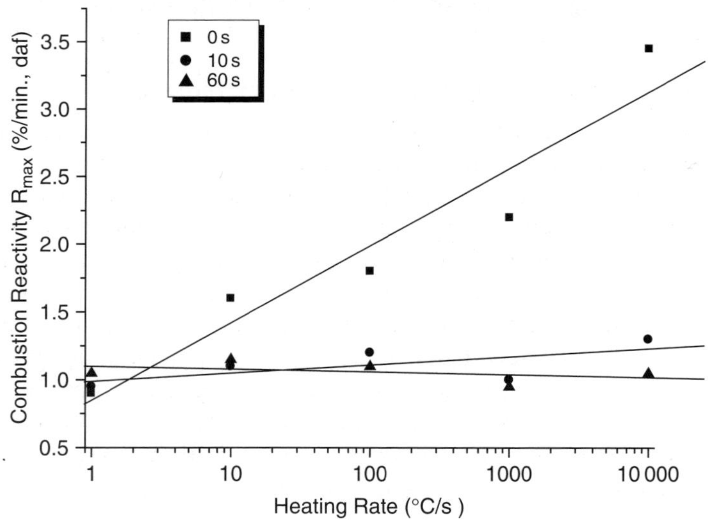

Figure 4.1 Relative char combustion reactivities of chars prepared in atmospheric pressure helium, in a wire-mesh reactor. Daw Mill (UK) coal. The initial pyrolysis experiments were conducted at heating rates between 1 and 10,000°C s^{-1} to 1,000°C, with holding at peak temperature for 0, 10 and 60 seconds. Rapidly heated chars lose over two-thirds of their reactivity after 10 seconds at 1,000°C. (Reproduced with permission: Fuel 79, (2000), 793; Copyright 2000 Elsevier).

holding time has relatively little effect, as most of the deactivation appears to take place during heatup. When heated at 10°C s^{-1}, the sample would spend 10 seconds between 900 and 1,000°C. Higher heating rates also tend to produce more porous chars, which present greater surface areas to diffused oxygen molecules during the combustion test.

These results imply that reactivity data obtained using slow-heating thermogravimetric balances are of questionable value, if results are to be interpreted in relation to fluidized or entrained bed reactors. Figure 4.1 also shows that char reactivity is a function of time at temperatures near or above 1,000°C. Unless a time dependence is built into reaction rate constants, results from kinetic schemes purporting to simulate the gasification or combustion of rapidly heated chars are thus likely to be in significant error. On the other hand, detailed time-*and*-temperature dependent kinetic constants would be difficult to produce to any degree of accuracy. Perhaps – just perhaps – experimentalists are indispensable after all.

Figure 4.1 also provided the clues for explaining how chars withdrawn from the spouted-bed of the British Coal air-blown gasifier could be very unreactive. The feed coal to this reactor was classed as "less than 3 mm" and the operating temperature range was mostly between 930 and 970°C. Experiments have been conducted in a laboratory-scale high-pressure fluidized-bed reactor at 1,000°C (described below), where conversions of the 106–152 μm size particles have been compared with conversions of 600–800 μm particles. The difference in conversion for a reaction time of 60 seconds was ~14% (about 72 for the smaller particles vs. 58% for the larger particles). Clearly, larger particles need more time to achieve total conversion – during which time the carbonaceous material rapidly loses reactivity at these relatively high temperatures. Some of the chars recovered from the British Coal ABGC reactor were thought to have spent over one hour within the reactor and their relative reactivities were found to have very low numerical values: between 0.25 and 0.5 on the scale presented in Figure 4.1 [Zhuo *et al.*, 2000a].

At Imperial College, the data shown in Figure 4.1 was considered sufficiently important to provide the starting point for a three-year study to examine the aging behavior of chars in a modified high-pressure fluidized-bed. It seemed in any case necessary to ascertain whether the observed phenomenon could be duplicated under different experimental conditions. A different operator, using a different batch of Daw Mill coal, undertook similar experiments. The procedure of Figure 4.1 was repeated at 3.5 bars in an atmosphere of CO_2. In this second set of runs, the heating rate interval was cut down to between 100 and 10,000°C s^{-1}; the slower heating rate runs take a long time to complete. Figure 4.2 does not therefore show the convergence of the reactivity lines at the low heating-rate end, as in Figure 4.1. However, the data were qualitatively similar to those of Figure 4.1. Heating at the highest available rate [10,000°C s^{-1}], the relative combustion reactivity was observed to drop to a quarter of its value in 10 seconds holding at 1,000°C, comparable to the drop observed in Figure 4.1.

4.2 The Design of High-Pressure Wire-Mesh Reactors

Three major high-pressure wire-mesh studies had been completed by the mid-1980s. The initial reactor constructed by Howard and co-workers [Suuberg *et al.*,1980] did not operate at heating rates below about 200°C s^{-1} and relied for tar yield determinations on

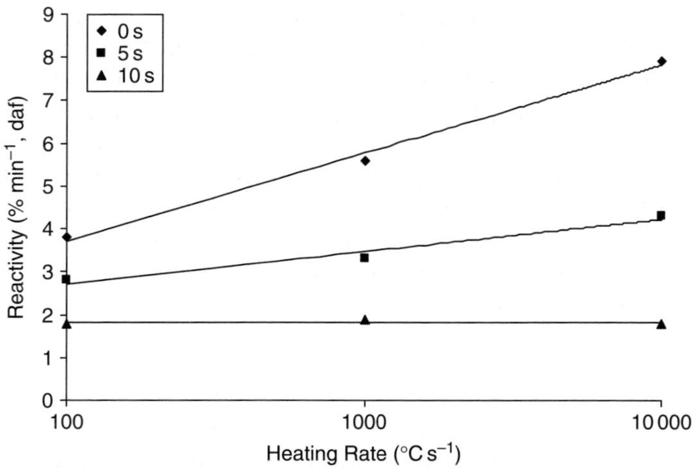

Figure 4.2 Relative char combustion reactivities of chars prepared in 3.5 bar CO_2 in a wire-mesh reactor. Daw Mill (UK) coal with 125–150 mm sized particles. The initial pyrolysis experiments were conducted at heating rates between 100 and 10,000°C s^{-1} to 1,000°C, with holding at peak temperature for 0, 5 and 10 seconds. Rapidly heated chars lost over two-thirds of their reactivity after 10 seconds at 1,000°C. [D. Peralta; unpublished work. Imperial College (2003)].

deposition onto reactor linings and in the gas filters. Recirculation of tar vapors around the mesh could not be avoided. van Heek and co-workers at Bergbau Forschung (now DMT) [Arendt & van Heek, 1981] similarly worked at rates above 200°C s^{-1} and relied on tar deposition on walls and a calculation involving butane concentrations in the product gas to account for *all tars* released from the sample. The work at Princeton University [Niksa *et al.*, 1982a, 1982b, 1984] including tar yield determinations in up to 25 bar H_2 by Bautista *et al.* [1986] relied on a design where the gas flow swept parallel to the surface of the mesh. Meanwhile, earlier work by Stangeby and Sears [1981a, 1981b] had demonstrated this configuration to give rise to partial tar cracking, particularly at higher heating rates. Initially, the motivation of all three groups for exploring high-pressure wire-mesh reactor designs was to examine the possibilities of hydropyrolysis as a process route for making liquids and substitute natural gas. None of these reactors went much above 1,000–1,100°C.

Hydropyrolysis was one of the research strands pursued in Europe and North America in the decade following the war of 1973 in the Middle East and the subsequent oil price increases. Pure high-pressure hydrogen was to be reacted with coal, to produce primarily methane and a tar. The suitability of the tars as a source of synthetic fuels and chemicals was to be explored. With the benefit of hindsight, it is not clear how this scheme was ever conceived as a process route with economic potential. In the first decade of the new millennium, we find ourselves in a quest for the "hydrogen economy", wishing for large amounts of pure hydrogen shorn of its "undesirable" companion, the carbon atom.

Nevertheless, in the mid-1980s, hydropyrolysis was investigated by British Gas and a consortium including several Japanese companies, led by Osaka Gas. At the time, the Japanese gas industry was looking into technically viable alternatives to the massive imports of LNG, which they later committed themselves to. A 2-ton per day pilot entrained flow reactor at Solihull (UK) was followed by a 50-ton per day facility in Osaka. The Japanese government, having initially pledged some \$125 million for this development, eventually abandoned the project early in the new century. Nevertheless, in the summer of 2001, Syngenta (formerly British Gas) was still putting out feelers for re-launching bench scale hydropyrolysis research. This occurred when natural gas prices had temporarily spiked to around \$10 per million BTU. At the end of the summer, however, gas fell back to about a third of that price and interest waned. Notwithstanding these economic ups and downs, hydropyrolysis was one of the technically most successful applications of the wire-mesh reactor configuration.

4.2.1 A gas-sweep facility for tar capture in the high-pressure wire-mesh reactor

We have seen in Chapter 3 how the design of the Imperial College wire-mesh reactor evolved along several parallel lines. First, novel instrumentation allowed heating rates as low as 0.1°C s^{-1} to be reached, giving access to a much wider range of heating rates than hitherto available. The data acquisition and control instrumentation required for producing smooth time-temperature ramps at slow heating rates had to be made in-house. The new system allowed directly observing the effect of heating rate on product distributions, during coal and biomass pyrolysis and hydropyrolysis [Gibbins-Matham & Kandiyoti, 1988; Gibbins & Kandiyoti, 1989a, 1989b; Gibbins *et al.*, 1989c]. From the start, water circulation was used for cooling the electrodes, to absorb the generated heat, particularly during slow heating experiments. The higher end of the heating rates was eventually extended to 10,000°C s^{-1}. We have also described in Chapter 3 how the sweep gas facility was used for carrying volatiles away from the heated zone. This facility served to direct tar vapors into a cold-trap for quantitative capture and eventual recovery. Initially, it was thought that the same configuration would serve equally during operation under high pressure. In fact, the reactor-base designs for the atmospheric and high-pressure reactors were similar, down to the O-ring seal design (see Figure 3.1). The substitution of a stainless-steel pressure head – tested to 300 bar – was thought sufficient to continue experimenting, much as had been done at atmospheric pressure. The first few commissioning runs quickly showed, however, that above about 10 bars, the gas sweep through the mesh gave rise to temperature instabilities [Gibbins & Kandiyoti, 1989b]. The explanation proved intriguing.

When operating at or near atmospheric pressure, flow distortions caused by the presence of the mesh and the sample particles do not give rise to detectable adverse consequences. However, at high pressures, gas densities, gas heat capacities per unit volume and gas-solid heat transfer coefficients all increase sharply. However, when local variations in gas velocities due to minor turbulence combine with greater rates of heat transfer under pressure, severe *lateral* as well as local (time-dependent) temperature oscillations are set up. The lateral oscillations were picked up by the two thermocouples placed on the mesh, about 1-cm apart. The magnitudes of these oscillations could be as large as ±150–200°C, sufficient for invalidating results from these experiments.

The solution adopted was to pass a "diffuse flow" through the cell and through and around the mesh. The quartz bell was also removed. In the absence of a sweep gas stream directing tar vapours into the traps, tar yields could not be measured reliably. The resulting high-pressure wire-mesh reactor was still able to operate at the pressures required for simulating the hydropyrolysis of coal and to determine total volatile yields, to within several percent accuracy [Gibbins, 1988]. On the other hand, process economics were said by British Gas to depend heavily on the yields and quality of tar from their entrained flow reactors. Accurate tar yields were therefore required, alongside the quantitative capture and recovery of the tars for structural characterization.

Eventually, the problem was solved by smoothing the flow of gas passing through the mesh [Guell, 1993; Guell & Kandiyoti, 1993]. The Reynolds number for hydrogen flowing through the 3 cm aperture in the brass plate underneath the mesh was calculated to be about 15 at atmospheric pressure. It was 950 for operation at 70 bar and 6850 at 150 bar, respectively. At first glance, therefore, the onset of turbulence would have been expected at pressures *above* 70 bar. However, flow through the wire mesh laden with dispersed sample particles appeared to disrupt streamlines, introducing imbalances in the flow crossing the sample-holding part of the mesh.

It was decided therefore to pass the sweep gas stream through a vertical bank of 3 mm-diameter tubes placed below the mesh, to reduce the characteristic flow diameter by a factor of 10. For hydrogen, calculations performed assuming an average gas velocity of 0.15 m s^{-1}, a flow diameter of 3 mm (one tube in the bank of tubes) and a temperature of 357°C, show Reynolds number values increasing from about 1.5 at atmospheric pressure to about 95 at 70 bar and to about 700 at 150 bar. The bank of tubes thus both reduced the Reynolds number and partly suppressed lateral velocity components, helping to even out heat transfer from the wire-mesh into the flowing stream of carrier-gas. The modification allowed safely passing a stream of gas through the sample holding mesh, much as in the atmospheric pressure instrument. It allowed operating the tar traps as originally intended, at pressures as high as 160 bar [Guell, 1993; Guell & Kandiyoti, 1993].

Figure 4.3a presents a schematic diagram of the high-pressure reactor, with the flow-smoothing section (Item 14) placed underneath the mesh. The significant difference between this design and the base-plate in Figure 3.1 was the positioning of the gas inlet beneath the mesh. The incoming flow was evened out by passage through two quartz sinter discs (Item 6), the space between them serving as a settling chamber. The bank of tubes (Item 14) used for smoothing the flow was later replaced by a "flame-trap matrix," which consisted of a rolled up sheet of finely corrugated stainless steel. This arrangement produced conduits of less than 2-mm diameter for the flow of carrier gas. These design modifications have allowed stabler equilibria to be established between the large energy inputs into the sample holding section of the wire-mesh (Item 16), and the large amounts of heat removed by the gas stream from that section, during operation at elevated pressures. Figure 4.3b shows the level of temperature stability, to within ±30°C, that was achieved with the introduction of the bank of tubes, during a run at the relatively extreme conditions of 80 bar H_2 at 850°C. The amplitude of the temperature oscillations was much reduced compared to carrier gas flow in the absence of the bank of tubes, but could not be eliminated completely.

The tar trap of the high-pressure wire-mesh instrument consisted of the "quartz bell" (Item 2 in Figure 4.3a) connected to a steel tube placed at the gas exit (Item 1 in Figure 4.3a), packed with strips of wire-mesh, to improve heat transfer and facilitate condensation. The

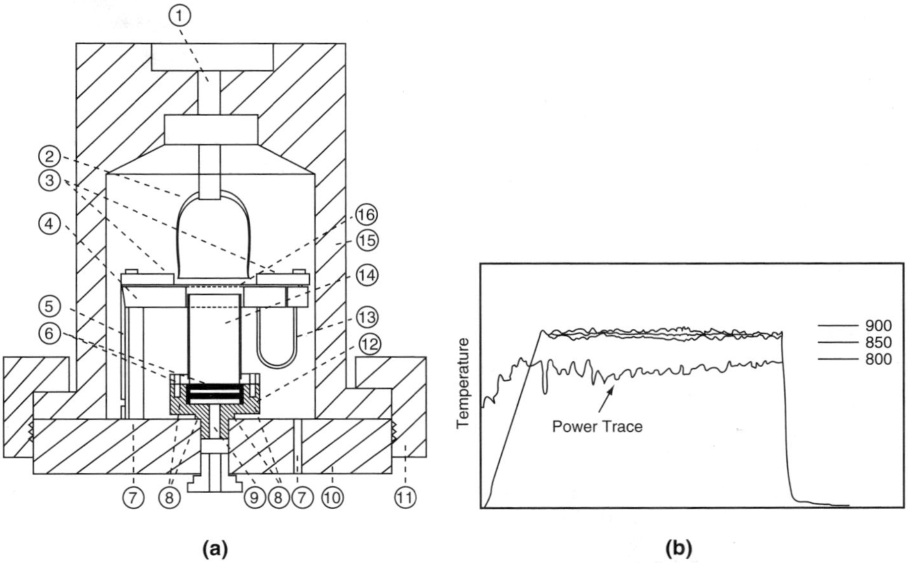

Figure 4.3 **(a)** The high-pressure wire-mesh reactor: (1) gas exit; (2) quartz bell; (3) electrode clamps; (4) mesh support plate; (5) current supply; (6) sinter disk; (7) support plate stands, hollow to allow water flow; (8) copper seals; (9) gas inlet; (10) base plate; (11) throw over sealing ring; (12) flow smoothing cell; (13) spring, hollow to allow water flow; (14) bank of tubes, later replaced by flame trap matrix; (15) pressure bell; (16) mesh. (Reproduced with permission: Energy & Fuels 1999, *13*, 122; Copyright 1999 Am.Chem.Soc.). **(b)** Time temperature history from the high-pressure wire-mesh reactor. Pressure 80 bar hydrogen. Heating at 1,000°C s^{-1} to 850°C with 10 s holding. (Reproduced with permission: Energy & Fuels 1993, 7, 943; Copyright 1993 Am.Chem.Soc.)

extension of the tar trap above the pressure case was sealed to the case by tightening a gland nut and cooled externally by liquid nitrogen. Initial attempts to weigh the tar trap before and after the experiment were foiled due to loss of steel mass as nuts were tightened and loosened to make and release the pressure seals. A procedure was then developed to wash the traps with a 4:1 volume/volume mixture of chloroform and methanol. The tar sample was recovered by evaporating the solvent. Details of the development have been given in the original publication [Guell & Kandiyoti, 1993].

4.2.2 High-pressure pyrolysis and hydropyrolysis of coals – General trends

Figures 4.4(a,b) show the variation of tar and total volatile yields from Linby (UK) and Pittsburgh No.8 coals in He and H_2, as a function of pressure. Samples were heated at 1,000°C s^{-1} to 700°C with 10 s holding. The Pittsburgh No.8 coal sample was drawn from the Argonne Premium Coal Sample set [Vorres, 1990].

As expected from trends observed in previous work at high-heating rates [Howard, 1981; Arendt & van Heek, 1981; Niksa *et al.*, 1982a, 1982b; Li *et al.*, 1991], tar and total volatile yields diminished rapidly between 1 and 10 bars, in hydrogen as well as in helium. The effect was first observed and discussed by Howard and co-workers [Howard, 1981].

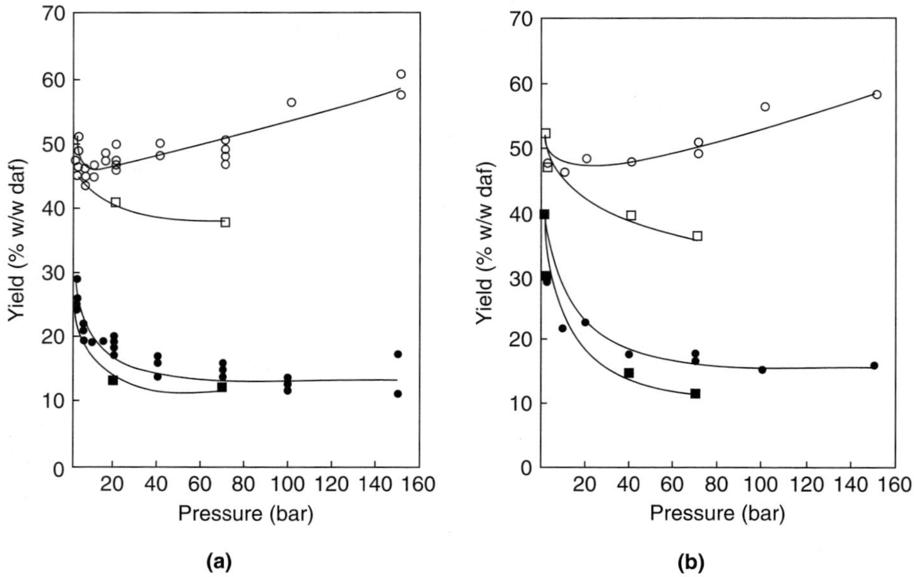

Figure 4.4 (**a**) Effect of pressure on pyrolysis and hydropyrolysis yields from Linby coal; heating at 1000°C s^{-1} to 700°C, with 10 s holding time: (○) total volatiles under H_2, (□) total volatiles under He; (●) tar under H_2, (■) tar under He. (**b**) Effect of pressure on pyrolysis and hydropyrolysis yields from Pittsburgh No.8 coal; heating at 1000°C s^{-1} to 700°C, with 10 s holding time: (○) total volatiles under H_2, (□) total volatiles under He; (●) tar under H_2, (■) tar under He. (Reproduced with permission: Energy & Fuels 1993, 7, 943; Copyright 1993 Am.Chem.Soc.)

Broadly, the decline appears related to higher ambient pressures physically suppressing the evolution of volatiles from coal particles.

To examine the behavior of coals more widely at these relatively low "high-pressures", the behavior of eight coals was examined at atmospheric pressure and at 2.5 bars (Table 4.1). All except one of the samples showed behavior similar to that in Figure 4.4, with tar yields declining more sharply than the corresponding total volatiles. All samples, except Tilmanstone, showed at least some particle softening during rapid heating (1,000°C s^{-1}), which is known to be enhanced by increases in ambient pressure. This is likely to impede the diffusion of potential volatiles moving outward within the particle, apparently allowing time for some cracking and char forming reactions to take place.

The only non-softening coal in the present set (Tilmanstone, UK; 91% carbon) showed a small *increase* in total volatiles alongside the drop in tar yield observed with the other coals. Repeated experiments have confirmed this result, which appears as the exception within this set of samples. This coal was not observed to soften or melt significantly during pyrolysis under these reaction conditions (1–2.5 bar; 1,000°C s^{-1}).

In Chapter 3, we cited the example of holding times that were too short for pyrolysis reactions to reach completion at 400 and 500°C. The data in Table 4.2 shows the relationship between weight loss and holding time at 2.5 and 70 bar under hydrogen. It provides satisfactory evidence of completion of hydropyrolysis reactions (in the form of tar

Table 4.1 The effect of a pressure increase from 1 to 2.5 bar on pyrolysis yields in helium (% weight/weight, dry ash free basis). Samples were heated at 1,000°C s^{-1} with 10 seconds holding at 700°C.

Coal	*Elemental carbon*	*Atmospheric pressure*		*2.5 bar*	
		Total volatiles	*Tar*	*Total volatiles*	*Tar*
Tilmanstone[a]	91.4	20.6	14.0	23.7	11.9
Taff Merthyr[a]	90.0	12.8	10.1	9.1	5.4
Thoresby[a]	84.0	46.8	33.6	42.6	27.0
Hem Heath[a]	83.9	44.6	33.4	42.2	27.4
Pittsburgh No8[a]	83.2	50.4	35.2	47.7	28.9
Longannet[a]	82.7	43.4	25.8	43.1	19.1
Linby	81.5	47.4	28.9	45.1	24.4
Gedling[a]	81.3	46.2	29.9	46.2	21.0

[a] Atmospheric pressure values obtained with 30 second holding.

Table 4.2 Effect of holding time and hydrogen pressure on pyrolysis yields (% w/w daf) from Linby coal heated at 1,000°C s^{-1} to 700°C.

Pressure (bar)	*Holding time (s)*	*Total volatiles*	*Tar*
2.5	0	30.6	20.3
2.5	1	40.0	27.1
2.5	10	41.9	26.4
70	0	23.2	9.9
70	1	41.0	17.4
70	10	50.3	16.6

release) at residence times of the order of 1 second at 700°C, while the steady increase in total volatiles shows the effect of hydrogasification reactions. Later on, we will use this information in selecting holding times for steam gasification experiments, which are normally carried out at higher temperatures and lower pressures (25–30 bar), but where tar yield determinations are difficult.

Figure 4.4a also shows that, between 20 and 150 bar, hydropyrolysis total volatile yields increased monotonically with pressure. Under these reaction conditions, intense methane formation has been widely reported, indicating the progress of hydrogasification reactions well beyond the completion of hydropyrolysis related volatile release [Howard, 1981; Arendt & van Heek, 1981]. The minimum in the hydropyrolysis total volatiles curve was first discussed by Howard and co-workers at the Massachusetts Institute of Technology. It appears at the cusp of two opposing trends, the one tending to suppress volatile release by the physical effect of high-pressure and the second, to enhance volatilization by increasing hydrogasification rates at higher pressures. However, a monotonic rise may be obtained (instead of the minimum) under two related conditions, namely, when the sample is sufficiently reactive and/or if the holding time is sufficiently long [Messenbock, 1998]. Despite important differences between *tar* yield determination procedures, the trends shown by the data in Figure 4.4, covering high heating rates, were in broad qualitative agreement with earlier work cited above. Data from slow heating experiments, however, told a different story.

4.2.3 Hydropyrolysis at slow heating rates

Figure 4.5 shows changes in tar and total volatile yields from Linby coal with increasing hydrogen and helium pressures during slow heating rate experiments (1°C s^{-1}) to 700°C with a 10 s hold. In contrast to the fast heating (1,000°C s^{-1}) results, these data show that after the initial drop in yields between 1 and 5 bar, there was a small but experimentally significant *upward* trend in the tar yield with increasing hydrogen pressures up to 70 bar.

By contrast, tar yields in helium tended to level off above 20 bar, at values well below the atmospheric pressure tar yield. As already explained, this is thought to be due to the physical effect of external pressure. Meanwhile, the *increase* in tar yields in the presence of H_2 indicates the effect of reactive interactions with the pyrolyzing mass. The differences between tar yields in hydrogen and helium increased with pressure. This latter trend clearly contrasts with that observed at high heating rates and requires clarification. Meanwhile, the longer exposure of char to hydrogen during these slow heating

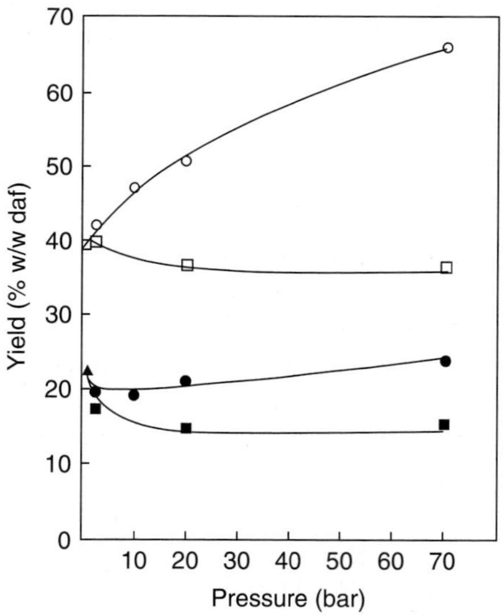

Figure 4.5 Effect of pressure on pyrolysis and hydropyrolysis yields from Linby coal heated at 1°C s^{-1} to 700°C, with 10 second holding at peak temperature. Total volatiles: (○) hydrogen, (□) helium. Tar: (●) hydrogen; (■) helium. (Reproduced with permission: Energy & Fuels 1993, 7, 943; Copyright 1993 Am.Chem.Soc.)

rate experiments gave total volatile yields that increased sharply compared to heating at 1,000°C s^{-1} (also see Figure 4.6b). There was no minimum.

Figure 4.6a shows the complex relationship that emerged between the effects of heating rate and pressure on tar yield. The data spanned heating rates between 1 and 2,500°C s^{-1} and pressures between 1 and 70 bar. At atmospheric pressure and 2.5 bar, tar yields were observed to increase with increasing heating rate. This trend was expected from findings outlined in Chapter 3 and could be repeated with nearly all vitrinitic coals and vitrinite concentrates tested. However, at 10 bar, the tar yield appears to be, within experimental scatter, relatively insensitive to changes in heating rate over the 1–1000°C s^{-1} range. Above 10 bar, tar yields were found to *decrease* with increasing heating rate. There was thus a reversal of the tar release trends with increasing heating rates at pressures above 10 bar.

Cross-reading the data, in the higher heating rate range, tar yields were found to drop rapidly with increasing pressure, as already observed in Figures 4.4 (a,b). Similar findings for high heating-rates had already been reported in previous work by Howard and co-workers and Niksa and co-workers, cited above. However, at heating rates below 10°C s^{-1} (1–10°C s^{-1}) tar yields are found to *increase* with pressure, after the initial drop between 1 and 5 bar. Overall, the reversal in the tar yield versus H_2-pressure trend is consistent with slow heating rates affording longer reaction times at temperatures relevant to tar release, allowing better contact between the pyrolyzing mass and the externally supplied hydrogen. As the heating rate is reduced, a larger proportion of tar precursors seem able to react with hydrogen and volatilize as tar.

By contrast, at low pressures, where the reactivity of hydrogen is not great, low heating rates would still lead to lower tar yields – as at atmospheric pressure. However, the volatile suppressing effect of high-pressure prevails across the heating rate spectrum. The highest tar yields were still observed during high heating rate runs at atmospheric pressure or at reduced pressures.

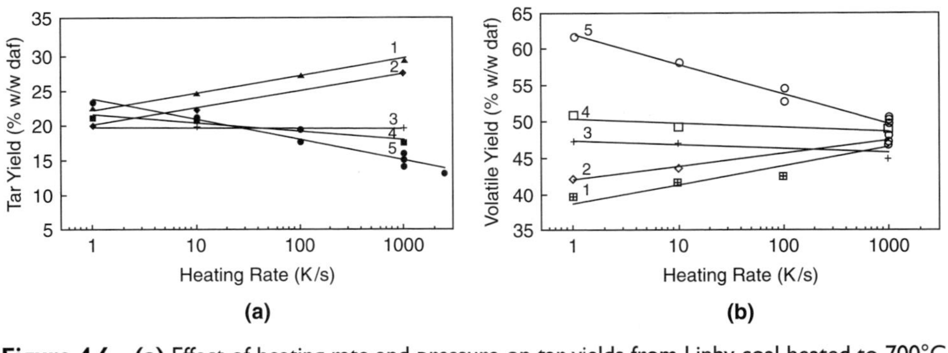

Figure 4.6 **(a)** Effect of heating rate and pressure on tar yields from Linby coal heated to 700°C with 10 seconds holding. (1) Atmospheric pressure helium; (2) 2.5 bar H_2; (3) 10 bar H_2; (4) 20 bar H_2 (5) 70 bar H_2. **(b)** Effect of heating rate and pressure on total volatiles from Linby coal heated to 700°C with 10 seconds holding. (1) Atmospheric pressure helium; (2) 2.5 bar H_2; (3) 10 bar H_2; (4) 20 bar H_2 (5) 70 bar H_2. (Reproduced with permission: Energy & Fuels 1993, 7, 943; Copyright 1993 Am.Chem.Soc.)

Figure 4.6b presents total volatiles data obtained during the same set of experiments. In the presence of hydrogen, sample weight loss has a component of volatile release due to hydropyrolysis and one due to hydrogasification. The two processes usually overlap to a certain extent, although tar release and char formation (hydropyrolysis) is viewed as preceding the char-hydrogen reaction stage (hydrogasification). These results show that total weight loss was greatest for the combination of high pressures and long exposures encountered during the slow heating rate experiments. Similar trends were observed during analogous experiments with the sample of Pittsburgh No.8 coal (not shown).

4.2.4 *Effect of coal thermoplastic behavior on hydropyrolysis tar yields*

Table 4.3 presents product distributions from the hydropyrolysis of two Southern Hemisphere coals as a function of pressure and heating rate. During slow-heating hydropyrolysis experiments at the highest pressure attempted, both coals gave greater tar yields than from fast heating rate experiments (1,000°C s^{-1}) at either atmospheric pressure or at 70 bar. Our previous experience suggested that it would be reasonable to expect the highest tar yields for any coal to be observed at low pressures and high heating rates. The indications provided by the two Chilean coals provided exceptions, which needed examining.

There was one obvious difference between these two coals and the other samples examined in this study. The Chilean coals showed no tendency to melt either during rapid heating or under high-pressure hydrogen. Slow heating rates appeared to provide greater opportunity for contact between tar precursors within the more open matrix of these non-melting coals and external hydrogen. The non-melting coals could thus offer greater surface area to ambient hydrogen and greater porosity for evolving volatiles.

Table 4.3 Effect of heating rate and pressure on hydropyrolysis tar and total volatile yields of Pecket and Catamutum (Chile) coals. 10 seconds holding at 700°C. All results are given as % weight/weight, dry ash free basis (daf)

Heating rate [°C s^{-1}]		*Hydrogen pressure*		
		2.5 bar	*20 bar*	*70 bar*
Pecket coal				
1,000°C s^{-1}	Total volatiles	57.1	69.9	74.3
1,000°C s^{-1}	Tar	15.5	11.1	11.1
1°C s^{-1}	Total volatiles	52.8	72.4	n.a.
1°C s^{-1}	Tar	11.3	14.5	24.1
Catamutum coal				
1,000°C s^{-1}	Total volatiles	54.3	n.a.	n.a.
1,000°C s^{-1}	Tar	20.4	n.a.	n.a.
1°C s^{-1}	Total volatiles	52.4	n.a.	83.3
1°C s^{-1}	Tar	9.7	n.a.	24.7

A short investigation of the relationship between the plastic properties of coals and the proportions of tars and volatiles released under hydropyrolysis conditions at different heating rates was then carried out. Table 4.4 presents tars yields for seven coals under the following conditions: (i) 1°C s^{-1} and 70 bar (in helium and H_2), (ii) 1000°C s^{-1} in atmospheric pressure helium, and, (iii) 1000°C s^{-1} and 70 bar (in helium and H_2). Of these samples, Linby and Pittsburgh No. 8 coals showed the greatest degree of melting when heated under pressure. At all pressures, Longannet, Taff Merthyr, and Tilmanstone displayed little softening during heating at 1°C s^{-1}, but showed limited evidence of melting when heated at 1000°C s^{-1}. The two Southern Hemisphere (Chilean) coals showed no trace of plastic behavior when heated at 1°C s^{-1} but limited deformation and agglomeration when heated at 1,000°C s^{-1}.

Compared to the Northern Hemisphere samples used in this study, atmospheric pressure tar yields from both Pecket and Catamutum were small relative to their corresponding total volatile yields (53.4 and 54.3% respectively). What makes these two coals atypical, however is the higher tar yields from slow heating hydropyrolysis experiments, compared to fast heating (1,000°C s^{-1}) atmospheric pressure runs. In this respect they stand out among the other samples, with the possible exception of the high-rank Tilmanstone sample. These findings may indeed correlate with the absence of plasticity of the samples upon heating. A more focused investigation starting with these findings would appear warranted.

Whatever the current level of industrial interest in hydropyrolysis *per se*, the methods outlined above have provided a platform for developing new wire-mesh reactor configurations suitable for testing the behavior of coal samples in CO_2 and steam-gasification. We will next describe a high-pressure wire-mesh reactor equipped with a steam injection facility.

4.2.5 Injection of CO_2 and steam in the high-pressure wire-mesh reactor

Initially, all research groups that attempted to develop high-pressure wire-mesh reactors were interested in investigating aspects of hydropyrolysis and hydrogasification. The initial designs were originally intended to work with ambient atmospheres of non-condensing

Table 4.4 Effect of heating rate and pressure on tar yields for coals showing different plastic behavior

Coal	*1°C s^{-1} 70 bar*		*1,000°C s^{-1} 1 bar*	*1,000°C s^{-1} 70 bar*	
	H_2	He	He	H_2	He
Tilmanstone	14.1	7.9	14.0	9.1	6.1
Taff Merthyr	8.9	n.a.	10.1	4.8	n.a.
Pittsburgh No8	29.2	20.5	35.2	17.4	11.4
Longannet	20.4	11.8	25.8	15.1	10.6
Linby	23.1	15.0	29.5	15.7	12.2
Pecket	24.1	9.5	14.9	11.1	9.3
Catamutum	24.7	11.6	20.4	n.a.	13.8

gases. They could all easily operate with helium and hydrogen. However, CO_2 and steam react with the SS 304 stainless steel mesh used in the pyrolysis experiments. A more resistant mesh material was required and molybdenum was selected as the mesh-material that was readily available and (reasonably) inert under these reaction conditions.

Moreover, CO_2 and steam have higher densities and heat capacities, requiring greater power inputs to attain a given temperature. They also have lower thermal conductivities, which makes operation more difficult. During experiments with CO_2, this combination of conditions led to overheating in parts of the mesh not directly cooled by the sweep gas stream. The layer of mica providing electrical insulation between the mesh and the support plate (between Items 4 and 16 in Figure 4.3) melted frequently, causing short circuits and destroying the sample holder. Whilst CO_2 gasification experiments at 1,000°C could still be undertaken below 30 bar and holding times shorter than 10 seconds [Lim *et al.*, 1997], design changes were required to insure reliable operation over a wider range of parameters.

The cooling water circulation pattern in the base-plate was changed next. This was done to suppress local overheating in parts of the wire mesh not swept by the gas stream, particularly the parts furthest away from the water-cooled electrodes. The new water conduit traced a broadly circular pattern around the aperture in the brass support plate [Figure 4.2 in Messenbock *et al.*, 1999]. In addition, the mica insulation was replaced by a 2 mm thick sheet of ceramic (Macor Machinable Ceramics; Goodfellows, U.K.), drilled through to open a 30 mm hole mirroring the shape of the support plate. These changes have allowed reliable operation for holding times up to 60 s at pressures up to 30 bar in CO_2. However, the Macor ceramic plates were brittle. The suppliers advise that "... over 200 distinctly shaped Macor® MGC parts can be found on America's reusable Space Shuttle Orbiter. Retaining rings of Macor® MGC are used at all hinge points, windows and doors ..." The plates were costly and after each fracture, the shards were fitted back together and re-used for as long as possible, eventually to be replaced by alumina plates of the same thickness. The alumina tiles worked well and retained their integrity over very large numbers of runs.

The basic wire-mesh design is quite useful when non-condensing gases are being used. However, steam injection inside the wire-mesh reactor vessel posed a new set of challenges, as all components except the mesh itself are normally kept at or near ambient temperature. Care was needed to avoid condensation on cold surfaces and dripping condensate from inside the (cold) high-pressure dome onto sample particles, the mesh and the electrodes. Prior to this effort, only DMT had attempted to introduce a limited amount of moisture into the system by heating the pressure casing to about 100°C.

In response to these constraints, the carrier-gas flow path of the reactor (Figure 4.3a), including the mesh and the sample itself, were preheated with a flow of helium, to about 300°C, in order to prevent condenzation. All components in the flow-path were re-designed to withstand the higher temperatures and insure the passage of steam through the flow-smoothing section, the mesh and the quartz bell above the mesh. Teflon seals and other temperature-sensitive components were replaced with others made of copper. The plastic insulation of the thermocouple wires was replaced with drawn glass capillaries [Messenbock *et al.*, 1999a, 1999b].

Figure 4.7 presents a schematic diagram of the wire-mesh reactor system equipped with a steam injection unit. The design aimed to introduce the minimum amount of steam

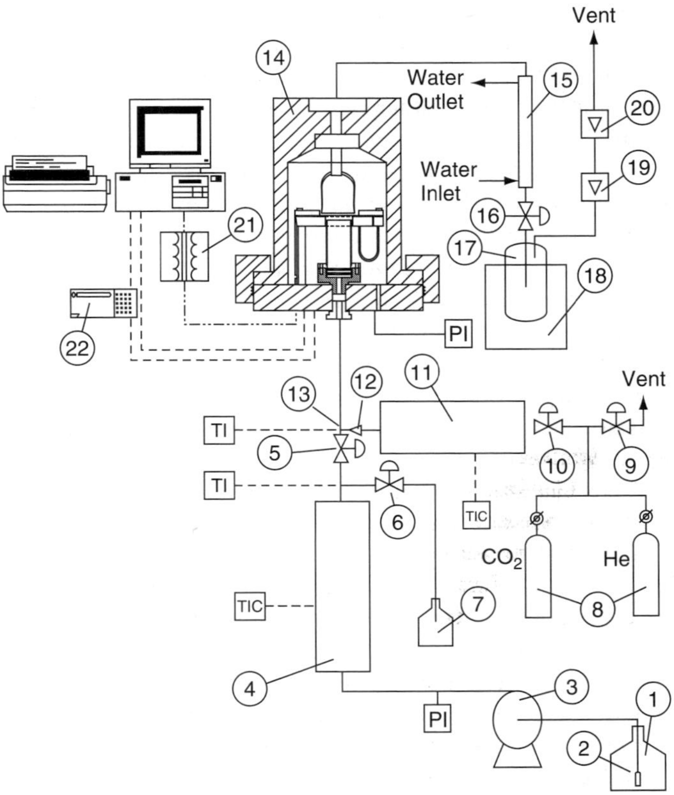

Figure 4.7 Simplified schematic diagram of the steam injection system. Steam is continuously produced and discarded through the by-pass. The steam flow-path is preheated to 300°C before electrical current and steam are simultaneously switched into the high-pressure cell. (1) water reservoir; (2) filter; (3) displacement pump; (4) steam heater; (5), (6) on/off valves; (7) steam bypass collector; (8) sweep gas; (9) safety valve; (10) flow control valve; (11) gas heater; (12) non-return valve; (13) mixing point of steam and gas; (14) wire-mesh reactor; (15) countercurrent condenser; (16) flow control valve; (17) water collector; (18) cold trap; (19, 20) flow-meter; (21) transformer; (22) watchdog. (P) pressure, (T) temperature, (C) controller, (I) indicator. (-) tubing, (--) thermocouple lines, (-··) electric current lines. (Reproduced with permission: Energy & Fuels 1999, *13*, 122; Copyright 1999 Am.Chem.Soc.)

into the reaction chamber during a given run. Prior to starting an experiment, the whole of the steam flow-path was preheated by a stream of helium (8), passing through a packed-bed heater (11). Meanwhile, water was pumped (3) from the reservoir (1) through a filter (2) into the steam generator operated at about 300°C (4), consisting of a 1-inch i.d. AISI 316 stainless steel tube, filled with 3–5 mm glass beads and heated with a resistance heater. Prior to initiating the experimental sequence, the steam was sent through the by-pass valve to a condenser (7).

Before triggering an experimental sequence, the reactor pressure was set by adjusting the pressure letdown valve (16). At that stage, the flow path would have been preheated, but the temperature of the mesh itself did not normally exceed 150°C. This was due to the cooling provided by water circulation through the electrodes and the support plate, which absorbed some of the heat input by the passage of pre-heated helium. Prior to initiating a run, therefore, the temperature of the mesh was electrically boosted to 300°C. Preheating to 300°C for short periods before a run does not appear to alter the sample behavior significantly. A *pyrolysis* run was carried out using the same temperature sequence as a steam-gasification run (but with no steam). At 1 bar and 10 s, the yields *and* the combustion reactivity of the char residue were similar to those from a pyrolysis experiment where the temperature had been directly ramped from ambient.

Steam was allowed to contact the coal sample only instants before the experiment was initiated. At "zero-time" the operator switched steam into the cell whilst simultaneously triggering the electrical current to start the heatup ramp. The temperature of the steam-helium mixture was independently monitored at the exit of the steam generator, where the two steams were mixed (13). The mixture temperature was found to vary between 260 and 270°C, well above the condensation temperature of steam at 20 bar (212.4°C), the maximum steam pressure used in this study.

During pyrolysis and CO_2-gasification experiments, the sweep gas was passed over the mesh at 0.1 m s^{-1}. A 80:20 steam: helium mixture was used in the steam gasification experiments and the mass flow rate of the gas was allowed to change with pressure, since the aim was to keep the gas velocity through the mesh constant. Samples were normally heated at 1,000°C s^{-1} to 1000°C, with the hold time at peak temperature altered between 0 and 60 s and the pressure from 1 to 30 bar. Temperature stability during steam injection was found to be comparable to levels of stability observed during experiments with any of the dry gases. At the exit of the reactor, the steam/gas mixture was cooled under pressure (15). The water was then collected in a cold trap (17) cooled by liquid nitrogen (18). The gas flow was measured by rotameters installed downstream of the condensers. Overall, sample weight loss in steam gasification has been determined with a repeatability of less than 1%, similar to the experiments in dry gases.

In many of the experiments described below, Daw Mill was used as the base-case test coal. This coal softens only marginally even at high heating rates and had been selected as the test coal in many UK gasification trials, including the British Coal Air Blown Gasification Cycle (ABGC) programe. Its composition is as follows (% w/w daf): C: 81.3; H: 4.8; N: 1.3; O: 11.5; S: 1.2; ash: 4.7.

Results from CO_2 and steam gasification of Daw Mill coal: Figure 4.8a presents total volatile yields from CO_2-gasification as a function of pressure (1–30 bar), with 0-second, 10, 20 and 60-seconds holding at 1,000°C. The conversions for "zero holding" and data from pyrolysis were close [Messenbock *et al.*, 1999a], showing no more than 1–4% conversion by gasification during the heatup stage, above the weight loss recorded by pyrolysis. For the longer hold-times (60 s), a sharp increase in weight loss was observed between 1 and 10 bar; conversions at 30 bar were found to be as high as 92%.

Figure 4.8a also shows that results from 10 s holding experiments traced a shallow minimum in the vicinity of 10 bar. Similar trends have been observed during hydropyrolysis experiments. In the lower pressure range, (1–10 bar), the physical suppression

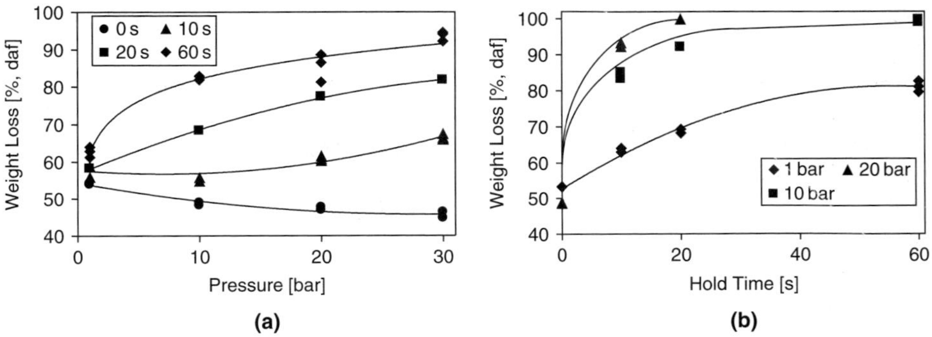

Figure 4.8 **(a)** Total volatile yields from CO_2-gasification **(b)** Total volatile yields from the steam gasification. Daw Mill coal, heated at 1,000°C s^{-1} to 1,000°C. (Reproduced with permission: Energy & Fuels 1999, *13*, 122; Copyright 1999 Am.Chem.Soc.)

of devolatilization by the external pressure appears to cause loss of volatile matter. At higher pressures, the reactivity of the gas appears to enhance increasing sample weight loss [Howard 1981]. Judging by the zero-hold-time results, the suppression of devolatilization due to pressure appears to take place (almost totally) before the end of the heatup period. Data from hydropyrolysis suggests that the effect is due mainly to suppression of tar evolution [Guell & Kandiyoti, 1993], the repolymerized tars depositing as relatively inert secondary char.

At this high heating rate, the gasification process appears to go through several *successive* stages. Volatile suppression mainly through tar repolymerization appears to be followed by gasification of the relatively inert secondary char and finally by the direct gasification of char residue [Guell *et al.*, 1993]. Furthermore, the results in Figure 4.8a show that the presence of the minimum at high heating rates depends not only on the pressure (related to the reactivity of the gas and the char) but also on the holding time at the given peak temperature. Experiments with 20 s holding did not show a minimum, suggesting that the secondary-char layer had already been consumed at the end of the 20 s period and that significant gasification of the char particle had already taken place.

For 60-second hold-time experiments, sample weight loss figures measured in a high-pressure fluidized-bed reactor were close to the results in Figure 4.8a [Megaritis *et al.*, 1998b]. Apart from the hydropyrolysis data already referred to above, there appears to be no other gasification related work carried out in wire-mesh type reactors to compare with these data. At the time of writing, comparable short-time resolution is not available in the vast amount of published data on the basis of TGA and DTA instruments.

Figure 4.8b presents changes in sample weight loss during *steam* gasification as a function of holding times up to 60 s at three different pressures (1, 10, and 20 bar). Differences between yields at 1 and 10 bar were large, qualitatively reflecting the results for CO_2 gasification. Differences between 0 and 10 seconds holding were again large, particularly at the higher pressures. At 20 bar, gasification of the sample (105–152 μm) was virtually completed after 20 s, leaving nothing but a dusting of ash particles on the sample holder.

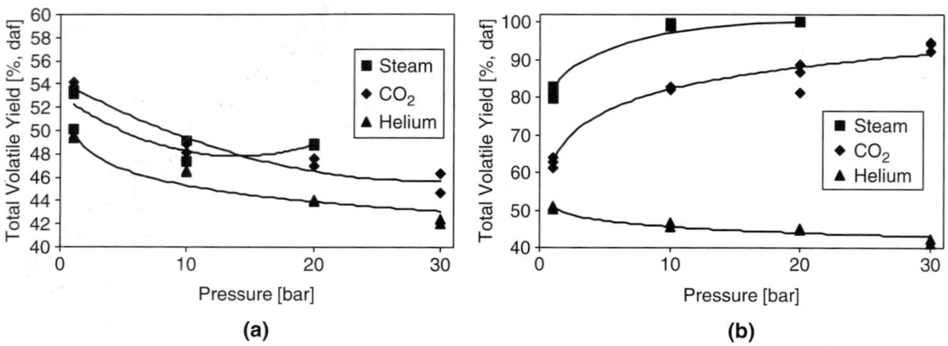

Figure 4.9 **(a)** Total volatile yields from the pyrolysis, CO_2 and steam gasification of Daw Mill coal, heated at 1,000°C s^{-1} to 1,000°C. 0 seconds holding time. **(b)** Total volatile yields from the pyrolysis, CO_2 and steam gasification of Daw Mill coal, heated at 1,000°C s^{-1} to 1,000°C. 60 seconds holding time. (Reproduced with permission: Energy & Fuels 1999, *13*, 122; Copyright 1999 Am.Chem.Soc.)

Figure 4.9a compares sample weight loss data from 0-second holding pyrolysis experiments in helium and gasification runs in CO_2 and steam. At atmospheric pressure, a difference of 3–4% was observed between the pyrolysis (helium) and gasification results, suggesting only a limited extent of gasification during heatup. Sharp drops were observed in sample weight loss between 1 and 10 bar. In helium and CO_2, the decline was monotonic for 0-second hold times over the 1–30 bar range. However, volatile yields in steam traced a minimum above 10 bar, similar to the one observed earlier for 10 s holding under CO_2 (Figure 4.8a). Chemically, the occurrence of the minimum coincides with the deposition and eventual consumption of the secondary char deposited during the pyrolysis stage. At longer hold times, the secondary-char layer appears to be consumed and conversions, due to direct gasification of the char particle increase monotonically. With 10 s holding at 1,000°C in *steam* [Figure 4.9 in Messenbock *et al.*, 1999a, not shown], the minimum is found to disappear. The effect appears associated with the greater reactivity of steam compared to CO_2 under similar experimental conditions. The occurrence of a minimum in conversion vs. pressure diagrams, previously observed by numerous researchers, is thus related to the reactivity of the gasifying agent *and* the holding time at peak temperature. Compared to experiments in CO_2, the minimum was observed at shorter holding times in the presence of more reactive steam. These data reflect the expected sharp difference in reactivity between steam and CO_2. At the higher pressures, after 60 sec holding at 1,000°C (Figure 4.9b), gasification in steam appears to be complete.

Some of these data reflect the rapid char deactivation discussed in an earlier section. In Figure 4.8b, the conversion line for atmospheric pressure steam tended to flatten out at longer hold times, showing practically no additional conversion between 40 and 60 seconds. Similarly, CO_2 gasification in Figure 4.9b does not appear headed for complete conversion and at 10 bars, the CO_2 gasification line flattened out at around 80% conversion [Figure 4.12 in Messenbock *et al.*, 1999a, not shown]. The effect

Table 4.5 Comparison of the "extent of gasification" for 0-second and 20-second holding at 1,000°C. Sample heating rate: 1,000°C s^{-1}

Gas	*Extent of Gasification (% w/w, daf)*		
	Pressure	*0-s holding time*	*20-s holding time*
Carbon dioxide	1	4.5	12.1
Carbon dioxide	10	2.0	24.4
Carbon dioxide	20	3.3	34.3
Carbon dioxide	30	3.3	41.9
Steam	1	2.8	20.6
Steam	10	1.7	48.3
Steam	20	4.8	54.4

appears due to prolonged exposure of the char to high temperature, in a medium not sufficiently reactive for total conversion to be achieved before the onset of deactivation.

The other important outcome of this investigation was the clear recognition that in both CO_2 and steam, only a small amount of gasification was observed during heatup. It then becomes relevant to define an "extent of gasification" by subtracting sample weight loss recorded during a *pyrolysis* experiment from the weight loss observed during a gasification run, performed under otherwise identical experimental conditions (heating rate, temperature, hold-time and pressure). With this definition in mind, the data from experiments just described can be recast into the form shown in Table 4.5.

As in the case of atmospheric pressure pyrolysis reactors, the strengths and weaknesses of high-pressure thermochemical reactors become clearer when results are compared using similar samples. In the next few sections, we will describe a high-pressure fluidized-bed reactor and return to comparing data obtained using these two reactors alongside the fixed-bed "hot-rod" reactor, described in Chapter 3.

4.3 The Design of High-Pressure Bench-Scale Fluidized-Bed Reactors

Experimental high-pressure fluidized-bed systems usually consist of a heated reactor body surrounded with thermal insulation to protect the outer "cold" pressure casing from elevated temperatures. This is done to distance the pressure containment problem from parts of the system where temperatures are high. At or near ambient temperatures, containment vessels operating at pressures required by gasification tests do not require exceptional alloys or very specialized designs. On the other hand, such assemblies usually turn out to be bulky. These larger rigs usually require several operators alongside relatively complex instrumentation. The high construction costs and accompanying expenses would, at least in part, explain the relatively small number of such units in existence.

4.3.1 A survey of small-scale high-pressure fluidized-bed reactors

Several, relatively large *laboratory-scale* high-pressure fluidized-beds designed for pyrolysis and gasification experiments have been reported in the literature. Morris and Keairns [1979] have described a reactor made of Inconel 600 (35 mm i.d., 330 mm long), placed inside a cold pressure casing, together with furnace heaters and surrounding insulation. The bed material (char) was fluidized with nitrogen. The coal sample was held in a horizontal tube attached to a solenoid valve and injected into the reactor using a small cylinder of high-pressure nitrogen. Gas analysis data from experiments up to 982°C and 10 bar have been reported for three different coals. From descriptions, the installation appears relatively large and seems to require significant maintenance. The schematic diagram of the apparatus in the original publication suggests that the feed tube did not extend into the fluidized-bed of solids. In these reactors, unless provision is made to introduce the sample directly into the fluidizing bed, the trajectories and temperature histories of sample particles remain ill defined. However, the authors have not signaled whether they considered this as a potential source of operating problems.

Adánez *et al.* [1985] have described another stainless steel fluidised-bed reactor (AISI 304; 40 mm i.d., 500 mm long) placed in a furnace. Instead of a coal injection system, the reactor was initially charged with 100–630 μm *char* particles and heated in nitrogen flowing at atmospheric pressure. When the intended reaction temperature was reached, the reactor was pressurized and reactant gases introduced. The temperature was determined by a single thermocouple placed inside the bed and the furnace temperature controlled manually using a variable voltage transformer. Gas pressure was regulated by means of a needle valve. Gasification experiments using a lignite have been reported (1,000°C, 25 bar).

Another fluidized-bed pyrolysis-gasification reactor (1,000°C; up to 25 bar) has been described by Hüttinger *et al.* [1988, 1994]. Few details of this reactor are available in the open literature. The reactor made of nearly pure alumina ("alsint") was housed in a furnace, within a bronze pressure casing. Coal samples of about ~200 mg were dropped by gravity from a syringe mounted at the top of the assembly, into the reactor freeboard. No attempt seems to have been made to guide the sample particles into the shallow (20 mm) fluidized-bed of 60–100 μm alumina particles.

A larger reactor with continuous feeding has been described by Sue-A-Quan *et al.* [1991; 1995]. The main reactor tube (Incolloy; 100 mm i.d.; 1,000 mm long) was centrally located and heated electrically inside a refractory-lined steel pressure shell of 305 mm diameter. The coal was fed at 2–5 kg h^{-1} by a star wheel feeder. Water, delivered by a diaphragm pump, was vaporized and superheated in coils immersed in a separate sand fluidized bed heated by a propane burner. Its final temperature was adjusted by heat exchange with product gas from the reactor. The reactor was operated at temperatures up to 900°C and 18 bars. From descriptions, the design and operation of the system appear complex.

4.3.2 Design of a simple bench-scale high-pressure fluidized-bed

Figure 4.10 presents the design of a high-pressure fluidized-bed, originally conceived for operation by batch sample injection. Compared to wire-mesh reactors, fluidized-beds are able to handle larger particle sizes and larger amounts of sample, providing greater

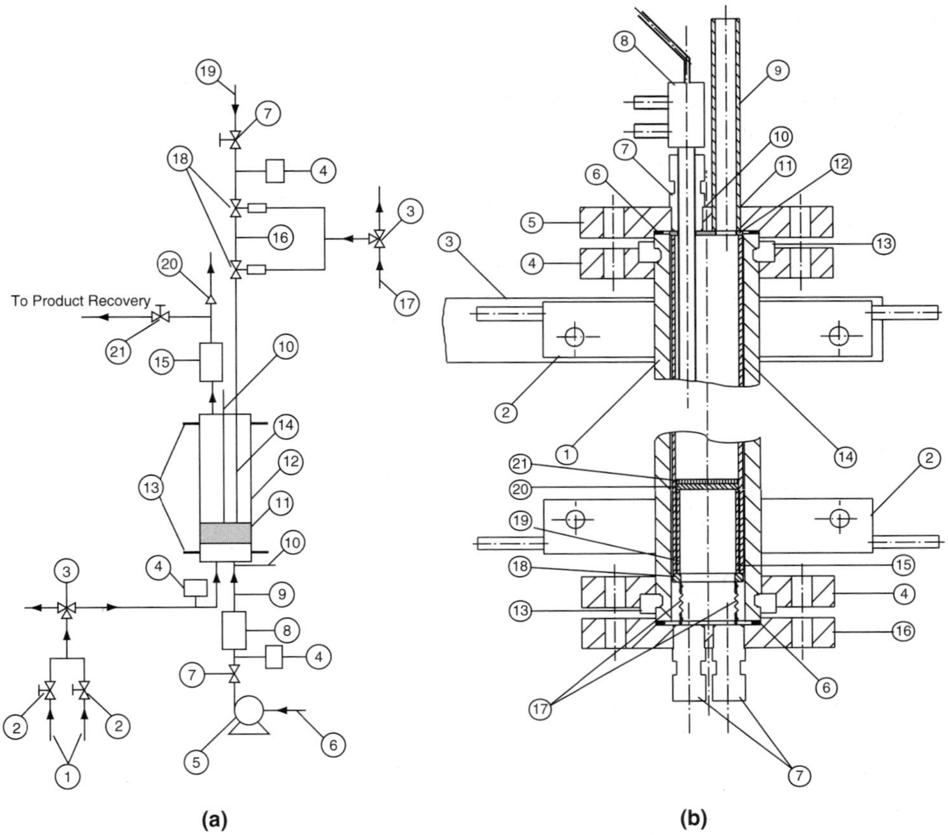

Figure 4.10 **(a)** Schematic diagram of the high-pressure fluidized-bed system. Below the reactor, a liquid metering pump (5) supplies the steam generator (8). Two non-condensable feed gas lines (1) may also be observed. Before injection into the bed, the sample (16) was held between two air-actuated valves (18). (1) High-pressure gas supply; (2) metering valve; (3) 3-way valve; (4) pressure transducer; (5) metering pump; (6) water supply; (7) valve; (8) steam generator; (9) heated line; (10) thermocouple; (11) sand bed; (12) reactor; (13) electrodes; (14) sample injection probe; (15) tar trap; (16) sample; (17) air supply; (18) air-actuated valves; (19) injection probe gas supply; (20) safety valve; (21) gas flow control valve.
(b) Main body of the high-pressure fluidized-bed reactor working at up to 1,000°C and 30 bar. (1) Reactor tube (2) Electrode (3) Copper bar (4) Flange (5) Flange (6) Copper sealing ring (7) Male weld connector (8) Sample injection probe (9) Gas exit line (10) Position of male weld connector for thermocouple (not shown) (11) Weld (12) Kaowool-paper sealing ring, wire-mesh plates (13) 'Half moon' positioning ring (14) Quartz tube liner (15) Distributor disk supporting quartz tube (16) Flange (17) Springs (18) Spring loaded ring. (b) Schematic diagram of the system, showing the two valves between which the sample was held for injection. Below the reactor, the steam generator is supplied by a liquid metering pump; two non-condensable gas lines may also be observed. (Reproduced with permission: Energy & Fuels 1998, *12*, 144; Copyright 1998 Am.Chem.Soc.)

quantities of chars and tars for subsequent characterization. The reactor shown in Figure 4.10 was simple to construct and capable of operation by a single researcher. It features direct electrical heating of the reactor body to avoid the use of a separate furnace.

Figure 4.10(a) presents a schematic diagram of the reactor system, showing the gas supply train (1), the reactor (12), the coal injection probe (14), the tar trap assembly (15), and the product recovery train. Water was supplied to the steam supply circuit by a calibrated high-pressure liquid metering pump (5). The steam generator (8) consists of an electrically heated stainless steel tube packed with ceramic spheres. The concentrations of the feed gas mixtures were controlled by metering valves (2) in the inlet line, where a pressure transducer (4) is positioned for tracking reactor pressure. Exhaust gases passed from the reactor, through a dryer into the product analysis train. The total flow of fluidizing gas was controlled downstream of the reactor system by a metering valve (21). The reactor was protected from overpressure by a safety valve (20) connected to the outlet line.

Various materials of construction have been considered for the reactor tube. Stainless steels could not be used because of unsatisfactory creep resistance properties above 600–700°C. Nimonic 105 and 115 alloys (Henry Wiggin Alloys) have adequate tensile properties for working at 1,000°C, but are difficult to machine, apart from not being available at the time, in the desired dimensions. Ni-Fe based Incoloy Alloy 800HT has high tensile strength and good creep resistance combined with resistance to high temperature corrosion. These properties make it useful for applications involving long-term exposure to elevated temperatures and pressures. Some of the Ni-Cr-Fe based heat resistant Haynes Alloys would also have been equal to this relatively exacting task [http://www.haynesintl.com/HTA.html]. The creep rupture limit of the reactor body designed for this work (34 mm i.d., 48 mm o.d; 504 mm long) was about 1,000 hours at 1,000°C and 40 bar. There was no need for a furnace or a "cold" pressure casing to separate the temperature maintenance and pressure containment functions. Instead, the reactor tube was attached directly to electrodes at the top and the bottom and served as resistance heater as well as pressure vessel.

The reactor tube was lined with a loosely fitting quartz tube, to limit corrosion attack by reactor contents and to block catalytic effects by the reactor walls. The initial design shown in Figure 4.10 was intended for operation as a bubbling fluidized-bed, equipped with a sintered quartz support plate. This is not dissimilar in conception from Tyler's atmospheric pressure quartz fluidized-bed (Figure 3.5a). Coal or biomass samples of 50–200 mg, held between two air-actuated valves, were injected batchwise ("single-slug" injection) through a water-cooled probe *into* the bubbling fluidized-bed of about 40 grams of acid washed sand. Exhaust gases were passed upward through a tar trap and dryer into the gas analysis stage. In addition to product gas analysis, the design of the reactor allows determination of tar/oil and char yields with a repeatability of ±2%. The water cooled sample injection probe, the tar trap assembly and other parts of the reactor system have been described in greater detail in the original publication [Megaritis *et al.*, 1998a].

Results from the high-pressure fluidized-bed reactor: Figure 4.11a presents total volatiles data from the pyrolysis of Daw Mill coal in helium. Experiments were conducted at 1,000°C, between 1 and 30 bar. Reactor cool-down was initiated after a holding time of 60 seconds at peak temperature. These results have been compared with high-pressure

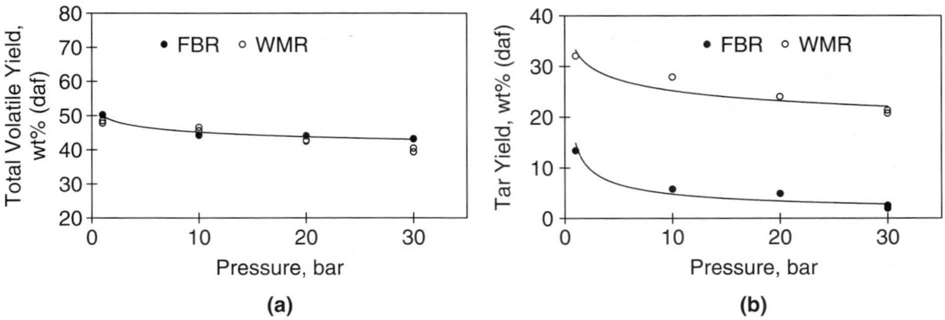

Figure 4.11 **(a)** Pyrolysis total volatile yields from Daw Mill coal as a function of pressure: fluidized-bed reactor: 1000°C, 60 s; wire-mesh reactor: 1000°C, 1000 K/s, 60 s. **(b)** Pyrolysis tar yields from Daw Mill coal as a function of pressure: fluidized-bed reactor: 1000°C, 60 s; wire mesh reactor: 1000°C, 1000 K/s, 10 s. (Reproduced with permission: Energy & Fuels 1998, *12*, 144; Copyright 1998 Am.Chem.Soc.)

wire-mesh reactor data from heating at 1,000°C s^{-1} to 1,000°C, with 10 s holding. We have already observed during experiments at temperatures of 700°C or above, described in Chapter 3, that all recoverable tars are released within the first second following the hea-tup ramp [Gonenc *et al.*, 1990; Li *et al.*, 1993]. At 1,000°C, the effect of differences in holding times between 10 and 60 seconds on *pyrolysis* (in helium) weight loss and tar yields may therefore be neglected.

The total volatile yields measured in the two reactors were similar; they both declined with increasing pressure as a result of the physical suppression of volatiles release. Previous work suggests this to be due to suppression of *tar* evolution [Guell & Kandiyoti, 1993]. However, the eventual fate of tars released in the two reactors was not the same, reflecting the differences in their configurations. Figure 4.11b shows that the proportions of tars recovered at the exit of the fluidized-bed reactor were considerably less than those from the wire-mesh reactor.

This was an expected result. In the wire-mesh reactor, the stream of carrier gas is intended to push tars released from coal particles rapidly out of the shallow reaction zone. In fluidized-bed reactors, on the other hand, tar vapors pass through the length of the freeboard before exiting from the reactor. In the case of the present reactor, before entering the cooled tar trap, volatiles must rise through a 280 mm freeboard (below the top electrode), where temperatures are high. At a distance 50 mm beneath the top electrode, the temperature of the freeboard was measured as 960°C. Below that point, the temperature was uniform at about 1,000°C. Clearly, tars are expected to thermally degrade and partially crack, both during passage through the bed of heated solids itself and *then*, during passage through the reactor freeboard [Stiles & Kandiyoti, 1989]. Previous work has shown that gas-phase tar *cracking* to lighter volatiles does not measurably alter total volatile yields [Tyler, 1979; 1980; Stiles & Kandiyoti, 1989].

During pyrolysis experiments, considerable agglomeration had been observed by visual examination of the bed material, although differences with results from the wire-mesh reactor were within experimental error. However, initial CO_2-gasification trials showed that sample agglomeration affected conversions measurably and adversely at pressures above 10 bars. In these initial experiments, sample conversion did not rise beyond the 50% level under 20 bar CO_2-pressure, i.e. not much above the weight loss observed during pyrolysis. The difficulty appeared due to agglomeration reducing sample surface area exposed to the reactive gas. Gasification reactions were slowed down, until the length of exposure at 1,000°C led to sample deactivation.

To reduce the extent of agglomeration (i) the design of the sample injection probe tip was altered to distribute sample particles radially outward, (ii) higher sample injection gas velocities were used, alongside (iii) larger, (150–300 μm) sand particles as fluidizing material, instead of the more usual 106–152 μm range. Some progress was observed. However, no further reduction in sample agglomeration was observed when a larger sand particle range (800–1000 μm) was tried. Similarly, mixing the coal particles with sand prior to injection was not found to reduce agglomeration significantly. However, less intense agglomeration was observed when smaller amounts of coal were injected. Sample weights of 50 rather than 200 mg were used during subsequent experiments.

Figure 4.12 compares total volatile yields from Daw Mill coal measured in the wire-mesh and fluidized-bed reactors, during CO_2-gasification experiments at 1,000°C. Extents of agglomeration were monitored by observing the sizes of particle clusters in the fluidized-bed material after experiments and by comparing data with the wire-mesh reactor. After introducing the corrective measures outlined above, results from the two reactors could be matched closely at 10 and 20 bar. However, at 30 bar, conversions in the fluidized-bed were found little changed from those at 20 bar, while the conversion in the wire-mesh reactor continued to rise. Experiments in steam were easier. Up to 30 bar (inclusive), total volatile yields from the two reactors were found to be well within experimental error when operating with *steam* [Zhuo *et al.*, 2000a].

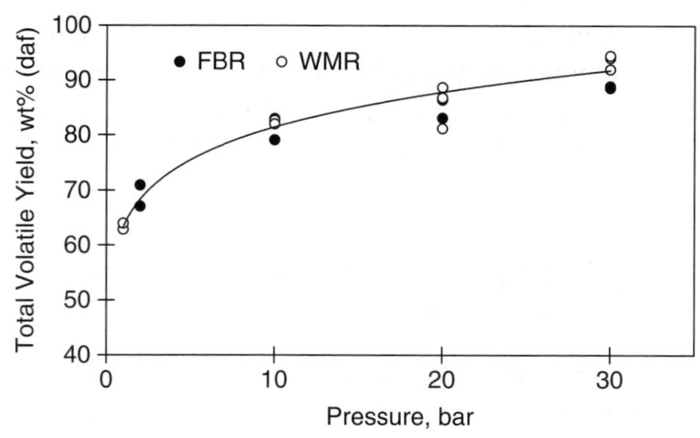

Figure 4.12 Total volatile yields from the CO_2-gasification of Daw Mill coal as a function of pressure. Fluidized-bed reactor: 1000°C, 60 s. Wire-mesh reactor: 1000°C, 1000°C s^{-1}, 60-seconds holding. (Reproduced with permission: Energy & Fuels 1998, *12*, 144; Copyright 1998 Am.Chem.Soc.)

These observations fit together reasonably well. We have seen that increasing the pressure of any gas will suppress tar evolution and enhance plasticity, which in turn promotes particle agglomeration. The deposited tar is expected to harden, forming relatively unreactive secondary char and block the pores of the solid matrix. However, very little agglomeration is possible in the wire-mesh reactor. The agreement between *pyrolysis* total volatiles from the two reactors suggests that pyrolytic volatile release was mostly completed *before* the plastic mass had set and hardened into a secondary char.

Meanwhile, extents of gasification in CO_2 would be more closely related (than in pyrolysis) on exposure of reactive surfaces to the ambient gas. Agglomerates in the fluidized-bed material indicate that the reactivity of CO_2 was not sufficient to break up the particle clusters, thus slowing down the gasification process long enough for the char begin to deactivate at 1,000°C. Increasing CO_2 pressures did not therefore significantly increase conversions. It is also clear from the wire-mesh reactor data (again, no agglomeration) that CO_2 is a less reactive agent for the gasification of residual chars than steam. The data also suggests that steam is able to scratch through the secondary char that held the agglomerates together. This would explain the agreement between the steam gasification data from the two reactors. Neither agglomeration nor deactivation took place to the same extent as in CO_2-gasification experiments.

However, an additional strand of information does not quite fit into this scheme. In parallel with these observations, Daw Mill chars from pyrolysis (in helium) and CO_2-gasification were compared by scanning electron microscopy (SEM). Both sets of chars were from experiments in the *wire-mesh* reactor. The chars formed during CO_2-gasification clearly showed greater fluidity. Ordinarily, the reverse would have been expected. Information provided by SEM is inevitably partial and a little subjective. If confirmed as systematic, however, this observation would suggest some chemical activity between the pyrolyzing plastic mass (mesophase) and CO_2, alongside the expected gasification reactions.

Returning to the design of the high-pressure fluidized-bed reactor, the next step in its evolution was the conversion of the system described in Figure 4.10 from batch mode to operation with continuous sample injection. The system was next reconfigured for studying changes in ammonia, NOx and HCN formation and release during air blown gasification [Paterson *et al.*, 2002; Zhuo *et al.*, 2002]. The new design and experimental results from it will be presented as a case study in Section 4.7, below.

4.4 Comparing Gasification Data from Reactors with Different Configurations

We have seen in Chapter 3 and Section 4.3, above, that comparison of results from different types of reactors may help examine the advantages and shortcomings of each particular configuration. In this section, results from the high-pressure wire-mesh reactor will be compared, first, with data from the fixed-bed ("hot-rod") reactor, described in Chapter 3. CO_2 and steam gasification data from three reactors will then be compared, using the same sample of Daw Mill coal. These findings will be evaluated against a background of attempting to narrow differences between bench-scale measurements and conditions in pilot or plant-sized equipment.

4.4.1 Comparison of CO_2-gasification in the wire-mesh and fixed-bed ("hot-rod") reactors

In wire-mesh reactors, coal particles go through a closely controlled time-temperature history. The sweep gas passing through the shallow reaction zone is intended to achieve near total absence of secondary reactions between released volatiles and the sparsely dispersed sample particles. The fixed-bed ("hot-rod") reactor also provides for a controlled time-temperature history and sweep gas passage through the fixed-bed of particles. However, heating rates in this reactor are limited by the thermal inertia of the system and radial heat transfer constraints to less than 10°C s^{-1}. Moreover, the minimum practicable sample size in this reactor is about 50 mg. Below this sample weight, experimental scatter due to errors in sample recovery grow to a level that obscures possible trends. A sample weight of 50 mg corresponds to a bed-depth of about 4 mm in this reactor when packed with coal [Gonenc *et al.*, 1990]; secondary reactions become inevitable. We have described this reactor in Chapter 3 and compared atmospheric pressure pyrolysis data from it with analogous results from the wire-mesh reactor.

Figure 4.13 compares total volatile yields from experiments at 1,000°C between 1–30 bar in the two reactors. In general, volatiles loss with increasing pressure is probably mostly through loss (re-polymerisation) of the heavier species in tars. In this sense, the steeper fall observed in the wire-mesh reactor data is likely to reflect the higher initial starting point, due to better general tar survival in that reactor. Furthermore, differences in conversion (total weight loss) between the two reactors increased with pressure. In contrast

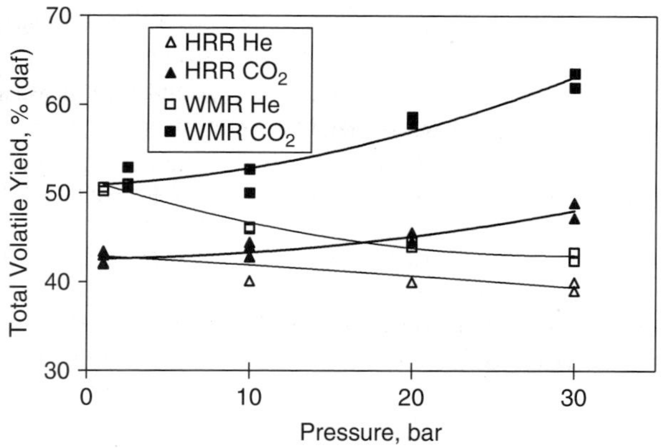

Figure 4.13 Comparison of total volatile yields from the high-pressure pyrolysis and CO_2-gasification of Daw Mill coal in the fixed-bed "hot-rod" (HRR) and wire-mesh reactors (WMR). Peak temperature: 1,000°C with 10 s hold in helium or CO_2. Heating rate 10°C s^{-1} in the fixed-bed ("hot-rod") and 1,000°C s^{-1} in the wire-mesh reactor. (Reproduced with permission: Fuel 1997, 76, 1327; Copyright 1997 Elsevier.)

to intimate gas-solid contact in the wire-mesh reactor, the stacking of particles in the fixed-bed ("hot-rod") partially blocks gasification reactions. This is probably made worse by the onset of plasticity and increasing tar deposition on heated chars [Lim *et al.*, 1997]. Experiments at the lower temperature of 850°C gave similar trends (not shown). However, in the fixed-bed ("hot-rod") reactor, relatively little evidence of gasification was found at this (lower) temperature. Sample weight loss in CO_2 was only marginally higher than that in He, even at 30 bar. By contrast, significant gasification was apparent in the wire-mesh reactor under the same conditions, with conversion in CO_2 exceeding that under helium by ~5 wt % across the pressure range.

The effect of heating rate on gasification conversions was also explored at 850°C, using two different heating rates in the wire-mesh reactor at 2.5 and 20 bar [Figure 4.3 in Lim *et al.*, 1997; not shown]. At 2.5 bar, the mass loss through gasification at 10°C s^{-1} was lower than that at 1,000°C s^{-1} but still higher than the pyrolysis yield at 1,000°C s^{-1}, reflecting the opposing effects of volatile loss through slow pyrolysis and weight loss through char gasification during the slow rise to peak temperature. At 20 bar, the greater char gasification rate was observed to reduce the difference between runs at 10 and 1,000°C s^{-1} in the wire-mesh reactor to within experimental error.

Figure 4.14 compares relative combustion reactivities of chars recovered from 1,000°C experiments in the two reactors. Both pyrolysis and CO_2-gasification chars from the fixed-bed ("hot-rod") reactor were less reactive than corresponding chars from the wire-mesh reactor. As outlined in Section 4.1.2, the slow ramp toward higher

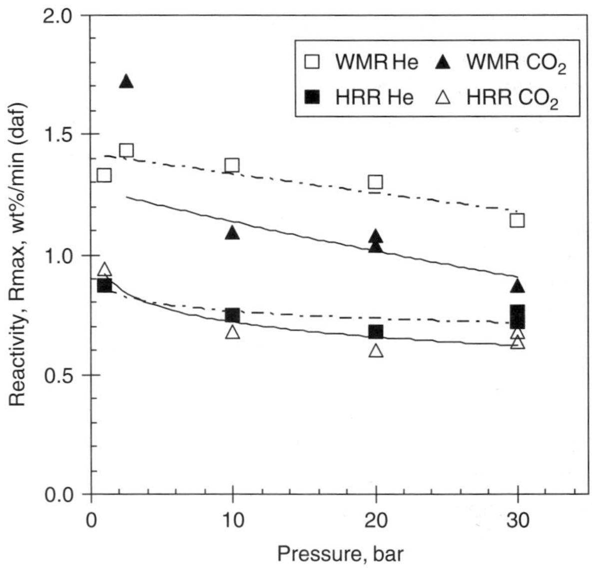

Figure 4.14 Comparison of relative combustion reactivities of chars from the high-pressure pyrolysis and CO_2-gasification of Daw Mill coal in the fixed-bed ("hot-rod") (HRR) and wire-mesh (WMR) reactors. Peak temperature: 1,000°C with 10 s hold in helium or CO_2. Heating rate 10°C s^{-1} in the fixed-bed ("hot-rod") and 1,000°C s^{-1} in the wire-mesh reactor. (Reproduced with permission: Fuel 1997, 76, 1327; Copyright 1997 Elsevier.)

temperatures in the fixed-bed ("hot-rod") reactor (10°C s^{-1}) would contribute significantly to char deactivation compared to the rapidly heated wire-mesh reactor chars (1,000°C s^{-1}). Moreover, we have seen how the fixed-bed ("hot-rod") reactor configuration allows more tar deposition on heated particles in the stacked bed. The amorphous char formed from deposited tar would be less reactive than the devolatilizing solid matrix [Guell *et al.*, 1993].

Comparing these results from fast-heating experiments in the wire-mesh and slow-heating in the fixed-bed ("hot-rod") reactor has a particular significance. The heating rate in the "hot-rod" reactor exceeds heating rates practicable in most thermogravimetric balances. Few thermogravimetric balances are equipped to operate at 600°C minute^{-1}. Moreover, volatiles are moved away more thoroughly and efficiently in the fixed-bed ("hot-rod") reactor than in thermogravimetric balances, since the latter can only accommodate a gas flow *around the furnace* within which the sample-bearing pan is held during experiments. In both respects therefore, the fixed-bed ("hot-rod") reactor is *better* equipped than most thermogravimetric balances in reflecting the conditions experienced by solid fuel particles in fluidized or entrained gasification. The data of Figure 4.14 shows that, in mimicking fuel behavior within these fast heating reactors, the fixed-bed ("hot-rod") reactor is clearly inadequate. It is important to view the utility or otherwise of high-pressure thermogravimetric balances in this perspective.

4.4.2 Internal consistency of gasification reactivities measured in bench-scale reactors

This section continues to explore the relevance of bench-scale experiments to the design and operation of pilot and industrial scale gasification. Within this framework, it is relevant to examine the internal consistency of gasification data generated in equipment and experiments with differing designs. The parameters to be compared include the pyrolysis yields, the char preparation conditions and how the latter affect the gasification reactivities of the chars.

Broad agreement was found between total volatile yields from pyrolysis and gasification experiments in the high-pressure wire-mesh and fluidized-bed reactors. Lower conversions were observed in the fixed-bed ("hot-rod") reactor, compared to the other two reactors. These differences are to be understood in terms of the basic design features of the equipment. The configurations of the wire-mesh instrument and the fluidized-bed reactor aim to minimize interference from neighboring particles with respect to volatile release and maximize contact between the sample particles and the reactive gas.

Oxygen blown gasifiers normally achieve reaction temperatures above 1,500°C (and pressures up to 30 bar), with consequent rapid and normally complete gasification of the coal. At these high temperatures, gasifier design becomes less sensitive to the reactivities of the feed coal and its chars, although we will see instances where establishing a hierarchy of coal (and char) reactivity may still be useful. Clearly, experiments at high-pressure and temperatures above 1,400°C are a great deal more difficult to undertake. Wire-mesh reactor experiments simulating conditions in oxygen blown gasifiers will be described later in this Chapter. In the present section, we will focus on experiments intended to simulate coal particle behavior in *air* blown gasification systems [Dawes *et al.*, 1991; Mojtahedi *et al.*, 1991;

Motter & Higginbotham, 1993; EPRI Report TR-103367]. Experimental conditions will therefore remain within the 1,000°C and 25–30 bar envelope.

The design of bench-scale gasification experiments: We have already discussed the significance of reaction conditions during pyrolysis on the structures and reactivities of product chars. One major source of uncertainty in data from bench-scale equipment arises from the way the particular gasification experiments have been designed. There are particular problems relating to what might be termed "two-stage" experiments, where gasification reactivities are measured using char samples prepared in a separate pyrolysis step. Often, the pyrolysis step is carried out under conditions not necessarily representative of the gasification process being simulated [cf. *e.g.* Goring, *et al.*, 1952; Sha *et al.*, 1990; Guo & Zhang, 1986; Haga & Nishiyama, 1988; Nozaki *et al.*, 1991; Ginter *et al.*, 1993; Meijer *et al.*, 1994; Yang & Watkinson, 1994]. These practices contrast with fuel particle histories in most modern pilot and industrial scale gasifiers, where the devolatilization and gasification of coal particles usually overlap and (quite clearly) take place under the same reaction conditions.

There is surprisingly little work reported in the literature on the relationship between char formation conditions and the *gasification* reactivities of chars [Laurendeau, 1978; Chitsora *et al.*, 1987; Guell & Kandiyoti, 1993; Guell *et al.*, 1993; Cai *et al.*, 1996]. Peng *et al.* [1995] have compared reactivities of directly gasified coals (a lignite, a sub-bituminous and a bituminous coal) with the reactivities of chars from the same coals, pyrolyzed previously under otherwise similar reaction conditions. The work was carried out in a thermogravimetric balance operated under atmospheric pressure steam-nitrogen mixtures (76 mol % steam), at temperatures between 1,000–1,400°C. Reactivities of chars generated during direct gasification experiments were reported to be up to six-times greater than those of corresponding chars prepared in the same TGA apparatus under nitrogen.

These experiments were reported to have been carried at heating rates that are surprisingly high for a thermogravimetric balance, estimated to be ". . . between 100 and 1,000°C s^{-1} . . .". They were reportedly achieved by lowering the sample-pan, laden with about 100 mg of 149–210 μm particles, into a heated TGA furnace. Such an experimental set-up, however, would lead to a number of potential uncertainties. It is possible that particles near the periphery of the pile were indeed heated rapidly, although the range of heating rates cited ("between 100 and 1,000°C s^{-1}") is indeed large. Furthermore, it is difficult to visualize how the uniformity of the temperature distribution and the constancy of the heating rate could be maintained *within* a pile of 100 mg of particles placed on the TGA-pan. The thermal conductivity of the pyrolyzing mass would be expected to govern the rate of heat transmission *within* the pile, instead of the externally imposed temperature gradient. Calculations for the fixed-bed ("hot-rod") reactor indicate that across a *radius* of 3 mm, heating rates greater than 10°C s^{-1} imposed at the external boundary would lead to severe temperature gradients [O'Brien, 1986]. Furthermore, experiments using samples of 50 mg (i.e. half that of Peng *et al.*, 1995) have shown that tar and volatile yields as well as char reactivities are affected by the stacking of coal particles. It is not clear how secondary char formation through tar deposition on pyrolyzing solid surfaces could be ignored in a 100 mg pile of sample particles. Clearly, secondary-char deposition would affect char reactivities.

Experiments have been outlined below, aiming to clarify some of the uncertainties in the use of single and two-step gasification tests. The effect of char formation conditions has been examined at pressures up to 30 bar, using the bench-scale reactors described in these two Chapters. "Extents of gasification" have been defined, by subtracting mass loss during pyrolysis in helium from mass loss during direct gasification, under otherwise similar reaction conditions. In turn, "Gasification reactivities" have been defined as the "Extents of gasification" divided by the hold time at peak temperature. "Gasification reactivities" calculated from CO_2-gasification data have been compared with conversions from two well defined, but distinct "two-step" procedures.

Experiments with untreated coal were carried out in the three reactors under helium and CO_2, at 1,000°C, with 60 s holding, at pressures between 1 and 30 bar. Heating rates in the fixed-bed ("hot-rod") and wire-mesh reactors were 10 and 1,000°C s^{-1}, respectively. No heating rates will be hazarded for the fluidized-bed experiments. It is assumed these were indeed high for most sample particles. The first set of *char* gasification experiments was carried out by pyrolyzing samples in helium in each of the reactors for 10 s at 1,000°C (1, 10, 20 and 30 bar). The carrier gas flows in the reactors were then switched to CO_2 for 60 s. The pyrolysis and gasification steps of these experiments were thus carried out at similar temperatures and pressures. For the second set of *char*-gasification experiments, a common sample of char was prepared in the fixed-bed ("hot-rod") reactor, by heating under atmospheric pressure helium at 10°C s^{-1} to 1,000°C with 60 s holding. The chars were sieved and dried (to drive away adsorbed moisture) prior to the gasification runs. Samples of char prepared in this way were gasified in each of the three reactors, under identical conditions to those used in the coal gasification runs (1,000°C for 60 s at 1, 10, 20 and 30 bar). *Combustion* reactivities of the residual chars recovered after these experiments have been determined in an atmospheric pressure thermogravimetric balance. In what follows, it is useful to remember that combustion and gasification reactivity *trends* for coal chars measured by TGA are nearly always similar.

Comparing gasification reactivities: Figure 4.15 presents ("one-step") CO_2-gasification reactivities of Daw Mill coal in each of the three reactors. Experiments were carried out between 1–30 bar with 60 s holding at 1,000°C.

Substantially lower conversions and reactivities were observed in the fixed-bed ("hot-rod") reactor, compared to the wire-mesh and fluidized bed reactors. The gasification reactivity in the fixed-bed ("hot-rod") reactor was also found to be less sensitive to pressure. The fluidized-bed and wire-mesh reactor data were found to be similar to each other, except at 30 bar where the reactivity in the wire-mesh reactor was higher, by about 5% min^{-1}. This difference is likely to be due to sample agglomeration, observed in the fluidized-bed at the highest pressures (see Figure 4.12 above).

In the first of the "two-step" sequences, the coal sample was first pyrolyzed *in-situ*, in each of the three reactors, under helium (10 s at 1,000°C), before switching over to CO_2 and holding for 60 s. This two-step procedure was found to reduce reactivities in the fluidized bed reactor by up to 4% relative to direct CO_2 gasification, but *qualitatively* the results did not differ much from Figure 4.15. However, significant differences were observed in both the wire-mesh and the fluidized-bed reactors between overall conversions from direct CO_2 gasification and those from the gasification of a sample of char pre-prepared in the fixed-bed ("hot-rod") reactor.

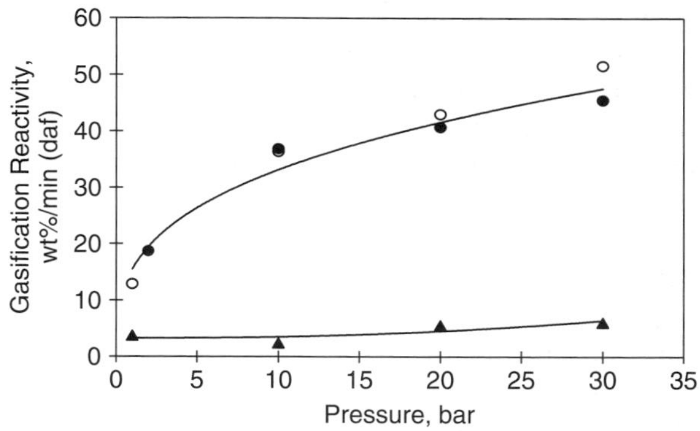

Figure 4.15 CO_2-gasification reactivity of Daw Mill coal as a function of pressure. (▲) Fixed-bed ("hot-rod") reactor: 10°C s^{-1} to 1,000°C with 60 s holding in CO_2. (○) Wire-mesh reactor: 1,000°C s^{-1} to 1,000°C with 60 s holding in CO_2. (●) Fluidized bed reactor: fast heating to 1,000°C with 60 s holding in CO_2. (Reproduced with permission: Fuel 1998, 77, 1411; Copyright 1998 Elsevier.)

Figure 4.16 compares results from the direct gasification of Daw Mill coal in the fluidized bed reactor and results from the two "two-step" procedures. The sharp drop in reactivity observed when using the char prepared in the fixed-bed reactor clearly illustrates the dangers inherent in working with chars from slow heating first-stage procedures.

In previous *hydro*pyrolysis work, the lower conversions observed in the fixed-bed ("hot-rod") reactor had been explained in terms of secondary char deposition (leading to low overall char reactivity) as well as poor char-gas contact [Gibbins *et al.*, 1991]. On

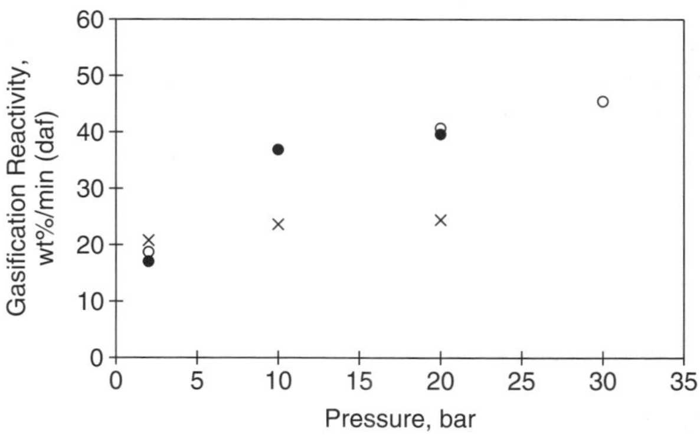

Figure 4.16 Gasification reactivities determined in the fluidized-bed reactor as a function of char preparation conditions. (○) Direct CO_2-gasification of untreated Daw Mill coal. (●) Daw Mill coal, successively pyrolyzed and gasified in the fluidized bed reactor. (×) CO_2-gasification reactivity in the fluidized-bed of char pyrolyzed in the fixed-bed ("hot-rod") reactor. Untreated coal: fast heating to 1,000°C, 60 s, CO_2. Pyrolysis-gasification in the same reactor: fast heating to 1,000°C, 10 seconds in helium followed by 60 seconds in CO_2. Pre-prepared fixed-bed ("hot-rod") pyrolysis char gasified by fast heating to 1,000°C, 60 s, CO_2. (Reproduced with permission: Fuel 1998, 77, 1411; Copyright 1998 Elsevier.)

their own, these data do not allow the two effects to be distinguished from each other. Data presented in Figure 4.16 indicate, however, that low conversions persisted when a coal *char* from the fixed-bed ("hot-rod") reactor was re-*ground* and gasified in the fluidized bed reactor. This suggests that char deactivation in the fixed-bed ("hot-rod") reactor plays the predominant role. However, experiments described in the present Chapter have indicated char deactivation (in the fixed-bed) to be primarily due to (i) secondary tar deposition superposed on (ii) char deactivation due to lower heating rates used in the fixed bed reactor (See Figure 4.1 above). Heating at 10°C s^{-1} allows long exposure times at temperatures near 1,000°C and the opportunity for the char to lose reactivity.

The magnitude of the error introduced by adopting this 'two-reactor, two-step' approach was about 20% per minute, for Daw Mill coal char tested at 20 bar. Sharper char deactivation would be expected with coals that soften on slow heating. It would also depend on the amount and properties of tars evolving from particular coals. Even when "only" aiming to establish *relative* reactivities within a suite of coals, therefore, this "two-reactor, two-step" procedure is likely to lead to erroneous results.

Gasification in Steam: Figure 4.17 presents conversions during the (one-step) steam gasification of Daw Mill coal in the same three bench-scale reactors. The experiments were conducted with 60 s holding at peak temperature (at 1,000°C). As in the case of gasification in CO_2, conversions increased monotonically with pressure. However, in the presence of 20 bar steam, nearly complete conversion of the (106–152 μm) sample was observed in the wire-mesh and the fluidized-bed reactors. These two reactors have been deliberately designed to segregate individual coal particles from one another and operate at high heating rates. In this case, agreement between them was within experimental error. In line with results from hydropyrolysis and CO_2-gasification experiments outlined above, lower conversions were obtained in the fixed-bed ("hot-rod") reactor.

Figure 4.17 clearly suggests, furthermore, an upper limit to the conversion of the char in the fixed-bed ("hot-rod") reactor, under these reaction conditions. Despite operating with a more reactive gas (steam), the maximum attainable conversion does not appear likely to reach completion, even at very long holding times. A similar effect was also observed in the operation of the British Coal ABGC pilot plant. In the preceding discussion, we have

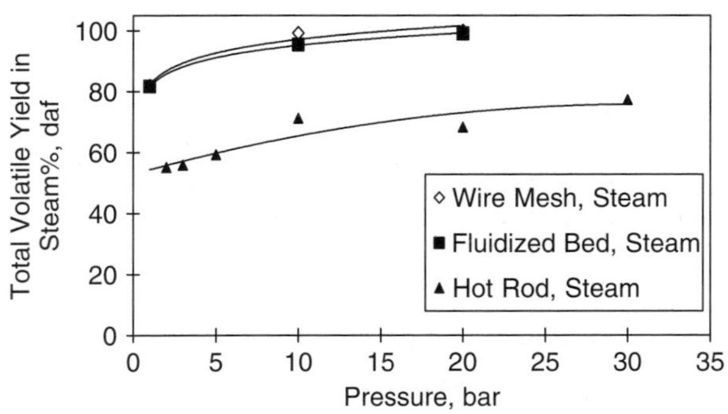

Figure 4.17 Weight loss during the steam-gasification of Daw Mill (UK) coal in the three reactors. 60 seconds holding at 1,000°C. (Reproduced with permission: Fuel 2000, 79, 793; Copyright 2000 Elsevier).

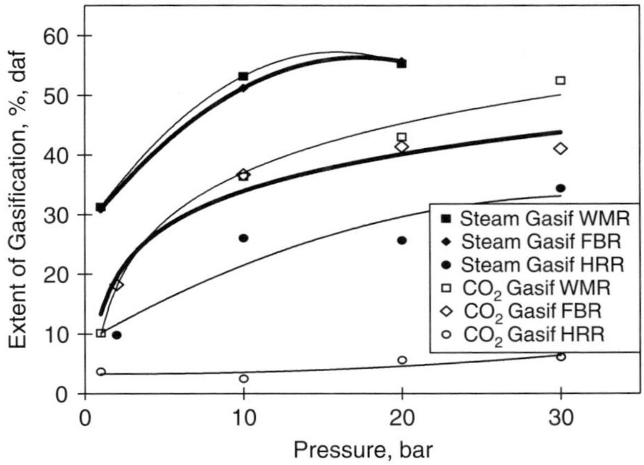

Figure 4.18 Comparison of "Extents of Gasification" in steam and CO_2. 60 seconds holding at 1,000°C. (Reproduced with permission: Fuel 2000, 79, 793; Copyright 2000 Elsevier).

partly attributed the low reactivity of the chars formed in the fixed-bed ("hot-rod") reactor to secondary tar deposition, as well as to poor gas-solid contact due to stacking of softened coal particles within the fixed-bed. Clearly, these factors operate alongside the char deactivation effect described in Figure 4.1 due to exposure at high-temperatures.

"Extents of gasification" in coal-steam reactions: The conversions (total volatiles) in Figure 4.17 correspond to the sum of weight loss through pyrolytic devolatilization and to weight loss due to reactions between the carbonaceous matter and steam. Figure 4.18 presents "extents of gasification" calculated by subtracting weight loss during pyrolysis [Figure 4.2 in Zhuo *et al.*, 2000] from conversions during steam-gasification (from Figure 4.17). For completeness, corresponding "extents of CO_2-gasification" have been included. Experiments carried out in the fluidized-bed reactor with 1/1 mixtures of steam and CO_2 gave results that were nearly indistinguishable from conversions in pure steam. In all reactors, the extent of *steam*-gasification was markedly greater than that in CO_2. Furthermore, char agglomeration during CO_2-gasification above 20 bar has been difficult to control and this was reflected in the lower conversion observed in the fluidized-bed at 30 bar – compared with the result obtained in the wire-mesh reactor. Measures adopted to control agglomeration in the fluidized-bed reactor have been described in Megaritis *et al.* [1998a]. Effects due to agglomeration were not found during operation with steam and were less pronounced during the pyrolysis experiments.

4.4.3 Comparison with pilot plant data

Figure 4.19 compares conversions of five different coals in a pilot-scale gasifier with direct CO_2-gasification conversions observed in the high-pressure wire-mesh reactor. The pilot reactor was the 200 kg hr^{-1} British Coal – Air Blown Gasification Cycle (ABGC) gasifier at Stoke Orchard (UK).

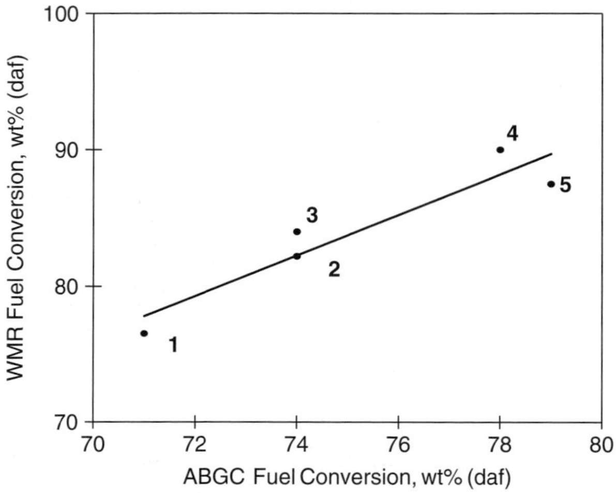

Figure 4.19 Comparison of fuel conversion in the high-pressure wire-mesh reactor and the ABGC pilot scale gasifier. Wire-mesh reactor: 1,000°C s^{-1} to 1,000°C with 60 s holding at 10 bar CO_2 pressure. ABGC gasifier: 970°C (approx.), 13 bar, air/steam. 1: Rietspruit (S. Africa), 2: Daw Mill (UK), 3: Drayton (Aus), 4: Illinois No. 6 (U.S.A.), 5: El Cerrejon (Colombia). (Reproduced with permission: Fuel 1998, 77, 1411; Copyright 1998 Elsevier).

The coals and their proximate analysis volatile matter contents (% w/w, dry basis) were as follows: Rietspruit (S. Africa; 28.7); Daw Mill (UK; 34.1); Drayton (Australia; 35.3); 4: Illinois No. 6 (U.S.A.; 41.2); El Cerrejon (Colombia; 37.8). They were selected as being representative of the range of the coals likely to be used by an eventual commercial gasifier. Detailed information on the compositions of the coals may be found in Table 4.6 [Megaritis *et al.*, 1998b]. The ABGC used a mixture of air and steam, with the feed coal crushed to "less than 3 mm diameter", compared to using pure CO_2 and the 100–150 μm particle size range in the wire-mesh reactor. It is not possible, therefore to seek an exact correspondence between conversions in the two reactors. Broadly, an estimated 30% of the original feed is converted to volatiles through pyrolysis and 10% by steam and CO_2-gasification. Another (estimated) 30% of the coal mass in the auto-thermal ABGC reactor is consumed by combustion, the energy being taken up by endothermic gasification reactions with steam and CO_2. Approximately 30% of the fuel charged is recovered as residual char. In the full-blown process, the chars would be combusted in a separate unit to raise additional steam.

Above 10 bar, conversions and "extents of gasification" determined in the wire-mesh and the fluidized-bed reactors were higher than conversions and estimated steam gasification levels in the pilot-plant gasifier. Results from experiments in CO_2 reflected the same trend although values were lower. It appears, the lower overall conversions observed in the pilot scale gasifier do not necessarily correspond to levels of reactivity that may be expected from the properties of the original feedstock. Instead, they represent consequences of long residence times in the reaction zone, in turn related to the particle size distribution. For equivalent grinding conditions, therefore, the order of reactivity within a particular set of coals may turn out to depend on the grinding properties of the coals.

Table 4.6 Proximate and ultimate analyses of the set of coals.

	Daw Mill[a]	*El Cerrejon*	*Drayton*	*Rietspruit*	*Illinois No.6*	*Daw Mill*[b]
Volatile Matter; (%, daf)	39.7	40.7	39.5	32.8	45.5	39.9
Ash (%, db)	14.1	7.2	10.7	12.5	9.5	4.4
Moisture (%, ad)	4.5	5.4	3.1	3.9	7.1	6.1
Swelling Number	1	1	2.5	1	6	1
Carbon (%, daf)	80.6	82.4	82.6	82.5	78.2	80.1
Hydrogen (%, daf)	5.4	5.8	5.7	4.9	5.6	4.7
Nitrogen (%, daf)	1.5	1.7	2.0	2.1	1.4	1.3
Sulfur (total) (%, daf)	2.0	0.8	0.8	0.7	4.4	1.1
Vitrinite (%Vol, mmf)	67	84	75	63	92	66
Inertinite (%Vol, mmf)	21	14	21	33	6	21
Liptinite (%Vol, mmf)	12	2	4	4	2	13
Mean Vitrinite Reflectance	0.60	0.72	0.65	0.73	0.40	0.60

[a] high ash Daw Mill sample; [b] low ash Daw Mill sample. (Reproduced with permission: Fuel 2000,79,109; Copyright 2000 Elsevier)

Despite these differences, a broadly linear relationship between the pilot-plant gasifier and the bench-scale reactor was observed in Figure 4.19, suggesting that data from the wire-mesh reactor may be successfully used to compare *relative* coal reactivities under pilot-plant gasification conditions. The conversions in the ABGC reactor were lower by about 10%. They provide an estimate for the char residue rendered inactive by a combination of larger particle sizes and longer residence times at high temperature. Thus, in addition to its ability to predict a correct order of reactivity for this set of coal samples, results presented in this section show the wire-mesh and the fluidized-bed reactors to have been useful in pinpointing reasons for the low reactivity of chars recovered from the spouted bed.

4.5 Case Studies: Factors Governing Coal Reactivity in Pyrolysis & Gasification

4.5.1 *Correlating results from coals and macerals*

The aim of the work was to evaluate the use of maceral analysis as a tool for predicting sample weight loss (conversions) during pyrolysis and gasification. Pyrolysis and

Table 4.7 Properties of the three maceral concentrates used in the calculation of predicted total volatile yields in Figure 4.20.(Reproduced with permission: Fuel 2000, 79, 109; Copyright 2000 Elsevier)

Maceral Group	Vitrinite	Liptinite	Inertinite
Maceral Type	Hand picked vitrain	Exinite	Semi-fusinite
Parent Coal (Seam)	Markham Main (Barnsley)	Peckfield (Beeston)	Roddymoor (Ballarat)
C (%, daf)	81	82	93
H (%, daf)	5.5	6.8	4.4
Vitrinite (Vol%, mmf)	98	3	10
Liptinite (Vol%, mmf)	1	92	0
Inertinite (Vol%, mmf)	0.3	5	90

gasification data on a suite of six coals and three maceral concentrates were collected using the high-pressure wire-mesh reactor (1–30 bar). The samples are described in Table 4.6. They correspond to the same suite of coals tested in the Air Blown Gasification Cycle (ABGC) pilot plant trials carried out at British Coal (Stoke Orchard) between 1993–97.

The choice of maceral concentrate samples was based on availability; the samples used in the study (Table 4.7) were derived from different coals. It was assumed from the outset that volatile yields from each coal in the series could be *calculated* from the weighted sum of (i) the maceral composition of the particular coal and (ii) the total volatile yields from the same three available maceral concentrates [Messenbock *et al.*, 2000]. The work was viewed as exploratory and likely errors inherent in the procedure were understood.

Predicting pyrolysis weight loss from coal maceral behavior? The pyrolysis and gasification experiments were run in helium and CO_2, respectively, at pressures of 1, 10, 20 and 30 bar. Samples were heated at 1,000°C s^{-1} to 1,000°C, with 10 seconds holding at peak temperature. The full set of results from these experiments has been presented in the original publication [Messenbock, *et al.*, 2000]. Weight loss measurements from the maceral concentrate samples were combined with the maceral analyses of the six coals, to *calculate* sample weight loss for the six "whole" coals. These calculated ("reconstituted") weight loss results were compared with the actual weight loss data from pyrolysis and CO_2-gasification experiments on the six coals.

Figure 4.20 shows that predicted values were only slightly lower than experimental values for volatile yields below ~45%. Above that value, the difference between the two sets increased, reflecting the systematic error due to the relatively high rank of the set of "constituent" maceral concentrates used for the purposes of this calculation. For the approximate calculation that this was meant to be, the predictive value of the correlation was considered as surprisingly good.

Added to those of Table 3.10, these results indicate that it is possible to predict *pyrolysis* total volatile and tar yields of individual coals reasonably well from (i) their maceral compositions and (ii) pyrolysis weight loss data from their (preferably own) maceral constituents. While it always makes sense to work with many more samples, the

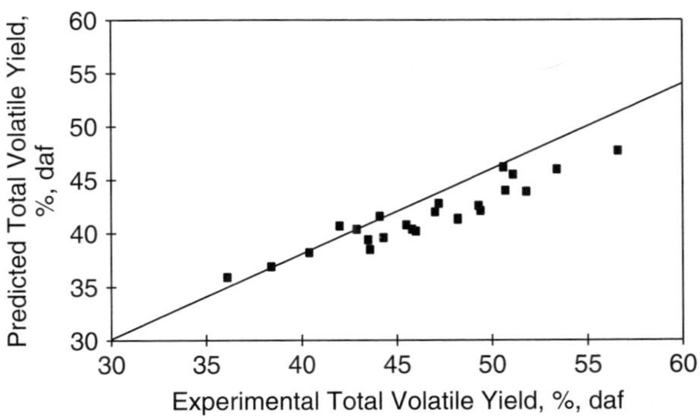

Figure 4.20 Correlation between experimental and predicted results for pyrolysis experiments in 1, 10, 20 and 30 bar helium. Samples were heated at 1,000°C s^{-1} to 1,000°C, with 10 seconds holding at peak temperature. (Reproduced with permission: Fuel 2000, 79, 109; Copyright 2000 Elsevier).

indications are that for middle rank bituminous coals, any synergistic effects between component macerals during pyrolysis are not experimentally detectable. This result is not surprising when taken together with findings outlined in Section 3.7.2. The commercially important question, however, is whether such a prediction method might be applied to coal conversions during *gasification*.

Predicting gasification weight loss from coal maceral behavior: Experimental and calculated CO_2 – gasification conversions of these samples have been given in the original publication [Messenbock *et al.*, 2000]. The attempted correlation (analogous to Figure 4.20) between predicted and experimental gasification conversions showed sufficient scatter to be virtually useless. However, in this comparison, the conversion had been defined as total weight loss. In other words, the "conversion" subsumed sample weight loss by pyrolysis *and* sample weight loss by gasification. When the predicted and experimental "extents of gasification" (see above) were plotted together, the correlation was found to be very weak indeed. In the interest of brevity, these diagrams have not been presented here.

Conclusion: Despite using a single set of maceral concentrates to represent the full spectrum of samples in the set, it appears possible to estimate the pyrolysis weight loss of coals from their maceral composition. This work covered a pressure range between 1 and 30 bar at 1,000°C and the results are in line with those from the atmospheric pressure maceral pyrolysis work described in Chapter 3. However, the method has been found *not* to enable prediction of the gasification performance of the same coals. Provisionally, it appears that gasification conversions were less dependent on original coal properties, than was found to be the case for pyrolysis weight loss.

Overall, gasification conversions appear less predictable. There was much scatter for middle-rank coals, probably due to a combination of factors including differences in char pore structure, catalytic effects of mineral matter components and differences in the inherent reactivity of the carbonaceous matter.

4.5.2 Correlating conversions with FT-ir spectra of coals

In a related study, possible correlations were examined between the FT-ir spectra of an array of coals and their pyrolysis and gasification conversions, as determined in the wire-mesh reactor. The study was undertaken using a proprietary software package "Quant+" for the correlations [Zhuo *et al.*, 2000]. The work originated in an attempt to explore reasons why coals of nearly similar rank gave different conversions during the operation of the British Coal pilot-scale air-blown spouted-bed gasifier [Gavin *et al.*, 1997]. In this type of process, conversions are usually incomplete and improving conversions of the feed coal in the gasification stage would have increased overall power generation cycle efficiencies.

Qualitative agreement has already been reported [Megaritis, 1998b] between CO_2-gasification conversions in the wire-mesh reactor and in the pilot-scale ABGC gasifier operated by British Coal. However, the distribution of particle sizes used in the pilot-scale gasifier ("less than 3 mm") was different from that in the wire-mesh instrument (106–152 μm). Coupled with the rapid deactivation of chars at temperatures around 1,000°C (in about 10 s), the longer times required for consuming these larger char particles appears to lead to the formation of significant quantities (~25% of the feed) of residual, relatively unreactive char [Zhuo *et al.*, 2000a]. Indeed, there are good reasons for linking the formation of relatively large amounts of unreactive char in pilot-scale air-blown gasifiers, to the use of relatively large feed coal particles.

As already outlined in Section 4.1, the kinetics of char *de-activation* as a function of time-at-temperature plays an important role in determining gasification rates. This factor is one of the often neglected complications involved in attempting to develop adequate model equations for gasification kinetics. Care was taken to reduce the effect of particle size differences for the purposes of this work, to isolate sample reactivity as a distinct parameter. This was done by consistently using the same standard size in the wire-mesh reactor (106–152 μm), as well as in the accompanying fixed-bed ("hot-rod") reactor tests.

The FT-ir spectrum of a coal is arguably a more complete structural description of individual coal samples than could be provided by a maceral analysis. The core of the study was an attempt to correlate the conversions of coals with their infrared spectra, with the aim of identifying structural features that related to the reactivity of coals. The commercial statistical data analysis package "QUANT+" [Perkin-Elmer (U.K.) Ltd., 1991] was used for the correlation work. Conversion (sample weight loss) data from pyrolysis and CO_2-gasification experiments were obtained using the high-pressure wire-mesh reactor. Similar correlations were attempted with analogous data from the fixed-bed ("hot-rod") reactor.

Table 4.8 presents the elemental analyses of the 26 coals used in the study. The samples were of diverse geological origins. Conversions of 23 of the coals were determined in the high-pressure wire-mesh reactor (see below), in helium and in CO_2. Analogous pyrolysis and CO_2-gasification data on a subset of 16 coals were acquired using the fixed-bed ("hot-rod") reactor: Taff Methyr, Tilmanstone, Emil Mayrisch, Santa Barbara, Heinrisch Robert, Candin, Point of Ayr, Thorsby, Hemheath, Bentinck, Longannet, Rietspruit, Gedling, Linby, Daw Mill, and Gardanne.

Table 4.8 Elemental Analyses of the set of 'calibration' coals (% w/w, dry ash free basis). (Reproduced with permission: Energy & Fuels 2000, *14*, 1049; Copyright 2000 Am.Chem.Soc.)

Coal	*C*	*H*	*N*	*S*	*O*
Taff Methyr (UK)	91.5	4.1	1.4	0.7	2.2
Tilmanstone (UK)	91.0	4.3	1.2	1.5	2.0
Emil Mayrisch (Germany)	89.2	4.4	1.4	0.8	4.1
Santa Barbara (Spain)	88.8	5.7	1.9	1.1	3.0
Heinrisch Robert (Germany)	87.7	4.9	1.2	0.9	5.2
Upper Freeport (USA)**	85.5	4.7	1.6	0.7	7.5
Candin (Spain)	84.6	4.8	1.7	1.2	7.7
Point of Ayr (UK)	84.5	5.4	1.8	1.5	6.1
Thoresby (UK)	84.0	5.3	1.8	1.0	7.9
WA1 (Australia)	84.0	4.0	4.7	0.3	10.0
Hemheath (UK)	83.9	5.4	1.8	0.8	8.1
Bentinck (UK)	83.5	5.6	1.7	2.3	6.9
Pittsburgh No. 8 (USA)**	83.2	5.3	1.6	0.9	8.8
Longannet (UK)	82.7	5.0	1.8	1.0	10.1
Lewiston-Stockton (USA)**	82.6	5.3	1.6	0.7	9.8
La Jagua (Colombia)	82.1	6.1	1.6	0.5	9.7
Rietspruit (S. Africa)	81.9	4.6	1.6	0.4	9.2
Gedling (UK)	81.3	4.7	1.5	1.0	11.1
Linby (UK)	81.0	5.3	1.7	1.0	11.0
WA2 (Australia)	80.8	5.1	1.9	0.3	11.9
Blind Canyon (USA)**	80.7	5.8	1.6	0.4	11.6
Daw Mill (UK)	80.1	4.7	1.3	1.2	11.5
Illinois No.6 (USA)**	79.6	5.0	1.4	4.5	9.5
Illinois No.6 (SBN)*	77.7	5.0	1.4	2.4	13.5
Fording Genesse (Australia)	74.3	4.4	0.7	0.5	20.1
Gardanne (France)	74.2	5.0	1.7	6.2	12.9

* Illinois No.6 (SBN) provided by Steinkohlebank Nederlands (SBN).
** Standard samples provided by the Argonne National Laboratories [Vorres, 1990].

The wire-mesh reactor experiments were carried out at 10 bar; samples were heated at 1,000°C s^{-1} to 1,000°C, with 20 s holding at peak temperature. In the fixed-bed ("hot-rod") reactor, experiments were conducted at 20 bar, heating 50 mg samples of coal at 10°C s^{-1} to 1,000°C, with 10 s holding at peak temperature. A superficial gas velocity of 0.1 m s^{-1} has been used in all the experiments. The FT-IR spectra of the samples were acquired as described by Li *et al.* [1994] and Madrali [1994].

The factor analysis and multiple linear regression methods used in the "QUANT+" correlation procedure have been described by Malinowski and Howery [1980] and

Weisberg [1985]. The first step in the procedure is to choose a set of "calibration" coals. Selected properties of each of these samples (e.g., carbon content, weight loss in pyrolysis, etc.) are measured. Their FT-IR spectra are acquired and stored. Relationships between the measured variables and the FT-IR spectra are then explored. The analysis procedure includes three sequential steps: calibration, validation, and prediction. An overview of the steps involved in the QUANT+ calculation is presented in Figure 4.21 [Perkin-Elmer, 1991].

Details of the construction of the "Model" involving calibration and validation have been presented in the original publication [Zhuo *et al.*, 2000b]. The aim of the calculation is to estimate the value of an individual property (e.g., elemental carbon content) of an unknown coal from its FT-IR spectrum, by using (i) already-measured properties of the "calibration" set of coals, and (ii) parts of the "model"'s segments of the FT-IR spectrum, statistically found to be significant in contributing to that particular property.

Cross Validation: A "cross-validation" procedure is then initiated to test the models generated. The procedure consists of excluding one coal sample at a time from within the calibration set and correlating the FT-IR spectra of the set of samples with their selected

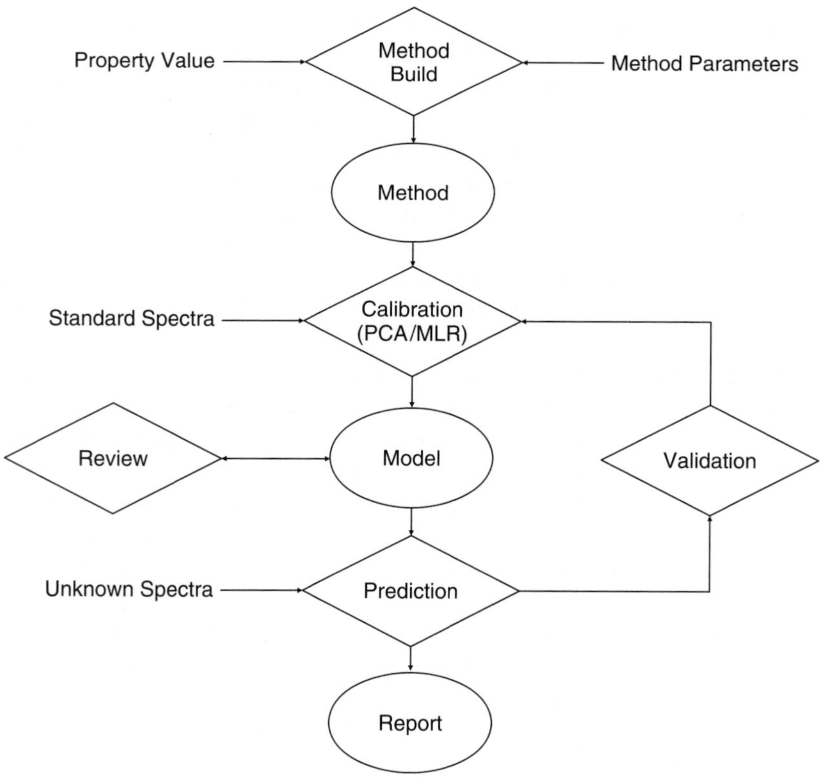

Figure 4.21 Overview of the QUANT+ correlation procedure Adapted from *Perkin-Elmer Ltd. QUANT+ User's Manual* (1991).

properties (minus the excluded sample). The next step is to predict the properties of the sample excluded from the calibration set during the calculation, by using the derived models (using (N – 1) calibration samples). The cross-validation procedure provides a measurement of average prediction error (standard error of prediction). These are summed and the "model" is then optimized to minimize the cumulative error of prediction. Only the factors (typically, segments of the FT-ir spectrum) with statistical importance are retained in the final regression model.

The final model is thus based on criteria minimizing the error of prediction. In what follows, we have confined ourselves to running the cross validation procedures against the existing pyrolysis and gasification data, i.e., taking one sample at a time as the "unknown", estimating the pyrolysis and gasification conversions of that sample, then moving on to the next sample.

Table 4.9 presents the total volatile yields during pyrolysis in He in the wire-mesh reactor, the total volatile yields from gasification in CO_2 and the "extents of gasification", calculated from (Total $\text{Volatiles}_{\text{gasification}}$ – Total $\text{Volatiles}_{\text{pyrolysis}}$) on a % weight/weight, dry ash free basis. Table 4.10 presents analogous results from experiments on a subset of 16 coals undertaken in the fixed-bed ("hot-rod") reactor.

Figure 4.22a showed excellent agreement between pyrolysis total volatile yields measured in the high-pressure wire-mesh reactor and weight loss values *predicted* by the correlation procedure. Despite the wide diversity in the geological origins of the samples, the calculation was shown capable of estimating the pyrolysis volatile yields of "unknown" coals, once their FT-ir spectra had been recorded and classified within the "library" of spectra. The level of agreement suggests that the initial structures of coals (as reflected in their infrared spectra) are closely related to their pyrolytic behavior. The method used in this work thus seems appropriate for estimating, for example, volatile matter yields of "unknown" power-station coals under pf-combustion conditions, provided a complete set of pyrolysis data are generated in a wire-mesh reactor, as the "calibration" set. The data set would need to be generated by pyrolysis experiments at atmospheric pressure, probably at some temperature at or above 1,500°C.

As in the case of the maceral-based correlation, however, the predictive value of the correlation procedure for conversions in CO_2-gasification experiments was poor (Figure 4.22b). Judging by this set of results, measuring FT-IR spectra of *coals* does not appear as a viable route for reliably predicting CO_2-gasification reactivities. Experimentally, the major part of the actual gasification process takes place between the reactive gas and the post-pyrolysis *char*. Several char properties relevant to gasification, such as surface area, porosity and detailed chemical composition and the occurrence of active sites, are all determined by the intervening pyrolytic stage. While the original properties of the coal are expected to influence parameters affecting gasification, these are mediated by the chemistry and physics of the intervening pyrolytic step. In fact the poor correlation of Figure 4.22b suggests that the intervening pyrolytic step played a predominant role in determining the gasification reactivity of the chars. When the same FT-ir based method was used to probe the relationship between extents of gasification and mineral matter types and amounts, once again, no *systematic* trends could be found [Lemaignen *et al.*, 2002].

Table 4.9 Total volatile yields from pyrolysis and CO_2-gasification experiments in the wire-mesh reactor and calculated 'extents of gasification'. (Reproduced with permission: Energy & Fuels 2000,*14*,1049; Copyright 2000 Am.Chem.Soc.)

Coal	*total volatile yields (pyrolysis in He) (% w/w, daf basis)*	*total volatile yields (gasification in CO_2) (% w/w, daf basis)*	*'extent of gasification' TV (gas) – TV (pyr) (% w/w, daf basis)*
Taff Methyr (UK)	14.3	16.0	1.7
Tilmanstone (UK)	20.4	24.4	4.0
Emil Mayrisch (Germany)	20.7	22.5	1.8
Santa Barbara (Spain)	37.9	54.9	17.0
Heinrisch Robert (Germany)	31.6	40.3	8.7
Upper Freeport (USA)	35.6	55.5	19.9
Candin (Spain)	40.9	49.8	8.9
Point of Ayr (UK)	41.8	48.5	6.7
Thoresby (UK)	42.2	49.7	7.5
WA1 (Australia)	20.7	32.5	11.8
Hemheath (UK)	44.2	81.0	36.8
Bentinck (UK)	42.6	56.6	14.0
Pittsburgh No.8 (USA)	47.5	58.8	11.3
Lewiston-Stockton (USA)	44.6	75.1	30.6
La Jagua (Colombia)	43.8	57.7	13.9
Gedling (UK)	44.0	49.5	5.5
Linby (UK)	45.9	95.6	49.7
WA2 (Australia)	41.2	52.7	11.5
Blind Canyon (USA)	50.9	74.3	23.4
Illinois No.6 (USA)	49.2	100.0	50.8
Illinois No.6 (SBN)	48.1	100.0	51.9
Fording Genesse (Australia)	39.7	82.1	42.4
Gardanne (France)	58.3	82.0	23.7

In attempting to probe sample behavior under process conditions, we are never free from considerations involving reactor design. Similar correlations were also attempted between the FT-ir spectra of a subset of 16 of the coals (Table 4.10) and their pyrolysis and gasification conversions in the *fixed-bed* ("hot-rod") reactor [Zhuo *et al.*, 2000b]. That the resulting correlation for gasification would be poor was to be expected. However, the correlation for pyrolysis total volatile yields was also quite poor, giving a correlation coefficient value of 56.6%, when using pyrolysis data from the fixed-bed ("hot-rod") reactor, compared to 97.7% for the wire-mesh reactor data. This result provides a large measure of justification for insisting on developing pyrolysis reactor designs, where product distributions are measured with as little interference as possible

Table 4.10 Total volatile yields from pyrolysis and CO_2-gasification experiments in the fixed-bed ("hot-rod") reactor and calculated 'extents of gasification'. (Reproduced with permission: Energy & Fuels 2000, *14*, 1049; Copyright 2000 Am.Chem.Soc.)

Coal	*total volatile yields (pyrolysis in He) (% w/w, daf basis)*	*total volatile yields (gasification in CO_2) (% w/w, daf basis)*	*'extent of gasification' TV(gas) – TV (pyr) (% w/w, daf basis)*
Taff Methyr (UK)	4.4	16.3	2.3
Tilmanstone (UK)	19.1	56.8	37.7
Emil Mayrisch (Germany)	17.5	21.8	4.3
Santa Barbara (Spain)	52.9	70.3	17.4
Heinrisch Robert (Germany)	25.0	90.7	65.7
Candin (Spain)	35.2	83.6	48.4
Point of Ayr (UK)	33.7	36.9	3.2
Thoresby (UK)	37.3	80.0	42.7
Hemheath (UK)	36.5	65.8	29.3
Bentinck (UK)	36.3	85.1	48.8
Longannet (UK)	37.3	37.9	0.6
Rietspruit (S. Africa)	34.5	60.9	26.4
Gedling (UK)	40.8	45.3	4.5
Linby (UK)	39.1	40.5	1.4
Daw Mill (UK)	40.0	45.1	5.1
Gardanne (France)	31.0	74.6	43.6

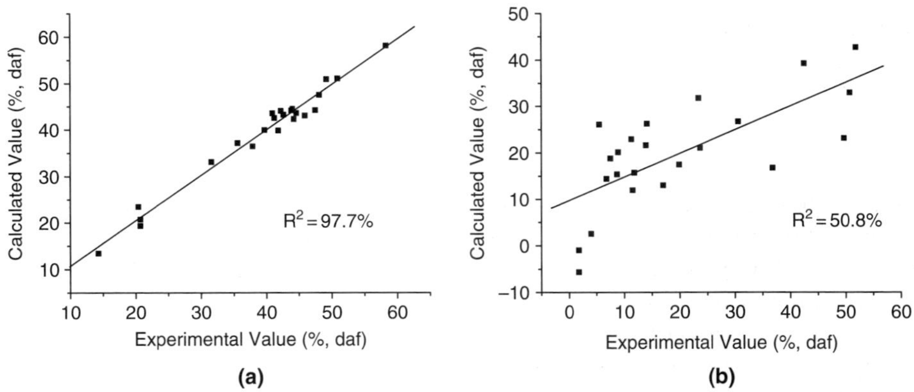

Figure 4.22 (a) Regression for Pyrolysis Total Volatile Yields; (b) Regression for the Extents of Gasification. All data from the high-pressure wire-mesh reactor (Reproduced with permission: Energy & Fuels, 2000, *14*, 1049; Copyright 2000 Am.Chem.Soc.)

from reactor related effects. This interpretation suggests, furthermore, that the statistical procedure used in this work is capable of leading to predictions of coal pyrolysis yields that may be perceived as physically meaningful.

Conclusion: We have been able to correlate the pyrolysis behavior of coals with their maceral contents and the pyrolysis yields of their constituent macerals. We have also been able to correlate pyrolysis yields with their FT-ir spectra. It would appear that original sample properties allow a level of prediction for weight loss during pyrolysis (total volatiles). However, it does not seem that the outcome of the "char gasification" stage of the process could be successfully correlated with the properties of the original coals.

4.6 Case Studies: Simulating Entrained-Flow Gasification in a Wire-Mesh Reactor

There are relatively few bench-scale experimental studies in the literature, on the assessment of coal performance under conditions that simulate conditions in entrained flow gasifiers. Mamori *et al.* [1998] used a drop-tube (entrained-flow) reactor at temperatures up to 1600°C. Other work has focused on pilot-scale rigs and their modeling [Brown *et al.*, 1998; Hara *et al.*, 2002].

Thermal Power Research Institute (Xi'An, Shaanxi Province) is developing a novel pilot-scale, dry feed, high temperature, high-pressure gasifier [Ren *et al.*, 2004]. This is being done in accordance with the strategic intent of the Chinese Government to develop dry-feed entrained flow gasification systems among their new generation of power technologies. In this section, we will outline the development of the high-pressure wire-mesh reactor (Section 4.2) for operation at temperatures up to 2,000°C and pressures up to 30 bar. The aim was to characterize coal particle behavior under conditions relevant to entrained-flow gasification and to determine the relative reactivities of fourteen Chinese coals, likely to be used under these reaction conditions [Peralta *et al.*, 2004; Wang *et al.*, 2005]. The work was done within the framework of a project in collaboration with TPRI, supported by the governments of China and the UK.

4.6.1 Extending the temperature range of the high-pressure wire-mesh reactor

The high-pressure wire-mesh reactor was modified to operate at temperatures up to 2,000°C and pressures up to 30 bar. The upgrade required only minor overhaul of the temperature control instrumentation. The Pt-PtRh pairs used up to about 1,500°C had to be replaced with higher-temperature thermocouples. The new thermocouples (type D) were alloys of 97% tungsten with 3% rhenium and 75% tungsten with 25% rhenium. This pair can be used at temperatures up to 2400°C. While the materials harden at the highest temperatures, this need not cause difficulties, so long as fresh thermocouples are used for each test. New ports for this set of thermocouples were installed on the reactor controllers, together with the corresponding thermocouple calibration data. In helium, the mesh could withstand temperatures up to 2000°C without physical damage. However, in CO_2, the mesh became brittle near 2000°C. At these higher temperatures, the holding time could be reduced, to protect the integrity of the mesh. On the other hand, the conversion in CO_2-gasification quickly ran to completion (~100%) at the highest temperatures. When operating in CO_2, therefore, the peak temperature could be reduced to 1500°C, which helped preserve the integrity of the mesh. An insulating sheet made of alumina was used to prevent the mesh from contacting the brass support plate and avoid a short circuit.

The heating rate used in this work was normally 1,000°C s^{-1}. In entrained flow gasifiers, rates are usually calculated to reach higher values. However, previous work with the wire-mesh reactor has shown that changes in heating rates above 1,000°C s^{-1} do not significantly impact on volatile release. Experiments at 1,000°C s^{-1} to 2,000°C and pressures between 10–30 bar were nevertheless difficult to conduct. Their successful completion reflects the considerable potential of the technique for studies under extreme conditions.

4.6.2 *Gasification reactivities of the set of Chinese coals*

Pyrolysis of the coals: Table 4.11 gives the compositions of the set of samples used in the study. They ranged from 75.3 (Niemeng) to 92.9% (Yangquan) elemental carbon content with (proximate analysis) volatile matter contents ranging from the mid-forties down to 9.7%.

The pyrolysis data for 1,500 and 2,000°C under 30-bar helium (1 s hold time) are presented in Table 4.12, showing weight losses larger than the proximate analysis volatile matter content. There are several reasons for this. The standard test is conducted at 900°C, with gram quantities of coal placed in a crucible and covered with a lid. Particle stacking in the crucible and the difficulty of volatile escape due to the lid gives rise to loss of volatile matter. This insures that volatile release during *any* wire mesh test, at or above about 700°C, exceeds the value from the standard proximate analysis determination [Gibbins *et al.*, 1990]. For some of the coals, there were significant differences between weight loss at 1,500 and 2,000°C.

It is well known that decomposition and volatilization of mineral matter in coals begins around 500°C with the escape of crystal water from kaolin and similar components [Taupitz, 1977]. Weight loss from mineral matter could be significant at the higher temperatures, particularly since some of the coals contain up to 30% ash. In one case (No.9; Huating), the ash content reached nearly 40%. Furthermore, the magnitude of the effect should vary with the composition of the mineral matter in the different coal samples. Analogous weight loss experiments have not been run with low temperature ashes to gage the magnitude of this effect. The uncertainty in the data is much reduced, however, when "extents of gasification" are calculated (Figure 4.23). This allows safely neglecting all weight loss from mineral matter with the possible exception of weight loss/gain due to mineral matter interactions with CO_2.

Table 4.12 shows that weight loss during pyrolysis varied widely, between 20 and 60% at 1,500°C, in rank ordered fashion. At 2,000°C, the rank ordering of sample weight loss was no longer discernible. At these higher temperatures, weight loss from the mineral matter would be expected to come into play as an interfering variable.

Gasification of the coals in CO_2: Extents of gasification have been estimated for each coal by subtracting the volatile yield, measured during the tests in He, from the total

Table 4.11 Compositions of the set of Chinese samples used in the study. 1. Yanzhuo Beishu; 2. Yanzhuo Yangchun; 3. Shen-Hua Houjitu; 4. Datong mixed; 5. Wujialu; 6. Shen-Mu mixed; 7. Shanxi Beigou; 8. Shen-Mu Daliuta; 9. Huating; 10. Neimeng; 11. Jiangyu; 12. Huangling; 13. Pingdingshan; 14. Yangquan (reproduced from Energy & Fuels 2005, 19, 2006 with permission. Copyright Am. Chem. Soc. 2005)

	Sample No.													
	1	*2*	*3*	*4*	*5*	*6*	*7*	*8*	*9*	*10*	*11*	*12*	*13*	*14*
Moisture % ar	3.7	3.6	7.6	4.0	7.4	6.2	1.4	9.8	1.1	11.1	1.2	2.0	1.2	0.9
Ash, % db	15.5	14.4	8.5	8.8	8.0	5.8	29.0	5.9	39.7	12.1	30.0	21.0	28.5	14.3
Volatile Matter	44.8	37.2	37.8	31.2	33.3	35.2	23.0	37.5	24.8	42.8	23.4	36.3	34.4	9.7
C	80.4	82.6	79.8	82.4	80.1	80.0	85.2	79.4	78.3	75.3	82.7	85.5	87.0	92.9
H	5.3	4.9	4.7	4.3	4.1	4.4	4.6	4.8	4.2	4.2	4.2	5.1	5.2	3.4
N	1.4	1.5	1.1	0.8	0.8	0.9	1.2	1.0	1.5	1.0	1.1	1.4	1.6	1.5
S	5.0	0.7	0.5	1.2	0.6	0.5	0.6	0.4	1.3	0.3	1.3	0.8	0.6	1.6

ar: as received; db: dry basis; all other data on dry ash free basis

Table 4.12 The Performance of the suite of coals in the wire-mesh reactor. Wang Baomin, Li Xiaoyu, Xu Shisen, Paterson, N., Dugwell, D. R. and Kandiyoti, R., *Energy and Fuels* 2005 (reproduced with permission from Energy & Fuels 2005, 19, 2006; Copyright Amer. Chem. Soc. 2005)

		Carbon Content %, daf	*Pyrolysis, 1,500°C, 30 bar %, daf*	*Pyrolysis, 2,000°C, 30 bar %, daf*	*Volatile matter %, daf*	*Gasification 1,500°C, 20 bar %, daf*	*Gasification 1,500°C, 30 bar %, daf*	*Extent of Gasification 1,500°C, 30 bar %, daf*
1	Yanzhuo Beishu	80.4	53.1	61.3	44.8	94.9	99.2	46.1
2	Yanzhuo Yangchun	82.6	48.8	51.9	37.2	85.4	88.6	39.8
3	Shen-Hua Houjitu	79.8	46	51.8	37.8	80.3	83.5	37.5
4	Datong mixed	82.4	40.1	43.1	31.2	62.2	69.3	29.2
5	Wujiala	80.1	44.6	46.2	33.3	89.7	93.2	48.6
6	Shen-Mu mixed	80	43.9	47.1	35.2	81.7	83.4	39.5
7	Shanxi Beigou	85.2	29.2	50.3	23	59.6	71	41.8
8	Shen-Mu-Daliata	79.4	49.4	50.7	37.5	81.9	85.5	36.1
9	Huating	78.3	33.3	54.9	24.8	78.1	88.7	55.4
10	Neimeng	75.3	58.8	59.8	42.8	100.9	103.2	44.4
11	Jiangyou	82.7	30.9	49.7	23.4	66	74.6	43.7
12	Huangling	85.5	44.8	54.7	36.3	71.8	83.2	38.4
13	Pingdingshan	87	44.6	65.5	34.4	76.1	82.8	38.2
14	Yangquan	92.9	18	22.2	9.7	31.5	36.2	18.2

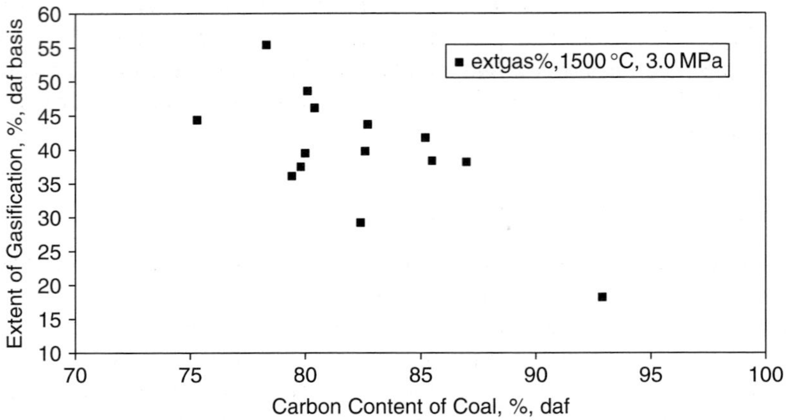

Figure 4.23 The "extent of gasification" in carbon dioxide as a function of carbon content of the coals. Wang Baomin, Li Xiaoyu, Xu Shisen, Paterson, N., Dugwell, D. R. and Kandiyoti, R., Energy and Fuels 2005 (reproduced with permission from Energy & Fuels 2005, 19, 2006; copyright Amer. Chem. Soc. 2005).

volatile yield during tests in CO_2. Values for experiments at 1500°C and 30 bar are presented in the final column of Table 4.12.

It may be noted that the pyrolysis experiments were run with a hold time of 1 s, whereas the gasification data was obtained with a hold time of 0.5 s. Initial tests, in CO_2, done with a hold time of 1 s at 1,500°C showed near complete conversion, which made it difficult to differentiate between the reactivities of these samples. Reducing the hold time to 0.5 s decreased the conversion and enabled differences between coals to be observed. A limited number of pyrolysis tests were repeated with a 0.5 s hold time, to check that it was valid to use the 1 s pyrolysis data set in the estimation of the extent of gasification after 0.5 s. The data obtained in He, at the shorter hold times, was virtually identical to that measured at 1 s and confirmed that pyrolysis was completed in less than 0.5 s, when samples are heated to these high temperatures.

The values for the extent of gasification, shown in Table 4.12, were in the 18 to 55% range. Figure 4.23 shows the extent of gasification plotted as a function of elemental carbon content on a dry, ash-free basis. The reactivity, as indicated by the extent of gasification, generally decreased with increasing C content, i.e. with increasing rank (maturity) of the coal.

While the general trend is clear in Figure 4.23, the diagram does nothing to provide *any* hope of serving as a predictive tool. As ever, diverse factors could be cited as contributing to the scatter, including the different effect of pyrolysis conditions on the particle morphologies for different coals. It appears that ordering the samples in some hierarchy of reactivity will inevitably require empirical means.

As expected, gasification rates at and above 1500°C are fast enough to gasify much of the coal. Coal reactivity is rarely considered an issue in oxygen-blown gasifiers, where complete conversion is achieved *provided sufficient residence time is allowed*. These data clearly show that the residence time in the gasifier that is required for complete conversion will vary with the rank of the coal. This has implications for the design of individual power plants, as the capital cost will be determined by the size and throughput of the gasifier. The aim would be to maximize the throughput for the minimum size of gasifier and therefore an understanding of the reactivity of the candidate coal(s) will be needed to optimize the design. The work outlined above indicates that the high-pressure wire-mesh reactor can provide a relatively low cost method for obtaining information that can be used to gain an insight into how the candidate coals will perform.

4.6.3 Relationship with pilot-plant design

The Thermal Power Research Institute in Xi'An (Shaanxi Province) is developing a new type of dry fed, two-stage entrained flow gasifier [Ren *et al.*, 2004]. In the first stage, approximately 80–90% of the coal is combusted and gasified in O_2 and steam. The exothermic reaction is sufficient to raise the temperature to over 1,800°C. The molten slag flows down the walls into a slag quench bath. The raw fuel gas from the first stage flows via a throat up into the second stage, where more coal and steam (but no oxygen) are injected. Sensible heat from the fuel gas is absorbed by endothermic steam gasification. It was estimated that the raw gas stream from the first stage would be cooled to approximately 900–1,000°C in the second stage. This is a benefit, since lower exit temperatures reduce the load on the raw gas cooler and enable this unit to be reduced in size, compared with a more conventional system. Lower exit temperatures also avoid sticky ash particles being released into downstream equipment, where they can form deposits.

Modeling studies have shown that the cold gas efficiency of the new concept is potentially 2–6% higher than a single stage system. However, we have already seen that temperatures around 900°C, expected in the *second* stage, do not give high conversions. *Some* carbon will therefore leave with the raw fuel gas. There appears to be every incentive to inject a more reactive coal into the second stage than need be injected into the first stage.

The pilot scale gasifier (24–36 t/d) is currently being constructed by TPRI. Initial studies were done with a single stage, 0.5-ton/day plant, which was then developed into the 2-stage version. Three of the coals that have been tested in the wire mesh reactor were also tested in the single stage 0.5 t/d plant. These were Huating, Yanzhou Yangchun and Yanzhou Beishu coals. The results obtained in the single stage plant and in the wire-mesh reactor are given in Table 4.13.

Table 4.13 shows similar orders of reactivity for the three coals in the pilot and laboratory scale reactors. The order of reactivity determined in the pilot-scale reactor was given by the cold gas efficiency [cf. Wang *et al.*, 2005] and by the CO content of the fuel gas, measured during tests at high fuel conversion. The reactivity order for the laboratory scale wire-mesh reactor was shown by the extent of gasification, measured after partial conversion of the fuel under standard test conditions. Compared to pilot scale results, numerically greater differences were found between the coals in the wire-mesh reactor. This is consistent with the laboratory scale results being obtained after partial fuel conversion (i.e. reaction times were short), where the differences in the reactivity will have had a greater influence on the results. Under the conditions in the pilot plant, a high fuel conversion was achieved and differences in the reactivity of the coals would be less apparent. However, these results show that the laboratory test can replicate the trend seen in the larger scale of operation.

Table 4.13 Performance of three coals in the single stage entrained-flow gasifier and in the high-pressure wire-mesh reactor. Wang B., Li X., Xu S., Paterson, N., Dugwell, D. R. and Kandiyoti, R., Energy and Fuels 2005 (reproduced with permission from Energy & Fuels 2005, 19, 2006; copyright Am. Chem. Soc. 2005)

	Huating	*Yanzhou Beishu*	*Yanzhou Yangchun*
Single Stage Gasifier Cold Gas Efficiency (1350–1400°C, 20 bar, in steam/air), %, LHV	79.1	77.2	74.8
CO, %, vol in fuel gas	55.1	52.2	51.7
Wire Mesh Reactor, Extent of Gasification (1500°C, 30 bar CO_2), %	55.4	46.1	39.8

The work has shown that the residence time in a commercial gasifier required for complete conversion, will vary with the rank of the coal. This has implications for the design of individual gasifiers, as the capital cost will be determined by the size/throughput ratio of the gasifier. Furthermore, the second stage of the reactor, described by Ren *et al.*, [2004] is expected to operate at lower temperatures, giving an exit gas temperature between 900 and 1,000°C. Clearly, the reactivity of the coals injected into the second stage would have a greater effect on overall conversions and on the amount of char in the exit gas.

4.7 Case Studies: By-Product Formation and Trace Element Problems in a Pilot Gasifier Processing Coal and Biomass

4.7.1 *Ammonia formation in a pilot-scale air blown spouted-bed gasifier*

High concentrations of NH_3 in fuel gas from gasifiers tend to enhance NOx forming reactions during subsequent combustion. In the operation of the pilot British Coal "Air Blown Gasification Cycle" (ABGC) gasifier, NH_3 concentrations in the product gas were found to be high, variable and worse, difficult to predict. The removal of NH_3 from fuel gas prior to combustion is technically feasible, but imposes an efficiency penalty on the process. The preferred alternative is to suppress NH_3-formation. That, in turn, requires investigating chemical pathways for its formation and decomposition.

However, open-ended investigations in pilot-plants can be expensive. On the other hand, bench-scale work lends itself to examining reaction parameters in isolation and experiments can be done quickly and relatively inexpensively. In order to clarify mechanisms and assess the impact of individual reaction parameters on NH_3 concentrations in the fuel gas, the laboratory-scale high-pressure fluidized-bed reactor described in Section 4.3 was converted from batch to semi-continuous operation.

4.7.2 *Re-designing the fluidized-bed for semi-continuous operation*

Changes to the design and operating procedures of the reactor have been described in detail by Paterson *et al.*, [2002] and Zhuo *et al.*, [2002]. Although the commercial-scale gasifier planned by British Coal was intended for operation in the 20–25 bar range, conditions in the pilot gasifier were typically 950°C and 13 bar. The "NH_3-project" undertaken in the bench-scale reactor mimicked conditions in the pilot reactor, to enable comparing data between the two systems. Since the gasifier contains oxidizing and reducing zones, the nitrogen chemistry in both types of environments had to be considered.

Figure 4.24 presents a schematic diagram of the new reactor system and the main components of the feeding circuit. As before, the main body consisted of a 34 mm (i.d), 504 mm tall Incoloy 800 HT tube that served as pressure shell and resistance heater. Unlike the batch bubbling-fluidized bed, the semi-continuous version of the reactor did not make use of a support plate. Instead, the design of the 28 mm (i.d.) quartz liner was altered to have a conical base, with a hole in the bottom to admit the spout jet. The steep angle of

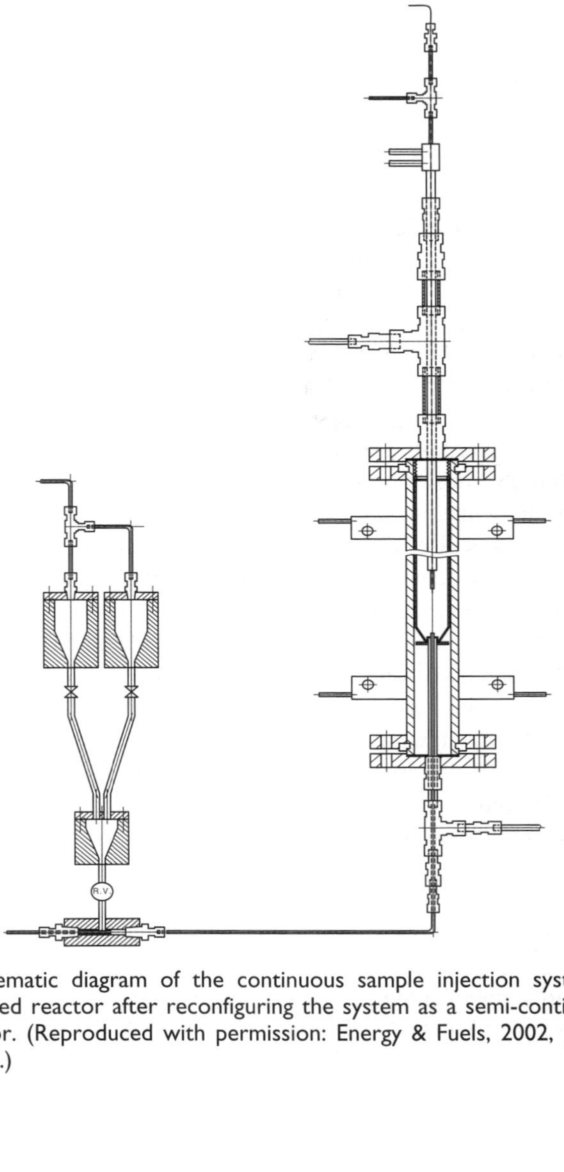

Figure 4.24 Schematic diagram of the continuous sample injection system and the high-pressure fluidized-bed reactor after reconfiguring the system as a semi-continuous, bench-scale spouted bed reactor. (Reproduced with permission: Energy & Fuels, 2002, *16*, 127; Copyright 2002 Am.Chem.Soc.)

the cone was similar to that of the base of the pilot scale gasifier. The coal and spout gas were fed by means of a spout jet, which fitted into the hole at the base of the quartz cone. This configuration was intended to qualitatively reproduce the hydrodynamic behavior inside the pilot scale gasifier.

The redesigned system continuously delivered 1–4 g min^{-1} solids to the reactor operating at up to 980°C and 30 bar. The system was capable of operating for periods of up to about 30 minutes, before ash sintering at the base interrupted experiments. A sampling probe attached to the top flange was installed to draw gas-samples directly from the spout jet. The probe also housed the exit tube, conveying gas from the reactor through a redesigned cooling section equipped with tar and steam traps. The gas then passed through a small cartridge filter to collect remaining particulates and through a needle valve, used for pressure letdown. As a safety measure, a small incinerator was installed to combust exit gases and vent into the laboratory extraction system.

Fuel feeding: Figure 4.24 shows the twin fuel hoppers, with capacities of 60 g coal each, feeding into a common hopper. The steep angles of the conical bases were intended to assist slippage and avoid bridging of fuel particles. Before a run, the hoppers were pressurized with nitrogen to approximately 0.2 bar above reactor pressure. From the common hopper, fuel was fed through a calibrated metering valve, driven by a variable speed DC motor. The metered fuel fell into an ejector, through which, the spout-gas mixture (air/nitrogen) was passed at high velocity. This entrains the fuel and conveys it through a 1.5 m long, 2 mm (i.d.) line to the spout jet at the base of the gasifier. The hopper and metering valve assembly were mounted on a frame that was vibrated by an eccentric motor, to avoid material bridging in the hoppers.

Details of the steam generator unit, fuel gas cleaning and analysis stages and the gas incinerator have been described in the original publications. The spout-gas temperature was measured using a thermocouple placed down the centerline of the gas-sampling probe, with its tip projecting by 5 mm into the bed. The design of the top flange and of the spout-gas analysis probe allowed vertical movement between experiments and served to extract gas samples from different heights within the submerged spout jet. It is in this region that the pyrolysis of fresh coal takes place alongside volatile and char combustion, releasing process heat. This is also where nitrogen oxides can potentially form from fuel nitrogen.

The sampling probe was designed in the form of three concentric tubes (cf. Figure 5 in Paterson *et al.*, 2002). The inner tube served as the gas outlet and thermocouple conduit, while the outer two tubes carried cooling water (or nitrogen) in and out, through a specially designed head. The height of the probe inlet above the spout may be changed between tests, by drawing the tube in or out, when assembling the top flange. Gaseous components analyzed included CO, O_2, NH_3 and NO. The collected NH_3 was determined by ion chromatography; the instrument measured NH_4^+ ion concentrations with a repeatability of ±2% of the measured value.

4.7.3 Operation of the bench-scale fluidized-bed gasifier

There was some debate on the choice of bed material. Sand had been previously used during batch operation, but would not have realistically simulated the high carbon content of the bed in the British Coal ABGC pilot gasifier. Using char from the pilot reactor was considered and rejected, as the sorbent content would have affected NH_3 release. Calcium oxide and its precursors have been noted to catalyze the formation of both NO*x* and NH_3 under conditions relevant to the gasifier [Lin & Johnssen, 1993]. Not enough Daw Mill char was available to use as bed material. A low volatile content fuel was necessary to start-up, in order to avoid the release of tar and other volatiles during heatup. In the event, a mixture of crushed *Coalite* (a commercially available "smokeless" fuel) and sand were used successfully. Coalite has a residual volatile matter content of about 9%, the majority of which was driven off during start-up.

Due to the continuous input of cold gas, there was only sufficient electrical power input to heat the reactor assembly to approximately 750°C at 13 bar. Instead of revamping the power supply, it was found that additional heat generation by combustion *in* the bed would allow operation at up to 950°C. When the temperature reached a plateau between 700–800°C, the fluidizing gas composition was changed to an approximate 20% air/80% N_2 mixture. The heat released by combustion raised the bed temperature toward the required level. Further increases in the proportion of input air were made to raise the bed temperature if required. The coal feed was then turned on and the rate adjusted to the desired value using the speed control on the calibrated valve.

In the laboratory-scale reactor, the fuel particle size was in the 200–300 μm range and the fluidizing velocity between 0.1 and 0.2 m s^{-1}, compared with values of up to 3,000 μm and 0.8 m s^{-1} in the pilot-scale gasifier, respectively. Typically, test durations were between 10–15 min. During this time, the gas *in* the spout, as well as the exit gas downstream of the gasifier, was analyzed. CO_2, CO, and H_2S concentrations in the exit gas were plotted against time to monitor reactor stability. There was no char off-take through the base, so that bed material built up in the gasifier during the test. Some sinter eventually formed on the base during most tests, blocking the spout entry and limiting the maximum run time to less than 30 min. Less sinter formed when steam was used as part of the fluidizing gas.

4.7.4 Nitrogen chemistry in the fluidized-bed

[Zhuo *et al.*, 2002] When fresh coal is injected into the spout jet of a gasifier, fuel-N compounds, in the form of pyrrolic and pyridinic nitrogen [Burchill & Welch, 1989], are released alongside other pyrolysis volatiles. At the temperatures (>900°C) and fuel rich conditions prevailing in the spout, fuel-N derived compounds break down, to release primarily HCN and smaller quantities of NH_3 [Baumann and Moller, 1991; Kanbara, *et al.*, 1993]. The escape of volatiles from fuel particles is normally retarded by the effect of elevated pressures. In fuel-rich conditions, this gives time for the primary HCN to be converted to NH_3, through reaction with hydrogen released during pyrolysis. The proportions of HCN and NH_3 thus depend on the type of reactor used and the conditions of the test. Not all of the fuel-N is released from coal by pyrolysis and the balance (somewhere between 40 and 60% of the total) remains in the char as stable compounds.

The NH_3 and HCN from pyrolyzing particles are released first into the oxidizing environment of the spout and then into reducing environments in the bubbling fluidized bed section above the spout and the reactor freeboard. In the oxidizing conditions of the spout, NH_3 and HCN can be oxidized to NOx and N_2O. HCN oxidation is more efficient at producing N_2O [Huulgard, 1993].

Under the oxidation/combustion conditions in the spout jet, char-N is thus converted to NOx, N_2O, or N_2. Raising the pressure increases the N_2O/NO ratio, because it decreases the concentration of radicals involved in NO formation [Haemaelaeinen & Aho, 1996]. Increased amounts of N_2O have also been noted in the transition region between oxidizing and reducing conditions [Hulgaard, 1992]. However, NOx and N_2O formed from char-N are likely to be reduced under the conditions of the bubbling fluidized bed above the spout to N_2 or NH_3.

In the fluidized bed, the NH_3 concentration tends toward gas-phase equilibrium according to the reaction, $2NH_3 \rightleftharpoons N_2 + 3H_2$. Thermodynamic modeling studies by British Coal have shown that *supra*-equilibrium concentrations were still present in the fuel gas at the point of measurement outside the pilot scale gasifier [Duxbury & Gavin, 1994]. This is consistent with other studies [Kilpinen *et al.*, 1991] and shows that equilibrium cannot be reached in the residence times allowed (up to 10 seconds) inside the reactor. During the gasification of residual char, the release of char-N may occur either as N_2 or as NH_3. The formation of HCN from the char-N, as precursor to NH_3 formation, is thought to be unlikely.

NH_3 production in the bench-scale fluidized-bed gasifier: When using fluidizing gas mixtures of air/N_2, the NH_3 concentrations in the exit gas may be understood in terms of the breakdown of volatile fuel-N compounds. Between 35–45% of the fuel-N was found in the chars recovered after the experiments. In the presence of air/N_2, residual char nitrogen does not appear to be released unless N-bearing parts of the solid matrix are consumed in combustion or gasification reactions.

Table 4.14 shows that the addition of steam resulted in large increases in the NH_3 concentration of the exit gas, compared with operation in air/N_2. The amounts of nitrogen in the ammonia were greater than the nitrogen content of evolving volatiles. Thus, much of the new NH_3 appears to have been formed by reaction with *char*-N. The formation of excess NH_3 from char-N in the presence of steam (or of hydrogen from steam decomposition) was reproducible. A large effect of steam to form ammonia had not been anticipated in these experiments, even though, the reactivity of char-N in the presence of high-pressure hydrogen had been noted during earlier hydropyrolysis experiments [Wu *et al.*, 1993].

The effect of steam on the char bed was examined further by varying the proportion of inlet steam during tests with a *Coalite* char bed, in the absence of the coal feed. The amount of input steam was varied from 0% (Test 61) to 10% (Test 56) and 16% (Test 55) by volume. The amount of NH_3 increased from 50 to 2300 vpm (volume parts per million) over this range, in proportion to the amount of added steam. Table 4.15 presents nitrogen analyses of selected residual char beds, from tests where coal was used as the feed. The data show that the presence of steam caused a reduction in the nitrogen concentration of final bed chars. This finding tends to confirm that steam was instrumental in removing some of the nitrogen from the char. The complexities of starting operations with a Coalite bed have been detailed in the original publications [Paterson *et al.*, 2002; Zhuo *et al.*, 2002]. However the overall effect of added steam on enhancing NH_3 formation, and in particular, its preferential reaction with char nitrogen is clear. Detailed mechanisms of these reactions are of interest and remain to be studied.

Table 4.14 The effect of steam input, operating temperature and coal/air ratio on NH3 concentrations in the product gas (Reproduced with permission: Energy & Fuels, 2002,16,742; Copyright 2002 Am.Chem.Soc.)

	Test No.	*% steam (by vol)*	*Pressure bar[a]*	*Cool feed rate $gmin^{-1}$*	*Coal: air ratio (mass)*	*Temp. °C*	*Dilution factor*	*NH_3 (adjusted for dilution by N_2) vpm, average*
No coal feed	61	0	12.5	0	–	850	6.06	49
	56	10.1	13.2	0	–	860	2.84	1297
	55	16.4	13.0	0	–	890	2.36	2294
Effect of % steam	48	6.0	13.3	2.1	0.32	795	2.59	2820
	62	14.3	13.1	2.2	0.28	800	2.21	5131
Effect of temp	48	6.0	13.3	2.1	0.32	795	2.59	2820
	43	6.1	12.6	2.0	0.25	837	2.20	1460
Effect of coal:air	43	6.1	12.6	2.0	0.25	837	2.20	1460
	49	6.5	13.5	2.2	0.46	841	3.10	3720
	47	6.3	13.6	2.2	0.80	830	4.78	5290

[a] Data corrected for start-up NH_3 formed fresh Coalite bed. Gas analysis adjusted for dilution by N_2.

Table 4.15 Nitrogen concentrations in final bed chars. (Reproduced with permission: Energy & Fuels, 2002, 16, 742; Copyright 2002 Am.Chem.Soc.)

Test no.	*Fluidizing gas (% vol)*	*N concentration in final bed char, % wt*
37	Air/N_2	0.8
39	Air/N_2	1.0
48	Air/N_2/steam (6%)	0.7
62	Air/N_2/steam (14%)	0.4

To summarize, these data show that during tests with steam, NH_3 in the exit gas is formed from both the breakdown of volatiles and from char-N. The proportion of the total NH_3 formed via each of these routes can be assessed from the results of tests with and without steam addition in Table 4.16.

The table shows data from tests with coal:air ratios of between 0.25 and 0.45. The NH_3 content measured at the reactor exit was found to increase substantially as the coal:air ratio was increased during tests in air/N_2 as well as in air/N_2/steam. Assuming that primary decomposition mechanisms remained unaltered by the presence of steam, it was calculated that the total NH_3 formed by pyrolysis (i.e., that released in the absence of steam) rose from 20 to 37% as the coal:air ratio was raised. The amount of NH_3 formed from char-N by the action of steam also increased with the coal:air ratio, but the proportion of the total NH_3 decreased from 84 to 67%. These data indicate that, during

Table 4.16 Proportions of NH_3 formed by pyrolysis and from char nitrogen. (Reproduced with permission: Energy & Fuels, 2002, *16*, 742; Copyright 2002 Am.Chem.Soc.)

Test no.	*Coal: air ratio (mass)*	*Steam addition % vol*	*Measured NH_3 vpm (% of total NH_3 during tests with steam)*	*Difference in NH_3 (with steam-without steam), vpm (% of total NH_3)*
43	0.25	6.1	1460 (100)	1173 (80)
18	0.24	0	287 (20)	
49	0.45	6.5	3720 (100)	2336 (63)
19	0.45	0	1384 (37)	

tests with steam in the air blown gasifier, the majority of the NH_3 is produced from char-N.

The relative proportions of NH_3 formed through (i) the decomposition of pyrolysis volatiles, and, (ii) by the reaction of steam with char-N varied with operating conditions. Daw Mill char was prepared during a separate test with the same reactor operated as a pyrolyzer. Several tests were done using initial bed material derived from the pyrolysis of Daw Mill coal, to seek further evidence about how char-N content affected the amount of NH_3 formed. The initial Coalite contained 2% nitrogen, whereas the Daw Mill coal contained 1.1% nitrogen. The coal char bed thus contained less nitrogen than the Coalite char bed and was expected to form less NH_3. Data obtained using the Daw Mill char bed is shown in Table 4.17, together with a set of data obtained with an initial Coalite char bed. Both tests used Daw Mill coal as feed.

Each initial bed contained a similar weight of char, although the Coalite char was mixed with an equal weight of sand. The data with the coal char bed were obtained with a higher coal:air ratio and with a higher proportion of steam in the fluidizing gas, which will have tended to raise the amount of NH_3 formed. However, the data in Table 4.17 clearly show that although the emission was high with the coal char bed, it was lower than with the Coalite char bed. This result confirms that char-N levels have a definite influence on the amounts of NH_3 formed during gasification.

Effect of temperature: The effect of temperature on NH_3 production was studied between 850 and 980°C, in the presence of sorbent, at a nominal pressure of 13 bar and

Table 4.17 Ammonia produced in the Daw Mill coal char bed[a]. (Reproduced with permission: Energy & Fuels, 2002, *16*, 742; Copyright 2002 Am.Chem.Soc.)

Test No.[a]	*% steam, vol*	*coal:air ratio*	*Temperature, °C*	*Pressure, bara*	*NH_3, vpm*
80	9.0	0.65	813	13.4	3475
82	14.2	0.72	825	13.7	3617
49	6.5	0.46	841	13.5	4249

[a] Tests 80 and 82, initial char bed produced by pyrolysis of Daw Mill coal. Test 49 initial bed formed from Coalite/sand mixture.

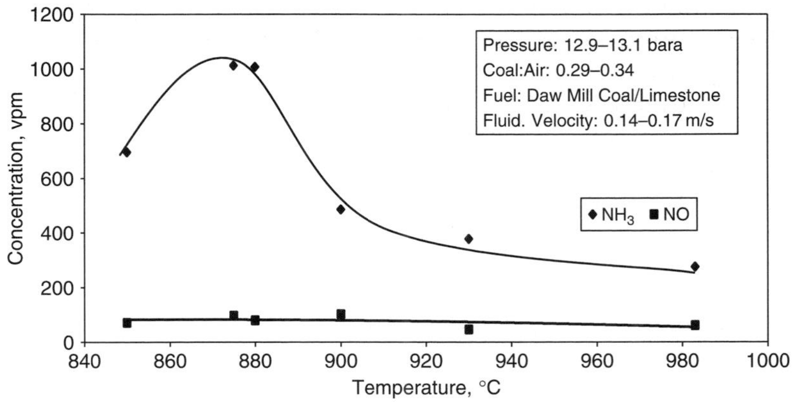

Figure 4.25 NH_3 concentration in the exit gas and the NO concentration in the spout as a function of the temperature, in the presence of sorbent. (Reproduced with permission: Energy & Fuels, 2002, *16*, 742; Copyright 2002 Am.Chem.Soc.)

a coal:air ratio of 0.3. Figure 4.25 shows the NH_3 concentration measured at the exit of the bench-scale fluidized-bed and the NO concentration in the spout, as a function of temperature. The NH_3 concentration peaked at approximately 880°C, indicating the temperature where NH_3-formation and destruction reactions were evenly balanced. In this context, NH_3-destruction is likely to result from decomposition to N_2 and H_2, as the composition tends toward gas phase equilibrium values.

Figure 4.26 showing the analogous peak in NH_3-concentration in the pilot scale reactor, confirms that trends observed in the bench-scale reactor reflected findings in the much larger gasifier. However, the peak occurred at approximately 940–950°C in the

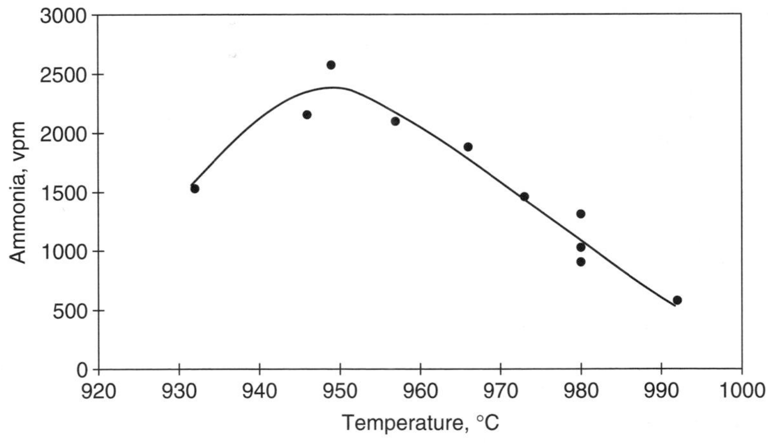

Figure 4.26 The Effect of Temperature on the Ammonia Concentration in the Fuel Gas from the Pilot Scale Gasifier. (Reproduced with permission: Energy & Fuels, 2002, *16*, 742; Copyright 2002 Am.Chem.Soc.)

pilot scale gasifier. The difference is attributed to the manner of temperature measurement in the pilot-scale gasifier. The temperature in the latter was the calculated mean of many readings, obtained by several thermocouples, set at different points throughout the bed, over an approximately 24 hour mass balance period. By contrast, the value for the laboratory scale rig was calculated as the average of readings from a single thermocouple over a 10–15 min test period. It is likely to reflect the temperature where NH_3-production and destruction reactions were balanced more accurately.

Summary: In addition to showing sharply rising NH_3 concentrations with increasing steam input, the bench-scale study allowed testing for the effect of several operating parameters on NH_3 formation: char-N content, reactor temperature and coal:air ratio [Paterson *et al.*, 2002; Zhuo *et al.*, 2002]. The effect of sorbent addition was also discussed in the original publications. In air and oxygen blown gasifiers, steam is used mainly to control the temperature of the bed. As we have seen, in air-blown gasification, steam injection leads to rising NH_3 concentrations in the product gas. The main control options to limit NH_3 concentrations in the product stream appear to revolve around the use of an alternative method of bed temperature control. This can be done by accepting higher bed temperatures, with the limit set by ash melting, or by operating at lower pressures within limits set by efficiency considerations for power generation in gas turbines.

4.7.5 Interaction between HCN and NH_3 formation during sewage sludge gasification

Analogous data on HCN formation during coal and sewage sludge gasification in the same reactor have been presented recently by Paterson *et al.* [2005]. Dried pellets of sewage sludge were gasified in air/N_2 and air/steam/N_2 mixtures to identify the effects of temperature and steam on HCN and NH_3 concentrations in the exit gas. Consistent with the work outlined above, higher concentrations of NH_3 (and HCN) were observed when steam was introduced into the spout-gas mixture. The mechanism of the effect is not obvious, but gives rise to increasing concentrations of potential NOx precursors. However, the concentration of HCN showed a somewhat different pattern to that of NH_3 formation.

As mentioned earlier, the bench-scale fluidized-bed reactor has no char discharge facility. Instead, the char bed builds up between the beginning and the end of a particular run. During the sewage sludge processing runs, measured HCN concentrations were observed to decrease as the experiment progressed and depths of char bed increased. The growing height of bed appears to have provided an effective environment for the reaction of HCN to form NH_3. This result was consistent with data obtained in the pilot-scale spouted-bed gasifier, where only low concentrations of HCN were measured in the exit fuel gas, after longer residence times in the char bed, compared to the laboratory scale reactor.

In bench-scale experiments, the concentration of HCN was observed go through a maximum at around 930°C and to decrease rapidly thereafter (not shown). Since tar release from the sample will have peaked at or before about 700°C, the increase of HCN with temperature indicates the intensity of tar cracking reactions taking place in contact with bed solids. The data shows that above about 930°C, its destruction by secondary reactions becomes increasingly rapid. NH_3 formation reactions would also be promoted by increasing amounts of molecular H_2 liberated through the cracking of the sewage sludge tars.

The results from sewage sludge gasification indicate that more complete tar cracking may lead to progressively greater proportions of H_2 in the product gas, possibly surpassing the already observed 25% level [Paterson *et al.*, 2005]. In this hydrogen-rich environment, very high (supra-equilibrium) NH_3 concentrations were reached at low temperatures, hovering above 9,000 vpm at about 780°C in the presence of steam. From these high values, NH_3 concentrations in the exit gas decreased with increasing temperature, both in the presence and absence of steam in the spout-gas. This appears to result from the faster approach to equilibrium values with rising temperature, effectively decomposing NH_3 to N_2 and H_2. The effect was sufficiently important to have masked the increase in NH_3 formation from HCN at higher temperatures.

The concentration of HCN formed from coal was also measured in experiments using the bench-scale spouted bed reactor, using air/N_2 mixtures. These tests required an initial char bed in the reactor (for operational reasons), and consequently, the effect of residence time in the char bed on HCN concentration could not be monitored. The concentration of HCN in the exit gas was observed to decrease at temperatures above 900°C. This is thought to be a result of the increased H_2 concentration released by pyrolysis, which would enhance the rate of decay of HCN to NH_3. The work has provided a useful insight into the reactions of fuel-N in the gasifier and has helped to explain why low HCN concentrations were measured in the raw fuel gas from the pilot-scale gasifier.

4.7.6 Trace elements in output solid streams during sewage sludge gasification [Reed et al., 2005]

Gasification and pyrolysis processes have attracted considerable interest from water utilities as technologies for sewage sludge disposal. The process has several advantages: waste volume reduction, destruction of pathogenic bacteria and energy recovery. However, elements such as Ba, Cu, Hg, Pb and Zn are present in sewage sludges at levels significant to the disposal of the residual solid streams from the gasifier. The behavior of these elements has been studied in the air blown laboratory-scale spouted bed gasifier [Figure 4.24], fueled by crushed dried sewage sludge pellets. Measurements of trace element concentrations by ICP-AES have been used to determine (i) the retention of selected elements in the solid streams, (ii) their relative depletion from the coarser bed residue and (iii) any enrichment in the fines carried to the gas cleaning system. The effect of gasifier bed temperature and type of sewage sludge has been investigated.

Results indicated that gasifier bed temperatures in excess of 900°C enhanced the depletion of Ba, Pb and Zn from the bed residue and their enrichment in the fines. Mercury and selenium capture required low temperature filters operating below 120°C. It was also found that thermodynamic modeling is not always helpful in predicting outcomes of experiments, due either to limitations of data-bases or problems arising from kinetic constraints. Parallel co-*combustion* experiments have suggested that the presence of chlorine somewhat enhances the volatility of lead and cadmium. This application was another instance of bench-scale experimental units having a role to play in estimating fuel behavior in pilot or plant scale equipment.

4.7.7 Calcium based liquid phase formation in pressurized gasifier environments [Paterson et al., 2001]

Limestone and dolomite are routinely used for retaining sulfur in air blown coal/waste gasification. An intractable deposit was observed to form, when *limestone* was used as sorbent during initial tests using a revised base design in the British Coal pilot scale gasifier. The deposit was found to be enriched in Ca and had formed via a melt, which eventually blocked the exit duct. An attempt was made to simulate the formation of melts by sorbent components under laboratory conditions. The detection of melting was done by impedance spectrometry. The technique measures the change in resistivity during the transition from a non-conducting solid to an ionic melt. The validity of the technique was demonstrated by reproducing melt formation that had previously been observed in the CaO/CO_2/steam system under the conditions of the CO_2-Acceptor Process. SEM/EDX and XRD were used to characterize the residues from the present tests.

$CaCO_3/CaSO_4$ was the only system tested that formed a melt under the experimental conditions corresponding to the pilot-plant trials. Melt formation was observed at 1,010°C at pressures in excess of 13 bar. The melting temperature was not affected by the presence of other Ca compounds (e.g. CaS) that would be present in the gasifier. It is thought that $CaSO_4$ formation occurred when CaS in the recirculating solids stream was exposed to the oxidizing conditions in the spout of the gasifier. The melt is thought to have formed by the interaction of $CaSO_4$ with as yet uncalcined $CaCO_3$ sorbent particles. Oxidation of sulfided limestone (and dolomite) and raising the temperature to above 1,010°C caused a melt to form and demonstrated the sequence of reactions. The properties of the melts produced from limestone and dolomite were different. The results also helped explain why dolomite did not give rise to deposits during pilot plant trials.

4.8 Case Studies: Synergistic Effects in Biomass Processing & Problems in Refining Biomass Pyrolysis Tars

Most of this Chapter has been devoted to explaining the multiple advantages of fast heating, short residence time bench scale reactors. In practice however, there are many cases where the use of a fixed-bed reactor is *required* by the course of the work or is made easier by it. It seems useful, therefore, to discuss two cases where the fixed-bed ("hot-rod") reactor was selected as the preferred reactor configuration.

4.8.1 Synergistic effects during the co-pyrolysis and co-gasification of coal-biomass mixtures

Co-processing coal with biomass? The costs of disposal by landfill for all types of waste have been rising. There are economic benefits in generating energy from domestic, industrial and agricultural wastes, as well as attenuating our reliance on premium fossil fuels. Furthermore, biomass and waste gasification/combustion may be viewed as CO_2 neutral.

However, the calorific and other properties as well as the supply of most biomass materials are subject to seasonal variations. Furthermore, the mass and energy densities of most biomass/waste are low, which reduces their economic transport radii. These factors provide a reasonable case for co-processing coal with biomass. On the other hand, incineration of coal with biomass-waste has not received wide acceptance due to unresolved technical as well as presentational problems regarding harmful emissions. The work described in this section was aimed at developing basic understanding of co-gasification of biomass with coal, in processes where emission controls are more easily worked into the overall design. Apart from *not* (yet?) being commercially available, however, the attractions of co-gasification are mitigated by the necessity of operating at higher temperatures than would have been necessary for biomass/waste alone.

Selecting a reactor vessel for coal-biomass co-processing: The research described in this section was undertaken in response to conflicting reports relating to the occurrence or absence of synergistic effects during coal-biomass co-processing. Sjöström *et al.* [1994] reported synergistic effects during the co-*gasification* of a Polish coal with silver birch wood. These experiments were carried out in a pressurized fluidized bed reactor, between 700–900°C at 4 bar of "oxygen enriched nitrogen". Sample was fed continuously at a rate of 27 $\sim$ 87 g min^{-1} into a sand fluidized bed. Char yields from fuel mixtures were lower than would have been expected from experiments on the coal and the biomass alone, whilst amounts of oxygen consumed and gas produced during co-gasification were higher than expected. It was suggested that the relatively high reactivity of wood and its chemical composition (high atomic H:C and O:C ratios) could lead to increases in fuel conversion. However, similar effects were not detected during experiments at the higher pressure of 15 bar. Only three co-gasification tests were carried out and it might have been premature to take these results as conclusive evidence for synergistic effects.

Working with an entrained flow reactor, Rüdiger *et al.* [1994] reported the absence of synergistic effects between various fuels. The findings were explained in terms of different temperature ranges at which the devolatilization of coal and biomass takes place, due to differences in particle sizes and chemical structures between the two fuels. Once again, problems in reactor selection appear to have impeded progress. The spatial separation of sample particles in entrained flow reactors does not appear to have been considered. Drop-tube reactors are instruments where the design principle *requires* sample particles to be segregated and are manifestly unsuitable for examining synergistic effects in blended feedstocks.

Pan *et al.* [1996] also studied the pyrolytic behavior of different mixtures of low rank coals with biomass in an atmospheric pressure thermogravimetric analyzer. Samples were heated at 100°C min^{-1} to temperatures between 110–900°C. No interactions were observed between the two fuels. In this apparatus, the *pyrolytic* behavior of blends in any proportion consisted of the additive behavior of the two individual samples. The absence of synergistic effects during the *pyrolysis* of maceral concentrates discussed in Chapter 3 would have led us to expect this last result. On the other hand, the work of Shafizadeh [1982] suggests that product distributions from the pyrolysis of intimate mixtures of cellulose and lignin (*e.g.* wood) show pyrolysis behavior that is not compatible with weighted sums of the constituent materials.

In the study summarized below [Collot *et al.*, 1999], tar and total volatile yields were measured during a set of co-pyrolysis experiments and co-gasification runs in CO_2. The experiments were carried out in two reactors: a fluidized-bed designed to provide near total segregation of sample particles and a fixed-bed reactor naturally providing intimate contact between neighboring fuel particles. Sample weight loss and tar yields from two sets of coal-biomass mixtures were determined singly and in mixtures, at 850 and 1,000°C and pressures up to 20 bar.

Results from coal-biomass co-processing experiments: Table 4.18 presents data from the co-pyrolysis of approximately 50:50 mixtures (by weight) of Daw Mill coal and silver birch wood in the "hot-rod" reactor at 1,000°C. The experiments were carried out at pressures of 5, 10 and 20 bar. Within experimental error, results from these and analogous experiments at 850°C (results not shown) gave total volatile yields matching values calculated on the basis of the *absence* of synergistic effects. However, experimentally determined tar yields were about 4% (w/w) higher than those calculated using tar yields from coal or biomass alone. The differences in tar yield were found to be systematic and somewhat larger than levels of experimental repeatability (±1%). As expected, analogous results from the fluidized bed reactor showed similar absence of synergistic effects. Table 4.19 outlines similar results for gasification in CO_2. Experimental total volatiles matched values calculated from the behavior of the individual fuels, whereas tar yields were again somewhat higher than expected.

Table 4.18 Pyrolysis total volatile and tar yields of Daw Mill coal and silver birch mixtures 1:1 obtained in the "hot-rod" reactor as a function of the pressure, at 1000°C with 100 seconds holding time. The heating rate was 10°C s^{-1}, and the carrier gas flow rate 0.1 m s^{-1}. (Reproduced with permission: Fuel 1999, *78*, 667; Copyright 1999 Elsevier.)

Pressure (Bar)	*Coal w/w%*	*Calculated Total volatile w/w,%*	*Experimental Total volatile w/w,%*	*Calculated tar w/w,%*	*Experimental tar w/w,%*
5	47.1	66.6	66.4	29.1	30.5
10	46.5	65.6	66.2	26.8	30.9
20	46.4	64.7	64.0	23.1	27.5

Table 4.19 CO_2-gasification total volatile and tar yields of Daw Mill coal and silver birch mixtures 1:1 obtained in the "hot-rod" reactor as a function of the pressure, at 1000°C with 100 seconds holding time. The heating rate was 10°C s^{-1}, and the carrier gas flow rate 0.1 m s^{-1}. (Reproduced with permission: Fuel 1999, 78, 667; Copyright 1999 Elsevier.)

Pressure (Bar)	*Coal w/w%*	*Calculated Total volatile w/w,%*	*Experimental Total volatile w/w,%*	*Calculated tar w/w,%*	*Experimental tar w/w,%*
5	53.0	72.8	74.8	21.0	26.1
10	50.7	74.2	76.9	20.0	22.9
20	48.4	76.4	78.3	19.1	25.0

Summary: Neither intimate contact between fuel particles (in the fixed-bed reactor) nor relative segregation (in the fluidized-bed reactor) of fuel particles during *pyrolysis* appears to affect total volatile release from mixed fuels – relative to results expected from independent sample behavior. Indications of differences in *tar yields* have been observed in the "hot-rod" reactor, where sample particles are mixed and closely stacked. Whilst differences were systematic and somewhat larger than levels of experimental repeatability, the effect was too small to provide effective information that might explain the results. Work with a different sample set suggests that even these weak trends are sample dependent. Co-gasification with CO_2 showed similar trends to those observed in pyrolysis. For a different sample set, *total volatile yields* were about 5% higher than would be expected in the absence of synergistic effects. Once again, the differences were too small to provide clues for explanations, and indeed, to constitute clear evidence of synergy. Apart from a possible catalytic effect of biomass ashes at high temperatures, observed synergistic effects were not of a magnitude that might have influenced process design.

4.8.2 How to process biomass pyrolysis tars?

Until about the last 70 years, the "destructive distillation of wood" was a principal source of methanol ("wood alcohol") and a host of other chemicals, creosote oils and tars. No longer economically viable, the process is otherwise known as wood pyrolysis followed by a fractional distillation. More recently, the thermal processing of agricultural, forestry and municipal solid wastes has been studied as a likely route for waste disposal coupled to energy and chemical feedstock production.

The rapid pyrolysis of finely divided wood particles gives liquid yields of 50 to 70% of the original biomass [*e.g.* see Scott & Piskorz, 1982a; 1982b; 1984]. These liquids are highly polar, very reactive and, clearly, are no longer economical to distil. During storage, they are corrosive, they separate into aqueous and organic phases and the organic phase gums up [Kirk *et al.*, 2001]. The problem is to find uses for these liquids, whose properties reflect the high oxygenate contents of the original feedstocks (20–45% w/w). Blending with kerosene has been found to ease storage problems. Exploratory catalytic hydrocracking work has shown the condensed liquids to be prone to high levels of coking and reactor blockages [Baker & Elliott, 1988]. Coke formation could be minimized by prior hydrogenation at lower temperatures in the presence of $CoMo/Al_2O_3$ or $NiMo/Al_2O_3$ catalysts, followed by hydrotreatment at higher temperatures. The two-step procedure reduced hydrogen consumption but generated large amounts of CO_2. Pindoria *et al.* [1998] have briefly surveyed research on the upgrading of biomass pyrolysis liquids. Overall, reported yields of lighter hydrocarbons were low and feedstock losses by coke formation relatively high.

In view of the reactivity of freshly condensed biomass pyrolysis tars/oils, it could be argued furthermore, that the work done during catalytic upgrading was partly an exercise in partially undoing the retrogressive reactions taking place during condensation and storage. An alternative route for upgrading pyrolysis-derived liquids involves passing pyrolysis volatiles over a catalyst bed prior to condensation.

Prior to the work described below, relatively few such experiments had been reported. They were mostly carried out at atmospheric pressure, where coke formation had once again taken an unacceptably high toll of the feedstock [Evans & Milne, 1988; Adjaye *et al.*, 1991; Bain *et al.*, 1993; Agblevor *et al.*, 1993; Diebold *et al.*, 1994a; 1994b]. This section describes results from a bench-scale, two-stage, fixed-bed reactor, designed to counter some of these problems. Mild hydropyrolysis in the first stage (~40 bar) was followed by a catalytic stage for the passage of freshly formed volatiles. The second stage operated at the same pressure but at (lower) temperatures more appropriate for catalytic hydrotreatment. The use of high-pressure hydrogen was intended to enhance liquids production in the hydropyrolysis stage and to minimize carbon lay-down within the catalyst stage. A schematic diagram of the reactor has been given in Figure 3.4b [Pindoria *et al.*, 1998].

Results from two-stage experiments: Products from the mild hydropyrolysis of Eucalyptus wood (*eucalyptus globulus*) shavings in the first stage (up to 500°C) have been catalytically hydrotreated in a second bed packed with H-ZSM5 as catalyst. The yields and changes in structural characteristics of product tars/oils were determined as a function of hydrogen pressure and *temperature in the second stage*.

In the absence of a second stage, tar yields from the first stage ranged from nearly 45 to slightly above 30% for operation at 10 and 40 bar-hydrogen, respectively. The use of the second stage diminished condensable yields progressively with increasing temperature. At 10 bar, tars went from about 27 to 22% as the temperature of the catalyst bed was raised from 300 to 400°C. At 40 bar, the condensables yield was about 23% for 300°C in the catalyst bed, declining to nearly 10% at 400°C. Thus increasing the severity of hydrocracking reactions led to lower liquids yields. Furthermore, when one batch of sample (about 50 mg) was hydropyrolyzed with a charge of fresh catalyst, between 40 and 46% w/w of the original sample was deposited as coke within the catalyst bed.

Taken together with amounts of volatiles entering the catalyst bed from the hydropyrolysis stage, the results showed that some of the materials, classed as light volatiles, were trapped within the catalyst matrix. TG-balance profiles from the burning off, in air, of carbonaceous material from once-used catalyst samples were consistent with removal by evaporation and/or combustion of relatively lighter species occluded in the Zeolite matrix, as well as the eventual combustion of insoluble and charred material.

The uptake of sample-derived material by the catalyst bed decreased sharply upon extended use of the catalyst. Qualitatively similar results were obtained during operation at 10 and 40 bar. Thus the first pass, when hydropyrolysis volatiles were contacted with fresh catalyst, appears to have largely consisted of a catalyst-passivation stage. However, the drop in sample uptake by the catalyst bed upon re-use did not result in the production of more liquid product but in a sharp increase in non-condensable light volatiles, suggesting that with the present combination of pressure, temperature and feedstock, H-ZSM5 acts as a more active cracking catalyst than an agent promoting hydrogenation and/or hydro-deoxygenation.

Observed shifts in liquid product characteristics were found to be relatively small and provided no evidence for significant molecular size reduction toward forming a sizeable distillate fraction. Despite the significant changes in tar/oil *yields* with reaction conditions, structural changes achieved during the hydrotreatment stage were relatively minor. The gas product has not been analyzed and may turn out to be more valuable than the tars. Overall, it was concluded that available reaction routes do not allow specific de-oxygenation pathways to predominate without disintegration of parent molecules to lighter volatiles. Contact with the H-ZSM5 catalyst under these conditions appears to lead to sharp transformations, forming lighter components (classed in this work as light volatiles or gas-make) and char, rather than progressive reductions in polarity and molecular mass. Finally, observed differences in product characteristics between fresh catalyst the first and second re-use were relatively minor.

Evaluation of results: Biomass pyrolysis liquids can be made in large volumes. For direct use as a fuel, the process may be conceived in ways that may be economically viable. However, the prospects for upgrading these fuels do not look so attractive. We have briefly reviewed work done elsewhere, to upgrade *condensed* biomass pyrolysis oils. The results were poor. We also attempted to upgrade the vapors of condensable materials – prior to condensation. In plain language, the results discussed in this last section mean that we have not (yet?) found a way of making a distillate fraction from biomass pyrolysis oils that could be used as a feedstock for making diesel or gasoline grade fuels. Even if the selected routes outlined had been proven technically attractive, the conditions selected here would be considered as reasonably expensive both in terms of raw materials used (notably hydrogen) and the use of high-pressure. It appears we have not reached the stage when we can entrust our future to the long heralded *bio-refineries* that environmental technologists promise will render fossil fuels redundant.

Closing remarks: Important residual problems must be resolved before co-pyrolysis and co-gasification technologies can be offered as commercially viable processes. Many of these phenomena can be effectively, rapidly and cheaply investigated at bench-scale. In these sections, several examples have been provided to show how laboratory-scale equipment can be used to determine primary product trends and provide basic diagnostics, in order to help resolve problems identified during plant or pilot scale design and operation. The studies have been carried out using a small array of thermochemical reactors, designed to simulate wide ranges of reactor conditions.

References

Adánez, J., Miranda, J. L., Gavilán, J. M. (1985) *Fuel*, 64, 801

Adjaye, J., Sharma, R. and Bakhshi, N.N., (1991) in "Energy from biomass and waste" vol XV, *ed D.L. Klass*, Institute of Gas Technology, Chicago Ill. USA, p 783

Agblevor, F.A., Rejai. B., Evans., R.J. and Johnson K.D. (1993) in "Energy from Biomass and Waste", vol XVI, *ed D.L. Klass*, Institute of Gas Technology, Chicago Ill. USA, p 767

Alvarez, T.; Fuertes, A. B.; Pis, J. J.; Parra, J. B.; Pajares, J.; Menendez, R., (1994) *Fuel* 73, 1358

Arendt, P (1980) *Ph.D. thesis*, University of Aachen, Germany

Arendt, P., van Heek, K.-H. (1981) *Fuel, 60*, 779

Bain, R., Diebold, J., Overend, R., Rejai, B., and Power, A.J.,(1993) in "Energy from Biomass and Waste" vol XVI, *ed D.L. Klass*, Institute of Gas Technology, Chicago Ill. USA, p 753.

Baker, E. G. and Elliott, D. C.(1988), in *Pyrolysis Oils from Biomass, ACS Symposium Series 376*, eds J. Soltes and T. A. Milne. ACS, Washington, p. 228.

Baumann, H. and Moller, P. (1991) *Erdol und Kohle, Erdgas, Petrochemie*, 44, 29

Bautista, J. R., Russel, W. B., Saville, D. A. (1986) *Ind. Eng. Chem. Fund.*, 25, 536

Bolton, C., Snape, C.E., O'Brien, R.J. and Kandiyoti, R., (1987) *Fuel*, 66, 1413

Bota, K. B.; Abotsi, G. M. K. (1994) *Fuel* 73, 1354

Brown, B.; Smot, L. D.; Smith, P. J.; Hedman, P. O. (1988) *AlChE J.* 34, 435–446.
Burchill, P. and Welch, L. S. (1989) *Fuel* 68,100
Cai, H., (1995) PhD Thesis, University of London
Collot, A-G., Zhuo, Y., Dugwell, D.R. and Kandiyoti, R.(1999) *Fuel* 78, 667
Dawes, S.G., Reed, G.P., Gale, J. and Clark, R.K. (1991) *Institution of Chemical Engineers Symposium Series No.123*
Dawes S G, Cross P J, Minchener A J. *UNECE Symposium*, Helsinki, "Development of the British Coal Topping Cycle", May 1993.
Dawes S.G., Mordecai, M., Brown, D., Burnard, G.K., "The Air Blown Gasification Cycle", Proc. 13th Intl. Conf on Fluidised Bed Combustion, Orlando, Florida, May 1995.
Diebold, J., Phillips, S., Tyndall, D., Scahill, J., Feik, C., Czernik, S., (1994a) *A.C.S. Div. Fuel Chem.* 39, 1043
Diebold, J.P., Bridgewater, A.V., Beckman, D., Elliott, D.C. and Solantausta, Y.,(1994b) in "Advances in thermochemical biomass conversion", *(Bridgewater, A.V., ed.), Blackie Academic & Professional, Glasgow* 2, 1325
Doering, E.L.; Cremer, G.A. (1994) *Proc. American Power Conference*, 56, 1686
Duxbury, J.; Gavin, D. MTDATA Studies (1994) *ETSU Report* R-018
EPRI Report No. TR-103367 (1993), 'Gasification of Pittsburgh No. 8 coal' in Rheinbraun's 'Pressurised High Temperature Winkler Plant', Final report
Elliott, M. M. (1981) Editor *Chemistry of Coal Utilization Second Supplementary Volume*, John Wiley, NY,
Evans, R.J. and Milne, T.A., (1988) "Pyrolysis Oils from Biomass", *Am. Chem. Soc. Symp. Series (Soltes, E.J. and Milne, R.J., ed.), 376*, Am. Chem. Soc., Washington, D.C., U.S.A. 311
Gavin, D.; Paterson, N.; Duxbury, J.; Maxwell, S.; Reed, G. P. (1997) Fuel Behaviour in the Air Blown Gasification Cycle, Report No. COAL R101, DTI Cleaner Coal Technology Programme
Gibbins-Matham, J. R., Kandiyoti, R. (1988) *Energy & Fuels*, 2, 505
Gibbins, J. R. and Kandiyoti, R. (1989a) *Energy & Fuels, 3*, 670
Gibbins, J. R. and Kandiyoti, R. (1989b) *Fuel, 68*, 895
Gibbins-Matham, J. R., King, R. A.V., Wood, R. J., Kandiyoti, R. (1989c) *Review of Scientific Instruments, 60*, 1129
Gibbins, J.R., Khogali, K. and Kandiyoti, R., (1990) *Fuel Proc Tech* 24, 3.
Gibbins, J.R., Gonenc, Z.S. and Kandiyoti, R., (1991) *Fuel* 70, 621
Ginter, D. M.; Somorjai, G. A.; Heinemann, H. (1993) *Energy Fuels* 7, 393
Gonenc, Z. S., Gibbins, J. R., Katheklakis, I. E. and Kandiyoti, R (1990) *Fuel* 69, 383.
Goring, G. E., Curran, G.P, Tarbox, R. P., Gorin, E., (1952) *Ind. Eng. Chem.* 1057
Guell, A.J., (1993) *PhD Thesis*, University of London
Guell, A. J., Kandiyoti, R. (1993) *Energy & Fuels*, 7, 943
Guell, A. J.; Cai, H.-Y.; Dugwell, D. R.; Kandiyoti, R. (1993) *Fuel Process. Technol.* 36, 259
Guo, C.; Zhang, L. (1986) *Fuel* 65, 1364
Haga, T.; Nishiyama, Y. (1988) *Fuel* 67, 743
Haemaelaeinen, J. P.; Aho, M. J. (1996) *Fuel* 74, 1377
Hara, S.; Inumaru, J.; Ashizawa, M.; Ichikawa, K. (2002) *JSME Int. J., Ser. B* 45, 518
Higginbotham, E.B., Motter, J.W., *13th EPRI Conf.: Gasification Power Plants*, San Francisco, CA (USA), October 1994.
Howard, J. B. (1981) *Chemistry of Coal Utilization Second Supplementary Volume* (ed. Elliott, M. M.), John Wiley, NY, p.665
Hulgaard, T. (1993) *AlChE J.* 39, 1342
Hulgaard, T. (1992) *Environ. Prog.* 11, 302
Hurt, R. H.; Sarofim, A. F.; Longwell, J. P. (1991) *Energy Fuels* 5, 290

Hüttinger, K. J. (1988) *NATO ASI Series. Series C: Mathematical and Physical Sciences* (ed. Yürüm, Y.), 244, 433
Hüttinger, K. J., Nattemann, C. (1994) *Fuel*, 73, 1682
Jahnke, Fred C., (1995) *Proc. American Power Conference* 57, 84
Jenkins, R.G., Nandi, S. and Walker, P.L., Jr. (1973) *Fuel* 52, 288
Kanbara, S., Takarada, T., Yamamoto, Y., Kato, K., (1993) *Energy Fuels* 7, 1013
Kilpinen, P.; Hupa, M.; Leppaelahti, J. (1991) Nitrogen Chemistry during Gasifications A Thermodynamic Analysis AAA-KTF/FKF-91/ 14 (Abo Akademi)
Kirk, D.W., Li, Z.R., Fuleki, D. and Patnaik, P.C. (2001) *Proc. ASME TURBO EXPO, June 4–7 2001, New Orleans, Louisiana*, Paper No. CBAF-2001-GT-006.
Koopmann, E.W.; Regenbogen, R.W.; Zuideveld, P.L., (1994) *VGB Kraftwerkstechnik* 74, 974
Lemaignen, L., Zhuo, Y., Reed, G. P., Dugwell, D.R. and Kandiyoti, R, (2002) *Fuel* 81, 315
Li, C.-Z.; Gaines, A. F.; Guell, A.; Kandiyoti, R. (1991) *Int. Conf. Coal Sci.* 72
Li, C.-Z.; Bartle, K. D.; Kandiyoti, R. (1993) *Fuel* 72, 3
Li, C.-Z.; Madrali, E. S.; Wu, F.; Xu, B.; Cai, H.-Y.; Guell, A. J.; Kandiyoti, R. (1994) *Fuel* 73, 851
Lim, J.-Y., Chatzakis, I.N., Megaritis, A., Cai, H.-Y., Dugwell, D.R. and Kandiyoti, R., (1997) *Fuel* 76, 1327
Lin, W. and Johnssen, J. E.(1993) *7th Int. Conf. Coal Sci.* (*Banff*) 554
Lowry, H. H (1963) Editor *Chemistry of Coal Utilization Supplementary Volume*, John Wiley, NY
Madrali, E. S. (1994) Ph.D. Thesis, University of London
Malinowski, E. R. and Howery, R. G. (1980) *Factor Analysis in Chemistry*; John Wiley, New York
Mamori, T.; Sagimori, K.; Baba, A.; Yatani, T.; Yasumi, T.; Inoue, H. (1998) (in Japanese) *Souken Houkoku 55*, 91 (ISSN: 0285-6697.)
Megaritis, A., Zhuo, Y., Messenbock, R., Dugwell, D. R., Kandiyoti, R. (1998a) *Energy & Fuels*, 12, 144
Megaritis, A., Messenbock, R. C., Collot, A.-G., Zhuo, Y., Dugwell, D. R., Kandiyoti, R. (1998b) *Fuel*, 77, 1411
Meijer, R.; Kapteijn, F.; Moulijn, J. A. (1994) *Fuel* 73, 723
Messenbock, R., (1998) PhD Thesis, University of London
Messenbock, R. C., Dugwell, D. R., Kandiyoti, R. (1999a) *Energy & Fuels*, 13, 122–130
Messenbock, R. C., Dugwell, D. R., Kandiyoti, R. (1999b) *Fuel*, 78, 781
Messenbock, R. C., Paterson, N., Dugwell, D. R., Kandiyoti, R. (2000) *Fuel*, 79, 109
Mojtahedi, W., Horrath, A., Salo, K., Gangawal, S.K., (1991) *10th EPRI Conference on Coal Gasification Power Plants*, Palo Alto, Ca., U.S.A.
Morris, J. P., Keairns, D. L. (1979) *Fuel*, 58, 465
Motter, J.W. and Higginbotham E.B. (1993) *12th EPRI Conference on Coal Gasification Power Plants*, San Francisco, Ca, U.S.A.
Muhlen, H.-J.; Van Heek, K. H.; Juntgen, H. (1985) *Fuel* 64, 944
Muhlen, H.-J.; Van Heek, K. H.; Juntgen, H. (1986) *Fuel* 65, 591
Niksa, S. J., Russel, W. B., Saville, D. A. (1982a) *Fuel, 61*, 1207
Niksa, S. J., Russel, W. B., Saville, D. A. (1982b) *19th Symp.(Int.) Combustion*, The Combustion Institute, Pittsburgh PA, 1151
Niksa, S. J., Heyd, L. E., Russel, W. B., Saville, D. A. (1984) *20th Symposium (Intl.) on Combustion*, The Combustion Institute, 1445
Nozaki, T.; Adschiri, T.; Fujimoto, K. (1991) *Energy Fuels* 5, 610
O'Brien, J. R. (1986) *PhD Thesis*, University of London
Pan, Y.G., Velo, E., Puigjaner, L.(1996) *Fuel* 75, 4
Paterson, N., Elphick, S, Dugwell, D.R. and Kandiyoti, R., (2001) *Energy and Fuels* 15, 894
Paterson, N., Zhuo, Y., Dugwell, D.R. and Kandiyoti, R., (2002) *Energy and Fuels* 16, 127

Paterson, N., Zhuo, Y., Dugwell, D. and Kandiyoti, R. (2005) *Energy and Fuels* 19, 1016
Peng, F. F.; Lee, I. C.; Yang, R.Y.K. (1995) *Fuel Process. Technol* 41, 233
Peralta, D., Li Xiaoyu, Xu Shisen, Paterson, N., Dugwell, D. R., Kandiyoti, R., (2004) Proc. "*2004 Intl. Hi-Tech Symposium on Coal Chemical Industry and Coal Conversion", 30–31 Oct. (2004), Shanghai, China*, pp. 136
Perkin-Elmer Ltd. (1991) "*QUANT +*" *User's Manual*
Pindoria, R.V., Megaritis, A., Herod, A.A. and Kandiyoti, R. (1998) *Fuel* 77, 1715
Piñon Pine Power Project near Reno, Nevada (1998) [http://www.netl.doe.gov/publications/press/1998/tl_pinon_ded.html;]
Reed, G.P., Paterson, N., Zhuo, Y., Dugwell, D.R. and Kandiyoti, R., (2005) *Energy and Fuels* 19, 298
Ren, Y., Xu, S., Xu, Y., Chen, C. and Xia, J., Proc. 2004 Intl. Hi-Tech Symposium on Coal Chemical Industry and Coal Conversion, 30–31 Oct. (2004), Shanghai, China, p. 172
Rüdiger, H., Greul, U., Spliethoff, H, Hein, K.R.G., (1994) in *APAS clean coal technology programme*, 1992–1994, Volume 3, C4
Scott S.C.; Piskorz J., (1982a) *Canad. J. Chemical Eng*. 60, 666
Scott S.C.; Piskorz J., (1982b) in "Flash pyrolysis of biomass" Chapter 23, Ann Arbor Science Pub., Ann Arbor Michigan
Scott S.C.; Piskorz J., (1984) *Canad. J. Chemical Eng*. 62, 404
Sha, X.; Chen, Y.; Cao, J.; Yang, Y.; Ren, D. (1990) *Fuel* 69, 656
Shafizadeh, F.(1982) *J. Anal. App. Pyrolysis* 3, 283
Shorter, V.R.; Marchese, M.R. (1988) *Proc. Energy Technology Conference* 15,247
Silveston, P.L. (1991) *Energy Fuels* 5, 933
Sjöström, K., Bjornbom, E., Guanxing, C., Brege, C., Rosen, C., Qizhuans, Y. (1994) in *APAS, clean coal technology programme*, 1992–1994, Volume 3, C3
Stiles, H.N. and Kandiyoti, R. (1989) *Fuel* 68, 275
Stangeby, P.C., Sears, P.L. (1981a) *Fuel*, 60, 131
Stangeby, P.C., Sears, P.L. (1981b) *Fuel*, 60, 125
Sue-A-Quan, T.A., Cheng, G., Watkinson, A.P. (1995) *Fuel*, 74(2), 159
Sue-A-Quan, T.A., Watkinson, A.P., Gaikwad, R.P., Lim, C.J., Ferris, B.R. (1991) *Fuel Proc. Tech.* 27, 67
Suuberg, E.M., Peters, W.A., Howard, J.B. (1980) *Fuel*, 59, 405
Shufen, L.; Ruizheng, S. (1994) *Fuel* 73, 413
Taupitz K.C. (1977) *Hydrocarbon Processing*, 219
Tyler, R.J. (1979) *Fuel*, 58, 680
Tyler, R.J. (1980) *Fuel*, 59, 218
Vorres, K.S. (1990) *Energy & Fuels*, 4, 420
Wang Baomin, Li Xiaoyu, Xu Shisen, Paterson, N., Dugwell, D.R. and Kandiyoti, R., *Energy and Fuels* 2005, 19, 2006.
Weisberg, S. (1985) *Applied Linear Regression*, 2nd ed.; Wiley: New York
Wu, F., Guell, A.J., Li, C-Z., Madrali, E. S., Cai, H.-Y.; Dugwell, D. R.; Kandiyoti, R. (1993) *Proc. ICCS*, Banff, 307
Yang, Y.; Watkinson, A.P. (1994) *Fuel* 73, 1786
Zhuo, Y., Messenböck, R., Collot, A.-G., Paterson, N., Dugwell D.R., Kandiyoti, R. (2000a) *Fuel*, 79, 793
Zhuo, Y., Lemaignen, L., Chatzakis, I.N., Reed, G.P., Dugwell, D.R., Kandiyoti, R. (2000b) *Energy & Fuels*, 14, 1049
Zhuo, Y., Paterson, N., Avid, B., Dugwell, D.R., Kandiyoti, R. (2002) *Energy & Fuels*, 16, 742

CHAPTER 5

Liquefaction: Thermal Breakdown in the Liquid Phase

In a book about fuel behavior and liquid hydrocarbon properties, coal liquefaction should have naturally taken center stage. This process was critical in supplementing the Central Powers' provisions of fuels and chemical feedstocks during two World Wars. It was much in evidence again as a likely process route for making transport fuels, after the oil shocks of 1973 and 1979. However, through the several cycles of price fluctuations, the price of oil has not remained sufficiently high long enough to give investors confidence that they would be able to recoup their investment [Jaffe, 1997].

At the time of writing, the nearly doubling of crude oil prices since July 2004 has not yet translated into a reconsideration of coal liquefaction as a viable option for making liquid fuels. The economics of synthetic fuels from coal and other starting materials may yet be revisited, however, if as at present, the price of a barrel of crude continues to hover above the lucrative US$55 mark. As explained in Chapter 9, the present high-price excursion is being driven by a number of structural elements. A combination of cost cutting by oil companies and OPEC's drive for high prices has largely wiped out commercial reserve oil inventories and marginal excess production capacity. The narrow balance between production and consumption levels was one important element in the recent upsurge in prices, where a succession of minor disruptions in supply have driven prices to progressively higher levels. If this price regime persists, we may expect that energy strategists will return to examine the economics of oil-from-coal projects as a potentially viable route for making transport and other fuels.

Returning to our present task, a short study of coal liquefaction will allow us to pursue the conceptual integration of sample characterization with reactor design. The approach is similar to that in the two previous chapters on Pyrolysis and Gasification. Moreover, examining liquefaction procedures will inevitably lead to a review of methods for liquid product characterization. We will need, therefore, to anticipate some of the analytical methods presented in Chapters 7 and 8, suited to the evaluation of these heavy products. Finally, the integration of product characterization with reactor design will lead us to attempt a unified understanding of successive thermally driven events that bring about thermal breakdown in pyrolysis and liquefaction. In Chapter 6, we will juxtapose data from pyrolysis and liquefaction experiments to highlight structural aspects of thermal breakdown that are common to both process pathways. We will show how this approach might impact on research in coke making, as well as on the design of liquefaction experiments.

Below, we will first outline general trends to be expected from coal liquefaction experiments. As before, we will closely observe how the interplay between reaction chemistry and reactor design can affect observations made during bench scale experiments.

5.1 Introduction: Coal Liquefaction

Limitations of space do not allow an exhaustive review of past liquefaction studies, going back over a century or more. Lowry [1963] and Elliott [1981] have provided excellent summaries of early work. A brief overview of technologies that have emerged during the last three decades and of research worldwide may be found in the 'Technology Status Report' by the UK Department of Trade and Industry [CCTP, 1999]. Kimber [1997] has reviewed the results of the British Coal Liquefaction Project. Short summaries of the process have been presented by Harrison *et al.* [1989] and Moore *et al.* [1989]. The massive amount of laboratory research carried out during the same period may be traced through successive proceedings of International Conference on Coal Science meetings. Furimsky [1998] has reviewed catalysts and reactors for hydroprocessing, while Mochida *et al.* [1998] have presented an overview of progress in coal liquefaction catalysts in Japan, where commitment to coal liquefaction technologies has been maintained longer than in Europe and the USA. A comparative analysis of costs of alternative coal liquefaction processes has recently been provided by Sun *et al.* [2005].

5.1.1 *Two stages in the solvent extraction of coals*

It is possible to distinguish between coal mass loss taking place prior to and following the onset of extensive covalent bond scission. Data from electron spin resonance spectroscopy [Fowler *et al.*, 1988, 1989a, 1989b] indicate that when coal is heated, spin concentrations of stable free radicals begin to rise from about 300–310°C and increase rapidly after about 350–375°C, providing – indirect – evidence for the onset of *extensive* bond scission at and above the latter temperature band. In view of experimental uncertainties [Fowler *et al.*, 1989a] and the likely distributions of bond strengths within complex coal structures, the 350–375°C mark must not be treated as a precise dividing line. Results presented below [e.g. see Figure 5.1] suggest that the characteristic temperature *band* may shift up or down by several tens of degrees, depending on the properties of individual coals.

Within this framework, the voluminous data on coal extraction obtained at boiling points of common solvents, such as chloroform and pyridine, may be safely viewed as reflecting dissolution in the absence of extensive covalent bond scission. Samples with good plastic properties normally release large amounts of extractable material (*e.g.* Pittsburgh No.8) while at the two extremes, lignites, sub-bituminous coals, semi-anthracites and anthracites give smaller extract yields. Extraction at lower temperatures may be carried out with several different objectives in mind: (i) to probe the relationship between solvent extractables content and the liquefaction conversions or pyrolysis tar yields of coals, (ii) to investigate relationships between amounts and structures of extractables and the coking behavior of coals, and (iii) to study the structures of coals and coal derived materials [*e.g.* cf. Lowry, 1963; Elliott, 1981].

A useful starting point for describing the solvent extraction of coals at room temperature is the work of Iino *et al.* [1988]. These researchers contacted an array of coals with a 1:1 mixture of CS_2 and 1-methyl-2-pyrrolidinone (NMP). Large extraction yields (30–66%) were reported for 29 of the 49 bituminous coals examined, with elemental C-contents ranging between 76.9 and 90.6% (daf basis). Extract yields were reported to trace a maximum for middle rank bituminous coals. The mixture of NMP with CS_2 has provided the strongest solvent combination yet, for extracting coals at room temperature. When temperatures are raised to levels *below* the onset of covalent bond rupture (below ~310–350°C), increased amounts of extract may be obtained, but a large proportion of the coal remains in solid form. In the past, several other solvents have been used for similar purposes. Methanol, chloroform, tetrahydrofuran (THF) and pyridine are among the most frequently mentioned.

During straightforward liquefaction experiments, the dissolution of products at temperatures below the characteristic band around 350°C may also be reasonably considered as taking place prior to extensive covalent bond scission. The proportions of coal extracted at temperatures up to about 350°C are not inconsiderable. In tetralin (tetrahydronaphthalene), a set of selected coals released extract that represented between 30 and 58% of the original mass (% w/w daf coal). These experiments were conducted by heating at 5°C s^{-1} to 350°C with 1600-seconds holding, in the "flowing solvent reactor" described below [Xu & Kandiyoti, 1996].

The foregoing does not pretend to contribute to debates on the description of coal structure in terms of a mobile and a macromolecular phase. Cumulative extract yields at temperatures up to 350°C change significantly depending on the nature of the solvent. In the flowing-solvent reactor, with a 400 s hold at 350°C, the weight loss from Point of Ayr coal changed from 24.6% in tetralin to 39.5% in quinoline [Xu *et al.*, 1994]. Multi-step extraction [Nishioka, 1991] or the use of more powerful solvents, such as NMP [Takanohashi & Iino, 1990] is likely to give higher extract yields prior to the onset of extensive covalent bond scission. It does not therefore appear possible to distinguish clearly between presumed distinct phases in coals.

Returning to liquefaction, once temperatures reach levels where extensive covalent bond cleavage is initiated, the remaining solid mass begins to break down into solvent soluble material. Evidence suggests that, unless hydrogen is quickly supplied to materials freshly released from the solid matrix, retrogressive recombination reactions may take place. The liquid products would then begin to increase in molecular mass and viscosity. In the laboratory, hydrogen donor solvents such as tetralin have often been used to quench free radicals and stabilize liquefaction products. At pilot or plant level, hydrogenated recycle liquids have been used for the same purpose. It is important to note that donor-solvents may not be the strongest solvents for coal-derived materials and strong solvents for coal-derived materials may not necessarily be good hydrogen donors. We will return to this point below (Table 5.3).

Figure 5.1 shows that between 30–40% of the mass of Point of Ayr (UK) and Pittsburgh No. 8 (US) coals may be extracted in tetralin, at temperatures up to about 350°C. For Point of Ayr coal, the amount of liquefaction product released rose sharply from about 360–375°C. This temperature range corresponds therefore to the onset of extensive depolymerization of the coal. In the case of Pittsburgh No.8 coal (Figure 5.1b), a sharp

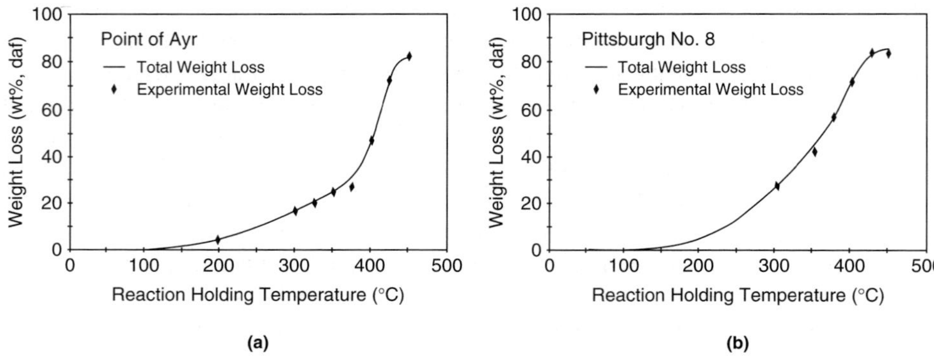

Figure 5.1 Sample weight loss from Point of Ayr and Pittsburgh No. 8 coals as a function of temperature in the flowing solvent reactor. Samples were heated at 5°C s^{-1} to 450°C with 400 s hold. Tetralin flow rate: 0.9 ml/s at 70 bar. (Reproduced with permission: Energy & Fuels 1996, *10*, 1115; Copyright 1996 Am.Chem.Soc.)

transition was not observed. As will be explained in the next Chapter, covalent bond cleavage reactions are thought to begin just above 300°C and intensify around 350°C. For some coals, however, the transitional temperature band for thermal breakdown may occur at higher temperatures (375–400°C) and in the unusual case of 'K-9 coal' described in Chapter 6, at about 425–450°C [Fukuda *et al.*, 2004].

The data in Figure 5.1 show, moreover, that up to 85–90% of the mass of suitable coals may be dissolved by the application of heat in the presence of hydrogen donor solvents. These extracts may be recovered in solution, but are not immediately usable. They are normally viscous and consist of large molecular mass materials that need to be hydrocracked and refined before use. Moreover, in pilot or industrial scale oil-from-coal operations, liquefaction of the coal is merely a first step. Reduction in molecular mass and of heteroatom content normally requires severe reaction conditions, with attendant high costs for producing saleable fuels and chemical feedstocks. Process stages designed for upgrading primary coal extracts thus tend to require a larger proportion of the total effort and expense. During pilot-plant tests at the British Coal Liquefaction Facility, hydrogen costs alone were calculated to make up as much as ~25% of total operating costs [Kimber, 1997].

5.1.2 A preliminary comparison of pyrolysis and liquefaction

When coals are heated to about 300°C in the absence of a solvent, small amounts of light gases are released, including hydrogen, methane, ethane, H_2S and possibly small amounts of light hydrocarbons [e.g. see Neuburg *et al.*, 1987]. At these temperatures, some of the potentially extractable material becomes mobile within the coal mass, although, in the absence of solvent, no mechanism exists for its removal, apart from small fractions that can evaporate below ~300°C.

When the temperature is raised above the 310–350°C threshold and covalent bond cleavage reactions set in, the underlying sequence of events during "dry" heating is not unlike that encountered in liquefaction. An array of free radicals is released within the particle, with a broad spread of molecular masses and chemical reactivities. If the coal mass softens, the viscous plastic phase that forms is thought to act as a hydrogen donor and exchange medium. However, left in close proximity (and in the absence of a donor solvent), the more reactive free radicals are likely to enter into retrogressive recombination reactions, forming fresh char, unless they are quenched by hydrogen released from within the coal mass itself. Small alkyl free radicals detached from aromatic and other structures may also perform a function similar to dissociated hydrogen in quenching reactive free radicals *or* they may evolve as light gases by scavenging hydrogen themselves.

As the temperature is raised further, some of the plastic material cracks, releasing tars and leaving a residue that eventually reverts to secondary char. We will present data in Chapter 6, showing that at or near 400°C, the solvent soluble plastic phase may persist for, at least, as long as 200 seconds. However, the re-formation of solids does not require extreme conditions and tends to reduce the amount of potentially volatile (or potentially soluble) coal derived material. Above 450°C, in the absence of solvent, the residual soluble materials within solid particles solidify rapidly. This is why much higher proportions of char are observed during the pyrolysis of a coal sample, compared to liquefaction in a donor solvent.

At bench scale, it is possible to delineate the lower-temperature "extraction" phase from "liquefaction" (depolymerization at higher temperatures) with reasonable clarity. It is also possible to observe when donor solvents quench free radicals. However, such observations require careful selection of reaction parameters and reactor design.

5.2 The British Coal Liquefaction Process

The investigation of coal liquefaction fundamentals outlined in this chapter were carried out with the British Coal Liquefaction Process as background, during a period when a 2.5 ton-per-day pilot plant facility was constructed, commissioned and operated at Point of Ayr, near Prestatyn in North Wales (UK). A brief description of the process, as laid out at the site would be useful for putting the parallel bench-scale research work in context.

The coal was slurried with a hydrogen-donor recycle solvent, pressurized to 20 bar, preheated to 410°C and fed to a digester with an average residence time of 1 hour. Up to 95% of the coal was dissolved. An "adequate" concentration of hydrogen donors in the recycle solvent insured a high degree of dissolution of the coal, without using externally added high-pressure hydrogen. The donor content of the recycle solvent could be "trimmed" by the use of saturates cracking or mild hydrogenation, in case of "over or under hydrogenation", respectively. "An excellent monitor of solvent quality during recycling is the solvent dissolving (the sample coal) . . . via a bomb test, picks up trends *before* (original authors' emphasis) plant performance is affected . . ." [Harrison, *et al.*, 1989].

After digestion, the product stream was cooled to 300°C and filtered before entering the ebullating bed catalytic hydrocrackers (210bar; ~420°C). The hot-filtration stage developed and successfully tested by British Coal was considered key to the technical and economic viability of the project [Kimber & Davies, 1988; Kimber, 1989]. The product stream from this stage was distilled to give three main classes of products: (i) propane and butane (approximating LPG), (ii) a naphtha fraction, boiling below 180°C, and (iii) 180–300°C boiling middle distillates. Some of the solvent was recovered and a by-product pitch stream boiling above 500°C was partly taken off but mostly recycled into the digestion stage. The process originally designed for a narrow rank range of British coals has been successfully tested with brown coals and lignites [Kimber, 1997].

It is also worth describing how other operational coal liquefaction processes relate to the British Coal "LSE" process. The Catalytic Two-Step Liquefaction (CTSL) process, supported by the US Department of Energy, is a leading US technology (Comolli & Zhou, 2000) and differs from the LSE process in that coal is reacted with process-derived recycle solvent and hydrogen in the presence of a supported Ni-Mo catalyst in two coupled reactors. The Ruhrkohle AG/VEBA OEL AG Kohleoel process similarly contacts coal with catalyst (here a disposable 'red mud' iron catalyst) in the presence of a recycle solvent and hydrogen, in a process based on one used commercially up to 1945 (CCTP, 1999). The 150-ton per day Japanese NEDOL process (Onozaki *et al.*, 2000) also involves mixing coal with solvent, hydrogen and a catalyst (here a synthetic iron-based catalyst); as in the LSE process, part of the reaction product is recycled as coal solvent after hydrogenation.

In this context, the simultaneous upgrading of coal and petroleum heavy ends ('co-processing' – see, for example, Bartle *et al.*, 1994), would seem attractive, at least in principle, since feedstocks could be varied in response to price trends, while favorable synergistic effects might operate. In fact, few convincing co-processing schemes have been proposed, largely because of the differing chemical natures of coal and petroleum resids, which necessitate markedly different processing conditions. The generally low concentrations of hydrogen donors in petroleum also militate against any synergy during co-processing.

5.3 On the Design of Bench-Scale Liquefaction Experiments

As in the case of pyrolysis and gasification, we will begin this chapter by an evaluation of conventional experimental techniques. This will lead us to a reactor design more amenable to showing the behavior of coals during thermal breakdown.

During the liquefaction process, materials extracted by the solvent and those detached from the solid coal matrix by the rupture of covalent bonds are released into the liquid phase. In the laboratory, coal liquefaction experiments are usually carried out in small batch reactors. These are normally small closed autoclaves or "bomb" reactors [e.g. see Clarke *et al.*, 1980]. Conventional autoclaves operating at several degrees per minute are of limited usefulness in determining rates of coal dissolution under liquefaction conditions, principally because the long heat-up and cool-down times imply that the substrate spends significant lengths of additional time in the vicinity of the peak experimental temperature. This creates uncertainties about the "reaction-time at temperature" measurement. The introduction of micro-bomb reactors has contributed considerably to the improvement of time resolution in liquefaction studies [Gorin, 1981]. However, micro-bombs are usually heated by immersion in a heated sand fluidized-bed. In this mode of heating, faster *initial* heating rates are observed when the reactor is immersed in a sand bed set at a higher temperature. The inevitable dependence of the heating rate on the target temperature gives rise to distinct time-temperature histories for every target temperature. On the other hand, accurate determinations of time-temperature histories of reactants and products are necessary for calculating kinetic constants. It is therefore desirable to perform liquefaction experiments in a manner that allows the ready and relatively reliable incorporation of sample time-temperature histories into the kinetic calculations. We will introduce one such scheme further on in this Chapter.

Another major problem with conventional batch (closed) liquefaction reactors concerns the fate of the products that are released into solution during successive stages of the process. In batch (closed) reactors, *all* reactor contents remain exposed to reaction conditions during the entire length of the digestion period, including unreacted coal and solvent as well as products freshly released from the coal. The digestion may last anywhere from 5–10 to 120 minutes. This procedure provides ample opportunity for liquefaction products to enter into secondary reactions with any of the materials in their environment. The range of products expected from the secondary reactions of primary coal dissolution products depend largely on the temperature and the local availability of hydrogen. They can form predominantly heavier products and char, or lighter liquid products and gas, the cracking possibly being aided by catalytically active solids. The sequence of reactive events within the closed reactor is too complicated to unravel once the scrambling of reactants and products has been allowed to take place. The mixture recovered from a batch reactor contains products of whatever secondary reactions have occurred, and perhaps some primary extraction products of the coal as well.

An investigation of the actual behavior of fuel particles during liquefaction requires that we obtain conversion data and information on extract compositions, structures and reactivities, in a manner that is free of the effects of secondary reactions. A measure of clarity in these experiments could therefore be achieved by distinguishing between coal mass loss and the subsequent reactions of coal extracts. This necessitates the removal of extracts from the reaction zone, as soon as they are released into solution. In this respect, the batch reactor configuration is entirely unsatisfactory. The challenge is not unlike the one faced in designing pyrolysis experiments, where it was desired to recover primary products not affected by secondary reactions and other reactor related effects. To this end, variants of the wire-mesh and of the fixed-bed "hot-rod" reactors described above have been tested in liquefaction mode.

5.3.1 The wire-mesh liquefaction reactor

Initially, constructing a wire-mesh reactor for liquefaction experiments appeared promising. The coal sample was loaded onto the mesh in a high-pressure wire-mesh chamber. Thermocouples were fastened onto the mesh as described in Chapters 3 and 4. The reactor was then filled with tetralin and the heating sequence triggered with a variable holding period at temperatures between 350 and 450°C. However, from the first test, sample particles showed evidence of swelling and coking. These were events expected at temperatures far higher than those measured by the thermocouples attached to the mesh. Discoloration of the mesh in the vicinity of coal particles also suggested that higher temperatures than those intended (~350–450°C, max.) had been reached by sample particles. An explanation was needed for the observed overheating.

When the horizontally positioned wire-mesh is heated, the solvent (in this case, tetralin) in the immediate vicinity of the mesh picks up heat and expands. Natural convection currents are thus initiated within the reactor. The solvent initially below the mesh sweeps through it, removing some dissolution products. The initially cool liquid removes vastly more energy from the mesh than was the case during pyrolysis in a gaseous atmosphere. This forces the control system to deliver large electrical currents, to keep pace with the pre-programed time-temperature ramp of the sample holder. It eventually became clear that the points where heat was not withdrawn from the mesh were those where solvent flow was blocked by the presence of sample particles. The resulting local overheating could not be suppressed even when the total sample was reduced to a single coal particle. This precluded the use of a wire-mesh type reactor for coal liquefaction. The attempt had to be abandoned.

5.3.2 The design of the "flowing solvent reactor"

The flowing solvent reactor described below was developed during the same campaign, by adopting a fixed-bed "hot-rod" configuration. However, several modifications of the original design described in Chapter 3 became necessary. A separate solvent preheating zone had to be worked into the design, as solvents sweeping through the reactor have higher heat capacities than the gas stream in the comparable pyrolysis/gasification reactor. Furthermore, the surge of released product occurring around 390–425°C necessitated dilution of the fixed bed of sample with sand, to avoid reactor blockages. The latter was not dissimilar to the arrangement Kershaw and Barras [1979] used in their hot-rod reactor constructed for hydropyrolysis experiments.

We have seen in Chapter 3 that the use of a continuous stream of gas to sweep products out of the reaction zone was a well-known feature of fixed-bed ("hot-rod") hydropyrolysis reactors. The concept appears to have found only limited application in liquefaction studies. The "flowing-solvent" reactor consists of a high-pressure tube within which a sample of coal is mounted in the form of a fixed-bed and swept by a stream of pressurized solvent. In liquefaction mode, this configuration has several advantages. Extraction and liquefaction products can be removed from the reaction zone within a matter of several seconds, avoiding the complicated mixtures formed during long digestions. Second, by pushing an excess of solvent through the bed of particles, it is possible to extract the coal particles against a virtually zero external product concentration in the solvent. This has the advantage of nearly standardizing extraction conditions and speeding up the extraction process. Furthermore, the use of a hydrogen-donor solvent is useful in blocking retrogressive secondary reactions of products released by covalent bond rupture. In other words, the major drawback of fixed bed reactors, when used in the pyrolysis mode (i.e. secondary reactions), may be by-passed to good effect when the reactor is swept with a solvent that is a hydrogen-donor.

There were several antecedents of the reactor constructed at Imperial College. Koll and Metzger [1978] installed a fixed bed reactor in a gas-chromatographic oven and swept it with a stream of supercritical acetone to remove the degradation products of cellulose and chitin. Squires and co-workers [Squires *et al.*, 1983; Aida *et al.*, 1985; Slomka *et al.*, 1985; 1986] used a similar reactor configuration, mostly for studying the supercritical extraction of coals with benzene and other light hydrocarbons. Interestingly, they attached an UV-absorption detector at the exit of the reactor to monitor changes in product composition. McPherson *et al.* [1985a; 1985b] used a fixed-bed reactor immersed in a sand-bath for heating. The coal bed mounted inside the reactor was swept with tetralin. These researchers reported on the morphology of solid residues and the reactions of tetralin during coal liquefaction.

The reactors mentioned in relation to these studies have all used various types of external heating: a GC oven and a sand fluidized bed have already been mentioned. Due to their high thermal inertias, these systems can only apply limited ranges of heating or cooling. In the work of Squires and co-workers, even with supplementary heating from electrical heating tape wound around the reactor, peak heating rates were limited to a maximum of about 2.5°C s^{-1}.

Not surprisingly, the flowing solvent reactor constructed at Imperial College was similar in conception to the "hot-rod" reactor constructed for pyrolysis and hydropyrolysis experiments (Figure 5.2). The basic device was a tubular reactor with an (empty) preheating zone making up the bottom section, while a fixed bed of coal was mounted above this zone, between two porous wire-mesh plugs. Both sections were heated directly, by

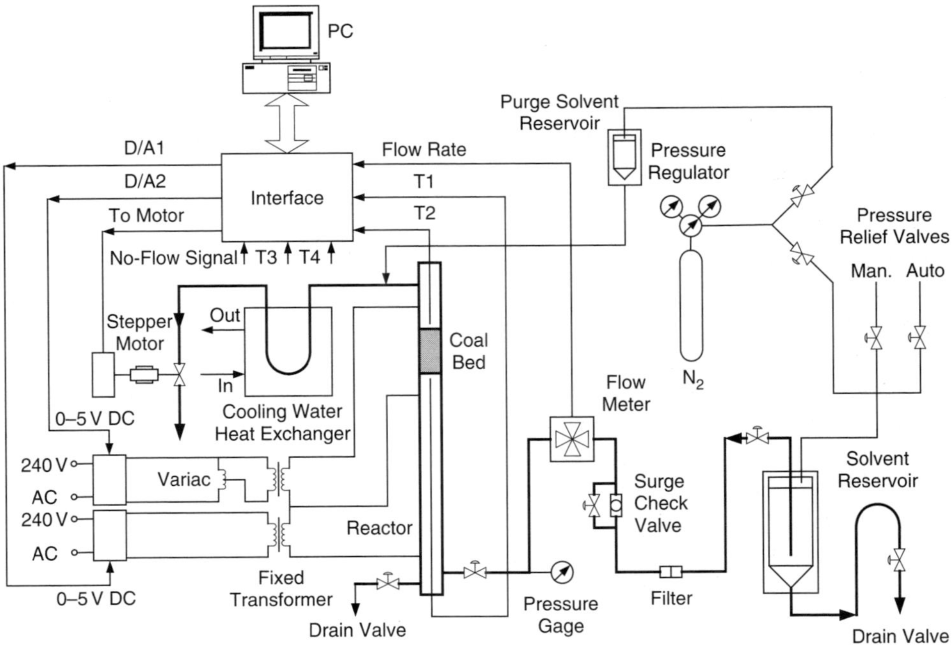

Figure 5.2 Schematic diagram of the flowing solvent reactor system. The thick line from the reservoir traces the solvent flow path. The solvent is forced through the fixed coal-bed and sweeps dissolved product into the heat exchanger. The letdown valve is attached to a computer-controlled stepper motor and serves to control the flow rate. (Reproduced with permission: Energy & Fuels 1996, *10*, 1115; Copyright 1996 Am.Chem.Soc.) to [Xu, 1995].

clamping copper electrodes and passing an electrical current through the reactor walls. With the thermal inertia of the system thus kept to a minimum, sample heating-rates well in excess of 10°C s^{-1} can be achieved. The temperature-time histories of the samples were pre-programed and controlled by an on-line computer. The original apparatus described by Gibbins & Kandiyoti [1990, 1991a, 1991b] was subsequently revised and equipped with a second-generation computer control system [Xu *et al.*, 1995; Xu & Kandiyoti, 1996].

Figure 5.2 presents a schematic diagram of the flowing-solvent reactor system designed and constructed at Imperial College. The reactor tube made of Haynes 230 alloy (Ni with 22% Cr, 14% W, 2% Mo, 3% Fe, 5% Co) had an outside diameter of 3/8-inch (0.95 cm), with 0.036-inch (0.091 cm) wall thickness and was capable of withstanding the intended pressure of 70 bar at up to 1,000°C. It was packed with a mixture of ~200 mg coal and 2,800 mg acid-washed sand (106–150 μm), mounted as an (approx.) 50 mm deep bed between wire-mesh plugs. The temperatures were measured by 1 mm diameter, mineral insulated sheathed thermocouples inside the reactor, positioned immediately above and below the bed.

The use of pumps was avoided, to maintain a steady flow of liquid. Instead, solvent was forced from a pressurized reservoir (0.9 ml s^{-1} at 70 bar) through a surge check-valve and flow meter, into the lower section of the reactor tube, which served as the solvent pre-heater stage. Power to both sections of the reactor was supplied by direct electrical heating and controlled separately. Temperature control in the pre-heating (lower) section was set for delivering solvent to the reactor (upper) section, at the temperature of the latter. Samples were usually heated at 5°C s^{-1} to the intended peak temperature and held for between 100–1,600 seconds. Figure 5.3 presents a typical time-temperature history, going through the heatup ramp, a 400 second holding period and a cool-down period. The power-input vs. time traces for the two reactor stages have been plotted

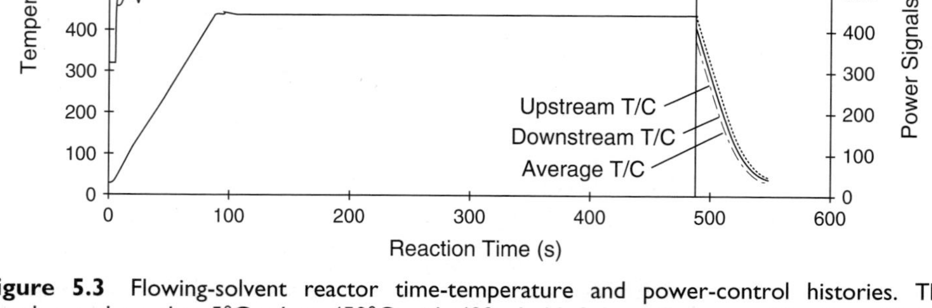

Figure 5.3 Flowing-solvent reactor time-temperature and power-control histories. The sample was heated at 5°C s-1 to 450°C with 400 s hold. Solvent flow rate: 0.9 ml/s at 70 bar. (Reproduced with permission: Energy & Fuels 1996, *10*, 1115; Copyright 1996 Am.Chem.Soc.)

against arbitrary units. At the end of an experiment, the reactor was rapidly cooled by the passage of cold solvent. Details of the heating and temperature control system, including the designs of the purpose built electronic hardware components have been described by Xu *et al.* [1995].

The reactor was operated without a gaseous medium (*e.g.* hydrogen) other than the N_2 used for pressurizing the liquid reservoir. A large excess of solvent (600–1,000 ml) is passed through the sample bed during a run, where no more than ~200 mg of coal is reacted. Products released by the coal are diluted by solvent and the mixture rapidly swept out of the reaction zone into a cooler-heat exchanger unit. The estimated time between the dissolution of product and its arrival in the cooling zone was estimated at between 6 and 10 seconds.

At the exit of the reactor, the product stream was cooled in the heat exchanger to near ambient temperature. This caused coal extracts to precipitate out of solution, which tended to block the letdown valve and slow-down the solvent flow-rate. To counter this effect, the computer-controlled pressure letdown valve at the exit of the reactor was pre-set to discharge at a fixed solvent flow-rate. The control system was designed to open the valve until the blockage was removed and then to restore the flow-rate to pre-set levels. The repeatability of sample weight-loss measurements was usually better than ±1.5%. Sample weight loss was not found to be sensitive to pressure changes between 50–100 bars and solvent flow rate changes between 0.9–2.4 ml s^{-1} [Gibbins & Kandiyoti, 1990, 1991b; Xu & Kandiyoti, 1996].

Had it worked as intended, the "liquefaction wire-mesh" mentioned above would have allowed the measurement of changes in conversion over a wide-range of heating rates. The thermal inertia of the flowing-solvent reactor system did not allow such wide variations in heating rates to be applied. In this respect, the system was similar to the fixed-bed ("hot-rod") reactor. Nevertheless, some variation in heating rate was possible. It was found that changes between 0.3–10°C s^{-1} had no effect on coal conversion, when operating with tetralin as the liquefaction medium [Gibbins & Kandiyoti, 1990]. In Chapter 6, we will return to the observed lack of sensitivity of coals to heating rates during liquefaction.

5.3.3 Liquefaction trends in the flowing solvent reactor

Figure 5.1 show the type of conversion vs. temperature data that are normally obtained with the flowing-solvent reactor. Each data point on the diagram represents the weight-loss from a single experiment (see Figure 5.3). A total of seven samples were drawn from the Argonne Premium Coal Sample (APCS) Program [Vorres, 1990] and one from the UK, a Point of Ayr coal sample from the Point of Ayr Coal Liquefaction Project (Table 5.1).

Table 5.2 presents experimental coal weight loss data as a function of temperature. Comparing results between 350°C (1600 s hold) and 375°C (400 s hold), all samples showed significant increases in weight loss over the interval, except for Point of Ayr (U.K.), Upper Freeport, and Pocahontas No. 3 coals. The data suggest that for these three coals, the transition temperature, T_d, between simple extraction and the depolymerization stage may be somewhat higher than 350°C. For these three coals, 375°C

Table 5.1 Elemental Analysis of the Argonne (APCS) Coals [Vorres, 1990] and Point of Ayr (U.K.) Coal [Xu, 1995]

Cool Sample	*C (daf)*	*H (daf)*	*N (daf)*	*S (daf)*	*O (by diff)*	*ash (dry basis)*
Beulah-Zap	72.9	4.8	1.2	0.7	20.4	9.7
Wyodak Anderson	75.0	5.4	1.1	0.5	18.0	8.8
Illinois No. 6	77.7	5.0	1.4	2.4	13.5	15.5
Blind Canyon	80.7	5.8	1.6	0.4	11.5	4.7
Pittsburgh No. 8	83.2	5.3	1.6	0.9	9.0	9.3
Upper Freeport	85.5	4.7	1.6	0.7	7.5	13.2
Pocahontas No. 3	91.0	4.4	1.3	0.5	2.8	4.7
Point of Ayr (U.K.)	84.5	5.4	1.8	1.5	6.8	9.6

Table 5.2 Primary liquefaction yields from eight coals in the flowing-solvent reactor. Samples were heated in tetralin at 5°C s^{-1}, to the holding temperature. Solvent flow rate: 0.9 ml s^{-1} at 70 bar (g). (Reproduced with permission: Energy & Fuels 1996, *10*, 1115; Copyright 1996 Am.Chem.Soc.)

Holding temperature (°C)	300	350	**350**	375	400	425	450	**450**
Holding time (s)	400 s	400 s	**1600 s**	400 s	400 s	400 s	400 s	**1600 s**
Coal sample	Weight loss (w/w, %, daf)							
Beulah-Zap	17.4	31.1	**40.6**	46.7	55.2	68.3	75.0	**84.0**
Wyodak-Anderson	25.0	35.0	**38.6**	45.2	62.0	73.1	82.0	**88.6**
Illinois No. 6	24.6	47.3	**54.7**	61.6	79.8	89.0	90.0	**94.7**
Blind Canyon	22.9	35.4	**37.9**	48.6	73.0	84.4	91.0	**92.0**
Pittsburgh No. 8	26.6	41.5	**55.2**	56.4	71.6	84.7	84.2	**89.0**
Point of Ayr (UK)	17.0	24.6	**36.0**	27.0	47.0	72.5	82.5	**84.0**
Upper Freeport	39.7	51.9	**57.9**	58.7	67.0	75.0	81.8	**86.0**
Pocahontas No. 3	9.0	24.0	**29.8**	27.3	32.7	39.8	43.5	**70.0**

was used as T_d, to delineate Stage A (extraction) and stage B (depolymerization) in the kinetic models, which we will outline below.

In Table 5.2, the large conversions of Upper Freeport coal at relatively low temperatures are consistent with the 59.4% extraction yield of this coal in a mixture of CS_2 and 1-methyl-2-pyrrolidinone at room temperature [Takanohashi & Iino, 1990]. In that series of experiments, the next largest extraction yield was reported as 39.2% for Pittsburgh No. 8 coal. At 450°C, the conversions of most coals for 400-second holding were close to the "ultimate" conversion obtained at the same temperature with 1600-second holding. Only the conversion of Pocahontas No. 3 coal increased quite substantially during holding between 400 and 1600 s, at 450°C.

5.4 Comparing Liquefaction in the Flowing-Solvent Reactor and a "Mini-Bomb"

The perceived advantages of the flowing-solvent reactor may be restated as the relative absence of retrogressive reactions leading to the largely intact removal of primary liquefaction products from the reaction zone. During the initial phase of the work on the flowing-solvent reactor, it was also anticipated that the new reactor would possibly give higher conversions and in any case show what effect changes in heating rates would have on coal behavior. As explained in Section 5.3, initial tests showed no differences in conversions for changes in heating rates between 0.3–10°C s^{-1}, when operating with tetralin as the liquefaction medium [Gibbins & Kandiyoti, 1990]. The proposition that higher conversions might be obtained was put to the test by parallel experiments in a batch reactor, to compare conversions and liquid product properties. The extractions were carried out using tetralin and 1-methylnaphthalene, the former a known hydrogen-donor and the latter, a good solvent but at best a poor hydrogen-donor [Gibbins *et al.*, 1991c].

The parallel "mini-bomb" reactor experiments were carried out at the Point of Ayr Coal Liquefaction Facility, using part of a common sample of Point of Ayr coal. A solvent to coal ratio of 4:1 was used in experiments between 385 and 460°C. The bombs were sealed under nitrogen at atmospheric pressure (cold). A heat up period of three minutes in the sand fluidized-bed was allowed for reaching the target temperature, where the bombs were held for contact times of 100, 400 or 1600 seconds. Following the digestions, contents of the mini-bombs were Soxhlet extracted with THF. Quinoline was used as the extraction solvent for a second set of runs for comparison.

5.4.1 Comparison of conversions in tetralin

Figure 5.1 shows the coal conversion trends observed in the "flowing-solvent" reactor as a function of increasing peak temperature. Substantial amounts of products were released during heatup, i.e. at 0-second contact time [see Figure 5.1 in Gibbins *et al.*, 1991c]. Results from the two reactors showed similar trends, although, at first glance, conversions in the flowing-solvent reactor appeared larger by about 10%. At the shortest contact time studied (100 s), Figure 5.4 shows that differences between conversions in the two reactors (solid lines) were systematic over the temperature range. However, conversions in the flowing-solvent reactor were then recalculated on the same basis as in the mini-bomb experiments.

On the basis of counting cold tetrahydrofuran solubles as product and considering hot-THF solubles as unconverted material, differences between conversions in the two reactors diminished significantly. The corrected results from the two reactors were quite similar, once allowance was made for minor variations between procedures employed in product isolation. These experiments have been described in detail by Gibbins *et al.* [1991c].

When the mini-bomb reactor contents were extracted with quinoline for 1 hour, instead of with cold THF, apparent conversions increased by about 10–20% and were less sensitive to the reaction temperature. This particular washing procedure was not

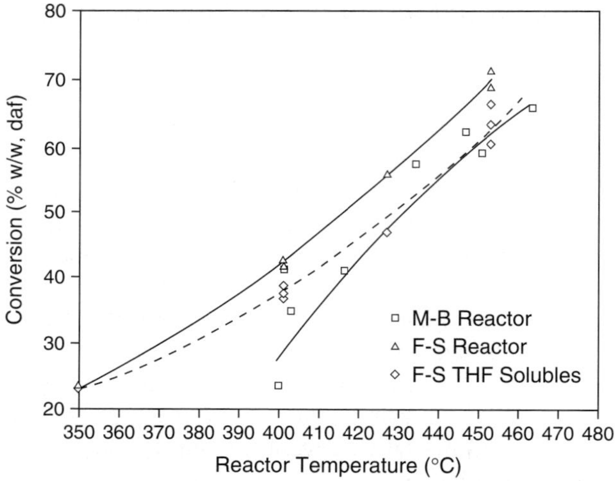

Figure 5.4 Comparison between overall conversions in tetralin, in the flowing solvent and mini-bomb reactors for 100 s holding at peak temperature. When procedures for calculating conversions were brought in line, differences between conversions in the two reactors were relatively minor. M-B: mini-bomb reactor; F-S: flowing solvent reactor; F-S THF solubles (broken line – – –): the recalculated conversion after the conversion was defined as cold-THF solubles as in the case of the "mini-bomb" experiments (Reproduced with permission: Fuel 1991, *70*, 380; Copyright 1991 Elsevier).

relevant to the pilot-plant operating conditions. It seems useful to note, however, that the selection of the washing procedure for residual solids actually determines what is meant by "conversion" for any given set of coal liquefaction experiments.

5.4.2 Comparison of conversions in 1-methylnaphthalene

We have seen that when using a tetralin-to-coal ratio of about 4:1, there are only minor differences between conversions observed in the "flowing-solvent" reactor and a small batch reactor. The next set of experiments was carried out to compare product distributions from the two reactors, when an essentially non-donor solvent was used as the liquid medium. Figure 5.5 compares conversions from the two reactors, using tetralin and 1-methylnaphthalene at 400 and 450°C. In the "flowing-solvent" reactor, conversions in 1-methyl naphthalene clearly increased with contact time at both 400 and 450°C. The trends broadly reflected those observed with tetralin, although conversions were lower by about 10–12% (Figure 5.5a).

On the other hand, Figure 5.5b shows major departures from trends observed when using tetralin, during runs with 1-methylnaphthalene in the "mini-bomb" reactor. Conversions at 400°C were lower by over 20% when compared with conversions in tetralin. They were also considerably lower than values from flowing-solvent reactor experiments using 1-methylnaphthalene. Furthermore, at 450°C, Figure 5.5b shows clear evidence for retrogressive char forming reactions in the mini-bomb. At contact times longer than 100 s, conversions diminished in the presence of 1-methylnaphthalene.

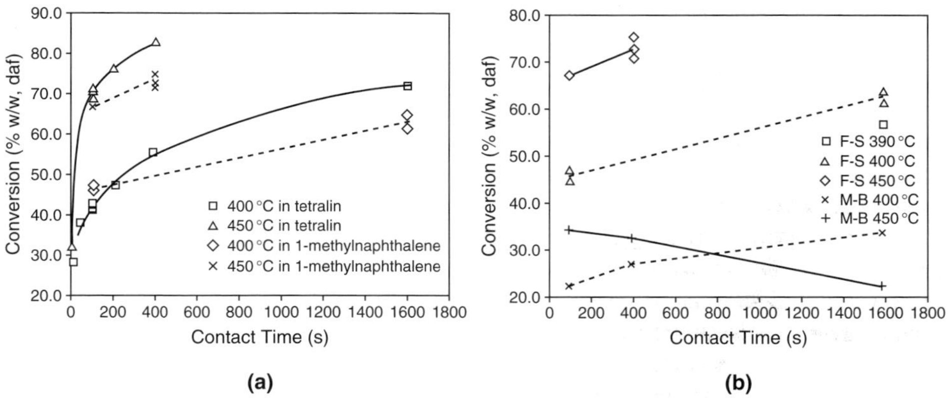

Figure 5.5 **(a)** Effect of solvent type on conversions in the flowing-solvent reactor. Heating at 5°C s^{-1}; solvent flow rate: 0.9 ml/s at 70 bar. **(b)** Flowing Solvent Reactor & Mini-Bomb reactor comparison using 1-methylnaphthalene as the liquefaction medium; solvent/coal ratio in mini-bomb reactor: 4/1 by weight (Reproduced with permission: Fuel 1991, *70*, 380; Copyright 1991 Elsevier).

Shin *et al.* [1989] have also reported small but identifiable increases in solid residue with increasing hold time, for three out of five Argonne coals liquefied in a micro-bomb using 1-methylnaphthalene as the liquid medium. The latter experiments had been done under hydrogen pressure, confirming the expected relative inertness of 1-methylnaphthalene regarding hydrogen transfer reactions.

The findings from the mini-bomb reactor were consistent with trends observed during earlier work in a larger batch "maxi-bomb" reactor [Clarke *et al.*, 1980]. While operating with a solvent-to-coal ratio of 4:1, extraction yields for Annesley (low rank UK bituminous) coal were observed to drop from 86% in tetralin to 48% in 1-methylnaphthalene. Experiments were also carried out with phenanthrene (3:1 solvent-to-coal), another non-hydrogen donor, but a good solvent for coal-derived materials. Sample weight loss for Annesley and Beynon coals initially increased, going through a maximum between 10–20 minutes contact time and then sharply declined.

Discussion of higher conversions in the "flowing-solvent" reactor: The contrast between results from the two reactors in 1-methylnaphthalene, at similar temperatures and contact times requires explanation.

In the flowing-solvent reactor, it is possible that impurities in 1-methylnaphthalene (~98% purity) could have been donating hydrogen to the coal. Due to the larger volumes of solvent used in this reactor – compared to the "mini-bomb" – a small concentration of donor-material passing through the reactor could indeed make a large difference. However, gas chromatographic analysis did not show any tetralin present at or above the detection limit of the instrument. No other donor materials were identified. The dimerization of 1-methylnaphthalene has also been offered as a possibility for hydrogen release, although dimerization products have not been identified in the reaction mixtures [Gibbins *et al.*, 1991c].

On purely reactor design considerations, the difference between data from the two reactors can be explained in terms of dilution in the "flowing-solvent" reactor. Nearly 1-litre of solvent is passed over the ~200 mg fixed-bed of sample during a single run. Fresh solvent thus continuously swamps the sample. Any dissolved coal derived species would be diluted by excess solvent sweeping through the fixed-bed, removing the products within ~10 s from the reaction zone into the cooling section. Clearly, in the absence of externally (i.e. solvent) supplied hydrogen, some 10% of the coal mass *does* undergo retrogressive reactions, compared to extraction in tetralin (Figure 5.5a). This appears due to fast retrogressive char-forming reactions taking place within the coal particles, i.e. before dilution by the solvent. Data for other coals showed similar small drops in conversion in the flowing-solvent reactor, when extracted in 1-methylnaphthalene for 100 sec hold time, compared to operation in tetralin. A 10% drop in conversion with Chriemhilt (Germany) coal was the largest drop observed among some 10 other coals tested [Gibbins *et al.*, 1994].

Based on these observations, coal liquefaction in the flowing-solvent reactor may be described, principally, in terms of two successive stages:

(i) Extraction of material soluble in the particular solvent, at up to the transitional temperature band prior to extensive depolymerization, followed by dilution and transport of product molecules out of the reaction zone,
(ii) Extraction of material released by the coal matrix during extensive depolymerization. Much of the product appears to be diluted and transported out of the reaction zone. However, in the case of non-donor solvents (e.g. 1-methylnaphthalene) some 10% of the product appears to enter into retrogressive reactions, apparently within the parent coal particles.

These product release data do not allow estimating the temperature for the onset of covalent bond scission, likely to take place at lower temperatures, but provide clear indications for the onset of generalized product release due to extensive bond rupture. We will return to this point.

We have been able to compare conversions from the two types of reactors. However, the results presented have not yet answered one of the questions raised at the beginning of this section regarding the effect of reactor design on product quality.

5.4.3 The effect of reactor design on product quality

Gibbins *et al.* [1991c] also compared size exclusion chromatograms (SEC) of products recovered in the two reactors. The eluent used during these characterizations was tetrahydrofuran (THF), subsequently shown to have several disadvantages when used for this purpose. The technique and its applications in characterizing coal and petroleum derived heavy hydrocarbons have been described in Chapter 8 of this book. However, it is useful to qualitatively review some of the observations made in the original publication on the comparison of the two reactors.

SEC chromatograms of products from different isothermal holding time runs and increasing reaction temperatures in the "flowing-solvent" reactor (in tetralin) were compared first. Progressively larger molecular mass material was released by the coal mass as hold time or peak experimental temperature was increased. The opposite trend was observed when products from the "mini-bomb" reactor were examined. Products became progressively lighter for longer exposure times and with increasing conversion. Since the *primary* products obtained in both cases were probably similar, these differences clearly reflect the effect of secondary reactions in the mini-bomb reactor. Not surprisingly, the "flowing-solvent" reactor gave larger molecular mass products under comparable holding times and similar temperatures. These observations were substantially verified by later work using NMP (1-methyl-2-pyrrolidinone), an altogether more appropriate eluent for SEC (see Chapter 8).

Figure 5.6 compares size exclusion chromatograms (SEC) obtained using NMP as eluent, of Point of Ayr coal extracts produced under four different sets of conditions. The samples shown in this diagram were prepared (1) in the flowing-solvent reactor, (2) in the mini-bomb (MB) using 10 minutes reaction time, (3) in the mini-bomb (MB) using 60 minutes reaction time. Sample (4) was the filtered extract from the Point of Ayr coal liquefaction pilot-plant (~1 hr average digester time) and was known to contain recycle solvent. Without going into details of size exclusion chromatography, it may be observed that the extract from the flowing solvent reactor gave more intense signal at short elution times. In SEC larger size (and mass) molecules are expected to elute at shorter retention times than smaller mass materials. The peak appearing at shorter times thus represents larger molecular mass material compared to the other samples. The diagram gives some idea of the extent of molecular size reduction, caused by extended residence times at reaction conditions in the mini-bomb reactor and in the pilot-plant digester [Zhang *et al.*, 1996a].

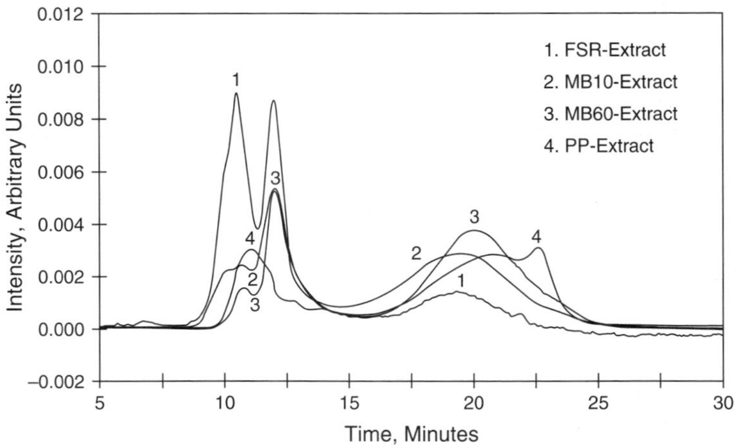

Figure 5.6 Size Exclusion Chromatograms (NMP eluent) Point of Ayr coal extracts prepared in different reactors, detection by UV-absorption at 450 nm. Samples extracted in (1) flowing-solvent reactor, (2) mini-bomb 10 minutes, (3) mini-bomb 60 minutes, (4) Point of Ayr liquefaction pilot-plant (~1 hr average digester time). (Reproduced with permission: Fuel 1996, *75*, 597; Copyright 1996 Elsevier).

5.5 Effect of Solvent Type on the Extent of Extraction

We have already observed in Figure 5.5a that in the "flowing-solvent" reactor, about 10–12% higher conversions were obtained in tetralin compared to liquefaction in 1-methylnaphthalene. We have also indicated that solvents could be classed according to their hydrogen-donor ability *or* their solvent power. Effective hydrogen-donors (e.g. tetralin) are not necessarily as powerful solvents for coal-derived materials as some non-donor solvents (e.g. quinoline, phenanthrene). We will next describe a set of experiments designed to distinguish between the effects of the two functions on the extent of extraction.

Experiments were undertaken with tetralin, quinoline, a mixture of quinoline and phenanthrene (both effective solvents for coal-derived materials) and finally hexadecane, a straight chain aliphatic compound, which is neither a hydrogen-donor, nor a known solvent for coal-derived materials. The Point of Ayr coal sample used in these experiments had already been extracted at 350°C, without observing the onset of extensive depolymerization reactions.

Table 5.3 shows that at up to ~350°C, both quinoline and the mixture of quinoline and phenanthrene could extract more of the coal mass than tetralin. The result is consistent with their known greater solvent power for coal-derived materials. Neither quinoline, nor the mixture of quinoline and phenanthrene could be said to have hydrogen-donor properties. The lower conversions than tetralin observed at 450°C using the two non-donor solvents were consistent with the results from using 1-methylnaphthalene, another non-donor solvent (Figure 5.5a).

Potentially important clues are provided by a comparison of 450°C data in tetralin and the two more powerful solvents. It seems safe to assume that, barring secondary

Table 5.3 Comparing Conversions: Liquefaction experiments with solvent flow rate of 0.9 ml/s at 70 bar (g). All data are given on % w/w dry ash free basis. (Reproduced with permission: Energy & Fuels 1994, 8, 1360; Copyright 1994 Am.Chem.Soc.)

	Heating Rate °C s^{-1}	Holding Time s	Ambient medium	Weight Loss 350°C	Weight Loss 450°C
POA Vitrinite	1,000	150	Helium	3.3	20.5 (2)*
	5	400	Tetralin	28.8**	77.6 (2)
	5	400	Q/P***	38.0	73.8 (2)
	5	400	Quinoline	—	72.7 (2)
	5	400	Hexadecane	12.5	27.3 (2)
POA Whole Coal	5	400	Tetralin	24.6	82.5 (4)
	5	400	Quinoline	39.5	74.7 (2)
	5	400	Hexadecane	—	24.0 (1)

* Number of repeated runs used for calculating the average value

** Holding time: 500 seconds. The weight loss from 100 s experiments under the same conditions was 29.2%, within experimental error.

*** Q/P: quinoline/phenanthrene (2.5:1 w/w) mixture.

effects, the rate of coal depolymerization is primarily a function of temperature alone. In other words, at equal temperatures and with comparable coal samples, we should expect approximately similar amounts of coal matter to detach from the parent solid matrix, irrespective of the surrounding medium (any solvent or gas). In the case of a gaseous medium, we have already seen that rapid char formation can be expected from the recombination reactions of reactive free radicals released by the coal matrix. That is why less volatile matter is released during pyrolysis, compared to product release during (most) attempts at coal liquefaction.

We have also observed retrogressive, char-forming reactions in 1-methylnaphthalene, in the "mini-bomb" reactor and less char formation in the flowing-solvent reactor. Furthermore, the rate of retrogressive reactions indicated by the downward slope of the conversion in Figure 5.5b were clearly rather slow. The amount of solids changed in a matter of *hundreds* of seconds. The conversion in the mini-bomb must have initially increased and reached a maximum somewhere between 0 and 100 seconds, before registering the observed reversal in conversion. At 400°C, longer contact times in the closed reactor would have been required for such a reversal in conversions to take place, if at all.

By contrast, less than 10 seconds elapses in the "flowing-solvent" reactor, between the dissolution of coal molecules in the solvent and the cooling (quenching) of the reaction stream in the heat exchanger (see Figure 5.2). Thus, in the flowing solvent reactor, not only was there insufficient time for extensive levels of these apparently slow retrogressive reactions to develop, but the vast dilution of products inherent in the design would tend to decrease the rate of retrogressive reactions even further. Nevertheless, about 10% more material must have re-combined in 1-methylnaphthalene, compared to liquefaction in tetralin, to produce insoluble material. It is likely that the retrogressive reactions leading to this decrease in conversion (by 10–12%) would have had to take place inside the coal particles themselves and not in the very dilute reaction stream.

On the basis of this evidence, it seems reasonable to postulate two distinct speeds for the occurrence of retrogressive reactions: (i) rapid within the coal particles, in the absence of donatable hydrogen, and, (ii) slower reactions of less reactive material, if present in a concentrated phase (as in Figure 5.5b). Clearly, in the flowing-solvent reactor, non-donor solvent carried dissolved material with ability to repolymerize. However, the level of dilution (200 mg in 700 to 1,000 ml solvent) appears to have blocked these reactions and to have carried the mixture from the reaction zone. In the case of tetralin, the hydrogen-donor ability of this solvent would readily quench the more reactive free radicals and serve to suppress the rapid as well as the slow retrogressive radical recombination reactions.

Taken together, these data provide strong indications for the occurrence of an initial burst of fast char-forming retrogressive reactions occurring in the absence of the hydrogen-donor species. The pyrolysis-based data of Fukuda [2002], discussed in Chapter 6, are consistent with the occurrence, in successive stages, of fast and slow recombination reactions, within the same coal mass. As an aside, it is also possible that at least part of the ~15–20% char residue left behind during liquefaction in tetralin of Point of Ayr coal may have formed as part of repolymerization reactions that were too rapid for the tetralin to be able to block. This type of residue being only fit for combustion or gasification, no work to examine its origins and formation appears in the literature.

The experiments described in Table 5.3 were designed to distinguish between the hydrogen-donor ability of tetralin and the high solvent power of essentially non-donor solvents, quinoline and phenanthrene. The latter had to be used mixed with quinoline, as it is a solid at room temperature and would not have flowed through the tubing upstream of the reactor. We have observed so far that coal liquefaction conversions (but not necessarily product quality) depend critically on the local availability of sufficient donatable/available hydrogen. It follows that in batch reactor experiments, the abundance of the donor-solvent, as well as its hydrogen-donating ability, should become critically important. Clarke *et al.* [1980] found conversions in the mini-bomb to remain fairly constant at and above donor-solvent:coal ratios of 4:1.

In the absence of both solvent power and of hydrogen donor-donor ability, Table 5.3 shows that extractions (in the "flowing-solvent" reactor) in hexadecane gave results closer to those obtained during pyrolysis in the wire-mesh reactor. Heating in helium at 1,000°C s^{-1} to 350°C with 150 s holding, the pyrolysis weight loss from Point of Ayr vitrinite concentrate was 3.3% (% w/w daf basis); at 450°C, the weight loss improved to 20.5%. This important result is a little counterintuitive; it might have been expected that mass transfer between a solid and a liquid (coal-to-hexadecane) would have been more efficient than mass transfer from coal to an inert gas stream. However, in the absence of solvent power for coal-derived materials, the presence of hexadecane appears to have made little difference from pyrolysis in *helium*. Results presented in Table 5.3 may therefore be understood in terms of the coal (or vitrinite concentrate) sample depolymerizing quite substantially during heating to (and at) 450°C. The outcome of the experiment then appears to depend on whether a large fraction of the depolymerized material can be chemically quenched by a hydrogen-donor or moved out of the reaction zone by dissolving in a powerful solvent.

In the case of liquefaction in hexadecane, the occurrence of extensive repolymerization was confirmed by the negligible amounts of extra material extracted from the solid residues, upon subsequent refluxing in a 4:1 mixture of chloroform and methanol. FT-ir spectra of hexadecane extracts showed greater aliphatic and hydroaromatic content than tetralin or quinoline extracts, suggesting a degree of preferential extraction [Xu *et al.*, 1994].

5.6 Flowing-Solvent Reactor: Successive Extract Fractions Released from Coal

Oil refineries routinely by-pass the problem of processing heavy residues by conversion to petroleum coke, in the process recovering some hydrogen and light hydrocarbons. The refractory nature of coal extracts has led to analogous questions about whether products released during shorter extraction times may be more amenable to subsequent upgrading [Moroni, 1991]. In any case, it rarely makes economic sense to go to great lengths to solubilize the last 10 or 20% of the coal mass to improve liquefaction conversions. Unless hydrogen is to be imported into the plant, some carbon must always be gasified to generate the requisite hydrogen and the solid residue from the digesters appears to be quite reactive for this purpose [Kimber, 1997].

At bench-scale, it is possible to attempt short duration experiments in micro-bomb reactors. However, in a batch reactor, only the cumulative extract mixture released between the beginning of the experiment and the endpoint is available for characterization. It is not possible to recover extract fractions released between specific time or temperature intervals. Furthermore, the length of heat-up and cool-down periods (order of minutes) introduces uncertainties in the time resolution of extracts released from the coal sample. The closed ("batch") reactor configuration also allows time for extraparticle secondary reactions of extracts to take place. On the other hand, the continuous flow of dissolved extracts out of the "flowing-solvent" reactor makes it naturally suitable for identifying and comparing successive product fractions released from coal during the liquefaction process. In the case of Point of Ayr coal, these fractions have been recovered and characterized. Once again, tetralin, quinoline and hexadecane were selected as solvents, based on their different hydrogen-donor abilities and solvent power.

However, the "flowing-solvent" reactor has its own shortcomings. While the recovery of successive product fractions from the reactor is straightforward, characterizing the products is made difficult by the relative abundance and the potential reactivity of the solvents. In earlier work, products from the thermal reactions of tetralin have been examined by gas chromatography – mass spectrometry. Tetralin-tetralin dimers and tetralin-naphthalene adducts, mainly of mass 258 and 262 u have been identified within the spectrum of products. The total concentration of this class of compounds did not exceed about 1% of the total solvent recovered from the "flowing-solvent" reactor. Greater concentrations of relatively stabler dimers and adducts, including naphthalene, butyl benzene and 1-methylindane, have been identified within products from a stirred micro-autoclave (residence time: 1 hour). These findings have immediate implications

Table 5.4 Temperature intervals corresponding to time-resolved product fractions. Samples were heated at 5°C s^{-1} to 450°C and held for 400 s. A Tetralin flow of 0.9 ml s^{-1} was maintained under 70 bar pressure. (Reproduced with permission: Energy & Fuels 1994, *8*, 1360; Copyright 1994 Am.Chem.Soc.)

Fraction number	*time interval (s)*	*Temperature interval (°C)*
1	0–70	ambient-350
2	70–80	350–400
3	80–90	400–450
4	90–100	450
5	100–140	450
6	140–190	450
7	190–490	450
8	490–	450-ambient

for the evaluation by GC-MS of coal liquefaction products prepared using tetralin as vehicle. Compounds similar to tetralin-derived by-products have been found in product solutions from coal liquefaction experiments [Brodzki *et al.*, 1994a, 1994b, 1995].

In assessing structural differences between successive extract fractions, therefore, effects due to products derived from the liquefaction solvent have to be eliminated or mitigated. The boiling points of tetralin-dimers and related adducts are above 350°C and it is not possible to remove them by distillation without unduly changing the sample itself. Furthermore preparing samples for characterization by distilling off the solvent, followed by drying, suffers from masking effects due to the presence of these larger molecular mass solvent by-products. However, unlike the coal sample – solvent residence times are constant in the "flowing-solvent" reactor. *Changes* in the structural features of materials identified in the reaction mixtures therefore can serve as a more reliable guide of changing extract properties in successive fractions recovered from the flowing solvent reactor. Detailed descriptions of successive extract fractions released from coal have been presented in the original publications [Xu *et al.*, 1994; Li *et al.*, 1995]. FT-infrared spectroscopy of product fractions showed decreasing extents of aromatic substitution and increasing product polarity with increasing extents of reaction.

In size exclusion chromatography, solvent-derived products appear at relatively longer elution times and could be blanked out. This enabled comparing the larger molecular mass fractions of the extracts. In both tetralin and quinoline, products showed increasing molecular mass distributions with the progressively increasing intensity of reaction conditions, i.e. peak reaction temperature and holding time at peak temperature. In both solvents, the size exclusion chromatograms of fractions released at 450°C indicated the presence of material at the high-molecular mass limit of the calibration of the SEC column. The findings suggested that, qualitatively, fractions released from coal during earlier stages of the liquefaction process probably required less severe catalytic hydroprocessing.

During efforts to characterize these extract fractions, a UV-fluorescence spectrometer was used for the first time, as detector in SEC, in series with a UV-absorption detector. The sensitivity of the UV-fluorescence spectrometer diminished with increasing sample molecular mass (i.e. at shorter elution times). Above about 3,000 u, where clear signal could be observed with the UV-absorption spectrometer, the quantum yields and detector sensitivity of the UV-fluorescence spectrometer were reduced to near zero [Li *et al.*, 1995].

5.7 A Two Stage Kinetic Model of Primary Coal Liquefaction

Overall, little effort has been spent examining the chemical kinetics of the thermochemical processes studied in this book. Partly this is because most of the effort has been spent on attempting to establish trends in sample behavior, free of interference from spurious variables. However, there are several further points that must be taken into account when considering the kinetics of the reactions of solid fuels (coals, biomass etc).

In all three types of processes that we discussed (pyrolysis, gasification and liquefaction), considerable weight loss takes place during heatup, which must be accounted for. It is also vital to consider that in gasification, a first major stage consists entirely of pyrolysis. In turn, a major part – but not all – of the pyrolysis phase is dominated by tar release. We have seen how the gasification reactivities of chars depend on the manner in which the pyrolysis step has been carried out. We have also seen that at peak temperatures, at or above 900–1,000°C, the gasification reactivities of coal chars rapidly change with time – of the order of 10 seconds. All these process stages represent different chemical reactions, each with their own kinetics – and assorted challenges in modeling the kinetics. It seems hasty therefore to proceed to kinetic calculations prior to identifying the critical rate determining processes and arrive at appropriate models. It is not, therefore, possible to view some kinetic models that have (even) reached the commercial stage, as much more than curve fitting exercises that work by means of manipulating adjustable parameters [Herod & Kandiyoti, 1994].

5.7.1 Elements of a two-stage kinetic model for coal liquefaction

Several strands of existing information must be taken into account in considering an even relatively simple global kinetic model for coal liquefaction.

(A) Reactor-Related Effects: When using tetralin as "vehicle", we have already observed that comparable coal conversions could be obtained in the "flowing-solvent" and "mini-bomb" reactors, even though, size exclusion chromatography showed evidence for extensive cracking of dissolved extracts in the "mini-bomb". On the other hand, different conversions were observed in the presence of the poor hydrogen-donor solvent, 1-methylnaphthalene. Shin *et al.* [1989] also observed retrogressive char formation in 1-methylnaphthalene during experiments in batch micro-autoclaves at lower temperatures. Reaction pathways in coal liquefaction depend, therefore, to some degree on interrelationships between the role of the reactor configuration and that of the solvent. It necessarily follows that the rates of these processes must depend on the same parameters.

Apparatus with specialized configurations such as the "flowing-solvent" reactor are not absolutely necessary for determining coal conversions independent of subsequent secondary reactions. When using tetralin as "vehicle", similarities between coal mass loss in the "flowing-solvent" and the "mini-bomb" reactors suggest that it may be possible to calculate coal dissolution (conversions) from batch reactor data, so long as an efficient hydrogen-donor solvent is used in sufficient abundance. In tetralin, solvent-to-coal ratios of about 4:1, or higher, would appear to provide the abundance of hydrogen-donor required for suppressing large-scale recombination reactions between coal-derived free radicals to the same extent as in the flowing solvent reactor. Ratios as high as 1:8 have been suggested by Hill *et al.* [1966].

Such *conversion* data could be considered relatively free of reactor-related effects. However, in batch reactors, diffusion of extract molecules out of coal particles takes place against a continuously rising concentration of extract molecules in the reaction mixture surrounding the particles. The *rates* of liquefaction (sample weight loss) are therefore affected by changing rates of outward diffusion of product molecules. For Point of Ayr coal, 10-min extractions in mini-bombs gave about 15% lower conversions than the flowing-solvent reactor (where coal was exposed to temperature for somewhat less than 500 s), while 60-min extraction results were comparable with conversions in the flowing-solvent reactor. In the latter reactor, the diffusion of extract molecules out of coal particles takes place against virtually zero concentration of dissolved coal extracts in the surrounding solvent phase, due to the continuous flow of fresh solvent. This reactor configuration therefore allows the calculation of kinetic constants independently of *external* (i.e. particle to bulk liquid) diffusion resistance effects.

It is important to note, finally, that in the absence of abundant H-donor solvent (e.g. in the presence of a poor donor solvent such as 1-MN), kinetic expressions describing coal liquefaction processes in batch reactors would need to reflect product loss through retrogressive radical recombination reactions, char formation, etc. In the flowing-solvent reactor, such retrogressive reactions would be minimized (but probably not totally eliminated) as observed in experiments with 1-methylnaphthalene, quinoline, and mixtures of quinoline and phenanthrene, for which conversions ~10% lower that those observed in tetralin have been obtained (Table 5.3).

(B) Processes preceding and following the onset of extensive covalent bond scission: We have already explained that it is possible to distinguish between coal mass loss taking place prior to and following the onset of extensive covalent bond scission. We will return to this point in Chapter 6 in some detail. Meanwhile, we will assume that dissolution of products at temperatures below the characteristic band around 350°C may be reasonably considered as taking place prior to extensive covalent bond scission.

It was recognized in early work that activation energies of processes taking place prior to the onset of extensive bond scission in coals are significantly lower than those taking place at temperatures where extensive depolymerization actually takes place [Hill *et al.*, 1966; Han & Wen, 1979]. Descriptions of coal dissolution in terms of single activation energies describing the succession of events in the entire temperature range from ambient to a peak temperature above 350°C would, therefore, mask a significant and (at least qualitatively) well recognized stage of the overall process. This frequently used assumption cannot but lead to the calculation of ill-defined kinetic constants and activation energies.

(C) Non-isothermal mass loss during heatup: It is generally accepted that when coal is pyrolyzed, sample weight loss during heatup is extensive, even during the application of rapid heating [Howard, 1981]. During experiments using a heating rate of 1,000°C s^{-1} to 700°C in a wire-mesh reactor, nearly all measurable volatile release was found to have taken place during heatup [Gibbins-Matham & Kandiyoti, 1988].

In coal liquefaction, lower temperatures are used and smaller proportions of total possible products are released during heatup; this proportion of extracts is nevertheless significant. In the flowing-solvent reactor, where rapid cool-down is made possible by allowing a flow of cold solvent to wash over the fixed bed of sample, experiments conducted in tetralin at 5°C s^{-1} to 450°C with "zero-second" holding at peak temperature gave about 20% weight loss for Pocahontas No.3 and about 30% weight loss for Point of Ayr coal [Xu *et al.*, 1994; Xu, 1995]. For most bomb reactors, average heating rates over the temperature interval are usually of the order of 2–5°C s^{-1}. Using an autoclave with a heatup period of 1.5–2 hours, Hill *et al.* [1966] reported that "80% of the total possible extraction" was achieved "before the system reached the reaction temperature". In such a reactor, the rates of heating would continuously change with rising temperature. Furthermore, sample weight loss during heatup must clearly take place in *non-isothermal* mode.

In the flowing solvent reactor, by the time the sample reaches 450°C in a stream of tetralin, depending on the type of coal, anywhere between 20 and 40% of a middle rank coal may be expected to have already gone into solution. Its neglect in the calculation of kinetic constants for coal liquefaction processes is not therefore warranted, irrespective of the duration of the heatup period. In coal liquefaction, events during heatup at temperatures between ambient temperature and 420–450°C thus encompass (i) stages of the process with different activation energies and (ii) significant levels of product release during heatup, taking place over a range of temperatures, at rates increasing with rising temperature.

In the next section, we will describe a set of equations enabling the calculation of reaction rate parameters for coal weight loss during coal liquefaction. The model explicitly accounts for product release during heatup, both before and after the onset of extensive covalent bond scission (around 350°C), leading to the calculation of different activation energies for the two stages. The calculations were based on a set of coal conversion data from the flowing-solvent reactor, with tetralin used as vehicle.

It must be noted in this connection that the use of higher heating rates would not significantly alter the amount of extract released during the overall process, provided a sufficiently long hold time is allowed. We have already described the relative insensitivity of liquefaction conversions to changes in heating rates between 0.1 and 10°C s^{-1}, in the presence of the efficient hydrogen-donor tetralin [Gibbins & Kandiyoti, 1991b]. Mechanisms underlying changes in yields with changes in heating rates appear to involve competition between recombination reactions and quenching of coal derived free radicals by locally available hydrogen [Li *et al.*, 1994]. In atmospheric pressure pyrolysis, the use of higher heating rates appear to lead to the release, in the form of volatiles, of some material which would otherwise have recondensed, leading to more repolymerized char residue [Gibbins-Matham & Kandiyoti, 1988; Li *et al.*, 1993].

In the presence of an efficient hydrogen-donor, the high conversions and the absence of a heating rate effect during liquefaction suggests that competition between recombination reactions and product release is not a predominant effect. In any case, changes in yields with increasing heating rates are not large. When, during a pyrolysis experiment, the heating rate is increased from 1 to 1,000°C s^{-1}, parallel increases in tar and total volatile yields rarely exceed 6–7% (w/w daf coal) [Li *et al.*, 1993]. Therefore, if significantly faster heating rates than those normally available to bomb reactors are used, we would expect outcomes that depend on the hydrogen-donor ability of the solvent. At "zero-holding", the amount of extractables released from coal up to the peak temperature *could* change as a function of heating rate. It would possibly increase by a small amount in the case of non-donor solvents, due to reduced time for retrogressive reactions. In the presence of a good hydrogen-donor, however, we would expect small reductions in conversion due to less time available for reaction [Gibbins & Kandiyoti, 1991]. It is, therefore, unlikely that very significant changes in product release can be achieved during heatup by changing the heating rate.

5.7.2 Description of the two-stage kinetic model of primary coal liquefaction: The Assumptions

1. The process is assumed to take place in two stages.
   (A) Stage A processes: This first stage consists of sample weight loss prior to the onset of extensive covalent bond rupture, taking place between ambient temperature and approximately 350°C. The model was formulated without explicit reference to the nature of physical and chemical phenomena occurring in this temperature interval. It is thought, however, that these processes include ordinary dissolution of more soluble (smaller molecular mass and/or less polar) species possibly occluded or held by weak interactions such as hydrogen bonding and van der Waals forces. In what follows, Stage A will be modeled, first, as a single activation energy, irreversible, first-order process and, second, as a set of parallel independent first order processes with a Gaussian distribution of activation energies, characterized by a standard deviation, σ_A [see Howard, 1981]. In the latter case, a common pre-exponential constant is used to characterize the set of parallel, independent reactions.
   (B) Stage B processes: The second stage is defined as the process of weight loss following the onset of extensive covalent bond rupture, broadly corresponding to coal weight loss above approximately 350°C. Once again, no explicit reference has been made to the nature of chemical and physical phenomena occurring in this temperature interval, apart from suggesting depolymerization due to covalent bond rupture. Changes in the structural characteristics of successive extract fractions released from coals described in Section 5.6 will not be addressed. Stage B will also be modeled, first, as a single, irreversible, first order process and, second, as a set of parallel independent first order chemical reactions with a Gaussian distribution of activation energies, characterized by a standard deviation, σ_B. A common pre-exponential constant will be used to characterize the set of parallel, independent reactions.
2. The coal is considered to behave as a homogeneous material. For ease of calculation, a single temperature is selected to distinguish between Stages A and B, defined as T_d. While this value was set at 350°C for most coals, in calculations about Point of Ayr, Upper Freeport, and Pocahontas No.3 coals, 375°C was found to be a more appropriate value.
3. The effect of intraparticle mass transport as a significant resistance or as a rate-determining step has been neglected. This assumption is treated as a first approximation and facilitates the development of the present model. The following experiments suggest that internal diffusion, may in the first instance, be neglected, at least for some coals. A set of parallel liquefaction experiments were carried out with a sample of larger particle size (250–500 μm) Point of Ayr (UK) coal over the 300–450°C range. Differences in weight loss were found to be within experimental error with data obtained using 106–150 μm particles [Xu, 1995]. In particular, no experimentally significant differences could be observed between the behavior of different sized particles at temperatures below the onset of extensive covalent bond scission. The evidence presented here is not thought to be conclusive and a separate study of the effect of intraparticle diffusion effects is needed.
4. For both Stage A and Stage B processes, the concentration driving force in all first order reaction terms has been expressed in terms of an "ultimate" (equilibrium)

weight loss value. For Stage A, equilibrium values have been estimated from sample weight loss determined by heating at 5°C s^{-1} to 350°C (or 375°C – depending on the coal – see below) with 1600 s holding, followed by washing the solid residue with a 4:1 chloroform-methanol solution at ambient temperature. "Ultimate" conversions at 450°C were similarly determined by 1600 s holding at peak temperature. Clearly, "ultimate" (equilibrium) weight loss may assume different values if the coal, the vehicle or the solid residue washing procedure is changed.

5.7.3 Description of the two-stage single-reaction model

Both stage A and stage B are modeled as single activation energy, irreversible, first-order processes. $x_A(t)$ is defined as the time-dependent weight loss variable for stage A. This variable has an experimentally determined ultimate (equilibrium) value of x_{mA}, defined as the conversion after 1,600 seconds at T_d – the transition temperature between the two weight loss regimes. The kinetic constants for Stage A processes at up to the temperature T_d were calculated first.

Sample weight loss due to stage B processes is defined as $x_B(t)$. The overall ultimate (equilibrium) conversion for stage B processes can be written in terms of the equation

$$x_m = x_{mA} + x_{mB}$$

where x_m can be experimentally determined as the *total* weight loss obtained by heating samples at 5°C s^{-1} to 450°C and holding for 1600 s. To account for weight loss during heatup, it is necessary to integrate mass loss over the heatup period as a function of changing temperature. With the initial (ambient) temperature defined as T_0 and the heating rate during the heatup period as k_h, the temperature T at any time may be calculated from

$$T = T_0 + k_h\, t.$$

The total weight loss "x(t)" at any time thus includes weight loss during heatup and during the subsequent holding period. When the holding temperature is less than T_d, the transition temperature, total sample weight loss is calculated using rate constants for Stage A processes only. When a run with holding temperature above T_d is being simulated, extraction from the two successive stages is calculated separately and added together. A set of three kinetic parameters must be determined for both stage A (e.g. x_{mA}, k_{0A}, E_A) and for Stage B (e.g. x_{mB}, k_{0B}, E_B) processes. Of these, x_m and x_{mA} are determined experimentally from 1600-second runs at T_d for the particular coal and at 450°C. x_{mB} is determined from the equation: $x_m = x_{mA} + x_{mB}$.

The first step of the calculation (i.e. for Stage A parameters) makes use of sample weight loss vs. temperature data at up to T_d (in Table 5.2). k_{0A} and E_A values were calculated using a two-dimensional surface-fitting non-linear regression algorithm and sample-mass vs. time expressions, described in the original publication [Xu & Kandiyoti, 1996]. The values of the starting parameters have not been found to influence the final values obtained in these calculations or in the multiple-reaction model calculations described below.

Table 5.5 Comparison of kinetic parameters of the set of coal samples in Table 5.1. Single-Reaction Model. (Reproduced with permission: Energy & Fuels 1996, *10*, 1115; Copyright 1996 Am.Chem.Soc.)

		Stage A			*Stage B*		
Coal Sample	x_m* (w/w, daf)	x_{mA} (w/w, daf)	k_{0A} (s^{-1})	E_A (kJ mol^{-1})	x_{mB} (w/w, daf)	k_{0B} (s^{-1})	E_B (kJ mol^{-1})
Beulah-Zap	0.84	0.41	1.52×10^2	55.4	0.43	4.12×10^6	124.0
Wyoming	0.89	0.39	0.50×10^2	47.2	0.50	1.67×10^7	130.4
Illinois No. 6	0.95	0.55	3.06×10^3	69.3	0.40	2.26×10^8	142.5
Blind Canyon	0.92	0.38	7.44×10^2	60.5	0.54	2.50×10^8	142.7
Pittsburgh No. 8	0.89	0.56	0.16×10^2	44.0	0.31	8.25×10^9	161.2
Upper Freeport	0.86	0.58	0.51×10^2	46.8	0.28	6.23×10^7	139.4
Pocahontas No. 3	0.70	0.30	2.05×10^4	80.5	0.4	9.40×10^7	150.0
Point of Ayr (U.K.)	0.84	0.36	0.21×10^2	35.0	0.48	4.65×10^{15}	238.0

* x_m: Total mass loss after 1600 s at 450°C

Results from the Two-Stage Single-Reaction Model: Table 5.5 presents kinetic parameters calculated using the single-reaction model [Eqs. 4 & 5 in Xu & Kandiyoti, 1996]. In Table 5.5, k_{0A} and k_{0B} denote the pre-exponential factors for Stage A and Stage B, respectively, while E_A and E_B denote the analogous energies of activation for the two stages. As expected from primarily desorption and diffusion driven phenomena, activation energy values for stage A processes, E_A, were found to be considerably smaller than those for stage B processes, E_B (Table 5.5). In showing different energies of activation for two clearly identifiable processes, this calculation justifies the effort put into the kinetic modeling. It shows that using a straight Arrhenius plot involving only final reaction temperatures against measured conversions is inherently erroneous.

With the exception of Pocahontas No. 3 coal, E_A values (Stage A) were found to vary within a relatively narrow band with no discernible pattern, suggesting a degree of similarity between rate-limiting steps in this temperature range. Most of the E_A values were low even for diffusion-limited processes. These energies of activation are attributable to processes involving desorption and dissolution of smaller molecular mass and/or less polar materials within the coal mass. It may be noted, however, that the lowest E_A value in Table 5.3 was found for Point of Ayr coal, the sample for which conversions were found *not* to change when the particle size was more than doubled (see above). The two elements of information appear to be consistent, in the sense of indicating weak resistance to the outward intraparticle diffusion of extract molecules.

Energies of activation for the higher temperature range (from T_d up to 450°C) Stage B processes, were found to lie between 124 and 238 kJ mol^{-1} (Table 5.5). This is the temperature range where extensive covalent bond scission is expected to contribute massively to the dissolution of the coal mass. The sharp difference between the ranges of E_A and E_B values found in these calculations confirm the validity and indeed the necessity of the added complications involved in introducing a two-stage model. Descriptions of coal liquefaction in terms of a single activation energy, expected to span the range of processes taking place between ambient and peak temperature, clearly conceal at least one distinct and important transition. Nevertheless, values at the lower end of the E_B range were smaller than would have been expected compared to values of bond dissociation energies and activation energies calculated in previous pyrolysis related work [Burnham *et al.*, 1989; Gavalas *et al.*, 1981].

5.7.4 Description of the two-stage multiple-reaction model

Early work by Jungten and van Heek [1970] has shown that when a process consisting of multiple parallel independent reactions is modeled in terms of a single reaction, the apparent activation energy of the imaginary lumped process turns out to be lower than the activation energies of the individual reaction pathways. To test the relevance of this proposition to coal liquefaction, both stage A and stage B of the liquefaction process were considered to proceed by means of multiple parallel independent irreversible first-order processes. The approach is adopted from the work of Howard and co-workers on volatile release during coal pyrolysis [Howard, 1981].

Once again the weight loss during a particular experiment includes weight loss occurring during heatup as well as the holding period. Weight loss during the rapid cooling period has been neglected. In the multiple-reaction model, in addition to the pre-exponential term and the mean activation energy, the standard deviation of the distribution of activation energies (σ) must also be calculated. As before, stage A parameters (k_{0A}, E_{0A}, and σ_A) were calculated first, using weight loss data obtained between ambient temperature and the transition temperature, T_d, for the particular coal.

To reduce the new three-dimensional nonlinear regression problem to a two-dimensional search, the third variable σ_A was preset at a series of fixed values ($E_{0A}/10$, $E_{0A}/25$, $E_{0A}/50$, $E_{0A}/100$ and $E_{0A}/200$) and the (k_{0A}, E_{0A}) values corresponding to the best fit were calculated using the same two-dimensional surface-fitting algorithm used for the single reaction model calculations. As before, when the peak experimental temperature exceeds T_d, weight loss due to stage B processes is added to that due to stage A processes (Eq. 15 & 16 in Xu & Kandiyoti, 1996). A similar algorithm was then used for calculating values of k_{0B}, E_{0B}, and σ_B. Listings of the codes have been given in Xu [1995].

Results from the two-stage multiple-reaction model: Table 5.6 presents kinetic parameters calculated using the multiple parallel independent reaction model. From a comparison of the parameters calculated from the two models (Tables 5.5 and 5.6), it may be observed that mean activation energies for stage A processes ($T < T_d$) calculated using the multiple parallel independent reaction model were only slightly larger than those calculated from the single reaction model. This result suggests that the number and nature of independent pathways involved in stage A processes are indeed probably fairly limited. The spread of activation energies, characterized by the standard deviations of the distributions, σ_A, were found to be correspondingly narrow. The steady decline of absolute σ_A values with increasing coal rank (with the exception of the two low-rank samples) may be interpreted in terms of the simplification of structural features with increasing coal maturation (also see below).

Activation energies for stage B processes, covering temperatures above the onset of extensive covalent bond scission ($T > T_d$), however, differ sharply from results calculated using the single-reaction model. The mean activation energy values were found to be

Table 5.6 Comparison of Kinetic Parameters of the Set of Coal Samples. Multiple-Reaction Model. (Reproduced with permission: Energy & Fuels 1996, *10*, 1115; Copyright 1996 Am.Chem.Soc.)

		Stage A				Stage B				Parameters from pyrolysis experiments*	
Cool Sample	x_m (w/w, daf)	x_{mA} (w/w, daf)	k_{0A} (s^{-1})	E_{0A} (kJ mol^{-1})	σ_A (kJ mol^{-1})	x_{mB} (w/w, daf)	k_{0B} (s^{-1})	E_{0B} (kJ mol^{-1})	σ_B (kJ mol^{-1})	E_0 (kJ mol^{-1})	σ_B (kJ mol^{-1})
Beulah-Zap	0.84	0.41	3.07×10^2	58.9	$E_{0A}/25$	0.43	2.23×10^{12}	200.0	$E_{0B}/25$	232	E0/38
Wyodak-Anderson	0.89	0.39	0.92×10^2	50.0	$E_{0A}/25$	0.50	4.32×10^{11}	188.4	$E_{0B}/25$	263	E0/45
Illinois No. 6	0.95	0.55	3.55×10^4	81.3	$E_{0A}/25$	0.40	2.32×10^{12}	194.1	$E_{0B}/50$	210	E0/45
Blind Canyon	0.92	0.38	9.41×10^2	64.5	$E_{0A}/25$	0.54	6.08×10^{13}	210.9	$E_{0B}/50$	196	E0/49
Pittsburgh No. 8	0.89	0.56	0.25×10^2	46.5	$E_{0A}/25$	0.31	1.97×10^{12}	191.0	$E_{0B}/50$	205	E0/54
Upper Freeport	0.86	0.58	0.93×10^2	49.5	$E_{0A}/25$	0.28	2.09×10^{16}	252.5	$E_{0B}/50$	262	E0/84
Pocahontas No. 3	0.70	0.30	3.70×10^4	83.6	$E_{0A}/100$	0.40	4.94×10^8	160.0	$E_{0B}/100$	222	E0/57
Point of Ayr (U.K.)	0.84	0.36	0.25×10^2	35.6	$E_{0A}/25$	0.48	1.97×10^{18}	275.0	$E_{0B}/100$	n/a	n/a

* from Burnham *et al.* [1989]

systematically greater, with an overall range between 160 and 275 kJ mol^{-1}. Pullen [1981] reported the activation energy for the bibenzyl cracking reaction as 201 kJ mol^{-1}, also quoted a somewhat higher value (235 kJ mol^{-1}) by Vernon [1980]. The E_{0B} values calculated for the present set of coals thus appear within the range that can be said to represent thermally induced bond scission [Nishioka, 1991].

In view of the multiplicity of parallel reaction pathways expected at temperatures *above* 350–375°C, it seems physically reasonable that the single-reaction model (Table 5.5) should underestimate the mean energy of activation for Stage B processes and that the *absolute* values of σ_B in Table 5.6 should be larger than σ_A values. The trend of decreasing σ_B values (Table 5.6) with increasing coal rank may be viewed in terms of increasing structural uniformity and possibly increasing degrees of cross-linking accompanying coal maturation.

In Table 5.6, the E_{0B} value for Pocahontas No.3 was lowest, while the value for Upper Freeport appears unexpectedly high. For these two coals, the values of E_{0A} and E_{0B} are probably related more closely than in other cases. The greater product release from Upper Freeport coal at lower temperatures has already been mentioned. The high E_{0B} value for temperatures greater than T_d may be viewed in terms of the more difficult (possibly the more polar or more densely cross-linked or more completely occluded) fractions of the coal remaining behind to be extracted. The opposite trend may be attributed to the case of Pocahontas No.3 coal. However, the high E_{0B} value found for Point of Ayr coal does not fit the same pattern. This last result does reflect, however, the consensus that this particular coal is not, after all, an "easy" one to liquefy.

For all coals in the study, good internal agreement was obtained between simulated results and data not used in the calculation of the kinetic constants [Xu, 1995]. If the kinetic procedure described in this study survives the test of time, the E_{0B} values calculated by this method may serve to assist in investigations of coal structure as well as help in determining the suitability of individual coals for conversion by liquefaction.

5.7.5 Comparison with previous studies of coal liquefaction kinetics

The present calculation takes into account weight loss during heatup, distinguishes between two distinct stages in coal liquefaction with different activation energies and shows that modeling the thermal breakdown step as a set of parallel independent reactions leads to apparently more realistic results. The latter assumption has been shown (Tables 5.5 and 5.6) to have a significant effect on calculated activation energies for stage B (thermal breakdown) processes. Clearly, a direct comparison with results of calculations from other work assuming (i) a single energy of activation and/or (ii) isothermal kinetics for the whole process is not entirely appropriate. The findings from this model need nonetheless to be placed in the context of previous work by comparison with activation energies calculated during different investigations.

Table 5.7 presents a short literature review of activation energies arrived at for coal liquefaction. The review is not meant to be exhaustive. A number of these investigations have reported activation energies that seem unacceptably low [e.g. Shin *et al.*, 1989; Curran, *et al.*, 1967; Brunson, 1979; Wiser, 1968; Cronauer *et al.*, 1978; Morita, 1979].

Table 5.7 Summary of results from a selection of previous studies on coal dissolution kinetics. (Reproduced with permission: Energy & Fuels 1996, *10*, 1115; Copyright 1996 Am.Chem.Soc.) [1. Wiser, 1968; 2. Curran *et al.*, 1967; 3. Cronauer *et al.*, 1978; 4. Weller (1951a); 5. Weller (1951b); 6. Morita *et al.*, 1979; 7. Hill *et al.*, 1966; 8. Shah *et al.*, 1978; 9. Gun *et al.*, 1979; 10. Shin *et al.*, 1989.]

Coal	*Solvent/Coal ratio*	*Temp (°C)*	*Definition of conversion*[a]	*Activation energy (kJ mol^{-1})*	*Ref. No.*
Utah HV	Tetralin (10:1)	409–497 (Pyrolysis) 350–450 (Liquefaction)	Wt. loss (pyrolysis) Benzene solubles	149 (2nd-order initial) 17.2 (1st-order later) 120.5 (2nd-order initial) 54.8 (1st-order later)	1
Pittsburgh seam (Ireland mine)	Tetralin (4:1)	324–387	Xylenol Cyclohexane, Benzene, cresol	E_a = 125.5 (fast) E_a = 159.0 (slow)	2
Belle Ayr (sub)	HAO[b] HPH[b]	400–470	Pentane, benzene, pyridine	Coal → product E_a = 70.0 (HAO) Coal → product E_a = 85.8 (HPH)	3
Pittsburgh seam (Bruceton mine)	none	400–440	Benzene	"Asphalt" hydrogen E_a = 150	4, 5
Miike	recycle stream MoO_3 catalyst	350–450 330–380	Hexane Benzene	Coal → asphaltene E_a (calcd) = 68.6 Asphaltene oil E_a (calcd) = 67.0	6
Utah Spring Canyon	Tetralin (10:1)	350–450	Benzene	E_a = 134 from 0 to 90% reacted ΔH = 155.6–358	7
Big Horn	Process (3:2)	413–440	Boiling ranges (Heavy oil > 343°C)	Coal → oil E_a = 230 Coal → furnace oil E_a = 169.5	8
Makum	Tetralin (1:1)	380–410	Benzene	Initial rxn = E_a = 330.5 Step 2 E_a = 196.6 Step 3 E_a = 146.4	9
Argonne PCS	1-MN[b]	375–425	Toluene THF	E_a = 18–112 E_a = 31.4–123.4	10

[a] Solid residue washed (extracted) in stated solvent.
[b] HAO, hydrogenated anthracene oil; HPH, hydrogenated phenanthrene; 1-MN, 1-methylnaphthalene
E_a denotes the energy of activation

Differences in model formulation outlined above (i.e. modeling coal liquefaction as an isothermal process and in terms of single reactions) may help clarify some of the reasons underlying the nature of these results. However, a small number of studies, using models similar to those used in the latter studies, have reported energies of activation similar to or greater than those found in the present work. These studies will be briefly reviewed.

Weller *et al.* [1951a, 1951b] worked with a rotating autoclave, which had a 1-hour heatup time. They used an isothermal kinetic scheme for their calculations, based on the peak temperature. Heating and cooling periods were assumed to add another 20 min to reaction time at peak temperature. During the catalytic "hydrogenolysis" in the absence of liquid solvent of an "anthraxylon" (vitrinite) fraction of Bruceton coal, "reaction was so rapid" that "no accurate value for the activation energy . . . can be deduced". Observing that "the simple Arrhenius relation . . . does not hold . . .", these authors calculated an activation energy of 150 kJ mol^{-1} for the hydrogenation of an "asphalt" fraction, from results at 430 and 440°C, "for which the greatest accuracy in *k*' was observed".

In an interesting early study, Hill *et al.* [1966] used a 1-litre autoclave into which Spring Canyon (Utah) coal was injected at the intended reaction temperature. The estimated 1–2 minute heatup time quoted in their paper corresponds to a heating rate of about 3.5–7°C s^{-1}. The authors state, "that at the initial stage of the experiment, the reaction is under diffusion control . . . this process has a very low activation energy." This sentence could be interpreted as referring to what we have described as the solvent extraction phase prior to the onset of covalent bond rupture. For purposes of the kinetic calculations, however, the coal was assumed in this study to have reached the reaction temperature instantaneously. No account was taken of possible successive stages and weight loss during heatup. They assumed a first-order dissolution process to take place in parallel with a second-order "extraction of interspersed materials" at the peak experimental temperature. Activation energies of 212 and 109 kJ mol^{-1}, respectively, were reported.

In a second model developed within the same report, the "first order reaction velocity constant" was found to vary with the fraction extracted at constant peak temperature. A gradual increase in the "enthalpy of activation" from 156 to 358 kJ mol^{-1} with increasing conversion (nearly 90%) was reported. The activation energy for the rate constant characterizing the *initial rate* at each temperature was found to be 134 kJ mol^{-1}. Despite the simplifications, the work clearly identified a sequence of process stages with corresponding distinct energies of activation. In this sense, it may be said to have foreshadowed the more detailed kinetic approach outlined above.

In modeling coal liquefaction and the subsequent reactions of extracts in a "segmented-bed" reactor, Shah *et al.* [1978] reported a coal-to-gas energy of activation of 357 kJ mol^{-1}. It is straightforward to show that if intervening steps within a set of *consecutive* reactions are ignored (and provided that pre-exponential constants are of comparable magnitude), calculated apparent energies of activation for the overall process may approach or even exceed the sum of the energies of activation for the intervening reaction steps. The same study found E_a (energy of activation) values closer to those in Table 5.7, of 230 and 169.5 kJ mol^{-1} for coal to "heavy oil" and coal to "furnace oil" conversion, respectively. In these calculations, isothermal kinetics and a single-stage coal dissolution step have been assumed. However, the results are more difficult to interpret than most since the presence of a preheater was indicated, operating at an unspecified temperature.

During their coal liquefaction experiments in tetralin, Gun *et al.* [1979] have also observed the reaction order and the energy of activation to change with time. These workers used a 2-litre stirred autoclave and assumed isothermal kinetics at peak temperature. The order of the reaction rate was reported to change from initial values of 1.0–1.2 to 2.0–2.2 and then to decrease to between 0.6 and 1.0. Activation energies during this three-step process were given as 332.5, 196, and 146 kJ mol^{-1}, respectively. Energies of activation were thus reported to have *decreased* with *increasing* reaction time. In view of the large thermal inertia of their reactor and the assumption that "experimental zero time" was assigned to the time when reaction temperature had been reached, it seems difficult to comment on these results in any detail.

Comparison with Kinetic Parameters Calculated from Pyrolysis Experiments: Similarities and differences between coal thermal breakdown during the initial stages of coal pyrolysis and liquefaction have been previously discussed [Li *et al.*, 1994; Zhuo *et al.*, 2003]. We will continue this discussion in Chapter 6. Generally, pyrolysis yields are even more sensitive than liquefaction conversions to reactor design. Not surprisingly, E_a values reported in the literature cover a very wide spectrum of values [Howard, 1981]. Total volatile yields from pyrolysis are also considerably lower than conversions normally expected from liquefaction. A comparison of energies of activation between pyrolysis and liquefaction experiments is only defensible (i) if it can be assumed that thermal breakdown constitutes the rate-limiting step in product release during pyrolysis and (ii) to the extent that stage B processes represent thermal breakdown in the liquefaction of coals. Within the present context, the first of these assumptions is not entirely justifiable. In addition to extensive recombination reactions, internal tar migration and tar release by mass transfer from external surfaces have been considered as significant resistances to product release from coal particles [Suuberg, 1985]. Furthermore, recombination reactions and diffusion limitations do not appear to affect the present liquefaction conversions, in any case not to the same extent. It may nevertheless be instructive to compare activation energies from the present liquefaction work and pyrolysis experiments from a study on similar coal samples in which a multiple parallel independent reaction model has been used.

The last column of Table 5.6 presents energies of activation and σ values (with units adapted to the present study) from the pyrolysis of the Argonne PCSP coals. A Rock-Eval reactor was used in these experiments [Burnham *et al.*, 1989]. The latter apparatus is not thought to be as free from extraparticle secondary reactions as "drop tube" or "wire-mesh" (heated-grid) instruments [Howard, 1981; Li *et al.*, 1993]. Comparing results with those from the present liquefaction study indicates somewhat higher pyrolysis E_0 values and some individual differences, notably for Wyodak and Pocahontas No. 3 coals. The results nevertheless appear to show the energies of activation from the two studies to be within the same range. σ values calculated from the pyrolysis data covered a narrower range, but progressively diminished with increasing rank, in line with the trend observed from liquefaction results in the present study.

5.8 Brief Overview of Liquefaction

In this Chapter, we have discussed the distinction between solvent extraction prior to and after the onset of extensive covalent bond rupture. For any given coal, the extractable material content prior to the covalent bond rupture stage is variable. It depends on the solvent used for extraction and the temperature. Our observations indicate that a coal is made up of materials on some sort of continuum of molecular masses. A more powerful solvent may extract more of the coal mass. On the other hand, a rise in temperature may change the rules. Some coal material may *become* soluble at these higher temperatures – without necessarily breaking covalent bonds. Rupturing covalent bonds would of course increase the amount of material extractable by a particular solvent.

In the next Chapter, we will review findings from electron spin resonance (esr) spectroscopy, showing the early onset of covalent bond scission reactions to occur as early as about 300–310°C. For most coals, the "esr" signal and extractable yields both increase rapidly within a band of temperatures (usually) around 350°C. This provides – indirect – evidence for *extensive* covalent bond scission, in other words, for the thermally induced breakdown (some would say depolymerization) of the coal.

During the latter process, the availability of a sufficient supply of hydrogen-donor solvent may quench reactive free radicals and chemically stabilize the extract. Alternatively, depolymerized material may be dissolved in a suitable solvent and carried away from the parent particle. In a batch reactor and using a relatively small amount of non-donor (albeit powerful) solvent, re-polymerization reactions have been identified by observing *increasing* char residues. If, however, the sample is heated in an inert gas, dry pyrolysis gives rise to greater extents of re-polymerization and the formation of larger char residues. A little surprisingly, the outcome turns out to be similar when coals are heated in inert liquid environments (e.g. hexadecane). Once again, we find that the outcome of an experiment – or of a processing step – is bound up not just with the chemical species involved but also by the configuration of the reaction vessel and relative abundance of the reagents.

To arrive at appropriate mathematical models for these processes, it is necessary to identify the rate determining steps, just like for any other process. The preceding analysis of coal liquefaction leads, in the first instance, to a two-stage process definition, each with its distinct kinetic constants and activation energies. The resulting model calculations gave different energies of activation. The magnitude of the energies of activation for the "extraction" phase turned-up values closer to those associated with diffusion and "E_a" values for "depolymerization" were closer to those associated with covalent bond rupture. There seems to be at least some semblance of correspondence between the model and the actual physical and chemical processes taking place.

In the next Chapter, we will attempt to draw together elements from our discussions on pyrolysis and liquefaction, to attempt a more detailed understanding of thermal breakdown in coals. In particular, we will attempt to explain how heating rates affect sample behavior during pyrolysis and seem less important in determining product distributions during liquefaction, when coal is heated in the presence of a donor-solvent. This will help us understand how the amount of extractable component in coals changes during heating at different rates and will tell us something of the behavior of poorly coking, as well as of premium coking coals.

References

Aida, T., Slomka, B., Shei, J. C., Chen, Y. and Squires, T. G. (1985) Am. Chem.Soc. Div. Fuel Chem. Prepr., 30(4), 274

Bartle, K.D., Bottrell, S., Burke, M.P., Jones, C., Louie, P.K., Lu, S.L., Salvado, J., Taylor, N. and Wallace, S. (1994) Int. J. Energy Res. 18, 299.

Brodzki, D., Djega-Mariadassou, G., Li, C-Z., Kandiyoti, R. (1994a) *Fuel* 73, 789

Brodzki, D., Abou-Akar, A., Djega-Mariadassou, G., Li, C-Z., Xu, B., and Kandiyoti, R. (1994b) *Fuel* 73, 1331

Brodzki, D., Abou-Akar, A., Djega-Mariadassou, G. and Kandiyoti, R. (1995) *Fuel* 74, 407

Burnham A. K., Oh, M.S., Crawford, R. W. Samoun, A. M. (1989) *Energy Fuels* 3, 42

Brunson, R. J. (1979) *Fuel* 58, 203

CCTP (1999). Technology Status Report 010; *Coal Liquefaction*, Cleaner Coal Technology Programme, Department of Trade and Industry/Pub URN 99/1120, London, UK.

Clarke, J.W., Kimber, G.M., Rantell, T.D. and Shipley, D.E. (1980) *ACS Symp. Ser. No. 139*, D.D. Whitehurst, Editor, 111

Comolli, A.G. and Zhou, P. (2000) 'Hydrocarbon Technologies Inc. Final Report to US DOE, AC22–92PC92148'. US Department of Energy, Pittsburgh, PA

Cronauer, D. C.; Shah, Y. T.; Ruberto, R. G. (1978) *Ind. Eng. Chem. Process Des. Dev*. 17, 281

Curran, G. P.; Struck, R. T.; Gorin, E. (1967) *Ind. Eng. Chem. Process Des. Dev. 6*, 167

Elliott, M. M. (1981) Editor *Chemistry of Coal Utilization Second Supplementary Volume*, John Wiley, NY,

Fowler, T. G., Bartle, K. D. and Kandiyoti, R. (1988) *Fuel* 67, 173

Fowler, T. G., Kandiyoti, R., Bartle, K. D. and Snape, C. E. (1989a) *Carbon* 27, 197

Fowler, T. G., Bartle, K. D. and Kandiyoti, R. (1989b) *Energy & Fuels* 3, 515

Fukuda, K. (2002) *PhD Thesis*, University of London

Fukuda, K., Dugwell, D.R., Herod, A.A. and Kandiyoti, R. (2004) *Energy & Fuels* 18, 1140

Furimsky, E. (1998) *Applied Catalysis A: General* 171, 177

Gavalas, G. R., Cheong, P. H-K., Jain, R. (1981) *Ind. Eng. Chem. Fundam*. 20, 113

Gibbins-Matham J. R. and Kandiyoti, R. (1988) *Energy Fuels* 2, 505

Gibbins, J. R. and Kandiyoti, R. (1990) *Fuel Proc. Tech*., 24, 237

Gibbins, J. R. and Kandiyoti, R. (1991a) *Rev. Sci. Instr*., 62(9), 2234

Gibbins, J. R. and Kandiyoti, R. (1991b) *Fuel*, 70, 909

Gibbins, J. R., Kimber, G., Gaines, A. F., Kandiyoti, R. (1991c) *Fuel* 70, 380

Gibbins, J.R., Gaines, A.F., Li, C. and Kandiyoti, R. (1994) *Intl. J. Energy Research* 18, 215

Gorin, E., (1981) in *Chemistry of Coal Utilisation-Supplementary Volume II* (M.A. Elliott, Ed.), J. Wiley, N.Y., 1485

Gun, S. R., Sama, J. K., Chowdhury, P. B., Mukherjee, S. K. and Mukherjee, D. K. (1979) *Fuel* 58, 171

Han, K. W. and Wen, C. Y. (1979) *Fuel* 58, 779

Harrison, J.S., Kimber, G.M. and Gray, M.D. (1989) *Intl. Conf. Coal Sci., Tokyo*, II, 655

Herod, A.A. and Kandiyoti, R. (1994) *Fuel* 73, 470

Hill, G. R., Hariri, H., Reed, R. I., Anderson, L. L. (1966) In *Coal Science*; Gould, R. F., Ed.; Advances in Chemistry Series 55; American Chemical Society: Washington, DC, p 427

Howard, J. B. (1981) in *Chemistry of Coal Utilization Second Supplementary Volume* (ed. Elliott, M. A.), Wiley, New York, p 665

Iino, M., Takanohashi, T., Ohsuga, H., Toda, K. (1988) *Fuel*, 67, 1639

Jungten, G. and van Heek, K. H. (1970) *Reaktion-ablaufe unter nichtisothermen Bedingungen*; Fortschritte der chemischen Forschung; Springer-Verlag: Berlin, 1970; Vol. 13, pp 601–699;

translated by Belov and Associates, Denver, CO, APTICTR-0776. Quoted by Howard [1981].

Jaffe, A. (1997), The James A. Baker III Institute for Public Policy of Rice University; Seminar Report "The political, economic, social, cultural and religious trends in the Middle East and the Gulf and their impact on energy supply, security and pricing" April 1997 http://www.rice.edu/energy/publications/docs/TrendsinMiddleEast_MainStudy.pdf

Kershaw, J. R., Barras, G. (1979) *ACS DFC Prepr.*, 24(3), 99

Kimber, G.M. and Davies, G.O., (1988) *Proc. Symp. Energy Production Processes, Inst. Chem. Eng.*, London.

Kimber, G.M. (1989) *Proc. Conf. Am. Filtration Soc.*, Pittsburgh (U.S.A.)

Kimber, G. M. (1997) *A history of UK coal liquefaction*, ETSU/DTI Report No. Coal R078

Koll, P. and Metzger, J., Angew. (1978) *Chem. Int. Ed. Engl.*, 17, 754

Li, C-Z., Bartle, K. D., Kandiyoti, R. (1993) *Fuel* 72, 3.

Li, C-Z., Madrali, E. S., Wu, F., Xu, B., Cai, H-Y., Guell, A. J., Kandiyoti, R. (1994) *Fuel* 73, 851

Li, C-Z., Wu, F., Xu, B. and Kandiyoti, R. (1995) *Fuel* 74, 37

Lowry, H. H (1963) Editor *Chemistry of Coal Utilization Supplementary Volume*, John Wiley, NY

McPherson, W. M., Foster, N. R., Hastings, D. M., Kalman, J. R., Okada, K., Heng, S. (1985a) *Fuel*, 64, 454

McPherson, W. M., Foster, N. R., Hastings, D. M., Kalman, J. R., Gilbert, T. G. (1985b) *Fuel*, 64, 457

Mochida I, Sakanishi K, Suzuki N, (May 1998) *Catal. Surv JPN* 2 (1): 17–30

Moore, S.A., Jones, M.A. and Kimber, G.M., (1989) *Intl. Conf. Coal Sci., Tokyo*, II, 663

Morita, M.; Sato, S.; Hashimoto, T. (1979) *Prepr. Am. Chem.Soc., Div. Fuel Chem.* 24, 63

Moroni, E. C. (1991) *Prepr. Pup.-Am. Chem.* Soc., *Div. Fuel Chem.* 36(2), 433

Nishioka, M. (1991) *Fuel* 70, 1413

Neuburg, H.J., Kandiyoti, R., O'Brien, R.J., Fowler, T.G. and Bartle, K.D. (1987) *Fuel* 66, 486

Onozaki, M., Namiki, Y., Ishibashi, H., Takagi, T., Kobayashi, M. and Marooka, S. (2000) *Energy and Fuels* 14, 355

Pullen, J. R. (1981) *"Solvent Extraction of Coal," IEA Coal Research: London*, p 65.

Shah, Y. T.; Cronauer, D. C.; McIlvried, G. G.; Paraskos, J. A. (1978) *Ind. Eng. Chem. Process Des. Dev.* 17, 288

Shin, S. C., Baldwin, R. M. and Miller, R. L. 1989 *Energy Fuels* 3,193

Slomka, B., Aida, T., Squires, T. G., (1985) *Am. Chem. Soc. Div. Fuel Chem.Prepr.*, 30(2), 368

Slomka, B., Aida, T., Chen, Y., Junk, G. A., Squires, T. G. (1986) *Am.Chem.Soc.Div. Fuel Chem.Prepr.*, 31(1), 238

Squires, T. G., Aida, T., Chen, Y., Smith, B. F. (1983) *Am. Chem. Soc. Div.Fuel Chem. Prepr.*, 28(4), 228

Q. Sun, J.J. Fletcher, Y Zhang and X. Ren (2005) *Energy & Fuels* 19, 1160

Takanohashi, T. and Iino, M. (1990) *Energy & Fuels* 4, 452

Suuberg, E. M (1985) in *Chemistry of Coal Conversion*; Schlosberg, R. H., Ed.; Plenum: New York, p 67

Weller, S.; Pelipetz, M. G.; Friedman, S. (1951a) *Ind. Eng. Chem.* 43, 1572

Weller, S.; Pelipetz, M. G.; Friedman, S. (1951b) *Ind. Eng. Chem.* 43, 1575

Vernon, L. W. (1980) *Fuel* 59, 102; quoted in Pullen [1981]

Vorres, K. S. (1990) *Energy & Fuels* 4, 420

Wiser, W. H. (1968) *Fuel* 47, 475

Xu, B., Madrali, E. S., Wu, F., Li, C-Z., Herod, A. A. and Kandiyoti, R. (1994) *Energy Fuels* 8, 1360

Xu, B. (1995) Ph.D. Thesis, University of London

Xu, B., Dix, M. and Kandiyoti, R. (1995) *Rev. Sci. Instrum.*, 66(7), 3966

Xu, B. and Kandiyoti, R. (1996) *Energy & Fuels*, 10, 1115

Zhang, S-F., Xu, B., Moore, S.A., Herod, A.A. and Kandiyoti, R. (1996a) *Fuel* 75, 597
Zhuo, Y., Herod, A.A. and Kandiyoti, R. (2003) "The thermochemical reactions of middle rank coals" in "Natural and Laboratory Simulated Thermal Geochemical Processes" (R.Ikan, Ed.) Kluwer Academic Publishers, Dordrecht Boston London, p 53

CHAPTER 6

Thermal Breakdown in Coals: Comparing Structural Changes in Pyrolysis and Liquefaction

Several of the initial, identifiable structural shifts that signal thermal breakdown in coals are common to pyrolysis and liquefaction. In this Chapter, we will try to set out what little we know about the similarities of the reaction pathways, when thermal breakdown in coals actually begins, how the paths begin to diverge, when and why heating rates affect product distributions and how retrogressive reactions actually work. We will present some early results from electron spin resonance spectroscopy (ESR), which provide valuable clues in these respects. We will adopt a primarily comparative approach that has provided useful insights in several previous attempts to improve our understanding of the complexities of thermal breakdown in coals [Li *et al.*, 1994; Zhuo *et al.*, 2003; Fukuda *et al.*, 2004]. In the process, we will try to answer some of the questions thrown up by research in modern coke making, as well as explain some findings relating to the now disappearing art of coal liquefaction.

6.1 Introduction

One obvious starting point for comparing thermally induced events in pyrolysis and liquefaction is an examination of larger char residues obtained during pyrolysis. To that end, we will first need to review some relevant findings from the previous three Chapters.

Raising the temperature to about 300°C does not appear to break many covalent bonds in coals. Results from solvent extraction experiments suggest that some of the smaller molecular mass materials, not covalently bound to larger molecules, are mobilized at temperatures up to 300–310°C. Evidence from ESR spectroscopy presented below suggests that some of these materials are free radicals and that the new freedom of movement allows some recombination reactions to take place, usually below 300°C.

While both pyrolysis and liquefaction processes are initiated by heating the coal sample, some differences may be observed early on. Many coals tend to swell in suitable solvents. We have also seen that large fractions of coals can be solvent-extracted, even at room temperature. The two phenomena are related; solvents that swell coals are often effective in extracting them. Furthermore, swelling appears to facilitate the movement of smaller, soluble molecules within sample particles. Above

300°C, thermally induced covalent bond scission is initiated and new extractable materials are released *within* the mass of coal particles. The proportions of the coal mass extracted in the presence of a suitable solvent prior to *extensive* covalent bond cleavage usually observed from about 350–375°C (see Figure 5.1) may be quite high.

During "dry" pyrolysis, these newly released extractables remain within the parent coal particles. We will show in Chapters 7 and 8 that "extractables" and coal itself, consist of molecules with a very broad and continuous range of molecular masses. Meanwhile, the data in Figure 6.2 (below) suggest that these materials can be remarkably stable and remain nearly unchanged as extractables over a time-scale of minutes, at temperatures as high as 400°C, and a little beyond. However, the trajectories of pyrolysis and liquefaction processes diverge in a major way once the temperature is raised above the level where covalent bond cleavage reactions give rise to extensive depolymerization. As already mentioned, the temperature band where these processes begin depends on the particular coal and is usually found somewhere around 350–375°C. We will encounter at least one exception, where the transition temperature occurred nearer 450°C.

Above the 350–375°C threshold, thermally induced depolymerization also takes place in "dry" pyrolysis. Tar/extract precursor free radicals are released from the solid matrix into the particle with increasing speed. Some of the lighter components are transported in bubbles through the plastic mass or diffuse through it to the external particle surfaces. These materials then come up against the solid-to-gas evaporation barrier [Suuberg, 1985; Suuberg *et al.*, 1981, 1985]. At or near the external surfaces, tar precursors may evaporate or repolymerize to a char or crack, producing some gas and smaller tar/extract precursors, which may in turn evaporate or undergo further pyrolytic degradation. With rising temperatures, the rates of cracking, evaporation and repolymerization all increase sharply. The competition between these reactions, partly governed by the rate of heating, goes some way toward determining the eventual product distribution.

We have also observed in Chapter 5 that a strong solvent that is not a hydrogen-donor may also remove considerable amounts of extractables, blocking or at least delaying some repolymerization reactions by diluting the extracts. We have seen evidence for this when extracting coals with 1-methylnaphthalene in the "flowing-solvent" reactor (Table 5.3). We have also seen, however, that in the absence of externally supplied hydrogen, even in excess solvent, some 10% of the coal mass *does* undergo retrogressive reactions, compared to extraction in a hydrogen-donor solvent (Figure 5.5a). In the "flowing-solvent" reactor, these repolymerization reactions appear to be taking place prior to removal by the solvent, from the particle into the bulk stream. Evidence for this is provided by the small amounts of insolubles carried by the process stream, normally far smaller that the 10% in question. This observation suggested that retrogressive char-forming reactions taking place within the coal particles must be relatively rapid. By contrast, retrogressive char formation reactions observed at 450°C in 1-methylnaphthalene in a mini-bomb reactor took hundreds of seconds (Figure 5.5a). It appears, we may thus be able to differentiate between fast and slow retrogressive reactions during coal liquefaction.

Compared to pyrolysis, removal of tar/extract precursors from parent particles into the surrounding solvent, presents a far less steep (solid-to-liquid) mass transfer barrier to overcome. However, not all classes of liquids perform the same function. Hexadecane is neither a hydrogen-donor *nor* a solvent for coal-derived material. Table 5.3 shows that in the case of this liquid devoid of hydrogen-donor and solvent properties, the mass transfer barrier between the solid and liquid phases was nearly as difficult to overcome as in the case of "dry" pyrolysis. The major distinguishing feature between pyrolysis and liquefaction thus turns out to be the *removal mechanism* from coal particles, of products already chemically released from the coal matrix.

So far, we have made a case for distinguishing between the chemical *formation* (and internal release) of tar/extract precursors and their removal from the coal particle. We now need to take a step back and explore the relationship between the *onset* of covalent bond scission and the internal *release* of tar/extract precursors. We will use data from the electron-spin-resonance spectroscopy of pyrolysing coals and from liquefaction experiments. These results will be used alongside data from the response of a weakly coking coal to changes in heating rate. We will compare the behavior of this coal with that of two prime coking coals under similar experimental conditions. The aim will be an attempt to improve our understanding of pre-pyrolysis phenomena in coals and the nature of retrogressive reactions during thermal breakdown.

6.2 The Electron Spin Resonance (ESR) Spectrometry of Thermal Breakdown

Coals ordinarily contain large numbers of free radicals. The population density of these free radicals increases with rank, typically from about 0.3×10^{19} spins gm^{-1} for a lignite to 1.7×10^{19} spins gm^{-1} in the case of an anthracite [Fowler *et al.*, 1989b]. Several early works on the electron spin resonance (ESR) spectrometry of carbonaceous materials [Singer, 1963; Retcofsky *et al.*, 1981] and their liquefaction behavior [Petrakis & Grandy, 1983] provide excellent background reading and are worth the reader's attention.

The earliest pyrolysis studies of coal by *in-situ* ESR spectrometry used vacuum cells placed within the spectrometer cavity [Austen, *et al.*, 1958; Smidt & van Krevelen, 1959], where volatiles were continuously removed from the reaction zone. However, later work studying free radical populations in pyrolysis was undertaken by heating coal samples in sealed (i.e. closed) quartz tubes [Petrakis & Grandy, 1983; Sprecher & Retcofsky, 1983]. These data showed maxima in spin concentrations above 400°C, suggesting that the populations of free radicals first increased with temperature and eventually declined at higher temperatures.

The Curie law is assumed in calculations of free radical concentrations. The procedure for determining the range of experimental variables over which assumptions inherent in the calculations of free-radical concentrations may be considered as valid is simple but exacting. The linearity of the temperature dependence of the ESR signal from pyrolysis *chars* must be ascertained. It turns out that the Curie law is poorly obeyed by samples pyrolyzed in sealed tubes. It may not surprise the reader that the problem appeared yet again to involve artefacts due to secondary interactions between evolved volatiles and residual solids, confined within the closed space of the sealed tubes.

To put this idea to the test, a shallow fixed-bed reactor was constructed of quartz, for carrying out pyrolysis experiments inside the spectrometer cavity. The design of this reactor deliberately emulated that of the fixed-bed geometry used in the hot-rod reactor (see Chapter 3). A continuous stream of inert gas was used to sweep evolving volatiles out of the reaction zone and out of the ESR cavity. The fixed-bed of sample was heated at 10°C min^{-1} to 480–500°C [Fowler *et al.*, 1987a, 1987b]. When chars from this open experimental system were examined, the temperature dependence of the ESR signal was clearly observed to be linear. Furthermore, for many of the coals studied, the maximum around 500°C, observed when using sealed tubes, had all but disappeared [Fowler *et al.*, 1987b; Gonenc *et al.*, 1988]. This suggests that the maxima observed in earlier work were artefacts of the closed cell configuration [Sprecher & Retcofsky, 1983]. We will need to return to this point below.

Figure 6.1 presents a schematic diagram of typical spin population vs. temperature curves observed, when coal samples were heated in a quartz fixed-bed reactor placed inside the cavity of an ESR spectrometer. In this work, spin populations (S) were defined as free radicals per gram of initial sample. Three distinct types of thermally induced processes have been identified.

Region I ($T_1 > T$): S increased to a relatively shallow maximum ($T_1 \sim 200$°C in Figure 6.1), due to the recovery of signal through desorption of gases adsorbed on

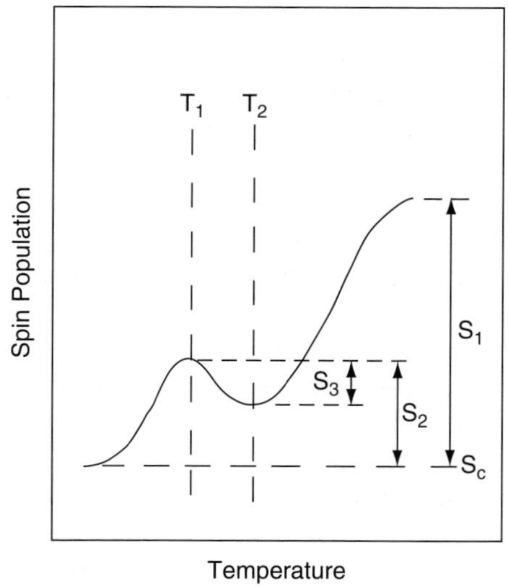

Figure 6.1 Schematic diagram of spin populations vs. temperature, defining parameters in Table 6.1. (Reproduced with permission: Carbon 1989, 27, 197; Copyright 1989 Elsevier)

Table 6.1 ESR parameters of coals given as "spin populations $\times 10^{-19}$" in the flow cell under slow sweep velocity conditions. (Reproduced with permission: Carbon 1989, 27, 197; Copyright 1989 Elsevier)

Coal	*Elemental C (%, daf)*	*Temp, T_1 (°C)*	*Temp, T_2 (°C)*	*Spin population S_C (g^{-1})*	*Spin population S_1 (g^{-1})*	*Spin population S_2 (g^{-1})*	*Spin population S_3 (g^{-1})*
Can[a]	54.2	250	310	0.3	0.8	0.46	0.12
Burning Star	75.5	220	310	0.8	2.9	0.54	0.23
Linby	83.0	205	310	1.07	3.26	1.19	0.33
Point of Ayr	85.4	220	325	1.37	1.98	0.58	0.3
Cortonwood[b]	87.2	250	340	1.36	3.21	0.67	0.23
Cynheidre[c]	95.2	–	–	1.72	–	–	–

[a] The three-region behavior not well developed
[b] Swelling forced part of the sample out of the cell
[c] Three region behavior not apparent with this coal

sample surfaces. These gases, primarily moisture and oxygen, would have been adsorbed by previous exposure to air.

Region II ($T_2 > T > T_1$): S decreased to a minimum ($T_2 \geq 300$°C in Figure 6.1). This decline is thought to be associated with recombination reactions, due to the thermally induced mobility of occluded material (i.e. extractables).

Region III ($T > T_2$): Above the temperature T_2, S climbed (slowly then fast) with rising temperature, signaling a monotonic increase in the stable free-radical population. Temperatures immediately above the minimum at T_2 are thought to mark the onset of covalent bond cleavage reactions.

As explained by Fowler *et al.* [1989a], it does not seem possible to observe (or quantify) the population of *reactive* free radicals. This is because their concentration at any time is small compared to the stable free radical populations in coals. Attempts to detect small differences between large actual and background readings would be lost in the experimental uncertainty of the measurement. Furthermore, the lifetimes of the more reactive free radicals are probably considerably shorter than scan times required by the types of spectrometers used in these studies (~30 s). What the ESR spectrometer *could* monitor with relative ease, however, were changes in the populations of stable free radicals. Explanations of how the free radical population calculation itself, in fact, assumes the population of free radicals to be a stable population have been given in the cited papers by Fowler *et al.*

Although the work was done some years ago, at the time of writing (Autumn 2005), the authors were unaware of recent work that might have overcome these experimental difficulties. It appears therefore that above T_2, increases in stable free radical concentrations represent the accumulating "debris" left over from already completed bond-scission reactions. Having removed evolving volatiles by means of the sweep gas from the vicinity of the fixed-bed of sample, the change in signal above T_2 (i.e. in Region 3) is presumed to reflect the changes in the concentration of stable free radicals embedded in the residual char matrix.

It was observed during these experiments that chars from *some* coals showed a drop in signal at temperatures above 400°C. In other words, the data showed an additional maximum, despite using the open flow-cell reactors. On the face of it, this maximum could have been interpreted as a *drop* in the stable free radical population. Experimentally, this maximum was always associated with an abnormal temperature dependence of the ESR signal. It was, eventually, shown to be an artefact associated with increasing electrical conductivities of some of the chars and unrelated to free radical populations.

During these experiments, it was also observed that increasing sweep velocities of inert gas forced through the quartz fixed-bed reactor gave rise to greater spin populations, from the start of Region III onward. In a parallel set of experiments undertaken in the steel "hot-rod" reactor described in Chapter 3, it was shown that tar yields increased with increasing sweep gas velocity. Taken together, the two results led to the conclusion that the rapid removal of pyrolysis tar vapors from the reaction zone caused increases in the spin population. The effect appears due to partial suppression of recombination reactions between tar molecules and the pyrolyzing char matrix, as increasing sweep gas velocities were instrumental in removing tar vapors from the reaction zone more rapidly and more completely [Gonenc *et al.*, 1988]. The direct observation of free radical recombination reactions within a pyrolyzing mass of coal is not an easy experiment to design. The experiment just described provided a second best: evidence for the *suppression* of recombination reactions between pyrolyzing coal particles within the ESR-cavity and evolved volatiles.

Temperatures of product evolution in liquefaction and pyrolysis: We have already observed that in liquefaction, extract release from coal particles is expected to increase sharply, between about 350 and 375°C, as a consequence of accelerating depolymerization reactions. On the other hand, in pyrolysis, tar precursors do not appear to overcome the evaporation barrier until about 400–425°C. Similarly, pyrolysis weight loss from coal samples in thermogravimetric balances usually begins well above 400°C, at times nearer 450°C.

Table 6.1 gives the temperatures and spin populations corresponding to Figure 6.1, measured for a small set of coals used in the study. The observed T_2 values (the minima) were well below temperatures where pyrolysis tars or liquefaction extracts are observed to be released from coals. It seems clear that coal samples climb a long way up the ESR curve in Region III (in Figure 6.1) before mass release is actually observed. This is consistent with the idea that more-than-one bond scission is necessary before large molecular fragments – attached to the solid matrix by more than one covalent bond – can be released, often as large free radicals. This description would explain the very gradual early increase in extractable material, released from the coal particles. The materials dissolving, say between 300 and 350°C, are probably mostly initially occluded materials. As shown in Table 5.3 materials released below about 350°C do not require a donor solvent for efficient removal from the reaction zone. What is needed up that point is a solvent capable of dissolving these coal derived materials.

In attempting to examine the evolution of pre-pyrolysis phenomena, we will next explore the state of coal particles in the 300–400°C temperature interval and the extent of extractables formation in "dry" heating mode. We will explore whether the way in which these extractables are formed is related to the heating-rate sensitivity of coals during pyrolysis.

6.3 Extractables as a Diagnostic Tool for Pre-Pyrolysis Phenomena

In a more recent study, Fukuda [2002] examined how the softening properties and swelling ability of certain coals might be improved by heating rapidly up to the softening temperature. The work aimed to examine mechanisms underlying pilot-plant findings, which showed that rapid preheating to ~400°C seemed to improve the coking properties of some ordinarily non-swelling and weakly coking coals.

6.3.1 *A new approach to coke-making*

The background to this work was an effort by several Japanese companies to develop a process aiming to reduce the amount of expensive coking coals used in coke-blends and to expand the range of coals that can be used in these operations [Sasaki *et al.*, 1998]. The process concept involves rapidly heating coal particles to ~400°C prior to injection into modified coke ovens, for conventional treatment from that point onward. At bench scale, Aramaki *et al.* [1996] have demonstrated the increased strength of resultant cokes, when weakly coking coals were "preheated" rapidly to temperatures up to 400°C. Tomita *et al.* [1996] also reported that the swelling ability and softening properties of coals could be improved with rapid "preheating" to their softening temperatures; however, the heating rates used in that work ranged between 5 and 500°C min^{-1}.

Useful summaries of the vast literature on conventional coal coking operations may be found in standard texts [e.g. Elliott, 1981; Speight, 1994]. The problem has drawn the attention of a distinguished line of researchers over decades. Central to this work is the recognition that the plastic properties of heated coals are closely related to their instantaneous contents of extractable materials. Early work by Brown and Waters [1966a; 1966b] showed that the inventories of chloroform extractables within coal particles correlated with their softening and agglomerating behavior. Using a melting coal (Pittsburgh No. 8), Howard and co-workers observed relationships between the duration of plasticity (at a given temperature), the extractable contents of the particles (at that temperature) and the temperature itself. They correlated the plasticity of a rapidly heated coal with the generation-destruction kinetics of (pyridine soluble) extractable material within the coal particles [Fong *et al.*, 1986a; 1986b]. Furthermore, some bituminous coals that remain morphologically unchanged when heated slowly (~1°C s^{-1}), have nonetheless been observed to soften and melt under rapid heating conditions (e.g. 1,000°C s^{-1}) [Hamilton, *et al.*, 1979, 1980; Gibbins-Matham & Kandiyoti, 1988]. Research on coke making is set to take advantage of such observations, to reduce the proportion of prime coking coals used in coke blends. These coals are usually more expensive and are becoming less available. Shifting the coke blend composition toward weakly coking coals is therefore commercially attractive.

6.3.2 The low temperature pyrolytic behavior of a weakly-coking coal

In the Fukuda [2002] study, experiments were designed for identifying (i) the range of "rapid" heating rates, which influence (promote) the formation and survival of extractables within the coal mass, and, (ii) the temperature interval in which fast heating is effective in causing discernible changes in the behavior of the coals. The atmospheric pressure wire-mesh reactor used for these experiments (see Chapter 3) was capable of multi-stage heating between pre-set temperatures.

Sample properties: Table 6.2 presents some properties of the three coals selected for Fukuda's study. Newcastle Blend coal (NCBC) was a weakly coking coal while Goonyella and K-9 were prime coking coals. As we will see below, K-9 was a blend with some rather peculiar properties. The 106–150 μm size range of all three samples was used for volatile and tar yield measurements in the wire mesh reactor. The choice of particle size range was partly determined by the smallest size of stainless steel mesh

Table 6.2 Characteristics of the three Australian coal blends used in the study. Legend: "TM": Total Moisture (%); "VM": Volatile Matter (%); "FC": Fixed Carbon (%); "CSN": Crucible Swelling Number (−); "Softening": Softening Temperature (°C); "Max. temp": Maximum Fluidity Temperature (°C); "Resol. Temp": Resolidification Temperature (°C); "MF: Maximum Fluidity". "NCBC": Newcastle Blend Coal

		NCBC	*Goonyella*	*K-9*
Elemental composition (% dry basis)	Carbon	83.6	87.7	90.7
	Hydrogen	5.6	5.0	4.6
	Sulfur	0.6	0.6	0.2
	Nitrogen	1.8	1.7	0.8
	Oxygen	8.3	4.8	3.3
Maceral composition	Vitrinites	72.6	61.5	82.3
	Liptinites	3.1	0.98	0
	Inertinites	26.8	37.9	20.8

	Proximate Analysis					*Gieseler Plastometer*			
Coal Brand	*TM*	*VM*	*FC*	*Ash*	*CSN*	*Softening*	*Max. Temp.*	*Resol. Temp.*	*MF*
NCBC	3.4	35.2	52.4	9.0	3.5	392	430	457	2.0
Goonyella	1.5	24.1	65.1	9.3	6.5	399	455	493	3.0
K-9	0.9	17.9	72.3	8.9	8.0	444	485	505	1.3

able to accommodate the easy-to-handle 50 μm diameter thermocouple wires. However, in cases where chars from coals ground to 106–150 μm were used for measurements of extract yields in NMP [1-methyl-2-pyrrolidinone], significant amounts of sample were observed to escape through the wire mesh (aperture size: 63 μm × 63 μm) during the extraction procedure. The problem was avoided by using a larger size fraction of 212–250 μm for extract yield determinations.

When NCBC coal was heated at 1,000°C s^{-1} to 400°C, about 20 s was required for volatile and tar release to die down. During this time, the weight loss increased from a little over 2% to about 11%. Rapid heating tends to require longer hold times at peak temperature for devolatilization processes to reach completion. This is because exposure to higher temperatures is short during rapid heat-up, compared to slow heating. Shorter hold times are needed at higher peak temperatures. For example, at 700°C about 1 second is needed for completing volatile release [Gonenc *et al.*, 1990]. To keep a safe margin during the low temperature runs described below, 30 sec holding was allowed for all experiments involving NCBC coal.

The effect of changes in heating rate on product distributions during atmospheric pressure coal pyrolysis at temperatures *above* the 400–500°C range is well-documented [Gibbins-Matham & Kandiyoti, 1988; Li *et al.*, 1993a; 1993b; also see Chapter 3]. Between 300 and 400°C, however, weight losses were small and did not exceed 12–14%. Greater weight loss could, nevertheless, be observed for rapid (1,000°C s^{-1}) heating compared to "slow" heating at 1°C s^{-1}. Tar yields were small but gave similar trends [Fukuda *et al.*, 2004].

Char extraction experiments: Figure 6.2 presents NMP-extract yields from NCBC chars, prepared using two distinct time-temperature programs. The upper line shows

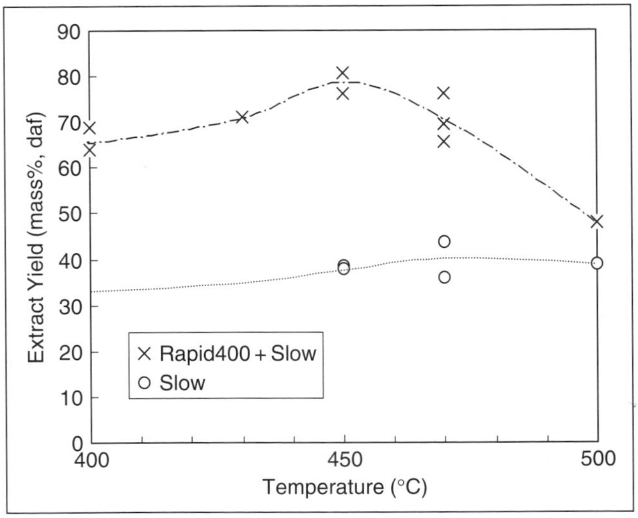

Figure 6.2 Comparison of NMP-extract yields between '*rapid400 + slow*' and simple '*slow*' heating (NCBC coal) (Reproduced with permission from Energy & Fuels 2004, 18, 1140; Copyright Amer. Chem. Soc 2004).

extract yields from chars heated at 1,000°C s^{-1} to 400°C, then heated to the peak temperature shown, at the rate of 1°C s^{-1}. This heating program was termed "rapid400 + slow". The "slow" heating pattern denotes heating the sample at 1°C s^{-1} to the peak temperature. At 400°C, the proportions of extractables were greater by a little over 25% when samples were heated rapidly (at 1,000°C s^{-1}) compared to heating at 1°C s^{-1}. The corresponding extract yields for chars heated *slowly* at 1°C s^{-1} to temperatures between 300 and 400°C were of the order of 33 ± 2%, significantly less than yields from rapidly heated samples. By contrast, the unheated coal gave ~35% extract. Extractable contents *increased* still further (upper line) when the temperature was then raised at 1°C s^{-1} from 400°C, reaching a maximum value of nearly 80% of the original coal mass between 450 and 470°C. Above 470°C, the plastic mass rapidly re-solidified.

Figure 6.3a shows the extent to which the increase in extractable content was found to be temperature dependent. The transition to high extractable yields occurred between 370 and 400°C. It was sharp and repeatable. The coal mass actually volatilized under these conditions does not exceed ~12–14% (Figure 6.7 in Fukuda *et al.*, 2004) and represents a significantly smaller proportion of the coal mass, compared with amounts of extractable material released within the particle. Figure 6.3b shows the extent to which this increase was heating rate, as well as temperature (Figure 6.3a) dependent. The transition between 500 and 1,000°C s^{-1} was, again, sharp and repeatable. Furthermore, when NCBC coal was heated in a thermogravimetric balance at 3°C s^{-1}, the onset of extensive weight loss occurred a little above 400°C, with maximum weight loss observed around 450°C. For most coals, this is a common temperature interval for massive weight loss in a TGA apparatus. Clearly, weight loss occurs at higher temperatures than many pre-pyrolysis phenomena including extractables formation and accumulation.

Above 450°C, tar release during pyrolysis is related to the competition between the escape of volatile species and their repolymerization reactions. When heated rapidly and

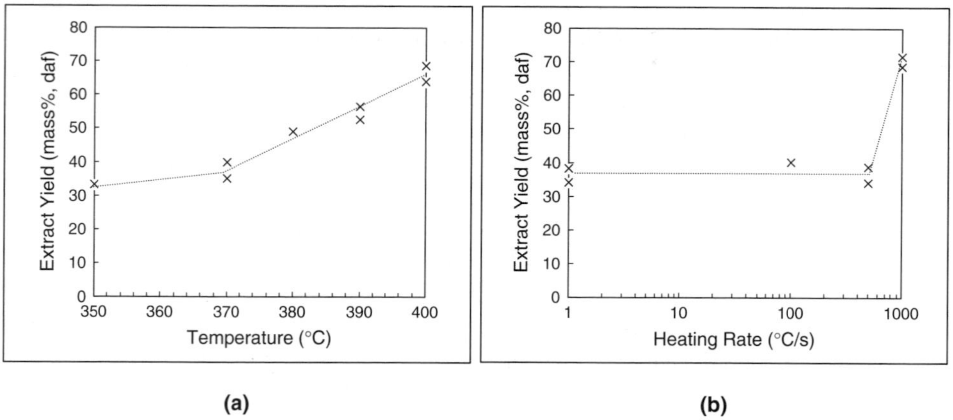

Figure 6.3 (**a**) Relationship between peak temperature and NMP-extract yield; heating rate: 1,000°C s^{-1}. (**b**) Relationship between heating rate and NMP-extract yield (370–400°C). (Reproduced with permission from Energy & Fuels 2004, 18, 1140; Copyright Amer. Chem. Soc 2004).

directly to temperatures above the 450–500°C level, many (but not all) coals and macerals give greater tar and total volatile yields compared to slow heating. The difference between yields at low (1°C s^{-1}) and high (1,000°C s^{-1}) heating rates might be as large as 6–8%, often mostly corresponding to the increase in tar yields due to fast heating (Chapter 3). With increasing heating rates, the sequence of pyrolytic events appears to get telescoped into a narrower time frame and shifted up the temperature scale. Therefore, when coals that show sensitivity to changes in heating rates are heated at 1,000°C s^{-1}, tar release appears to reach completion at temperatures nearer 700°C, compared to ~550–600°C for slow heating, say, at 1°C s^{-1}.

Figures 6.2 and 6.3b show that the proportion of the original coal mass released within the particle as NMP-extractables is a direct function of the heating rate. When coal particles are rapidly heated to 700–1,000°C, their greater content of extractables translate into a larger pool of tar pre-cursors. In previous work, Gray [1988] has ascribed increased tar yields obtained at high heating rates to greater tar survival through the explosive ejection of tar precursors (i.e. extractables). The quick rise of the internal pressure due to volatiles evolution within coal particles is probably instrumental in enhancing this outward bulk flow. When coupled to the "explosive ejection" model, the larger pool of extractables identified in rapidly heated particles help explain why the extra volatiles identified during fast heating consist mostly of tars. The flow would quicken the escape of the greater pool of tar precursors and reduce the probability of their undergoing secondary (retrogressive) reactions. However, how the excess extractable material is formed during rapid heating remains to be explained.

6.3.3 The link between findings from ESR spectroscopy and extract yield

At this point, some data directly linking changes in ESR spin populations with extract yields would have completed the picture. We would have been able to gage how far up the temperature and spin population scales we would have to climb before observing the formation of extractable material by the break-up of the solid matrix. The one study the authors are aware of, which makes a direct link between ESR spin populations and extractable yields is that of Neuburg *et al.* [1987]. However, the solvent used in that study was chloroform, a less polar and less powerful solvent compared to NMP. The latter was adopted as standard solvent several years later. The coals examined in the Neuburg *et al.* study were Linby (low-rank bituminous) and Cortonwood Silkstone (good coking) coals. The ESR behavior of these samples has been summarized in Table 6.1.

In the Neuburg *et al.* [1987] study, extraction yields from Linby coal showed a shallow minimum around 350°C, rising steadily thereafter. However, extracts from Cortonwood Silkstone coking coal increased steadily after 300°C. In other words, extract yields from both samples qualitatively shadowed their ESR spin population curves. The use of chloroform (rather than NMP) in these experiments does not allow detailed comment or direct comparison with the NCBC data of Figures 6.2 and 6.3a. Nevertheless, the direct observation of parallel increases in extractable yields and free radical populations with increasing temperature is useful in conceptually linking the two sets of results. Much of what was already glimpsed by ESR spectroscopy as taking place

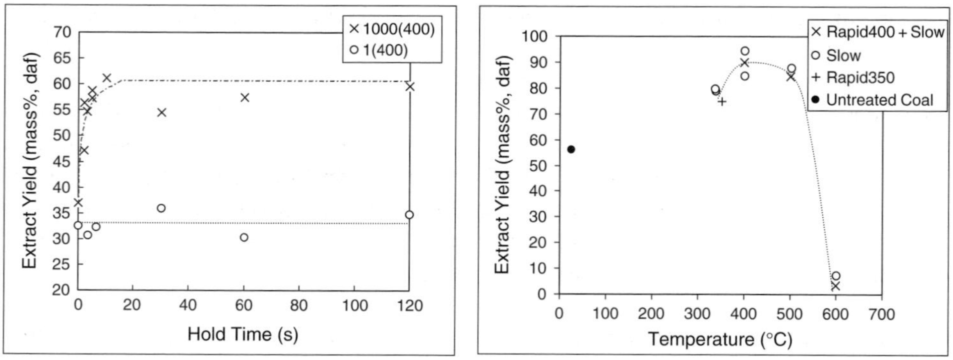

(a) (b)

Figure 6.4 (**a**) Effect of holding time at 400°C on NMP-extract yields for NCBC samples heated at 1 and 1,000°C s^{-1}. (**b**) Relationship between NMP-extract yield and temperature for samples of Goonyella (coking) coal heated at 1 and 1,000°C s^{-1} to 400°C, followed by heating at 1°C to a variable peak temperature up to 600°C (30 s holding at peak temperature). (Reproduced with permission from Energy & Fuels 2004, 18, 1140; Copyright Amer. Chem. Soc 2004).

within (several other) coal samples between 300 and 400°C could in fact be read into the NCBC data on changes in extract yields. However, that is as far as we can go without fresh ESR data.

Figure 6.4a indicates that the internally released extractables were remarkably stable for as long as 120 s. It would have been interesting to continue the experiments for longer times. It is clear, however, that the repolymerization reactions of these extractables proceed remarkably slowly at 400°C. Meanwhile, Figure 6.4b indicates that Goonyella (prime coking) coal showed no discernible heating rate sensitivity regarding the formation of extractables when heated at 1 and at 1,000°C s^{-1}. This finding clearly contrasts with the behavior of NCBC coal. Let us recall that NCBC only softens when heated rapidly (and *not* when heated at 1°C s^{-1}) while Goonyella becomes fluid when heated, irrespective of the heating rate. We will return to this point in the next two Sections.

6.3.4 On the chemical stability of extractables at 400°C

Figure 6.4a presents extract yields from NCBC coal as a function of holding time at 400°C. For the sample heated at 1,000°C s^{-1}, the extract yield climbed during the first 5 seconds, probably reflecting continuing extract formation (depolymerization) reactions *after* the sample had reached 400°C. For heating at 1°C s^{-1}, extract formation appears to have been completed by the time the sample had reached 400°C. At this temperature, the depletion (i.e. repolymerization) rate of NMP-extractables appears negligible over the time interval of 120 s.

The stability of extractable materials within coal particles for up to 2 minutes (and probably beyond, had the experiment been continued) might, in fact, have been expected from the data of Fong *et al.* [1986a]. These authors reported a depletion rate for pyridine extractables for the higher temperature interval of 600–800°C, characterized by the reaction rate constant

$$k = 1.9 \times 10^{10} \exp(-21{,}200/T) (s^{-1}).$$

The depletion rate for extractable materials calculated for 400°C by this equation is almost negligible. It also appears that depletion by volatilization of the extractables is not a significant factor at 400°C. During the first few seconds after reaching 400°C, the relatively small loss by devolatilization appears to have been made up by new extractable formation.

The stability of the extractables during at least 2 minutes would allow time for agglomeration within the coke ovens, which the coal particles would next be fed into. Thus, from these data, it seems reasonable to link the observed increase in extractables content arrived at by rapid heating, to improved coke strengths via the related increase in thermo-plasticity.

The pyrolysis of coals that form a plastic phase upon heating has been the object of previous studies. Howard and co-workers have reported parallel pyrolysis experiments in a high heating-rate plastometer and a wire-mesh pyrolysis reactor. The viscosity of the "plastic" phase of Pittsburgh No. 8 coal was monitored using the plastometer. In parallel, pyrolysis chars prepared at similar temperatures in the wire-mesh reactor were rapidly quenched with liquid N_2 and extracted with pyridine. In the viscometer, the low viscosity phase identifiably overlapped with the time-temperature interval during which the sums of evolved volatiles and recovered pyridine-extracts (from the wire-mesh reactor) went through a maximum [Fong *et al.*, 1986a; 1986b]. These experiments provided a clear link between the *decrease* in the amount of pyridine soluble material within the solid substrate as a function of increasing temperature and increasing time *at* temperature (in the 600–800°C range). The latter effect coincided with the increasing viscosity of the mass of coal during resolidification. According to results from Gieseler plastometry (Table 6.2), NCBC coal displayed melting behavior between 392 and 457°C. Clearly, Gieseler plastometers do not allow high heating rates. Once again, however, the temperature of maximum extractable content corresponded to the temperature of maximum thermo-plasticity, to a reasonable approximation.

6.4 Hydrogen Donors in Coals: Liptinites and Others

Developing ideas originally formulated by Brown & Waters [1966a, 1996b] and Wiser [1968], Neavel [1981] has summarized a consensus view on plasticity in coals as representing a transient hydrogen donor process. Within this framework, the coal itself is viewed as supplying the solvating and hydrogen-donating vehicle. The hydrogen-donor ability of the plastic phase is held to reside in the hydroaromatic component. During pyrolysis, the extent of softening and the magnitude of tar yields are thought to depend on local hydrogen availability and to be "directly proportional" to hydroaromatic hydrogen content.

In this description, the pyrolytic process prior to evaporation of tars may be viewed as a softening stage, an (almost) internal liquefaction step. This image of hydrogen donation by hydroaromatic groups during pyrolysis must owe something to what we know of liquefaction in tetralin. On raising the temperature further, this stage would be followed by the usual parallel evaporation, charring, and cracking reactions already described. Of course, resolidification is largely suppressed during liquefaction, provided the bulk of the plastic phase is dissolved in the available liquid medium. When the liquid medium is a hydrogen-donor, it also serves as a source of hydrogen for quenching and stabilizing the free radicals formed by bond scission. The Introduction to the present Chapter has already rehearsed the role of non-donors and inert liquids such as hexadecane. The properties of the fluid medium within which the coal is heated thus appear to provide the critical difference between pyrolysis and liquefaction.

Some of these ideas are central in considering the behavior of coals that do not *ordinarily* soften, such as Linby or NCBC coals. We have already observed in Chapter 3, that no strict dividing line exists between melting and non-melting coals. Some coals, like Linby (UK), and like NCBC (Australia), do not melt when heated slowly ($\sim$1°C s^{-1}) but do soften when heated rapidly (1,000°C s^{-1}). These low rank bituminous coals appear to be transitional between sub-bituminous and high volatile bituminous coals on the one hand and the readily melting and swelling middle rank coals (including coking coals) on the other.

Unfortunately, we do not seem to have a direct, easy to use and reliable method for determining hydroaromatic content in coals and coal derived materials. Hydroaromatic carbon content has been reported to range between $\sim$10 to 15% of total carbon for coals containing 80–83% elemental carbon, possibly corresponding to between 40–50% of the total aliphatic carbon content. Hydroaromatic content in coals is thought to decrease with increasing coal rank [Reggel *et al.*, 1968, 1971, 1973; Snape *et al.*, 1985].

Figure 6.5 presents plots of pyrolysis tar yields against (a) FT-ir derived aliphatic/aromatic hydrogen ratios (H_{ali}/H_{aro}) and (b) ^{13}C-nmr derived aliphatic (15–37 ppm) band

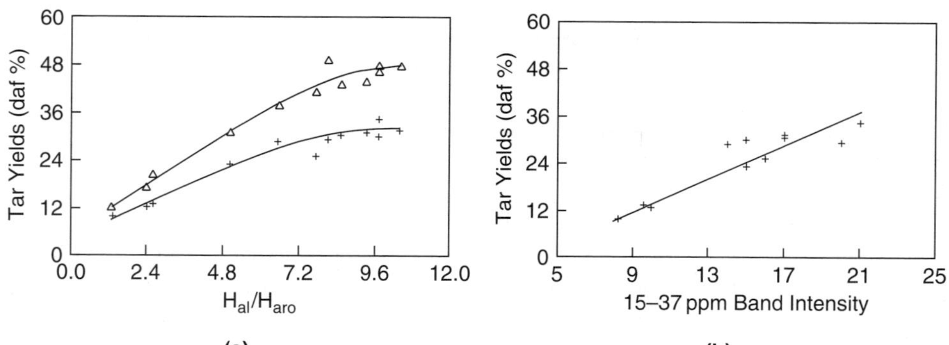

Figure 6.5 (**a**) Pyrolysis yields as function of FT-ir-derived aliphatic: aromatic hydrogen ratio for a rank-ordered series of Northern Hemisphere coals: + tar, △ total volatile. Pyrolysis in atmospheric pressure helium at 1000 K/s to 700°C with 30 s holding time. Non-melting coals: Taff Merthyr, Emil Mayrisch, Tilmanstone; melting coals: Heinrich Robert, Santa Barbara, Longannet, Candin, Bentinck, Thoresby, Gelding, Linby, Illinois No.6. (**b**) Pyrolysis tar yields as a function of ^{13}C-n.m.r-derived 15–37 nm aliphatic band intensity for a rank-ordered series of Northern Hemisphere coals. Coals and pyrolysis conditions as in (a). (Reproduced with permission: Fuel 1994, 73, 851; Copyright 1994 Elsevier).

intensities, respectively. These data, collected for a rank ordered set of Northern Hemisphere coals, showed reasonably smooth trends of increasing yields with increasing aliphatic content. Clearly, both sets of data lump signal from alkyl and alicyclic/hydroaromatic structures. However, in view of the performance of hexadecane during liquefaction (Table 5.3), it seems reasonable not to assign much of the changing activity to alkyl structures.

It now seems reasonable to assume (but difficult to *prove*) that internal hydrogen transfer from hydroaromatic species contributes significantly to the softening behavior of coals. This assumption would, at least, lend us a language in which we can begin to explain the behavior of coals like Goonyella or indeed the behavior of liptinites and heating-rate sensitive vitrinites, described in the "Pyrolysis" Chapter of this book (Tables 3.6–3.8).

Among much other detail, Figure 6.6 shows that the difference between liptinite conversions in pyrolysis and liquefaction were small compared to the magnitudes of analogous differences in conversion for the other macerals. What we know of liptinites suggests that they contain larger proportions of donatable hydrogen compared to vitrinites and inertinites from the same coal. A little counter-intuitively, there is also evidence from liquefaction experiments showing that liptinites generally dissolve more *slowly* at higher temperatures and less completely at lower temperatures [Neavel, 1981; Joseph *et al.*, 1991; Gaines *et al.*, 1991; Li *et al.*, 1991]. Their molecular mass distributions tend to be broader, although these data obtained using tetrahydrofuran (THF) as eluent [Li *et al.*, 1993a; 1993b], should be reworked using NMP as eluent (also see Chapter 8). For the two predominantly vitrinitic coals considered in Figure 6.6, results for "whole coals" were quite similar to those of their respective vitrinite concentrates. In contrast to the liptinites, the vitrinite (and whole coal) pyrolysis conversions were considerably smaller

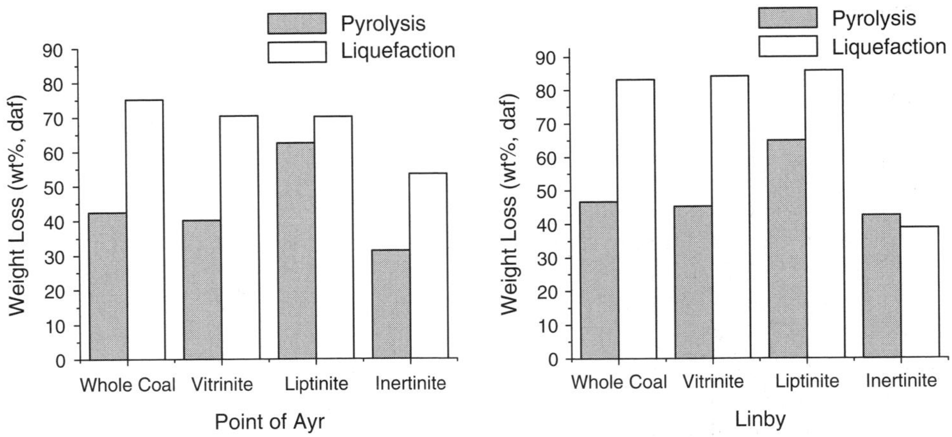

Figure 6.6 Sample weight loss in pyrolysis and liquefaction of whole coals and maceral concentrates. Pyrolysis: 1000°C s^{-1} to 700°C with 30 s holding time in atmospheric pressure helium. Liquefaction: 5°C s^{-1} to 450°C with 400 s (Point of Ayr) or 100 s (Linby) holding time. Tetralin flowing at 0.9 ml s^{-1} under 70 bar pressure. (Reproduced with permission: Fuel 1994, *73*, 851; Copyright 1994 Elsevier).

than those in liquefaction. It was interesting to note, however, that during liquefaction in tetralin – with a glut of donatable hydrogen available in the flowing-solvent reactor – vitrinite concentrate conversions were comparable to those of the liptinite concentrates.

6.5 Overview of Thermal Breakdown in Coals

We have already established that pre-pyrolysis phenomena such as covalent bond cleavage (which produces rising ESR spin populations) and internal extract release are concurrent phenomena. We have also observed that both effects need to be fairly well advanced before sample weight loss can proceed to any significant extent. However, we have no data from fast heating experiments in the ESR spectrometer cavity; experimental difficulties appear as insurmountable now as they did in the mid-1980s. The experiments described in these Chapters may be regarded as exploratory. As ever, we would need to experiment with more coal samples to confirm these observations. However, the emerging picture is a coherent one. Data presented in Chapters 3 and 6 show that pyrolysis tar yields from transitional coals such as Linby and NCBC are highly dependent on the heating rate. Linked to this general observation, we can see that the build-up of extractables within particles of "transitional" coals is bound up with higher heating rates, which favor the survival of a greater proportion of extractables within the coal particles.

In Section 3.2.2, we put forward two mechanisms in attempting to explain the sensitivity of pyrolysis tar yields of some coals to changes in the heating rate. The first, involved greater tar survival through rapid ejection of tar precursors and is straightforward. We also suggested that sample-derived hydrogen and hydrogen-rich light hydrocarbons may be incorporated into the pyrolyzing mass more efficiently during fast heating, to quench and stabilize tar precursor free radicals. Fast heating tends to telescope the sequence of events into a narrower time frame and shift the temperature scale upward. During fast heating, hydrogen known to be released within pyrolyzing solids from about ~300°C during slow heating [*e.g.* cf. Neuburg *et al.*, 1987] would remain in contact with the pyrolyzing mass up to higher temperatures during fast heating. We do not have precise models about the mechanism by which the alicyclic/hydroaromatic components donate their hydrogen. It appears consistent with the data in hand to argue that evolving and/or donated hydrogen (from alicyclic/hydroaromatic species) reacts with, or is donated to, some of the internally released free radicals. Such reactions would tend to block some of the char forming recombination reactions. Consistent with this view, we have seen how NCBC coal internally retained a larger proportion of extractable material intact, during heating at 1,000°C s^{-1} (~61%) compared to 1°C s^{-1} (~33–34%). Between 25 and 28% more of the mass of coal appears to have survived as "extractables" in the case of fast heating. Furthermore, these extractables were chemically sufficiently stable to survive intact for at least 120 seconds, when the experiments were interrupted – in retrospect an unfortunate saving of time.

As a corollary, there appears to be a whole class of *reactive* free radicals internally released by the coal, which if not quenched by whatever happens during fast heating (we *think* they are quenched by internally generated or donated hydrogen and light hydrocarbons) would undergo *immediate* recombination reactions. That would give rise to low extractable accumulation observed during slow heating of NCBC coal. According to our model and observations, a small but critical amount of hydrogen would escape during slow heating.

It may be noted from Figure 6.3a above, however, that the difference in extract yields is manifested from "0 seconds holding" onward. In other words, the reactions that have served to quench and stabilize between 25 and 28% of the coal mass, causing them to remain as extractables during heatup, must have been relatively rapid. They must have been completed during heatup at the rate of 1,000°C s^{-1}. Conversely, the char formation reactions occurring during slow heatup to 400°C (i.e. at temperatures below 400°C) must have had low activation energies.

These reactions probably involved the more reactive of the coal free radicals. This observation is in glaring contrast to the demonstrated chemical stability of extractables, at temperatures between 300 and 400°C. Figures 6.3a and 6.3b clearly show extractables remaining intact for over 120 s. Returning to Figure 5.5b, it seems clear that the slow rate of solid *char* formation (drop in conversion) observed at the higher temperature of 450°C also shows the progress of *slow* recombination reactions.

On the other hand, we found a large difference in conversion at 100 s between the flowing-solvent reactor and the mini-bomb, when both were operating at up to 450°C in 1-methylnaphthalene. We think we can explain the difference by sample dilution in the flowing-solvent reactor, without being able to say much about the speed of the recombinations. However, the difference of slightly over 10% between operation in tetralin and in 1-methyl naphthalene in the flowing solvent reactor would have taken place under nearly identical dilution conditions. The recombinations involving about 10% of the coal mass are likely to have been rather rapid.

Taken together, these data provide evidence for the occurrence of both fast and slow retrogressive repolymerization reactions. Stated more simply, during the thermal breakdown of coals, some of the free radicals formed are more reactive than others. For transitional coals, the effect appears closely related to the sensitivity of pyrolysis yields to changes in heating rate. Compared with premium coking coals, it seems likely that these transitional coals are those marginally deficient in donatable hydrogen and indeed, probably more richly endowed in hydrogen scavengers such as sulfur and oxygen. Table 6.2 of this Chapter clearly shows lower oxygen contents for the premium coking coals.

So far, relatively little has been said about the molecular structures and mass distributions of tars and extracts that we have encountered. The next two Chapters will attempt to address some of the issues involved in estimating the molecular masses and unraveling the structures of these complex molecules.

References

Aramaki, T.; Arima, T.; Yamashita, Y.; Inaba, A. (1996) *Tetsu to Hagane* 82, 5, 34

Austen, D. E. G., Ingram, D. J. E. and Tapley, J. G. (1958) *Trans. Faraday Soc.*, 54, 400

Brown, H. R., Waters, P. L. (1966a) *Fuel*, 45, 17

Brown, H. R., Waters, P. L. (1966b) *Fuel*, 45, 41

Elliott, M. A., (1981) Ed. *Chemistry of Coal Utilization, Second Supplementary Volume*; Wiley: New York

Fong, W. S., Khalil Y. F., Peters W. A., Howard J. B.(1986a) *Fuel*, 65, 195
Fong W. S., Peters W. A., Howard J. B. (1986b) *Fuel*, 65, 251
Fowler, T. G., Bartle, K. D., Kandiyoti, R. (1987a) *Fuel*, 66, 1407
Fowler, T. G., Bartle, K. D., Kandiyoti, R. (1987b) *Carbon*, 25, 709
Fowler, T. G., Bartle, K. D., Kandiyoti, R. (1989a) *Energy and Fuels*, 3, 515
Fowler, T. G., Kandiyoti, R., Bartle, K. D., Snape, C. E. (1989b) *Carbon*, 27, 197
Fukuda, K., (2002), PhD Thesis, University of London
Fukuda, K., Dugwell, D.R., Herod, A.A. and Kandiyoti, R. (2004) *Energy & Fuels* 18, 1140
Gaines, A. F., Li, C-Z, Bartle, K. D., Madrali, E. S., Kandiyoti, R. (1991) *Proc. Intl. Conf. Coal Sci.*, Newcastle-upon-Tyne, UK, 830
Gibbins-Matham, J. R. and Kandiyoti, R. (1988) *Energy Fuels* 2, 505
Gonenc, Z. S., Fowler, T. G., Kandiyoti, R. Bartle, K. D. (1988) *Fuel* 67, 848
Gonenc, Z. S.; Gibbins, J. R.; Katheklakis, I. E.; Kandiyoti, R. (1990) *Fuel* 69, 383
Gray, V. R. (1988) *Fuel* 67, 1298
Hamilton, L.H., Ayling, A.B., Shibaoka, M. (1979) *Fuel* 58, 873
Hamilton, L.H. (1980) *Fuel* 59, 112
Joseph J.T.; Fisher R.B.; Masin C.A.; Dyrkacz G.R.; Bloomquist C.A.; Winans R.E. (1991) *Energy and Fuels* 5, 724
Li, C-Z., Gaines, A. F., Kandiyoti, R. (1991) *Proc. Intl. Conf. Coal Sci.*, Newcastle-upon-Tyne, UK, 508
Li, C.-Z.; Bartle, K. D.; Kandiyoti, R. (1993a) *Fuel* 72, 3
Li, C.-Z.; Bartle, K. D.; Kandiyoti, R. (1993b) *Fuel* 72, 1459
Li, C-Z., Madrali, E. S., Wu, F., Xu, B., Cai, H-Y., Güell, A. J., Kandiyoti, R. (1994a) *Fuel*, 73, 851
Neavel, R. C. (1981) *Coal Science, Volume I* (ed. Gorbaty, M. L., Larsen, J. W., Wender, I.), Academic Press, New York, 1–19
Neuburg, H. J., Kandiyoti, R., O'Brien, R. J., Fowler, T. G., Bartle, K. D. (1987) *Fuel*, 66, 486
Ohtsuka, Y., Wu, Z., Tomita, A. and Itagaki, S. (1996) *Tetsu to Hagane*, 82, 5, 28
Petrakis, L. and Grandy, D. W. (1983) *Free radicals in coals and synthetic fuels*, Elsevier
Reggel, L., Wender, I., and Raymond, R., (1968) *Fuel* 47, 373
Reggel, L., Wender, I., and Raymond, R., (1971) *Fuel* 50, 152
Reggel, L., Wender, I., and Raymond, R., (1973) *Fuel* 52, 162
Retcofsky, H. L., (1981) *Coal Science Volume I*, (ed. Gorbaty, M. L., Larsen, J. W., Wender, I.) Academic Press, NY
Sasaki, M., Komaki, I., Matsuura, M., Sato, K., Fukada, K. (1998) *Proc. ICSTI Ironmaking Conference*, 803
Singer, L. S. (1963) *Proc. 5th Carbon Conf. 1961*, Pergamon, NY, 37
Smidt, J. and van Krevelen, D.W. (1959) *Fuel*, 38, 355
Snape, C.E., Ladner, W.R. and Bartle, K.D. (1985) *Fuel* 64, 1394
Speight, J. G. (1994) *The Chemistry and Technology of Coal, 2nd Ed.*; Marcel Dekker: New York
Sprecher, R.F. and Retcofsky, H.L., (1983) *Fuel* 62, 473
Suuberg, E. M., Unger, P. E. (1981) *18th Symposium (Intl.) on Combustion*, The Combustion Institute, Pittsburgh, PA, 1203
Suuberg, E. M., Unger, P. E., Lilly, W. D. (1985) *Fuel*, 64, 956
Suuberg, E. M. (1985) *Chemistry of Coal Conversion* (ed. Schlosberg, R. H.), Plenum Press, NY, 67
Wiser, W. (1968) *Fuel*, 47, 475
Zhuo, Y., Herod, A.A. and Kandiyoti, R. (2003) in "*Natural and Laboratory-simulated Thermal Geochemical Processes*" (R.Ikan, Ed.) Kluwer Academic Publishers, Dordrecht Boston London, p. 53

CHAPTER 7

Analytical Techniques for Low Mass Materials: Method Development

In this book, we focus on the examination of solid fuel behavior and liquid fuel properties, emphasizing the conceptual integration of sample characterization, reactor design and evaluation of product distributions. In the next two chapters, we will present a systematic overview of methods for analyzing smaller molecular mass materials and procedures for characterizing high mass samples not amenable to analysis by GC-MS or probe-MS. We will attempt to show how properties of fuel-derived samples need to be matched to the capabilities of the range of available analytical techniques. In the present Chapter, we will describe the use of chromatographic, mass spectrometric and other supporting methods for the *analysis* of materials, broadly below molecular masses of about 500 u. Recent developments in size exclusion chromatography and MALDI-mass spectrometry will be outlined in Chapter 8, where we will describe methods allowing realistic estimates of molecular mass distributions of heavy hydrocarbon liquids.

Methods described in this Chapter have been in use and under development since the invention of chromatographic methods. The introduction of thin layer (planar) chromatography is attributed to the work of Michael Tswett (1872–1919) [see Lederer 1994 for a brief biography]. Gas chromatography dates from the work of Martin and Synge on partition chromatography, for which they received the Nobel Prize in Chemistry in 1952 (see Bartle and Myers [2002] for a brief history). Other forms of chromatography were devised from these developments. The essence of chromatographic separations is a bed of particles forming a chromatographic separation medium, due to particular chemical or physical surface properties, and a gas or liquid medium to carry the sample through the separation bed. Liquid chromatography and supercritical fluid chromatography were later developments that took advantage of improved solvent properties, compared with gas chromatography that depends only on the volatility of the analytes for its operation.

Thin layer chromatography, also called "planar chromatography", is probably the simplest of these methods. The principles of operation serve as a reminder of how chromatographic systems operate in general. TLC involves a layer of particles of relatively uniform size (usually silica, alumina or cellulose) stuck onto a backing plate of glass or metal. A complex mixture applied to the plate in a suitable solvent as a spot or as a band, is dried and the chromatogram developed by allowing solvent to spread along the plate as a sharp front from one edge toward the opposite edge of the plate. The solvent front should proceed in parallel to the edge from which it was applied for good results to

Table 7.1 Chromatographic methods in coal science

Mobile phase	*Method*	*Spectroscopic method*	
Gas	GC	*Coupled* MS (IR, UV)	*Off-line*
Supercritical fluid	Supercritical Fluid Chromatography	MS, IR, UV Fluorescence	
Liquid	Thin Layer Chromatography	MS, UV, UV-Fluorescence, HPLC, (SEC)	
	HPLC	MS, IR, UV-Fluorescence	NMR

be achieved. A selection of solvents can be used in any order so long as the previous solvent is dried-off before the next one is applied. Components of the sample usually have different affinities for the separating medium. For example over a silica bed, the sample mixture would separate according to the polarity of the solvents used. The technique is particularly useful with samples containing heavy material. Larger and more polar molecules in the initial sample mixture may be isolated by their lack of mobility, nearer the origin, after smaller molecules are removed by appropriate solvents. Some smaller volatile components may be lost during solvent evaporation. However, many of the components covered by the current discussion can be isolated according to molecular mass and polarity, by use of an appropriate solvent sequence. The reader may wish to refer to Touchstone [1992] and Hamilton and Hamilton [1987] for a self-learning text.

TLC is also the chromatographic method that involves the lowest cost to set up. Column chromatography involves more cost than TLC because the volumes of solvent are greater. Gas chromatography is relatively expensive both in instrumentation and in the cost of (expendable) columns. HPLC is even more expensive to set up because the pumps and detectors as well as the (expendable) columns involve capital expenditure. SFC is probably the most expensive version of chromatography. Once these chromatography methods are interfaced with mass spectrometry however, the costs of installation and maintenance as well as the need to have an operator who understands the mass spectrometric instrumentation, become relatively high.

The chemical complexity of coal and other fuel-derived materials requires detailed characterization of their constituents to be carried out by a combination of separation and spectroscopic methods [Bartle, 1989]. Given the wide molecular mass (MM) range, a variety of chromatographic methods is required to separate coal derivatives with gas, supercritical-fluid, or liquid mobile phases. Many of these have been used in tandem with a spectroscopic identification method, either coupled or off-line (Table 7.1).

7.1 Gas Chromatography

7.1.1 *Capillary–column GC*

For the relatively small molecules present in coal products, ranging from permanent gases to hydrocarbons and their derivatives with molecular masses up to approximately 400, gas chromatography (GC) is most often the method of choice [Lee *et al.*, 1984, Bartle, 1985].

The complexity of fuel derivatives demands the greatest possible resolution capability. In this respect, GC using packed columns even as long as 20 meters may fall short of the separation power available on capillary columns. Fused-silica columns are used almost universally for these samples and if surface activity is controlled by silanization treatments, trace compounds can be eluted as sharp peaks. For most analytical work, columns 10–25 meters in length are suitable, with internal diameters of 0.2–0.3 mm (although see Section 7.1.3). Column performance may be readily evaluated from the separation of isomer pairs such as anthracene/phenanthrene, and chrysene/benzanthracene.

Capillary column GC has been widely used in the analysis of coal derived products. The non-polar or slightly polar stationary phases – methylsiloxanes (0V-1, 0V-101) or 'small content' phenyl siloxanes (SE-52 and SE-54) have generally been used, especially after free radical cross-linking, to improve thermal stability, although high-phenyl content siloxanes and more polar phases such as Poly S-170, are also used. The latter was especially useful in separating the dibenzofluoranthene/dibenzopyrene (MM 302) isomers [Borwitzky *et al.*, 1978]. Liquid crystal blends have been observed to give remarkably selective separations, but their use is limited by low thermal stability [Bartle 1985]. Poly-(mesogenmethyl) siloxanes are gum phases which show high column efficiencies and stabilities but retain a high selectivity for polycyclic aromatic hydrocarbon (PAC) isomers of coal origin (Figure 7.1). They have a wide nematic temperature range (70–300°C)

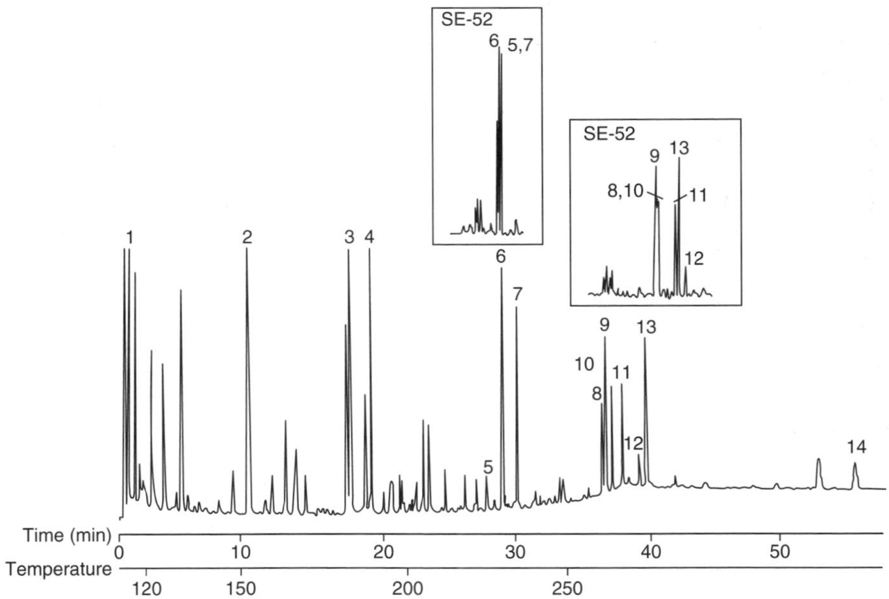

Figure 7.1 GC chromatogram of coal tar on a 20 m long column coated with a mesogenic stationary phase. Inset are portions of the chromatogram on an SE-52 methylphenyl stationary phase. Peak assignments: 5, triphenylene; 6, benz[*a*]anthracene; 7, chrysene; 8, benzo[*j*]fluoranthene; 9, benzo[*b*]fluoranthene; 10, benzo[*k*]fluoranthene; 11, benzo[e]pyrene; 12, perylene; 13, benzo[*a*]pyrene and 14, indeno [1,2,3-*cd*] pyrene. Reproduced from Lee *et al.*, 1984 with permission, copyright Wiley.

[King *et al.*, 1982]. The shape selectivity of smectic liquid-crystal GC phases has proved useful in the separation of a number of PAC isomer groups present in standard reference materials, e.g. those with molecular mass 228, 252 and 278.

Figure 7.1 shows components of the tar volatiles identified using a 20 m long column coated with a mesogenic stationary phase. The analysis showed signal up to benzopyrenes, $C_{20}H_{12}$ of mass 252 u, after which peak intensities fall toward zero. The molecular mass range of this tar is likely to extend much further than this mass but the higher mass components were presumably too involatile or decayed to char in the injection system and were not observed. Although the molecular structures of the components detected are known, it is not clear if the structures of the components not detected could be predicted from those known, simply by the addition of benzo groups and ethylene bridges. This problem is considered further in Chapter 8 for a coal tar pitch, where we will see that evidence from elemental carbon and hydrogen contents makes it most unlikely that the structures of larger molecules could be predicted in this way. Where tars are produced in the relative absence of secondary reactions, as in a wire-mesh reactor (see Chapter 3), the tar sample dried at 50°C showed few low-mass components that could be detected using GC methods. The technique would be of little use for the structural analysis of such samples, apart from demonstrating the absence of low mass components.

A similar result was arrived at when liquefaction extracts from a "flowing-solvent" reactor (Chapter 5) was examined by GC-based techniques. Once again, this was a reactor designed to suppress secondary reactions of coal extracts released from coal particles. While, therefore, GC appears of little use for analyzing even small fractions of primary tars and extracts released from coals, it is a powerful technique for analyzing coal-derived samples which have been thermally or chemically degraded, as well as for biomass, petroleum and other fossil fuel derived materials. As described in Section 7.6, GC with mass spectrometric identification is often the final step of combined procedures in which open-column LC and HPLC are used to separate coal liquids, with hundreds of compounds being identified. What must always be remembered, however, is that the proportion of the total sample that shows up in the analysis may be unknown.

Capillary-column flame-ionization chromatograms may be used for identifications, after GC-MS calibration, by making use of retention indices – systems of reproducible retention parameters. The use of a retention index, I, based on a homologous series of retention standards is useful, in handling variations in chromatographic conditions and column film thickness. For PAC, the Kovats system, which uses *n*-alkanes as reference standards, is less reliable than that of Lee *et al.* [1979], which is based on the internal standards naphthalene, phenanthrene, chrysene, and picene:

$$I_x = 100\,Z + 100\,(T_{RX} - T_{RZ})/(T_{RZ+1} - T_{RZ}),$$

where T_{RX} is the elution temperature of compound x in the linear temperature programed GC, and Z and Z + 1 are the number of rings in the bracketing standards.

The original system [Lee *et al.*, 1979] has been critically evaluated by Vassilaros *et al.* [1982] and the effects of variations in programing rate, column internal diameter and initial temperature have been discussed. Values of "I" for 310 PAC compounds have been listed by Romanowski *et al.* [1983], with standard deviations (generally less than

0.10 unit) and 95% confidence limits. This system has the advantage that most of the reference standards are generally present in coal-derived oils. Lai and Song [1995] have described similar retention indices for coal- and petroleum-derived liquid fuels.

The high concentrations of nitrogen and sulfur containing compounds makes selective detection especially useful in GC analysis of coal liquids to show dual trace nitrogen – (nitrogen and phosphorus detector-NPD) and sulfur (by flame photometric detector-FPD) selective chromatograms of enriched fractions. The increased number of structural isomers created from each aromatic molecule by incorporating a nitrogen atom (as pyrrolic or pyridinic nitrogen) makes the identification of particular isomers difficult without a specific detector or without a selective fractionation to isolate basic or neutral nitrogen compounds. Concentrations of individual components tend to be around 1% of the intensity of unsubstituted aromatics. Limited use has also been made of infrared and UV detection [Lee *et al.*, 1984]. A recent development is the use of positive and negative ion electrospray mass spectrometry at high resolution, for the qualitative detection of N and O containing compounds in liquids without chromatographic separation (see Section 8.6.9).

The wide volatility range of coal-derived mixtures means that care is necessary to exclude discrimination against higher molecular mass compounds when using split-less injection. Only cold on-column injection completely avoids fractionation during sample vaporization and is more reproducible, since the liquid sample is injected directly into the column. The following ratios of pyrene to benzo[*a*]pyrene were measured for a coal tar oil with different injection methods: split, 5.6; splitless, 4.2; cold on-column 3.4 [Bartle, 1989].

7.1.2 High-temperature GC

The increase in thermal stability necessary to extend appreciably the working temperature of stationary phases was achieved by *in situ* free radical cross-linking. Such columns are stable at temperatures above 300°C [Bartle, 1985] and allow the elution of compounds with molecular mass somewhat above 450 u in coal tar and combustion products. For example high temperature GC-MS of a fraction separated from coal tar by liquid chromatography allowed the presence of compounds containing 8–9 aromatic rings and molecular masses up to 456 to be determined [Romanowski *et al.*, 1983]. Similar results were obtained in a comparison of 5% diphenyl-substituted polysiloxane and biphenyl-substituted silarylene-siloxane polymers. The latter showed high separation efficiency, but retention times for PAC were longer [Bemgard *et al.*, 1993]. Examples of the use of HT-GC-MS for analyzing a coal tar pitch, a petroleum residue and a low temperature coal tar are presented in Figures 7.7, 7.15 and 7.16, respectively.

Distillation data are of obvious importance in the characterization of petroleum feedstocks and coal liquids, and are commonly obtained by simulated distillation (SIMDIST) by GC, rather than by actual distillation tests [Abbott, 1995]. The underlying principle is that hydrocarbons and their heterocyclic isosteres elute from a column with a non-polar stationary phase, during a temperature-programed run, in a sequence of increasing boiling temperatures. Calibration with a set of *n*-alkanes enables retention times to be converted to a boiling-temperature scale.

Such procedures should be readily applicable to coal-derived oils of low and intermediate molecular. Two standardized American Society for Testing Materials (ASTM) GC methods for SIMDIST of petroleum fractions designated ASTM D2887 and D3710 have been used for this purpose. ASTM method D5307 was established in 1992 for crude oil boiling ranges. However, these methods are only applicable to materials boiling up to 538°C because they use packed columns which are susceptible to stationary-phase loss at high temperatures. Thermally stable capillary columns have been used for SIMDIST of higher-molecular mass oils [Trestianu *et al.*, 1985]. Reddy *et al.* [1998] have shown that an upper temperature, equivalent to an atmospheric boiling point of 847°C, could be achieved by using a temperature program to 425°C, eluting alkanes in excess of n-C_{92} .

Higher resolutions and better reproducibility were obtained with capillary columns than with packed columns, and hydrocarbons boiling up to 750°C have been successfully eluted using this technique. However, the use of fused-silica columns with stationary phases composed mostly of siloxane or even silylarene is inevitably limited, while the polyamide coating for fused silica cannot repeatedly withstand temperatures greater than 430°C. After extended use at high temperature, the material becomes brittle. Although the problems have been resolved using aluminum coated [Lipsky & Duffy 1986] and stainless steel [Takayama *et al.*, 1988] capillaries, the high temperature exposes analytes to the possibility of thermal degradation. For example, there is no guarantee that complex molecules such as hydroaromatics do not decompose at these high temperatures [Bartle *et al.*, 1992]. Indeed, the evidence from comparison of probe-ms and GC-MS of coal tar pitch fractions indicates that the polycyclic aromatics of mass greater than 278 u are present and detected by probe-ms but not detected by GC-MS in similar abundance.

7.1.3 Fast GC

Recent advances have emphasized "Fast GC", reducing GC analysis time by the use of small diameter columns and fast temperature programing. Column lengths may also be cut by using selective stationary phases to reduce the number of theoretical plates required for given separations [Cramers *et al.*, 1999]. In comparison with analysis on more conventional 250 μm i.d. capillaries, separations are more rapid and efficient on 100 μm i.d. capillaries (Table 7.2) [Boden *et al.*, 2002]. Optimization of injector volumes is necessary, and the column oven must be capable of temperature programing rates up to 20°C min^{-1}. Even faster analysis may be achieved if hydrogen is employed as mobile phase.

7.1.4 Two-dimensional gas chromatography

In conventional capillary GC of complex mixtures such as coal oils, many compounds co-elute at a single given retention time. Single-column separations, based e.g. on vapor pressure must therefore be interpreted with care. However, in comprehensive two-dimension GC (so-called GC × GC) such overlapping peaks in the first dimension are subjected to a second separation with a different basis (e.g. polarity or polarizability). The technique is capable of achieving much greater resolution of components from neighboring compounds [Phillips & Venkatramani, 1993]. In practice, fractions from the

Table 7.2 Comparison of properties of capillary GC columns in PAC analysis[a]

Parameter	*Fast GC*	*Column Standard GC*	*Fast GC*
Column length/m	10	30	20
Internal diameter/μm	100	250	100
Film thickness/μm	0.1	0.25	0.1
Theoretical plates per m	8600	3300	8600
Total theoretical plates	86000	99000	172000
Relative column efficiency	0.93	1	1.32
Relative analysis time[b]	0.38	1	0.55

[a] From Boden *et al.*, 2002
[b] For benzo[*ghi*] perylene

first column are introduced into a second column by modulated injection. Figure 7.2 shows a schematic of the joining of the two columns with the modulation mechanism; the computer software needed to handle the data is available, but the combination of GC × GC with a time-of-flight mass spectrometer provides a good solution when examining mixtures of unknown compounds.

Column 1 is typically a 20 m low polarity phase column, while Column 2 is typically a more polar phase column of short length (1–2 m), joined by a sleeve. In operation, sample and carrier gas flow through column 1 while points A and B are alternatively frozen and heated on a, say, 10 sec cycle. Thus, when A is frozen, material condenses at A while material previously condensed at B is heated and evaporates into Column 2. While the chromatogram in column 2 develops, point B is frozen, point A is heated and material moves to point B. The cycle is repeated resulting in pulses of material at 10 sec intervals entering Column 2. Because of the different polarities of the two columns, co-eluting mixtures of aliphatic and aromatic molecules behave differently on the second column than on the first with aromatics delayed more than aliphatics. The second column must produce a gas chromatogram every 10 sec compared with a total elution time for a chromatogram from the first column of 0.5 to 1 hour. This is achieved by a combination of faster linear flow through the second narrow column combined with a much shorter length. An example is given in Figure 7.9, below.

A second advantage accrues in GC × GC: in many regions of a single-column GC chromatogram of coal or petroleum oils no one compound is present at sufficient concentration to be detected above the 'baseline'. This is due to the elution of many (literally hundreds) of

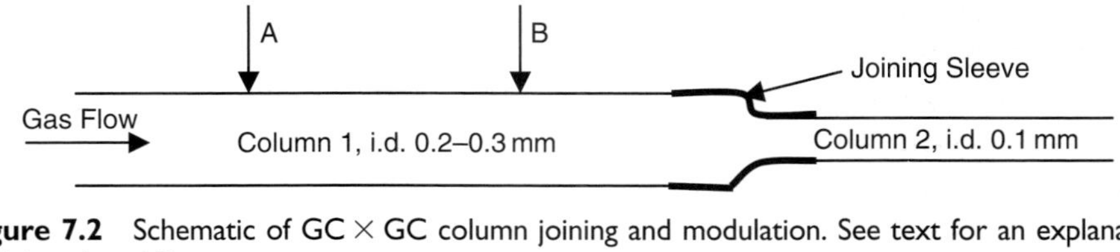

Figure 7.2 Schematic of GC × GC column joining and modulation. See text for an explanation of Points A and B.

low-concentration species. This situation is shown below, in Figures 7.15a, 7.15b and 7.16b. However, in GC × GC the high speed of analysis in the secondary GC column also brings about a considerable improvement in peak amplitude, and hence in sensitivity [Lee *et al.*, 2001]. Consecutive fractions of the primary peak are processed through the modulation to produce a series of much sharper peaks, which then pass into the detector. The periods between individual modulation pulses are then mapped by linear interpolation, resulting in a marked increase in peak amplitude. Reviews of the method have been given [Shellie *et al.*, 2003, Marriott *et al.*, 2004].

Coal oils contain very many individual compounds (as many as 10^6 in coal tar) but a limited number of chemical classes which impose impossible demands on any single chromatographic technique. They are ideally suited to analysis by GC × GC. An example of this method applied to a coal extract is discussed below (section 7.6.1 b, Figure 7.9).

7.2 Supercritical Fluid Chromatography (SFC)

7.2.1 SFC of coal derivatives

The application of GC in the analysis of coal derivatives is inevitably limited by the low volatility of PAC of relatively modest molecular mass at temperatures at which chromatographic columns can be routinely operated (about 350°C) without significant degradation. Applications of GC are also limited by the possibility of decomposition of the constituents of coal oils held at high temperatures for periods of minutes typical of GC analysis. Chromatography with a supercritical fluid as mobile phase (SFC) should in principle extend the molecular mass range amenable to high resolution analysis [Shariff *et al.*, 1997].

Above its critical point, substances have density and solvating power approaching that of a liquid but viscosity similar to that of a gas and diffusivity intermediate between those of a gas and of a liquid. Hence, above the critical temperature, fluid properties become very favorable for use as a chromatographic mobile phase. Extraction and solvation effects allow the migration of materials of higher molecular mass than in more usual gas chromatography. The low viscosity means that the pressure drop across the column is greatly reduced for given flow rates. High linear velocities can be achieved and high diffusivity confers very useful mass-transfer properties, so that higher efficiencies in shorter analysis times are possible than can be achieved in HPLC. Furthermore, the density of the supercritical fluid and hence the solubility and chromatographic retention of different substances can be easily changed by varying the applied pressure. The analogue of temperature programing in GC and gradient elution in HPLC is SFC with pressure programing – slowly increasing the mobile phase density and decreasing solute retention. There are advantages in increasing the density in an asymptotic program to keep the retention differences approximately constant between members of homologous series. The much-reduced operating temperatures in SFC compared with GC allow high-resolution chromatography to be applied to mixtures, which would normally be separated by HPLC, but only after considerable investment of time to determine the appropriate mobile phase composition.

In general, the separating power of SFC is midway between that of GC and HPLC. The separation achieved by SFC on a 34 m × 50 μm internal diameter capillary column falls little short of that obtained by GC with the same stationary phase on a 20 m × 300 μm i.d.

capillary for a coal tar mixture. Standard HPLC columns are suitable for SFC, and many applications in the separation of polycyclic aromatic compounds have been reported [Shariff *et al.*, 1997]. Efficient and rapid analysis of coal-derived oils is possible with CO_2 as mobile phase at low temperatures. The SFC chromatogram of an anthracene oil (within 5 minutes) on a conventional HPLC packed octa-decylsilane bonded-phase (ODS) column gave a separation of benzopyrenes and benzofluoranthenes similar to that obtained by capillary GC of the same mixture [Bartle *et al.*, 1987]. The sixteen toxic PACs on the list of the United States Environmental Protection Agency containing 2–6 rings were separated within 6 minutes by SFC with a CO_2/methanol mobile phase with simultaneous temperature, pressure and composition gradient [Heaton *et al.*, 1994]. These PACs are as follows in ascending order of mass (with molecular mass in u): naphthalene (128), acenaphthylene (152), acenaphthene (154), fluorene (166), phenanthrene (178), anthracene (178), fluoranthene (202), pyrene (202), benz[*a*]anthracene (228), chrysene (228), benzo[*b*]fluoranthene (252), benzo[*k*]fluoranthene (252), benzo[*a*]pyrene (252), benzo[*ghi*]perylene (276), indeno[1,2,3-*cd*]pyrene (276), and dibenzo[*a,h*]anthracene (278).

The greater permeability of capillary columns allows column lengths to be greatly increased, with subsequent high efficiency. The advent of new column technology has made small-diameter fused-silica columns for SFC widely available. The stationary phase, usually a polysiloxane, must be subjected to free radical cross-linking to prevent extraction by the mobile phase. Capillary columns offer a number of further advantages in SFC; high sensitivity, the maximum use of density programing and compatibility with a variety of detectors including the universal flame ionization detector. The use of the last section of the capillary as a detector can help eliminate band spreading. Capillary columns also permit low flow rates of flammable and therefore dangerous mobile phases, such as *n*-pentane (critical temperature 196.6°C) can be used as the mobile phase which is able to elute larger coal derived molecules than can be eluted using CO_2 alone. UV absorption or fluorescence detection is used to give chromatograms (e.g. Figure 7.3) in which molecules beyond ovalene (MM 398) are eluted with high resolution [Bartle, 1989].

The selectivity of the fluorescence detector in capillary SFC can be exploited (Figure 7.3) to identify specific compounds in heavy coal derived oil. The regular groups of peaks extending to high molecular mass in this chromatogram indicate how only certain isomers may be present in major concentrations even for molecular masses in the asphaltene range (Bartle, 1989). However, the upper mass range of the chromatogram still falls short of the upper mass range of the actual sample. The components detected probably correspond to 20–30% of the whole sample. The remaining 70–80% of the pitch may be presumed to remain in the injection system of the chromatograph.

Simultaneous density-temperature programing may be carried out in capillary-column SFC to achieve variations in selectivity, making use of an increase in solubility and diffusion in the supercritical fluid with temperature. For example, on a liquid-crystal stationary phase, isothermal density programing at 90°C gave acceptable resolution of PAC up to indeno [1, 2, 3–*cd*] pyrene on the basis of molecular shape (e.g. length to breadth ratio), but higher mass PACs gave very broad peaks. On the other hand, slow simultaneous density temperature programing gave better resolution for the compounds eluting beyond coronene (MW 300) with good peak shapes because of increased solute diffusion in the mobile phase [Kithinji *et al.*, 1990].

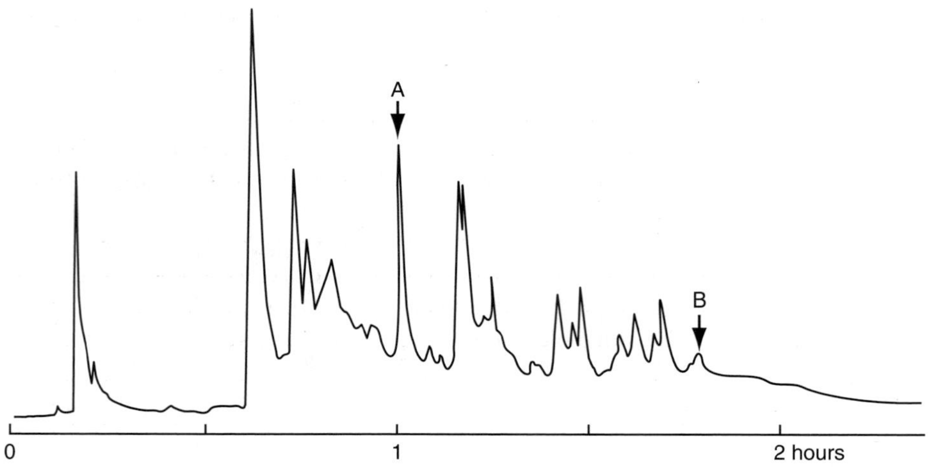

Figure 7.3 SFC chromatogram of coal tar pitch on 10 m long capillary column. Mobile phase pentane at 220°C with fluorescence detection. Peak assignments: A, coronene; B, ovalene. Reproduced from Bartle, 1989 with permission, copyright Elsevier.

Fractionation of complex coal-derived mixtures is also possible by elution with supercritical CO_2 from columns packed with an NH_2-modified stationary phase. Fractions are collected in pressurized vessels, according to the number of aromatic rings, with much 'cleaner' separation than is possible in preparative HPLC [Campbell & Lee, 1986].

7.2.2 *SIMDIST by SFC*

The greater solubilities of PAC in supercritical fluids at much lower temperatures than in gases have led to a number of attempts to replace GC by SFC for simulated distillation (SIMDIST). It was early noted that capillary SFC under mild conditions (column temperatures as low as 100°C) could be used to characterize heavy petroleum oils with true boiling temperatures extending beyond 700°C, thus eliminating possible sample decomposition [Shariff *et al.*, 1997]. The accessible boiling range has been extended to 750°C, but problems of sample capacity and loadability have been cited. More recently, the composition of high-boiling temperature mixtures has been determined by SFC using packed capillary columns [Shariff *et al.*, 1994]. 250 μm i.d. columns were packed with silicas, bonded with different functional groups: C_{18}, C_8, C_6 and C_1. SIMDIST was carried out at 120°C with carbon dioxide as mobile phase with pressure programing from 100 to 350 atmospheres, using flame ionization detection. Calibration was carried out with *n*-alkanes (polywax polyethylene) standards. At the limit of the technique, using the hexylsilyl bonded silica, hydrocarbons boiling beyond 800°C (n-C_{130}) can be eluted and detected. Excellent agreement was obtained with SIMDIST by high-temperature GC.

It has been pointed out that there may be discrepancies between true boiling points of PAC and those calculated on the basis of a calibration of an SFC method with *n*-alkanes [Shariff *et al.*, 1997]. Deviations of boiling temperatures in simulated distillation are tabulated in Table 7.3. The greatest deviation for SIMDIST is by GC, but the deviation is

Table 7.3 Deviations between the true boiling points and the predicted boiling points of polycyclic aromatic hydrocarbons

		Deviations (°C)			
Compound	*True boiling point (°C)*	*High temp. GC**	*SFC (open tubular)**	*SFC (packed octylsilyl)*	*SFC (packed hexylsilyl)*
Naphthalene	218	0	+24	+25	+19
Fluorene	293	−11	+20	+18	+7
Phenanthrene	338	−19	+4	+12	+2
Pyrene	393	−25	0	+5	0
Chrysene	447	−38	−14	+2	−2
Benzo[e]pyrene	493	−48	−11	−4	−10

* Values from Shariff *et al.*, 1994.

reduced for SIMDIST by capillary column SFC and is minimized when packed capillary SFC is applied, especially for aromatic compounds containing three or more aromatic rings. This procedure was found to be particularly effective in determining the progress of the co-refining of coal and petroleum from the boiling temperature distribution of products [Bartle *et al.*, 1994].

7.2.3 Aromatic content of fuels by SFC

A niche application of SFC is the rapid determination of the content of aromatics in fuels, with high precision and reproducibility. The standard method (ASTM Method D5186-91) replaces the old fluorescent indicator adsorption (FIA) method, and employs a packed silica column, of length 25 cm, with CO_2 as mobile phase and an FID as detector – clearly not possible in a liquid chromatographic separation. A hydrocarbon group separation on a packed capillary column may be used to determine contents of mono-, di-, and polynuclear aromatics in coal-derived fuels (Table 7.4) [Shariff *et al.*, 1998].

Table 7.4 Aromatic contents of coal liquefaction recycle solvents by packed – capillary column SFC[a]

		Mono- [%]	*Di- [%]*	*Polyaromatics [%]*	*Total aromatics [%]*	
					SFC	*FIA*
Recycle solvent	1	15.1	4.1	<1	19.2	21
	2	17.1	5.4	1.0	23.5	28
	3	28.4	10.6	4.2	43.2	46
	4	31.1	15.5	8.9	55.5	58
	5	67.3	17.4	3.4	88.1	88

[a] from Shariff *et al.*, 1998.

7.3 High Performance Liquid Chromatography

Since its inception in the early 1970s, HPLC has been widely used in the separation of PAC. The small particle sizes, necessitating the pumping of the liquid mobile phase through the column, result in much more efficient separations than can be achieved on open columns [Wise, 1983]. Unless microcolumns with very small internal diameters are used, the high separation efficiency of capillary GC and SFC is not achieved in HPLC. However, HPLC does offer a variety of stationary phases capable of providing unique selectivity both for functional group types and for difficult-to-separate isomers, because of interactions of the solute with both stationary and mobile phases. HPLC thus provides a useful fractionation technique and the means of high-resolution analysis for compounds with molecular masses up to 600 [Bartle *et al.*, 1981]. In principle, the greater solubilities of PACs in liquid solvents should extend the applicable molecular mass range over those achievable in GC and SFC.

Unusual among HPLC techniques, size exclusion chromatography (SEC) is an example of HPLC where separation is achieved through penetration of the porosity of the packing. Smaller molecules penetrate the porosity of the column packing and are delayed to a greater extent than larger molecules, which elute earlier. Interaction of sample with column packing surfaces is not sought, indeed is deemed undesirable. As will be discussed below, SEC has been found useful in resolving aliphatic as well as aromatic species (also see Chapter 8).

Columns used for HPLC are normally packed with particles of uniform size, often with a chemically bonded surface modifier, such as amino-groups for normal phase or alkyl groups for reverse phase operation. These surface bonded groups allow sample to adhere to the surface until a sufficiently polar solvent can displace them, allowing a more effective separation of chemical types. Column technology is driven by the need to carry out specific separations of critical components, needed in pollutant detection. Column performance will thus be classified on the basis of the separation of particular isomers or compound types, possibly PAC standards. However, it will still be the responsibility of the experimentalist to risk time and resources to see what can be achieved in the separation of coal tars more complex than the standard mixtures.

7.3.1 Normal phase HPLC

In normal phase HPLC, when silica or amino-group bonded phase columns are used with an alkane as eluent, saturated hydrocarbons exit the column first, followed by aromatics. Polar compounds are eluted by back-flushing. An alternative procedure involved solvent programing to a more polar solvent with elution of some more polar components [Herod *et al.*, 1988] as determined by LC-MS. Class fractionation of coal liquids by functional group has also been investigated for a wide variety of other normal-phase columns (chemically bonded NO_2, CN, diol, sulfonic acid), and NH_2 and NO_2 found to be the most selective for fractions containing hetero-functions [Bartle *et al.*, 1979].

Excellent resolution of PAC from azaarenes has been reported on a nitrophenyl phase [Bartle *et al.*, 1979]; about 70% of a coal tar could be analyzed with this phase. Normal phase HPLC is commonly used to separate PAC on the basis of ring number (Figure 7.4),

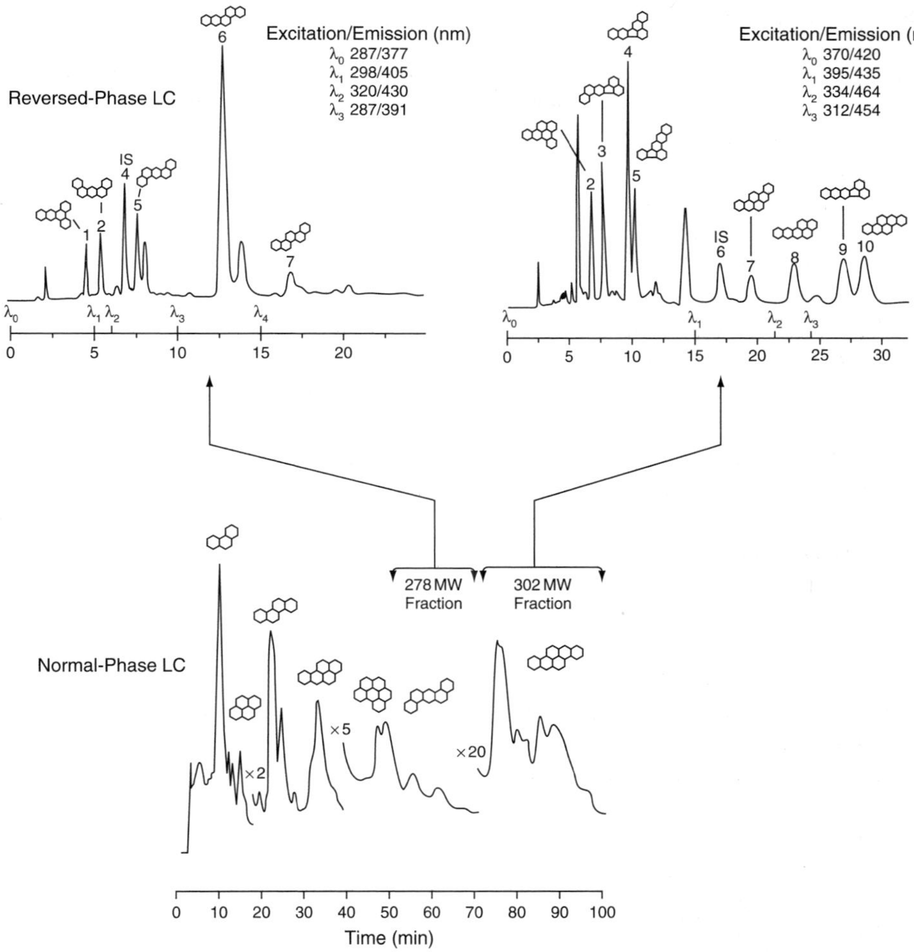

Figure 7.4 Sequential normal phase (lower) and reverse phase (upper) HPLC chromatograms of structurally isomeric 5 and 6 ring aromatic compounds in coal tar. Reproduced from Wise, 2002 with permission, copyright Taylor and Francis.

with alkyl derivatives eluting with the parent. Such separations usually precede analysis by another chromatographic method, or by a spectroscopic procedure. The elution profiles of coal liquid distillates and residues during normal phase (*n*-heptane mobile phase) semi-preparative HPLC and analysis of fractions by mass spectrometry showed that separation had been achieved on the basis of the number of double bonds [Boduszynski *et al.*, 1983].

The data of Figure 7.4 shows that the chromatographic intensity fell away markedly as the 5- and 6-ring aromatic components eluted. It seems unlikely that the sample did not contain larger aromatic systems. Clearly, elution using an alkane cannot possibly elute all the aromatics of a coal tar since many of the higher mass aromatic components are

Table 7.5 HPLC retention characteristics[a] for polycyclic aromatic hydrocarbons

Compound	*Normal phase*[b]		*Reverse phase*[c]
	Silica	*NH_2 column*	*ODS*
Naphthalene	2.00	2.00	2.00
Acenaphthene	2.30	2.10	2.64
Fluorene	2.25	2.55	2.70
Anthracene	2.95	2.94	3.20
Phenanthrene	3.00	3.00	3.00
1-methylphenanthrene	3.26	3.02	3.39
3-methylphenanthrene	3.24	3.12	3.29
9-methylphenanthrene	3.18	3.02	3.34
Benzo[*a*]fluorene	3.95	3.51	3.79
Pyrene	3.06	3.37	3.58
1-methylpyrene	3.23	3.46	3.98
2,7-dimethylpyrene	–	3.47	4.42
Fluoranthene	3.42	3.51	3.37
Benzo[*c*]phenanthrene	–	3.64	3.63
Benz[*a*]anthracene	4.00	4.00	4.00

[a] values of log retention index based on naphthalene, phenanthrene and benz[*a*]anthracene standards (from Wise 1983)
[b] *n*-hexane mobile phase;
[c] 85% acetonitrile in water mobile phase

not soluble. Solvent programing to introduce a more polar solvent might improve the situation. Table 7.6 (below) indicates that significant proportions of three coal tars were not soluble in pentane.

7.3.2 *Reverse phase HPLC*

High resolution HPLC is generally achieved on reverse-phase columns, such as octadecylsilane (C_{18}) with a polar solvent as mobile phase (methanol, acetonitrile etc. in water), often with its composition changed continuously during the run, termed "gradient programing". Increasing the proportion of organic solvent decreases retention. Methyl and other alkyl derivatives are usually separated from parent compounds (Table 7.5). Numerous separations of PAC and polycyclic aromatic sulfur heterocycle (PASH) fractions from coal have been reported [Wise, 1983], which show the excellent selectivity for the separation of PAC isomers and alkyl derivatives. The great advantage of reverse phase packings is their compatibility with a variety of mobile phases and with gradient elution. Selective separations are achieved on the basis of the length-to-breadth ratio of solutes: the more nearly linear molecules are retained longer [Wise, 1985].

Table 7.6 Fraction weights recovered from column chromatography using a sequence of solvents

	Petrox flash column residue	*Pitch*	*Coal digest*	*Low Temp. Tar*
*Fractions**	%	%	%	%
Pentane 1st	50.67	3.77	12.41	10.5
Pentane 2nd	17.96	17.38	29.64	25.71
Toluene	5.72	26.52	10.12	19.55
Acetonitrile	1.74	5.20	3.14	13.72
Pyridine	2.52	15.29	17.50	4.01
NMP	6.16	15.09	8.54	4.25
Water	11.26	2.84	5.08	6.64
SUM	**97.03**	**86.09**	**86.43**	**84.38**

Data from Al-Muhareb *et al.* 2005. *Pentane 1st and 2nd fractions are from successive elutions of the columns with 50 ml each of pentane.

Not all C_{18} stationary phases provide the same selectivity for PAC separations. The selectivity depends on the method of stationary phase preparation, and there are clear differences between separations achieved on 'monomeric' (prepared by reaction of monofunctional silanes) and 'polymeric' (prepared using trifunctional silanes) C_{18} packings. The separation of PAC isomers is vital in studies of coal-derived mixtures. Thus, the selectivity of 'polymeric' phases is especially marked e.g. in the separation of isomers with MW 302. Other factors influencing the separation include the pore size of the packing and the phase coverage, as well as column temperature during HPLC [Wise, 2002].

A major advantage of HPLC in the analysis of coal liquids is the availability of sensitive and selective detectors. The UV detector is universal for PAC. The sensitivity and selectivity may be increased by monitoring at specific wavelengths for given compounds. For example, benzo[*a*]pyrene exhibits nearly maximum absorbance at 290 nm with very little interference from perylene [Wise, 1983]. A scanning UV detector or photodiode array detector allows the possibility of identifying chromatographic peaks either from complete spectra or absorbance ratios at several wavelengths. Compositional changes in coal liquids as a function of process conditions have been demonstrated by this approach [Klatt, 1979].

Fluorescence detection provides unique selectivity in the identification of individual compounds separated by HPLC. For example, six isomeric PACs of molecular mass 252 (perylene, benzopyrene etc.) are generally found in coal-derived mixtures and are not completely resolved in HPLC with ODS columns, but all may be determined by varying the emission and excitation wavelengths [Bartle *et al.*, 1981]. Combination of HPLC with mass spectrometry has provided the most useful results, particularly in the analysis of higher molecular mass constituents of coal liquids. For example, HPLC-MS allowed identification and qualification of PAC with masses up to 450 with an acetonitrile-dichloromethane mobile phase gradient [Caslavsky & Kotlarikova, 2003].

Microcolumn HPLC has been proposed as a method of improving the efficiency of LC separations and has proved capable of resolving many constituents of coal-derived materials containing between 5 and 9 rings. The columns are typically 200 μm i.d. and, when packed with 3 μm C_{18} particles, can generate over 200,000 theoretical plates, although very long analysis times are necessary (Figure 7.5) [Novotny *et al.*, 1984].

7.4 Combined Chromatographic Methods

The coupling of chromatographic techniques can provide particular advantages in the analysis of mixtures as complex as coal derivatives. The most powerful of these comprises one or more separations by liquid chromatography into fractions based on chemical type, followed by a high resolution analysis of the constituents of the fraction by capillary GC, usually with mass spectrometric identification. Alternatively, two sequential liquid chromatographic dimensions may be employed [Saini & Song, 1994].

Fully integrated analytical methods based on this strategy can improve existing knowledge concerning the composition of coal extracts and liquefaction products. In one example of such applications, a coal extract was fractionated by first separating aliphatic and PAC fractions using adsorption chromatography on neutral alumina [Campbell & Lee,

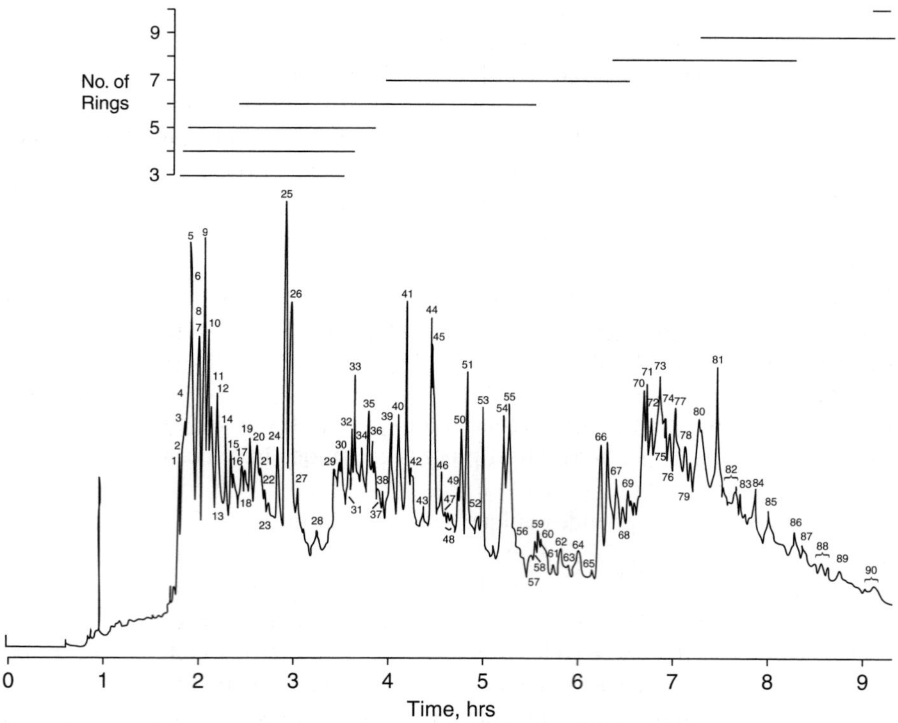

Figure 7.5 Microcolumn HPLC chromatogram of solvent-refined coal fuel oil. Reproduced from Novotny *et al.*, 1984 with permission, copyright Amer. Chem. Soc.

1986]. The aliphatic fraction was separated into branched/cyclic alkanes and *n*-alkanes using a 5Å molecular sieve. The neutral PAC fraction was further fractionated by normal-phase HPLC into sub-fractions based on aromatic ring number. The numerous constituents of these subfractions were identified by capillary column GC-MS.

In another application, capillary GC-MS of fractions based on compound type has allowed the detailed analysis of polycyclic aromatic sulfur compounds, carbazole, amine, azaarene, nitrile, hydroxyaromatic, hydroaromatic, *n*-alkane and branched/cyclic alkane fractions of coal derivatives [Chang *et al.*, 1988]. While such detailed investigations of coal extract compositions are noteworthy, however, the proportion of sample detected is not clear. It seems likely that the coal extract left a considerable quantity of material on the alumina chromatography column and the GC-MS analysis failed to detect the entire range of aromatics isolated from column chromatography.

A direct reverse-phase HPLC analysis provides precise, reliable quantitative results for only 10–12 major PAC in complex mixtures of fossil fuel origin. This limitation has been partially overcome by using a multidimensional HPLC procedure in which a normal phase separation on an aminopropyl phase is used to produce fractions based on the number of aromatic carbon atoms. This is followed by analysis of these fractions by reverse phase HPLC with fluorescence detection to quantify the various isomers. The separation achieved by this approach identified PACs with molecular mass 278 (dibenzanthracenes and dibenzophenanthrenes) and 302 (dibenzofluoranthenes and dibenzopyrenes) in a coal tar standard reference material [Wise 1985]. Off-line HPLC uses relatively large volumes of solvent and requires intermediate re-concentration steps prior to GC analysis. The use of microbore HPLC columns, with internal diameter 1 mm, can reduce the volume of solvent containing the various separated fractions so that these are contained within 100 μl of HPLC eluent [Davies *et al.*, 1987]. The fraction solutions may be transferred directly to an analytical GC capillary column from an automatically and pneumatically operated 10-port valve via a 25 m long retention gap of deactivated but uncoated fused-silica capillary. The retention gap provides a zone for solvent evaporation and solute refocusing by the solvent effect.

On-line HPLC-GC is a rapid and highly reproducible method for the analysis of complex samples such as coal liquids [Kelly & Bartle, 1994]. Thus, coupled normal phase HPLC-GC was applied in the analysis of lignite liquefaction products using two approaches. In the first, aminosilane/silica columns were used in series to provide an aliphatic/aromatic separation and a ring size separation of the aromatic fraction. To resolve alkenes from alkanes, a silver-impregnated silica stationary phase located after the silica column was used to retain the alkenes selectively.

7.5 Unified Chromatography

We have already observed that many mixtures of interest in coal or petroleum chemistry are made up of constituents with a very wide range of volatilities that are difficult to analyze by a single chromatographic procedure. GC, SFC and HPLC may be applicable to different series of compounds and separations may be carried by any of the above techniques, either alone or in sequence on the same instrument: the unified chromatograph [Tong *et al.*, 1993]. The appropriate mobile phase is supplied to either a small-diameter capillary column or a micro-packed column. Detection is either by flame ionization or UV absorption. Independent optimization of mobile phase (helium and carbon dioxide respectively) flow rates is made possible in GC and SFC. Thus, light oils (GC), heavy oils (SFC) and pitches (SFC and HPLC) may be analyzed on the unified chromatograph [Bartle *et al.*, 2000]. Figure 7.6 shows the chromatogram of a coal tar recorded on this instrument, with simultaneous mobile phase composition and pressure programing. Chester and Pinkston [2004] have described the role of SFC in unified chromatography.

In the example shown (Figure 7.6), the largest aromatic detected is of mass 278 u and is still within the range of gas chromatography using normal GC-columns and equipment. Clearly, the use of SFC and unified chromatography has some way to go before extending its range to higher molecular mass aromatics. At the time of writing, the approach seems to be at the stage of initial development and not yet ready for routine application to the analysis of coal or petroleum-derived samples.

7.6 Mass Spectrometric Methods

General descriptions of mass spectrometric procedures and ionization methods are available [Chapman, 1993, Hamming & Foster, 1972] as are methods for interpreting mass spectra [Chapman, 1993; Hamming & Foster, 1972; McLafferty & Turecek, 1993; Eight Peak Index, 1992; Watson & Sparkman, 2005]. Applications of mass spectrometric methods to petroleum chemistry and coal have been described [Genuit, 2005; Herod, 2005]. In what follows, we propose to limit the discussion to aspects of mass spectroscopy relevant to samples derived from the several major classes of fuel already described.

Mass spectrometric methods most suitable for the detection of low mass components of complex mixtures are based on vaporization of the sample in a gas stream or under vacuum. In these applications, the method of introduction of the sample to the mass spectrometer becomes more important than the type of mass spectrometer (magnetic sector, quadrupole, ion trap, time-of-flight, cyclotron or some combination of different sectors). The mode of operation of the mass spectrometer is one of repetitive scans over a range of masses. The frequency of the scans is adjusted, such that the period of one scan is appreciably smaller than the rate of change of sample concentration at the sample inlet of the mass spectrometer. During input from a gas chromatograph, the mass spectrometer must be capable of several complete scans over the duration of the chromatographic peak of a pure compound, to minimize distortion of the mass spectrum. This may be arduous for fast gas chromatography but can be achieved by time-of flight instruments as used in two-dimensional GC-MS. Under such conditions, the determination of nominal mass (measurement of the mass to charge ratio of each ion to the nearest integer mass) is often adequate.

In two-dimensional GC-MS, the second GC column can effect a separation of components co-eluting from the first column by use of a second column with different separation properties. An alternative method of using the mass spectrometer is to utilize multiple ion detection with accurate mass measurement for the detection of specific compounds of known mass spectrum. In this method, the mass spectrometer steps in sequence over a set of predetermined accurate values of m/z characteristic of the analyte. The concept of accurate mass is based on the ^{12}C relative mass (mass 12.00000) scale of isotopic

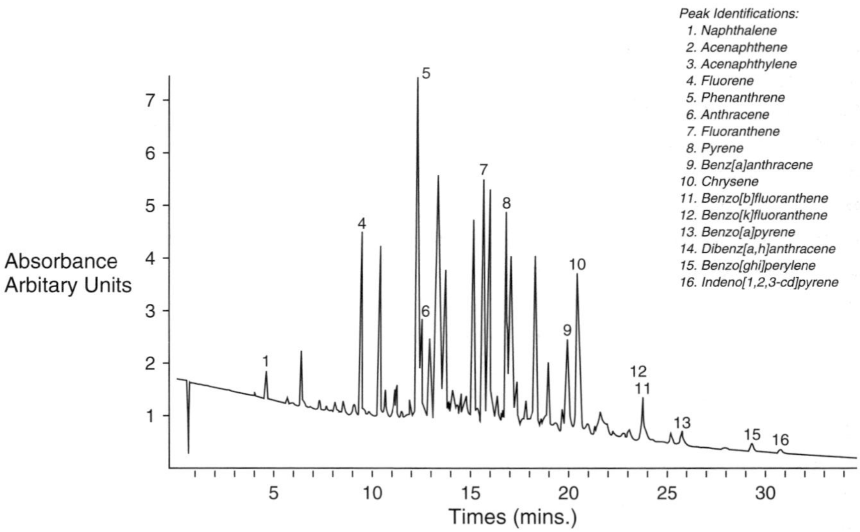

Figure 7.6 SFC chromatogram a coal tar oil on a 30 m long packed capillary column in a unified chromatograph. Mobile phase carbon dioxide with programed methanol content and pressure programing. Detection by UV absorption. Reproduced from Bartle *et al.*, 2000, with permission copyright Amer. Chem. Soc.

atomic mass with ^{1}H of mass 1.00782, thus operating with mass measurement to at least a few parts per million of the mass of the ion. The atomic composition of the ion may then be calculated. Whereas a double-focusing mass spectrometer used to be required, in the past, to achieve accurate mass measurement, it is now possible to use FT-mass spectrometers and time-of-flight mass spectrometers equipped with advanced electronic data systems.

When the sample for analysis can be entirely vaporized and held in a reservoir and leaked slowly into the mass spectrometer with an unchanging composition, repetitive scans over a wide mass range using accurate mass measurement becomes possible. This combination can give a wealth of information on the small molecules since their atomic compositions can be determined directly from the accurate masses of their molecular ions. The method can also be applied to samples partially volatile in the reservoir.

Methods of ionization commonly used in the mass spectrometry of volatile molecules include electron ionization, chemical ionization and field ionization. Electron ionization involves the production of electrons in the mass spectrometer ion source with energies sufficient to induce ion formation in neutral gas-phase molecules of sample, through electron repulsion. Because molecular ion formation and fragmentation of molecular ions are functions of the energy of the ionizing electron, it is common to use relatively high energy electrons, of 70 eV, to approach ionization and fragmentation patterns which are independent of the ion source configuration. The configuration of the ion source has some effect on the relative ion intensities and fragmentation pattern which can be seen, for instance, in the relative intensities of the molecular ions of alkanes compared with their most intense fragment ion, often m/z 43, 57 or 71. An alternative approach is to select an electron energy low enough to avoid ionizing aliphatic molecules but to form relatively intense molecular ions from aromatics without extensive fragmentation. Aliphatic molecules have ionization potentials of around 14 eV while aromatic molecules have ionization potentials of around 7–8 eV and by selecting an ionising energy of about 10 eV, it is possible to focus on forming only aromatic molecular ions. This method has been used to examine the molecular ions of aromatics in coal liquids with accurate-mass mass spectrometry to give semi-quantitative analyses of the more volatile fractions.

Chemical ionization methods involve the production of a reagent gas mixture in the mass spectrometer ion source, using electron ionization of a relatively high-pressure gas such as methane. Ions form through electron ionization and ion-molecule reactions to give a set of ions which cannot react further with the reagent gas, but which can react by proton transfer to analyte molecules of lower ionization potential than the reagent gas itself. For methane, the set of ions formed that can react further include $CH_5{}^+$ and $C_2H_5{}^+$. Field ionization is a method of ionization, which avoids extensive fragmentation. The ionization is effected by passing volatile sample through an intense electrical field, produced by applying a voltage of several kV to a sharp edge such as a razor blade or a sharp point. In one ion source, an array of points was used to improve ionization efficiency by passing the sample through the array of points into the ion source.

There are many possible combinations of sample introduction system, method of ionization, mass separation technique and detection. In practice, however, each mass spectrometer comes equipped with a system of mass separation and detection that is fixed by the manufacturer. There may be several options of sample introduction and ionization methods, but they will be limited by the pumping capacity of the vacuum system of the mass spectrometer and the geometry of the ionization source region. There might not be sufficient space around the ion source to fit all the sample inlet systems desired. The solution to this problem is to gain access to several mass spectrometers with the required attributes of sample introduction system and ionization method.

In the following sections, examples of the different sample introduction methods are discussed.

7.6.1 Gas chromatography

GC-MS: This method involves the use of one GC column with the effluent passing into the mass spectrometer through some interface held at an elevated temperature. The components of highest mass eluting from the column are meant to neither condense nor thermally crack in the interface. For many interfaces, the operating pressure at the column outlet is at reduced pressure because of proximity to the mass spectrometer vacuum. Under these conditions, the column performance may differ somewhat from operation at atmospheric pressure with a flame ionization detector or with a mass spectrometer with an atmospheric pressure inlet system.

We have already discussed how the more volatile components of coal-derived liquids can be identified by GC-MS. The proportion of sample that can be identified by this technique is not readily quantifiable. Menendez *et al.* [2002] have suggested that this could be as high as 20% of a coal tar pitch, with the upper limit normally around m/z 300–350 corresponding to coronene and benzocoronene. Components described for creosotes and coal tars range from indan and naphthalene to benzo[*ghi*]perylene (creosote) and to coronene (tar) and consist mainly of unsubstituted polycyclic aromatics (PACs) [Howsam & Jones, 1998]. Nitrogen and sulfur derivatives formed by replacement of aromatic [–CH=] by [–N=] or replacement of [–CH=CH–] by [–S–] were shown to be present in coal liquids [Herod 1998]. However, these compounds are found at lower relative intensities than PAC and need to be concentrated before analysis.

Figure 7.7 shows a chromatogram of the pentane soluble fraction of a coal tar pitch [Herod *et al.*, 2005]; identities of the main peaks are given in the Figure caption. This sample is the second pentane soluble fraction of Table 7.6 below. The reason for examining this sample by GC-MS was to look for aliphatic components that might have survived the coke oven and subsequent tar distillation processes usual for isolating the pitch. In the event, all of the components detected in the chromatograms were aromatics and no aliphatics were evident.

Figure 7.8 shows the chromatogram of a Polywax sample (a mixture of even-carbon number polyethenes) extending up to C_{56}, on the same column as used for Figure 7.7. Although known to be present from probe-mass spectra, aromatics above m/z 350 decayed or otherwise failed to pass through the column under conditions able to pass much higher mass alkanes. This inability to pass through normal or HT-columns appears to be a problem of the higher aromatics. Extrapolation of the structures known in pitch to components at higher masses by sequential addition of benzo groups, does not produce molecules within the C/H ratios of tars [Parker *et al.*, 1993] and therefore some structural changes must occur at higher masses.

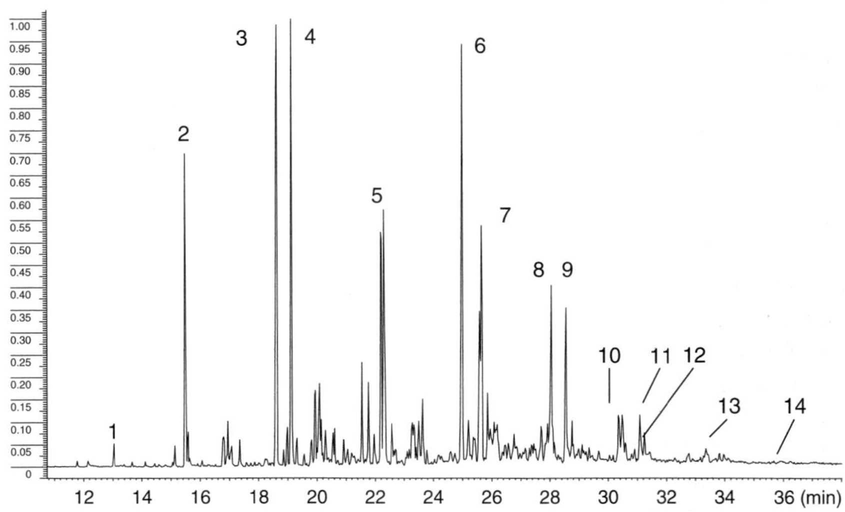

Figure 7.7 HT GC-MS of second pitch pentane soluble fraction. Peaks are: 1-m/z 166 fluorene; 2-m/z 178 phenanthrene; 3-m/z 202 fluoranthene; 4-m/z 202 pyrene; 5-m/z 228 chrysene isomers; 6-m/z 252 benzofluoranthenes; 7-m/z 252 benzopyrenes; 8-m/z 276 and 278 indenopyrene and dibenzanthracene; 9-m/z 276 benzo[*ghi*]perylene; 10- and 11-m/z 302 dibenzopyrene isomers; 12-m/z 300 coronene isomers; 13-m/z 326 rubicene isomers; 14-m/z 352 tribenzopyrene isomers. From Al-Muhareb *et al.*, 2005.

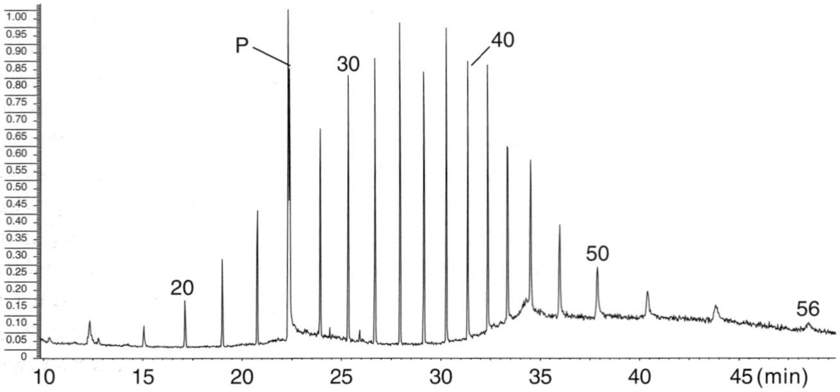

Figure 7.8 High-Temperature GC-MS chromatogram of Polywax 500. The peaks are of even-carbon number alkanes, C_nH_{2n+2}, carbon numbers as shown above peaks (P is a phthalate impurity coeluting with C_{26} alkane). From Al-Muhareb *et al.*, 2005.

GC-MS has been applied to tars and extracts from wide ranges of coals, kerogens and macerals, as well as from biomass and waste materials. Some examples are given here. An oil shale from Mongolia [Avid *et al.*, 2004] gave a tar after pyrolysis and gasification at 800°C consisting of polycyclic aromatics, phenols, aza-aromatics, furans, thiophenes and alkylated aromatics up to benzopyrene isomers. Sewage sludge gave a tar after pyrolysis and gasification in a bench scale spouted bed reactor [Adegoroye *et al.*, 2004] that consisted of polycyclic aromatics and their aza-derivatives with some furans and carbazole, up to mass 228. Pyrolysis tars from animal bones (at 700°C) and milk casein (at 550°C) in a retort were shown by GC-MS to contain alkanenitriles, alkanes and amides (bone tar) and alkyl benzenes, phenols, alkanes and alkylnitriles (casein tar) [Purevsuren *et al.*, 2004a,b]. Both sets of products were quite unlike coal or petroleum derived tars. The pyrolysis of wood and lignin samples gave tars that in GC-MS showed components based on the three oxygenated structural groups typical of biomass. Derived from the terpenes, these are syringol, guaiacol and coniferol [Evershed *et al.*, 1985, Robinson *et al.*, 1987, Guell *et al.*, 1993a,b; also see Section 3.10]. P-coumaryl alcohol from grasses extends the spectrum of materials identified toward more aromatic, less oxygenated molecules, as the pyrolysis temperature is increased [Bocchini *et al.*, 1996].

Wood waste tars have been examined by Evans and Milne [1987a,b] and Challinor [1995]. Wood and cellulose tars from the atmospheric pressure wire-mesh reactor described in Chapter 3 have been analyzed by GC-MS [Fraga *et al.*, 1991]. The proportion of the tar detected by GC-MS cannot be determined from these published data. Total ion chromatograms of wood and biomass tars from GC analysis differ for the different feedstocks and the molecular structures of the main components have not always been determined from the mass spectra [Fraga *et al.*, 1991; Pindoria *et al.*, 1997b], in part because libraries of mass spectra do not contain the required compound spectra. Many of the compounds generated by the wire-mesh and hot-rod reactors are unlikely to be in the mass spectral libraries because the pyrolysis conditions do not match those used for the more usual wood pyrolysis products, such as the Stockholm tars [Evershed *et al.*, 1985, Robinson *et al.*, 1987].

The role of GC-MS in the petroleum industry has been described by Genuit [2005]. Components of petroleum products identified range from alkanes, through mono- to hexa-cyclo alkanes and aromatics from benzenes to chrysenes, including benzo-, dibenzo- and naphthobenzo-thiophenes to nitrogen substituted aromatics. Extensive alkylation with up to C_{10} groups has been encountered in all cases. The vacuum residues of petroleum crudes are extremely complex and it is generally beyond the capability of GC-MS to resolve most of the components of the mixture. Techniques such as hydrocarbon type analysis have been developed to simplify the analysis. GC-MS chromatograms of two Petrox Refinery (near Concepcion, Chile) flash-column residue fractions are shown in Figures 7.15 a,b below. An underlying broad peak of unresolved material may be observed in each chromatogram of the pentane soluble material.

Two-dimensional GC-MS: This method was developed during the past ten years and relies for its operation on access to a fast scanning mass spectrometer such as a time-of-flight instrument. The reproducibility of the method appears to be good [Shellie *et al.*, 2003, Marriott *et al.*, 2004]. The gas chromatographic part of the method has been discussed above (section 7.1.4). The GC system consists of a long capillary GC column (20–30 meters in length) with a phase able to elute components in approximately boiling point order, in the range of polymethyl silicone. The sample is injected into the first column as normal, with temperature programing of the oven to volatilize the heavier components of the sample. At the column exit, the second column is joined, consisting of a short length (say 1 meter) of narrow bore capillary, low phase film thickness and of different polarity than the first column (e.g. a poly-phenyl methyl silicone). As described above, the end section of the first column is operated by alternately freezing and thawing the components eluting in real time, such that an injection of thawed sample is made into the second column every few seconds (typically 5–8 seconds). The second column is set at higher temperature than the first (by say 10°C) and programed at the same rate of temperature rise. The column phase types described would be suitable for coal-derived materials, petroleum residues and biomass tars. Ideally, however, the operator of the system should be familiar with the compound types to be expected in the samples.

The elution time on the second column is arranged to be the same time as the frequency of injection. The mass spectrometer must be able to collect spectra sufficiently fast to characterize all the components eluting from the second column after a single injection. The process is then repeated for every subsequent injection into the second column. Using proprietary software, the mass chromatograms corresponding to the separate injections into the second column may be plotted orthogonally to the direction of the chromatogram emerging from the first GC column. This constitutes the second dimension and allows components co-eluting from the first column to be separated by the second column. The mass spectra are linked to the separated spots and can be manipulated in the normal ways to show specific ions or particular fragment ions or molecular ions. In 2-D GC-MS, the spectra of individual components are much less prone to overlap from other components. For complex mixtures, individual mass spectra are much cleaner than spectra from one-dimensional GC-MS.

An example from using the method is shown in Figure 7.9 for a coal derived process stream from the Point of Ayr coal liquefaction pilot plant: the feed to the hydrocracker

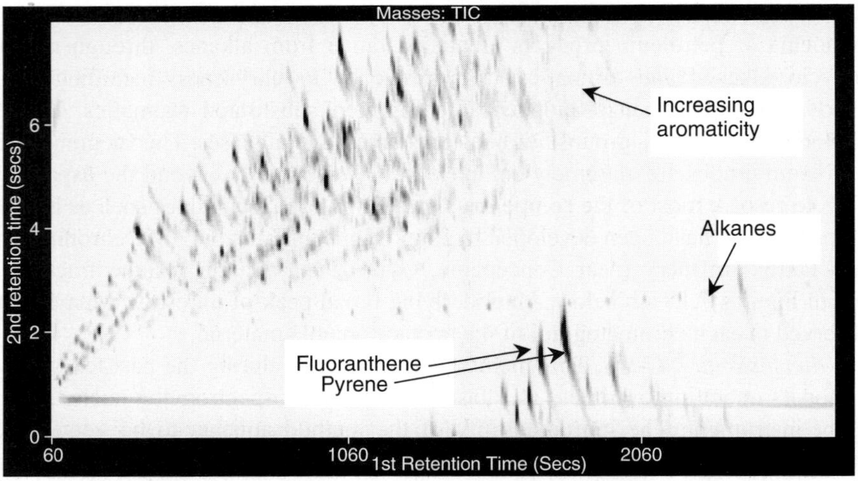

Figure 7.9 Point of Ayr coal liquefaction plant, feed to hydrocracker by 2-D GC-MS. From Hamilton *et al.*, 2005.

[Hamilton *et al.*, 2005]. Differences attributable to the process can be distinguished between the feed to and product (not shown) from the hydrocracker. These relatively small molecules are mostly associated with the process-derived recycle oil and are involved in the production and destruction of hydrogen-donor species and their isomerization into non-hydrogen donor structures.

In Figure 7.9, the series of *n*-alkanes may be clearly noted. The scatter of points of the more aromatic types indicates a very complex mixture of hydroaromatics. Pyrene and fluoranthene are shown and their position on the diagram indicates that these components exceeded the elution time allowed for the second column (8 sec) and appeared at very early elution times in a later scan. In the Figure, the spot density increased with the concentration of the component. The method has not been used extensively for coal tars but an example has been shown for a petroleum fraction, to show the presence of benzothiophenes that co-eluted with naphthalenes [Genuit, 2005]. At the time of reading, 2-D GC-MS must be considered as a promising technique in the process of development, rather than a routine analytical tool.

Pyrolysis GC-MS: The pyrolysis GC-MS method described here involves placing a pyrolysis stage in the GC carrier gas stream. Volatiles may be fed into GC-columns from several types of pyrolysis stages. These include resistively heated filaments such as platinum, laser pyrolysis and Curie point pyrolysis stages, where ferromagnetic metals or alloys are heated to a particular temperature by inductive heating, by exposure to a high frequency field in an induction coil. The mass range limitations of the pyrolysis products are the same as for GC-MS. Involatile material remains behind as char in the pyrolysis stage, or deposits onto the inlet system or parts of the column. The method is useful in identifying structural elements of involatile samples that cannot themselves pass though the GC-column. The technique has been used in investigations of coal tars and petroleum residue fractions as well as for biomass tars that do not give a direct response in ordinary GC-MS.

In pyrolyzing polymers, it is expected that they thermally break down into their characteristic monomers to reveal the chemical nature of the polymer. In the samples of interest for fuel characterization work, the notion of monomers is not useful. Instead, the method aims to identify pyrolysis fragments from large complex molecules of unknown but probably irregular structures. We will describe further on in this Chapter some examples where no detectable fragment molecules were released by samples being studied.

PY-GC-MS of coals indicates the main volatile fragments to be aliphatic rather than aromatic in nature, with biomarker molecules from original depositional material in evidence [Scott & Fleet 1994]. Experiments with Curie point pyrolysis – GC has shown how the aromatic groupings and aliphatic materials released by several coals change with coal rank [Tromp *et al.*, 1988, 1989]. Under mild pyrolysis temperatures, homologous series of alkanes from C_{12} up to C_{30} were observed alongside a range of aromatics up to phenanthrenes. Phenols up to xylenols were detected at high pyrolytic temperatures.

Results from the pyrolysis GC-MS of coal tar and extract fractions have been widely reported [Herod *et al.*, 1999, Islas *et al.*, 2000, 2002, 2003a, Song *et al.*, 1992]; identified components closely resembled pyrolysis products from unfractionated liquids or the parent coals. However, only those low-mass fractions soluble in acetonitrile or THF gave significant amounts and numbers of individual products. Fractions of material insoluble in acetonitrile and pyridine gave virtually no signal. This suggests fundamental structural changes in material with increasing molecular masses (as identified by size exclusion chromatography and MALDI-MS) and increasing polarity. It appears these largest molecules either form char at the pyrolysis stage and/or their pyrolytic fragments condense somewhere within the GC-system.

However, parallel increases of molecular masses and associated increasing sizes and complexities of PAC groups observed for coal liquids, are not applicable to samples of different origins. Solvent extracts of a Baltic amber were examined by pyrolysis-GC-MS [Islas *et al.*, 2001]. The components detected in pyrolysis products from the pentane-solubles, toluene-solubles, the NMP-extracts and the NMP-insoluble residue were all relatively similar to each other and to the products from the pyrolysis of the whole amber. Despite the wide spread of molecular masses, units making up the amber structure appear not to differ greatly from one another. This similarity of products from the different solubility fractions contrasts markedly with observations on coal tar derived solubility fractions displaying very different properties. In this sense, the behavior of amber is more closely related to those of polymers with known regular structures.

Pyrolysis-GC-MS can thus be used to investigate whether high mass, relatively intractable molecular units can be readily broken up into smaller structural units or whether these structural units themselves evolve and become more intractable with increasing molecular mass and polarity. We have observed how the pyrolysis of toluene, acetonitrile, pyridine and NMP insoluble fractions of coal liquids and a petroleum vacuum residue (Table 7.6) did *not* give molecules detectable by GC-MS. The larger mass molecules from coal-derived fractions appear therefore to have different structures from smaller molecules ordinarily identifiable by GC-MS. The *characterization* of these larger molecular mass materials is the subject of Chapter 8.

7.6.2 Heated glass inlet

Another sample introduction system into mass spectrometers is the heated glass inlet, where the sample is evaporated without fractionation. This allows hydrocarbon-type analysis through the summation of ions characteristic of various hydrocarbon types. For example, a distillate liquid sample may be evaporated completely into a reservoir inlet system, to give a constant pressure in the mass spectrometer ion source. Low ionizing voltages may be used to suppress the formation of doubly charged ions and the fragmentation of aromatic molecules by loss of alkyl side chains. The ionization potentials of aliphatic molecules are higher than those of aromatics; at 10 eV, any ions observed are normally molecular ions of aromatic molecules. In high-resolution mass spectrometry, the definition of aromatic types by accurate molecular ion mass provides a formidable, semi-quantitative analysis tool.

The concept of the Z number from the general hydrocarbon formula C_NH_{2N+Z}(N.S.O) is essential for the understanding of type analysis results. N.S.O represents the possible inclusion of nitrogen, sulfur or oxygen atoms and does not alter the Z number. It reflects the total number of rings and double bonds in a structure defined by the carbon number N, and is used as an index of hydrogen deficiency, compared with the corresponding alkane, where $Z = +2$. The number of heteroatoms N, S and O indicate that their structural configuration is not specified; they may be present in rings or as parts of pendant groups [Aczel, 1972, Aczel & Lumpkin, 1979, Herod, 1989]. Distillate samples can be analyzed by GC-MS and the data evaluated to give the equivalent of type analysis. Methods were initially developed by ASTM for petroleum middle distillates using 70 eV electron impact (ASTM 1972). Applications of hydrocarbon group type analysis methods using both complete evaporation (with electron ionisation at 70 and 10 eV) and GC-based methods in petrochemistry have recently been described by Genuit [2005].

The method is not now much used because 1) quantification was not accurate since assumptions had to be made concerning the make-up of the structural isomers of each type of compound based on available standard materials and 2) relatively high mass resolution was needed for the low eV work, requiring a double-focusing mass spectrometer. Such instruments are not generally available for this kind of work and would be very expensive to acquire and set up for routine analysis.

7.6.3 Solids probe

Solids probe mass spectrometry may be considered as an extension of hydrocarbon type analysis into the volatility region beyond that accessible by the heated glass inlet system. It is a method for detecting components not amenable to gas chromatography due to high polarity or low volatility (Section 7.6.6 below). Distillation of the sample into the ionizing region with increasing probe temperature permits the examination of a series of components with increasing molecular mass. High or low mass resolutions and ionizing voltages may be used. In terms of scanning a range of masses, the use of nominal mass rather than accurate mass permits faster scans over a wider range and may be necessary to evaluate the components of a complex mixture. Since very small sample quantities are required, the possibility exists of repeated runs with different mass spectrometric conditions to investigate aspects of the sample.

In this configuration, samples may be part pyrolyzed if the maximum probe temperature is high enough. The presence of char on the probe after the experiment is an indication of incomplete volatility. The mass range of SEC fractions of the asphaltenes (benzene insolubles) of a hydropyrolysis tar [Herod *et al.*, 1993] extended to m/z 500. Figure 7.10(a) shows a probe-mass spectrum at 10 eV summed over a range of scans and showing only molecular ions in the aromatic fraction of a hydropyrolysis tar. The spectrum shows a very complex mixture. For instance, at m/z 322, at least seven components can be seen in the cluster of peaks. These range from m/z 316 to 328 and correspond to aromatics of decreasing Z number. At m/z 316, we can propose a structure derived from benzofluorene (m/z 216) by addition of two benzo groups, as $C_{25}H_{16}$. Thus, the other components of the group may correspond to molecular ions of $C_{25}H_{18}$ (m/z 318), $C_{25}H_{20}$ (m/z 320),

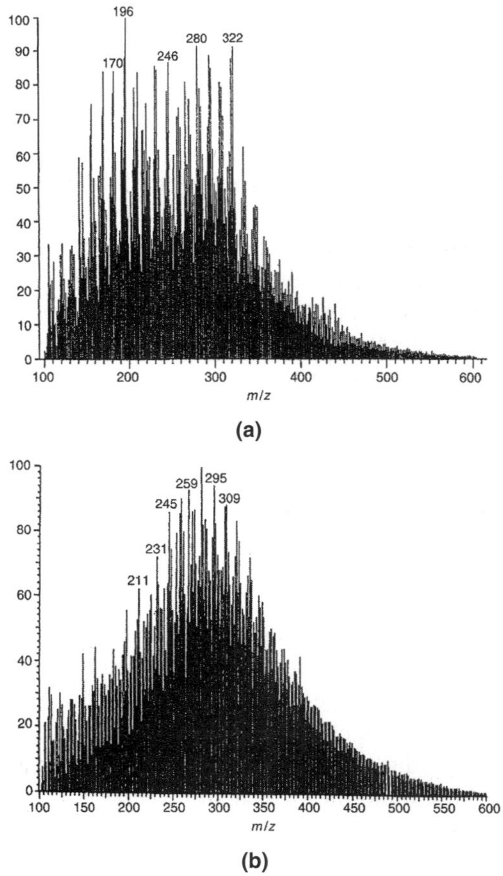

Figure 7.10 **(a)** Probe-mass spectrum at 10 eV of a hydropyrolysis tar from coal **(b)** Chemical ionization mass spectrum of the same tar using isobutane gas. Reproduced from Herod *et al.*, 1987 with permission, copyright Elsevier.

$C_{25}H_{22}$ (m/z 322), $C_{25}H_{24}$ (m/z 324), $C_{25}H_{26}$ (m/z 326) and $C_{25}H_{28}$ (m/z 328). Accurate mass measurement would be required to verify these suggestions. A pyridine-insoluble fraction of a coal tar pitch [Begon *et al.*, 2000] gave no significant ions when examined by probe mass spectrometry with 70 eV ionization, because the lighter materials had been removed. However, after catalytic hydrocracking, a wide range of aromatic molecular ions was observed, typical of a coal-derived material. This evidence was taken to show that the molecules of the pyridine-insolubles were very large and broken down by hydrocracking into relatively small molecules.

A sample of Baltic amber has been pyrolyzed in the wire-mesh reactor described in Chapter 3 [Pipatmanomai *et al.*, 2001 and Chapter 3] and extracted to give separated solvent extracts [Islas *et al.*, 2001]. The product pyrolysis tar and solvent extracts were examined by probe mass spectrometry. The mass spectra of the pyrolysis tar and the solvent extracts contained many common ion fragments but the mass range of the spectra of the solvent extracts extended to m/z 700 and indicated homologous series of components in the mass range above m/z 500. Thus, these probe-MS spectra showed material far beyond the range that is normally covered by GC-MS.

Probe-mass spectra of sample spots recovered from thin layer (planar) chromatography plates have shown sample aromaticity increasing with decreasing mobility in solvents like pentane, toluene and pyridine. Increasing solvent polarities were required to move components with increasing polarity and increased heteroatom content. In another application, fractions of coal tar pitch collected from an analytical SEC column as successive 1 min time fractions and examined by probe-MS showed little signal for the early eluting fractions. Components of the early-eluting fractions were of too large molecular mass (and too involatile) to ionize by this method. Ions characteristic of pitch were observed only in the last few fractions [Lazaro *et al.*, 1999].

Desorption chemical ionization provides a method of evaporating material from a heated probe by a relatively gentle ionization method which may be selective for particular types of compounds, depending on the ionization reagent gas used. The limitation is the volatility of sample components. Figure 7.10(b) shows the DCI mass spectrum using isobutane reagent gas, of the same aromatic fraction as in Figure 7.10(a), showing a different mass range than the low voltage probe, possibly because the relative sensitivities of aromatics and alkylated aromatics differ in the two ionization methods.

7.6.4 Field ionization

Field ionization mass spectrometry depends on the tunnelling of an electron from a molecule in the gas phase on passage through a steep electric field, to an electrode. The sharp edge can be a razor blade edge or sharp points held at several kV to a counter electrode. The molecule loses one electron in the electric field, to produce a molecular ion. The method was designed to minimize the fragmentation of sample molecules.

The method detects those molecules which are volatile in vacuum or which form on pyrolysis in vacuum. High resolution FIMS may allow determination of atomic compositions of peaks, but the complexity of mass spectra observed for asphaltenes and preasphaltenes tends to limit the unambiguous use this technique. In addition, the relative intensity of odd-mass peaks to even-mass peaks in spectra of asphaltenes tends to be greater than in spectra of aromatics or oil fractions, indicating that the molecular complexity has increased by the introduction of heteroatoms, further reducing the ability of high resolution mass measurement to give unambiguous results. The upper mass limit of the technique, defined by the requirement of volatility under the operating conditions, appears to be about m/z 1200, but that need not correspond to the largest masses present in any given sample. Mass spectra of aromatics extended to m/z 1100 [Herod *et al.*, 1993]. FIMS and pyrolysis FIMS have been used to analyze coal liquids [Malhotra *et al.*, 1993, Cagniant *et al.*, 1992). Type analysis of a base oil using FIMS showing types ranging from alkanes to benzodicycloalkanes (Z number range from +2 to −10) has been presented by Genuit [2005].

The best application of FIMS is probably to distillate samples likely to be entirely volatile in the vacuum of the mass spectrometer. The technique has been used to examine the volatile components of coal liquids, as described above, but the limitations are clear. Examination of the sample holder after an experiment can be useful in indicating the proportion of sample that remained involatile.

7.6.5 Isotope ratio mass spectrometry

This method of operation requires a mass spectrometer set to measure two ions of different mass, m/z 44 $^{12}CO_2$ and m/z 45 $^{13}CO_2$. An oxidation stage is necessary to convert the carbon of the analyte to carbon dioxide and if the sample elutes from a chromatography column, the oxidation must be rapid and complete. The ratio of ^{13}C to ^{12}C in coal liquids has been measured by GC-MS to show that the ratio differs from that of petroleum liquids and the source of environmental pollution from the two types of material can be distinguished [McRae *et al.*, 2000]. The method has not yet been applied to the intractable fractions of coal tars or vacuum residues. It may be useful in revealing differences in isotopic compositions between such fractions and provide clues relating to the origins for large and small molecules. The method has not reached the stage of routine operation for characterizing relatively heavy fractions but clearly presents novel analytical options.

7.6.6 Tandem mass spectrometry

Tandem mass spectrometry is a method in which an ionization source and a mass selective device, such as a quadrupole or magnetic sector is used to select a particular ion from the initial ion source for study in one or more subsequent analytical mass selection detectors.

The aim of such work may be to determine the characteristic fragments from the selected ion after collision-induced dissociation or the fragments resulting from the specific loss of a neutral fragment. The possibilities of this instrument configuration have been summarized by Chapman [1994]. A schematic of the tandem mass spectrometry experiment, or MS-MS, is shown in Figure 7.11. Sample introduction by probe or reservoir is needed to maintain constant pressure. Preasphaltenes from liquefaction extracts of coals [Wood *et al.*, 1985] have been examined to identify homologous series of hydroxy- and dihydroxy-aromatics that would not pass through a gas chromatography column.

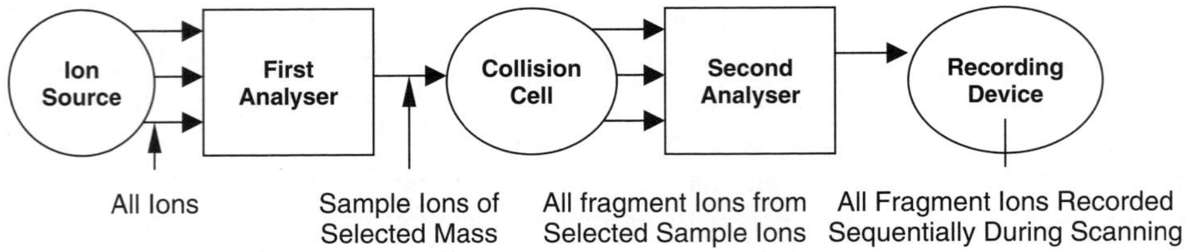

Figure 7.11 Schematic of a tandem mass spectrometry experiment (MS-MS), after Chapman [1994].

The tandem mass spectrometry approach is one able to pick out specific components or types of component in complex mixtures. It is necessary to understand what is being looked for at the outset so that the instruments may be set up to search for either the neutral mass loss characteristic of fragmentations of a chemical type or the specific ion fragments that may identify a compound class. The experiment can be achieved using quadrupole analyzers or even an ion trap mass spectrometer by using appropriate software normally supplied with the instrument. However, given the complex mix of chemical types found in coal and petroleum derived samples, the scope for experimentation is almost endless. The method is a research tool rather than a routine analytical method as far as fossil fuel derived liquids are concerned.

7.6.7 *MALDI and LD-mass spectrometry*

MALDI and LD-MS are methods normally used to generate molecular ions from large molecular mass, thermally labile molecules, described in Chapter 8. However, MALDI-mass spectrometry can be used at low laser power to produce mass spectra of small aromatic molecules, equivalent to a probe-mass spectrum, showing aromatics in acetone and pyridine-soluble fractions of pitch up to m/z 400 [Millan *et al.*, 2005]. Small aromatic molecules have also been ionized by laser-desorption mass spectrometry [Herod *et al.*, 1994]. These applications of MALDI- and LD-MS approximate to the solids probe method described above. The low-mass spectrum of a pyridine-soluble fraction of a coal tar pitch is shown in Figure 7.12.

7.7 Aliphatic Materials from Coal and Petroleum

7.7.1 *Introduction*

When coal is pyrolyzed and the products examined by GC-MS, the major components detected are *n*-alkane series and multicyclic terpane structures [see for instance, articles in Scott and Fleet 1994]. Alkanes and alkenes up to about C_{33} were observed in pyrolysis and hydropyrolysis tars from coals [Snape *et al.*, 1985]. Aliphatic materials are thus known components of coals, with methane (firedamp) as a volatile and dangerous hazard of coal mining [ICS Ref Lib 1920]. Higher alkanes can be evaporated at low temperatures from coals (and peat) when the latter are used as a gas chromatographic column packing [Herod *et al.*, 1983, 1991]. The gas products from the high-temperature coking

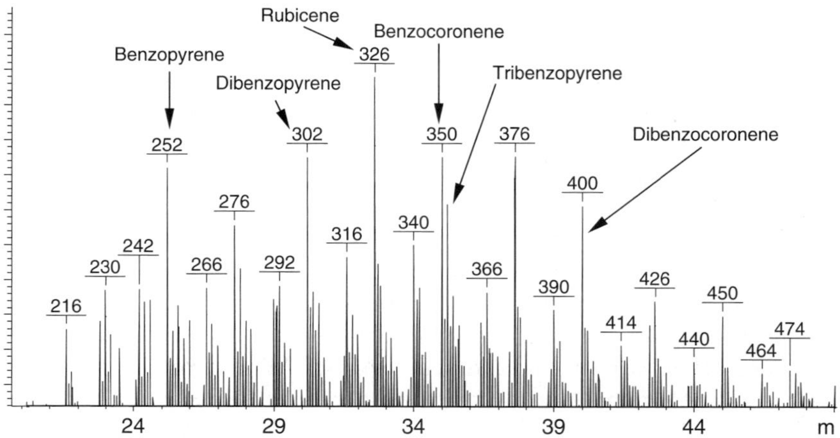

Figure 7.12 LD-MS of pitch pyridine soluble fraction of pitch at low laser energy showing low mass aromatic molecules. Reproduced from Millan *et al.*, 2005 with permission; copyright John Wiley & Sons.

of coals [Owen 1979] include methane and small alkanes (17% weight of dry coal), often considered to result from the pyrolysis of alkyl groups attached to the coal structure, with liquor (2.5%), light oils (0.8%) and tar (4.5%) as the other components of the volatiles leaving behind the coke product (75%). Alkanes and cyclic aliphatics (mono-, di-, tri- and tetra-cyclics) have been detected in a saturate fraction of a coal liquefaction recycle solvent [Wilson *et al.*, 1987] and by GC-MS of a coal digest [Islas, 2001]. Series of alkanes have also been detected through LC-MS work on hydro-pyrolysis tars [Herod *et al.*, 1987, 1988] showing alkanes up to C_{60} together with cyclic alkanes including pentacyclic triterpanes.

Alkanes have been detected by GC-MS in coal-liquefaction fractions [Walton, 1993, Herod *et al.*, 1995a], in a low-temperature coal tar [Islas, 2001], but not in a coal tar pitch from high-temperature coking. Molecular mass ranges of technical waxes and paraffins have been compared using SEC (with o-dichlorobenzene as eluent), SFC and MALDI-MS and found to agree reasonably well up to a mass (M_n) of about 2,000 u [Kühn *et al.*, 1996] but the methods have not been applied to coal waxes.

7.7.2 SEC methods for alkanes in coal liquids and petroleum fractions

Investigations and applications of size exclusion chromatography (SEC) have involved the solvent 1-methyl-2-pyrrolidinone (NMP) in order to examine the polar and large sized molecular material of coal-liquids, petroleum residues and humic acids [Herod *et al.*, 1995b]. In that work, we were able to examine material insoluble in pyridine that could not be examined by SEC using a solvent less powerful than NMP. However, NMP is a poor solvent for aliphatic materials and the examination of petroleum vacuum residues using NMP, only allowed the examination of aromatic materials with aliphatic material remaining insoluble.

Accordingly, a method for the examination of mixtures of aliphatics and aromatic materials was sought. A solvent that dissolves both aliphatics and aromatics was considered desirable as eluent in SEC. Tetrahydrofuran (THF) can dissolve both types of material and in principle, SEC could be done with THF as eluent. Since aliphatic materials do not absorb UV-light, the tandem use of UV-absorbance and a universal "evaporative light scattering" detector would actually provide an element of distinction between aromatic and aliphatic species. However, fractions containing larger aromatics are only partially soluble in THF. This is easily ascertained by observing increasing pressure drops across guard columns as sample progressively precipitates out from THF onto column packing. Furthermore, due to the relatively low solvent power of THF, the true size exclusion mechanism for aromatics is partially lost during the size exclusion chromatography of some of the heavier coal and petroleum-derived materials. It turns out THF causes the elution of aromatics after the permeation limit defined by polystyrene standards; this will be discussed again in Chapter 8 [Herod *et al.*, 1995b]. In the absence of solvents able to carry all classes of materials through the SEC column, possibilities of analyzing different classes of compounds separately had to be considered.

Carbognani [1997] performed SEC using toluene as eluent, in silica columns operated at 45°C. The injection valve and transfer lines were held at 60°C. This system would probably elute *most* (but not all) aromatics together with the aliphatics. Large alkanes ($>C_{30}$) are soluble in hot toluene but not in cold toluene. Thus, toluene might enable a level of isolation of alkane concentrates. Another alternative was to use heptane as eluent.

In contrast to the application of NMP as eluent in SEC, which tended to minimize the surface interaction between aromatics and the polymeric material of the column packing, the use of heptane as eluent tended to maximize the surface interaction between aromatics and the polymer packing. The elution of aromatics was delayed until well after the permeation limit for aliphatics, allowing a complete separation of the types and an unambiguous identification of aliphatic material.

Accordingly, heptane was selected as eluent in the SEC of aliphatic materials. The SEC column was calibrated using standard alkanes with carbon numbers 13, 14, 16, 20, 22, 25, 30, 40, 44, 50, 60 and branched-C_{19} as well as Polywaxes 500, 655 and 1000 (consisting of even-numbered-carbon alkanes from about C_{20} to C_{100} in different ranges) with detection by an evaporative light scattering detector. Pentane and toluene soluble fractions isolated by column chromatography from a petroleum residue, a coal tar pitch, a coal digest and a low temperature tar, where aliphatics were concentrated, were examined [Islas, 2001, Islas *et al.*, 2003b]. High temperature GC-MS using a 25 m HT-5 column was used to identify some of the aliphatic standards.

Figure 7.13 presents elution times of standard *n*-alkanes from the SEC column. The elution times of the alkanes up to C_{50} appears to follow a linear trend against ln (MM). Larger alkanes do not appear to be readily soluble in heptane. Detection of the alkanes was by ELS since they do not absorb UV-light.

The four "polywax" samples were examined and their ELS chromatograms are shown in Figure 7.14. These chromatograms show slight shifts in elution times from Polywax 500 to 655, but very little of the Polywax 1000 sample dissolved in heptane. The range of alkanes in the Polywax samples given by the suppliers was: Polywax 500 C_{20-60}; Polywax 655 C_{40-60}, Polywax 1000 C_{40-80}. No aliphatic materials eluted later than the small-molecule

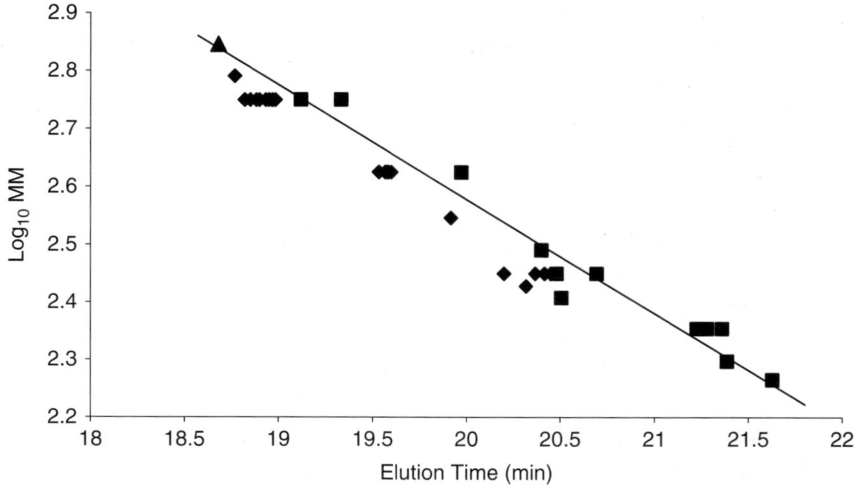

Figure 7.13 Calibration of the "Mixed-E" column using alkane standards up to n-C_{50} (triangle); squares and diamonds represent calibration at different times. Detection by an evaporative light scattering detector. From Al-Muhareb *et al.*, 2005.

permeation limit, at about 24 min for this column. However, the small aromatics, benzene and toluene, did not elute before the permeation limit, and indeed, did not appear until about 45 min later as a broad peak. Similarly, polystyrene standards eluted much later than the permeation limit. The exclusion limit of the column has not been defined using alkanes, but the colloidal silicas of diameters 9, 12 and 22 nm *all* eluted at about 11 min.

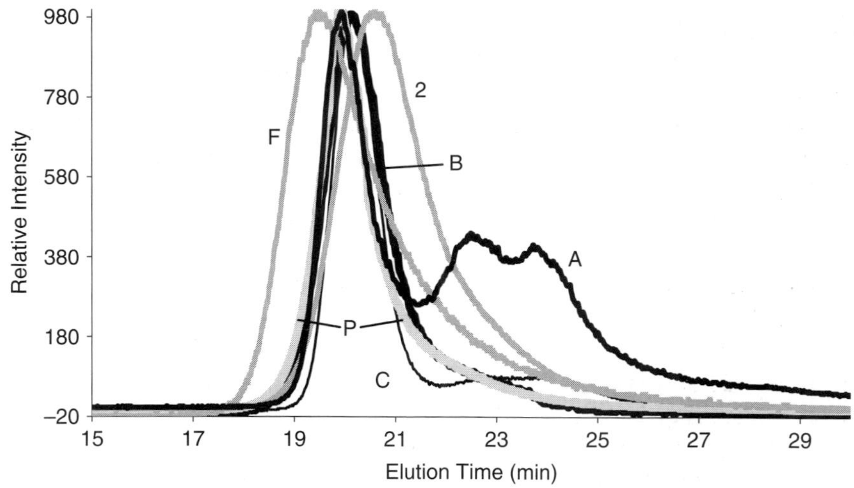

Figure 7.14 Heptane soluble fractions of several vacuum bottoms on Mixed-E, ELS detector. Samples are A-vacuum bottoms A, B-vacuum bottoms B, C- vacuum bottoms C, F- Forties vacuum residue, P- Petrox residue, 2-residue sample 2. From Al-Muhareb *et al.*, 2005.

When, the pentane and toluene soluble fractions from the fractionation of the Petrox petroleum residue were examined by using *n*-heptane as eluent, only the pentane soluble fractions gave signal in SEC. The analysis by GC-MS however showed a familiar picture. The first pentane solubles fraction showed a series of *n*-alkanes superimposed on a broad peak of aliphatic material with no discernible molecular ions, the broad peak presumably reflecting the presence of branched and cyclic aliphatics. The second pentane fraction showed only the broad peak of aliphatic material with some alkyl-substituted phenanthrenes. The analysis of the toluene solubles fraction by GC-MS gave no signal, confirming the presence of larger mass materials than could be handled by the GC columns. The mass balances for the fractionation of four samples by column chromatography using pentane, toluene, acetonitrile, pyridine, NMP and water are shown in Table 7.6. Figure 7.14 presents the SEC chromatograms of several vacuum bottoms obtained using the mixed-E column with detection by ELS. Differences between samples were evident in the ranges of alkanes contained in the samples. However, the range of elution times for these samples did not exceed the range of the calibration samples. It is unlikely, therefore, that the solubility of the sample in heptane was anything near complete. The peaks in Figure 7.14 for samples A and C from 23 to 26 min may correspond to multicyclic alkanes since these elution times correspond to alkanes smaller than C_{12} and such small alkanes do not register with the ELS detector, being too volatile for the system.

7.7.3 HT-GC-MS of petroleum and coal liquid fractions

Figure 7.15(a,b) presents chromatograms of two pentane soluble fractions of the Petrox refinery flash column residue, examined by HT-GC-MS. The first pentane fraction gave a series of alkanes from C_{16} to C_{44}, superimposed on a broad underlying peak of aliphatic material. There was no light material in this sample of flash

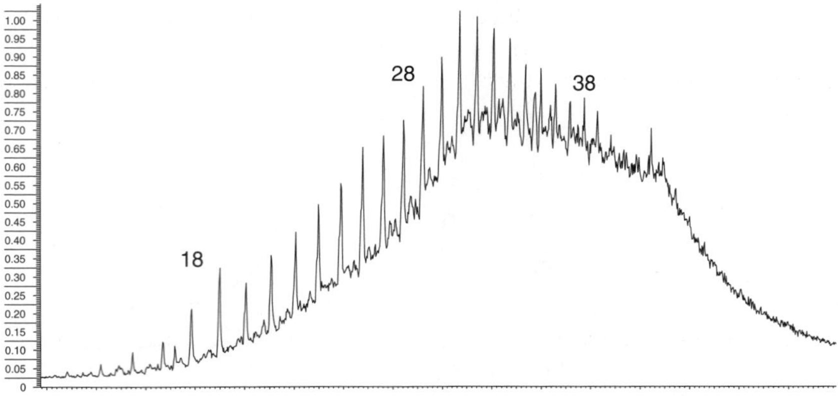

Figure 7.15a HT-GC-MS of first pentane-soluble fraction of Petrox residue; carbon numbers of *n*-alkanes shown and the unresolved hump corresponds to branched alkanes. From Al-Muhareb *et al.*, 2005.

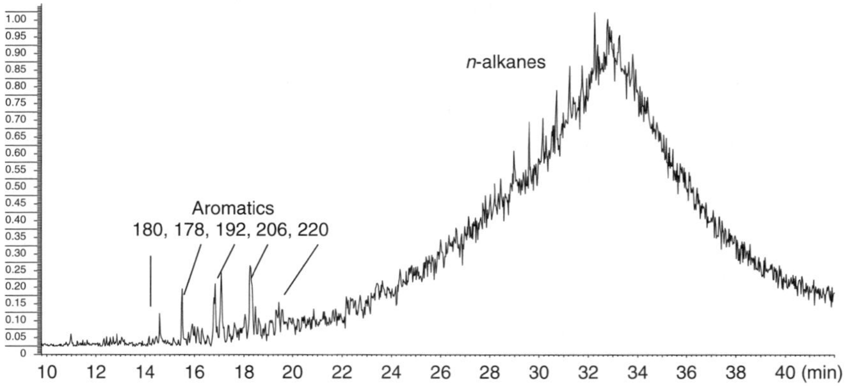

Figure 7.15b HT-GC-MS of second pentane-soluble fraction of Petrox residue; the aromatics are m/z 180 alkyl fluorene, m/z 178 phenanthrene, m/z 192, 206 and 220 methyl, C_2-and C_3-alkyl phenanthrenes respectively while the unresolved hump corresponded mainly to branched alkanes. From Al-Muhareb *et al.*, 2005.

distillation residue. The second pentane fraction showed very few alkanes and some aromatics but the underlying unresolved peak of aliphatic material shifted to later scans than that observed for the first fraction. The toluene fraction showed very little signal, indicating that the fraction contained very little material able to pass through the high temperature column.

Heptane extracts of several petroleum vacuum residues also gave aliphatic chromatograms in SEC, and these chromatograms are shown in Figure 7.14. The Figure shows differences amongst the residues, but there is no resolution of the individual alkanes. When one asphaltene sample (the heptane-insoluble asphaltene from Maya crude) was extracted again with heptane, giving a small quantity of heptane-soluble material from the mainly heptane-insoluble fraction, no aliphatics were observed to pass through the SEC column. This shows that there were no more alkanes in the asphaltene samples within the range of the method, that is smaller than C_{50} alkane.

Aliphatics in coal-derived samples: SEC using heptane as eluent has indicated that both pentane soluble fractions of the low temperature tar (Table 7.6) contained aliphatics, which presented peaks in the retained region. The toluene soluble fraction gave no signal in this column. The coal digest gave a similar distribution of SEC peaks. In contrast, the pitch fractions showed no aliphatic signal in SEC with heptane as eluent. When analyzed by GC-MS, the first pentane fractions of the low temperature tar and the coal digest gave series of alkanes. Chromatograms for the two fractions of the low temperature tar are shown in Figures 7.16a,b. In contrast, the pitch gave no aliphatic peaks in GC-MS, even though the range of aromatics detected was up to m/z 352, tribenzopyrene. The range of alkanes observed in the Polywax 500 sample was up to C_{56}. It appears then that in two of the coal-derived samples (low temperature tar and coal digest), GC-MS could detect the presence of aliphatic species when there was an SEC peak in the retained region – during SEC with n-heptane as eluent. Thus, although the SEC method can provide an unambiguous indication of the presence of alkanes in a

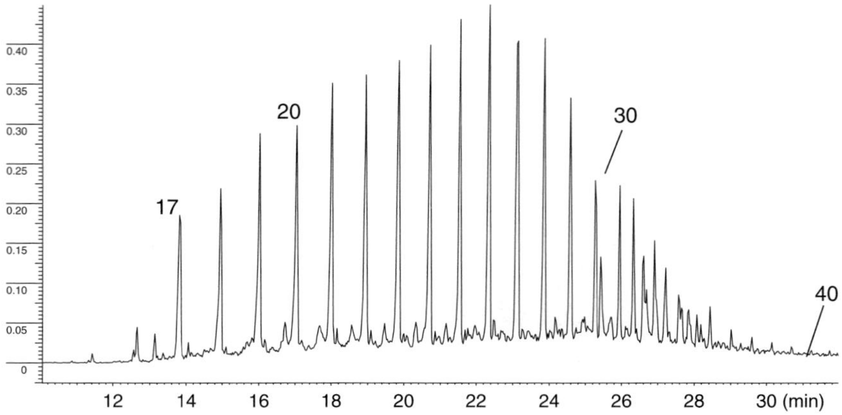

Figure 7.16a HT GC-MS of first pentane-soluble fraction of low temperature tar; *n*-alkane carbon numbers as marked. From Al-Muhareb *et al.*, 2005.

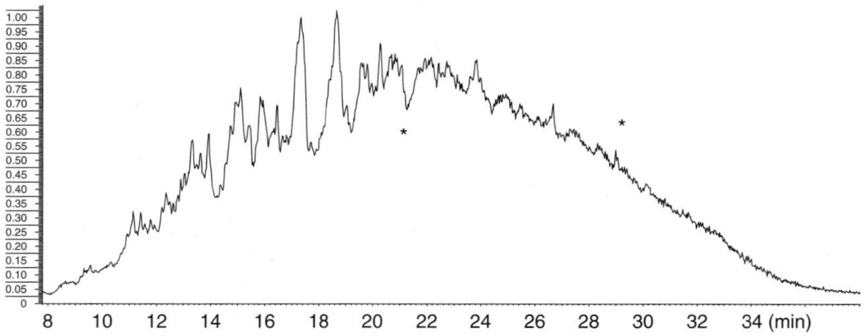

Figure 7.16b HT GC-MS of second pentane-soluble fraction of low temperature tar; all the components were aromatic with no detectable alkanes. From Al-Muhareb *et al.*, 2005.

complex mixture, high-temperature GC-MS must remain the method of choice for the examination of alkanes, particularly when molecular mass measurements are required.

Figures 7.17a,b show mass spectra taken from the chromatogram of Figure 7.16b to indicate the complexity of the mixture at the points marked by stars. The underlying unresolved peak of overlapping components is clearly displayed by the spectra and prevents a meaningful discussion of the sample. Coal tar pitches as in Figure 7.7 present a much simpler problem of resolution of components by GC-MS. The spectra of Figure 7.17 indicate that the low temperature tar contains many overlapping types of aromatic molecule even after simplification by fractionating by column chromatography. The spectra are difficult to interpret because the fragmentation patterns cannot be deconvoluted. Two dimensional GC-MS could, in principle, be used to resolve the overlapping spectra if it was considered necessary to identify all the minor components. The complexity of the whole sample can be illustrated by comparing mass spectra from direct mass spectrometry methods, such as shown in the complex mass spectra of Figure 7.10 for an unfractionated hydropyrolysis tar.

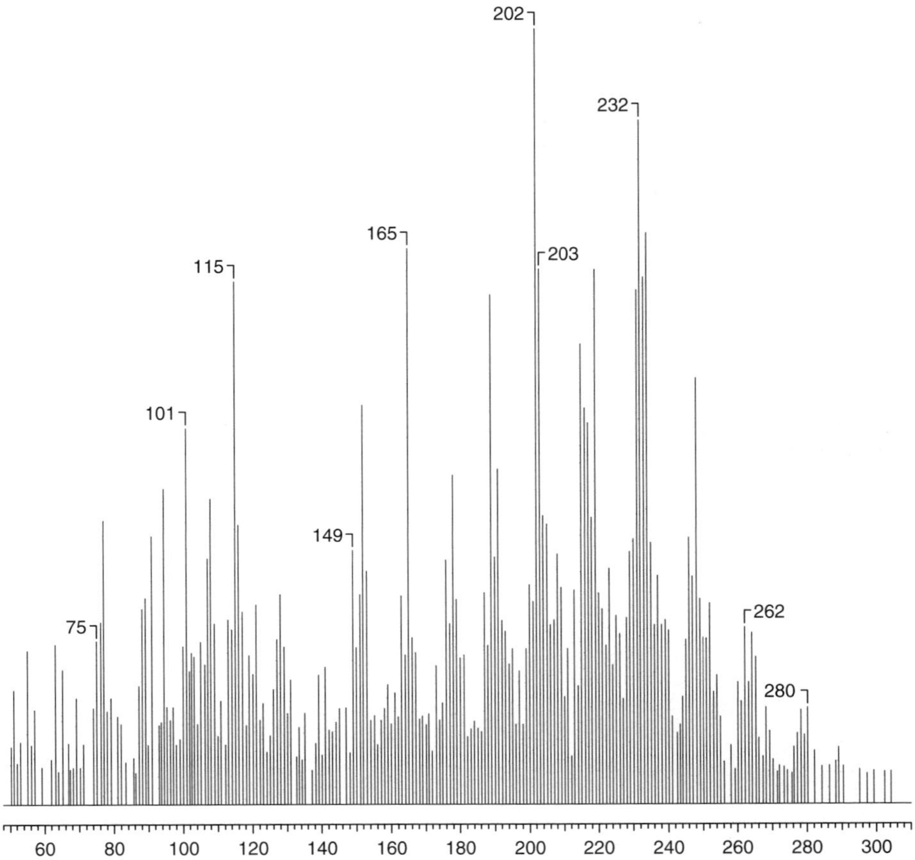

Figure 7.17a Mass spectrum from the chromatogram of Figure 7.16b at about 22 min. The major components of the spectrum in the valley shown have molecular ions from m/z 200 to 230 with minor components to m/z 300. While some of these components might be known, the overlap of individual mass spectra in this mixture is too complex for meaningful identification.

7.8 Conclusion

GC remains pre-eminent in the chromatographic analysis of lower molecular mass complex materials derived from fossil and biomass fuels. It gives the greatest resolution. Recent advances in column technology have both increased analysis speed and extended the molecular mass range of possible analyses, while GC × GC has the potential to offer unparalleled resolution. The inherent limitation of GC, however, is the insurmountable problem that, at temperatures at which higher molecular mass compounds have sufficient volatility to pass through a column, both column and analyte may be thermally degraded. However, the method will remain important for the determination of those lighter PACs considered to indicate the carcinogenic properties of coal-derived liquids.

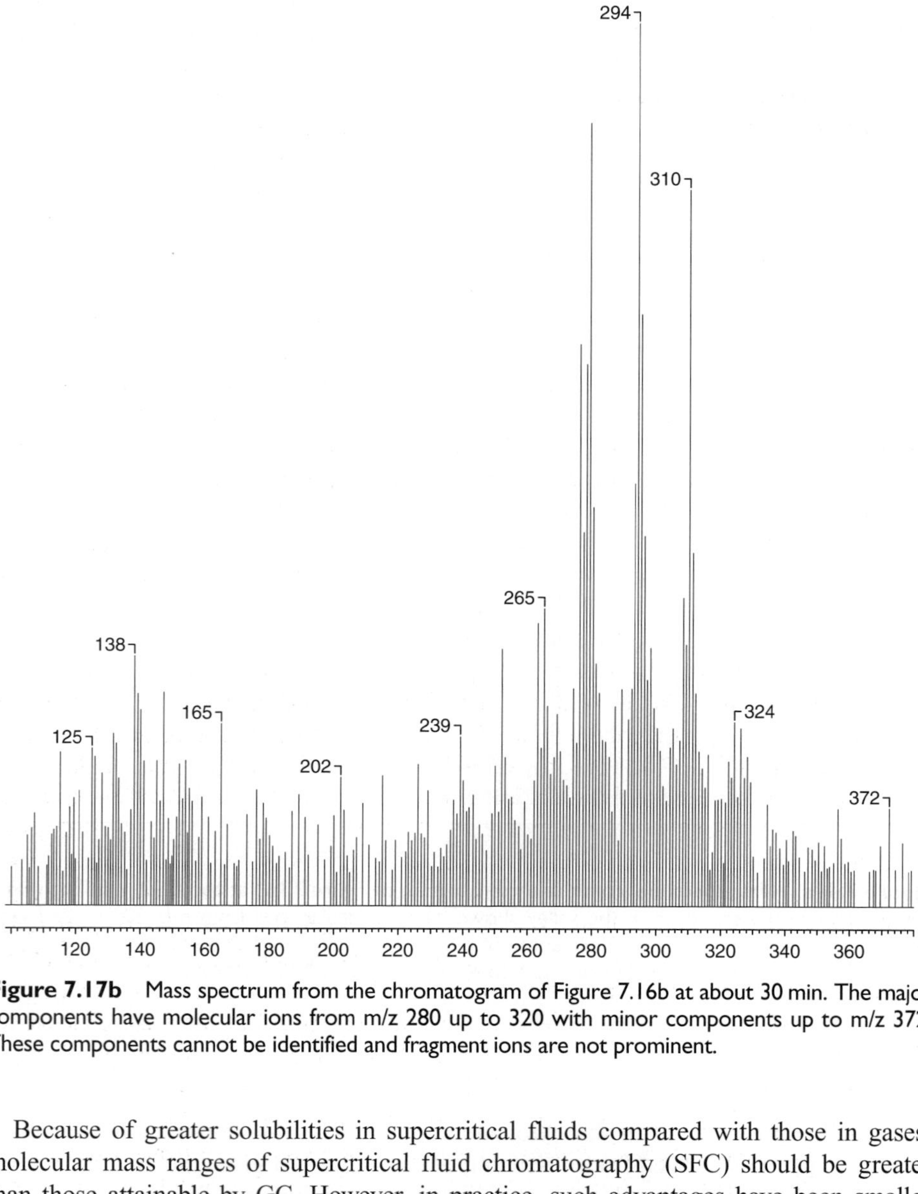

Figure 7.17b Mass spectrum from the chromatogram of Figure 7.16b at about 30 min. The major components have molecular ions from m/z 280 up to 320 with minor components up to m/z 372. These components cannot be identified and fragment ions are not prominent.

Because of greater solubilities in supercritical fluids compared with those in gases, molecular mass ranges of supercritical fluid chromatography (SFC) should be greater than those attainable by GC. However, in practice, such advantages have been smaller than anticipated for PAC. The accessible molecular mass range has increased by approximately 200 mass units. Separations are, of course, carried out at much lower temperatures in SFC than in GC (raising the temperature reduces the fluid density at a given pressure, and reduces solubility) so that thermal decomposition of analytes is not an issue. Furthermore, because of the greater diffusivity of supercritical fluids in comparison with gases, SFC analysis is more rapid than HPLC. Future developments may require mobile phases other than carbon dioxide. Nor has HPLC greatly extended the available molecular mass range of PAC analysis. Higher molecular mass PACs are not soluble in the hydrocarbon solvents used in normal phase HPLC as is conventionally practiced. However, some advances have been reported by using more polar solvents such as dichloromethane. New stationary phases are also required for which partition on to the column packing is less marked. Similar considerations apply to reverse-phase HPLC.

In our experience, none of the gas or liquid chromatography methods has given sufficient resolution to allow unambiguous identification of many of the numerous components present in the complex coal or petroleum-derived liquids we have examined. GC-MS is directly useful for the estimation of the individual PAC molecules that are generally considered to present a carcinogenic hazard, such as the EPA 16, mentioned above. Also, the method quickly provides indications of the probable processing temperature of coal tars. The higher the temperature, the fewer the peaks and these tend to be unsubstituted aromatics with few alkyl derivatives. Tars produced at lower temperatures tend to show an underlying unresolved peak of material, as in Figures 7.15a,b and 7.16b above and the spectra of Figure 7.17. In addition, the data of section 7.7.3 on the search for alkanes using HT-GC-MS broadly indicated that only those components soluble in pentane could pass through the GC column. Therefore all the material of these samples insoluble in pentane but soluble in toluene, acetonitrile, pyridine or NMP, could not be detected by GC methods. Mass balances for four samples shown in Table 7.6 above reinforce this argument while the data of Table 2.6 indicates that the proportion of coals as extracts amenable to GC study is limited to less than 5% for the 8 coals cited.

In these circumstances, two methods can give significant information: GC-mass spectrometric detection and GC $\times$ GC with mass spectrometry. HPLC methods have given molecular mass information over wider ranges of mass than by GC, but compounds larger than those commonly available as standards are difficult to identify. In our work, the low-mass MALDI-MS spectrum gave mass values (m/z) as shown in Figure 7.12. The structures likely to represent those mass values were guessed from masses of smaller standards using some simple rules. These include: 1) the addition of a fused benzo group adds 50 mass units; 2) the addition of a methyl group adds 14 mass units; 3) the addition of an ethylene bridge adds 24 mass units and 4) the addition of a ethane bridge adds 26 mass units. Molecular ions of homologous series are separated by 14 mass units. This does not add up to positive identification of molecular types for which accurate mass measurement of molecular ions is necessary. The variety of structural isomers making up each of the particular molecular ions cannot be determined by MALDI-MS and some chromatographic separation would be needed to achieve resolution of structural isomers. In Figure 7.12 for instance, the peak for m/z 252 can be assumed to correspond to a collection of 7 isomers.

Mass spectrometric methods give valuable information on molecular mass and molecular type either when used directly in sequence with GC, SFC or HPLC, or when used off-line in the analysis of fractions prepared by other methods. These might include distillation, various chromatographic methods or by use of different methods of ionization. The limitations applicable to mass spectrometric methods are generally those associated with the thermal volatility of sample. Volatility does not appear to be a problem in the case of LD-MS at low laser power. In general, off-line mass spectrometry is able to generate molecular ions for coal and petroleum derived materials in excess of those achieved by using chromatographic methods on-line with mass spectrometry.

In view of these considerations, a chromatographic approach based on a separation principle other than partition is required for the higher molecular mass constituents of coal derivatives. High-performance size-exclusion chromatography (SEC) satisfies this criterion since, molecules of different sizes are separated according to their degree of penetration into the pores of a gel packed into a column in the form of small-diameter spheres. Elution is in reverse sequence of molecular weight. The gel is typically a three-dimensional network of a cross-linked polymer or co-polymer, with controlled porosity, such as polystyrene-divinylbenzene.

Choice of mobile phase is vital in size exclusion chromatography. The work described in this Chapter has shown that heptane can be used to elute aliphatics with enhanced interaction of column packing and aromatic solutes, thereby allowing the unambiguous identification of aliphatics in the presence of aromatics. In contrast, recent work has shown how 1-methyl-2-pyrrolidinone is preferable to the earlier-used toluene and tetrahydrofuran solvents because of its greater solvent power and its capacity to prevent intermolecular association of solutes and interaction between column packing and aromatic solutes. The application of this method to larger molecular mass materials is described in the next Chapter.

References

Abbott, D.J. (1995) in Chromatography in the Petroleum Industry, E.R. Adlard (ed.) Elsevier Amsterdam p 41.

Aczel, T. (1972) *Reviews Anal Chem, 1*:226.

Aczel, T. and Lumpkin, H.E. (1979) Chapter 2 in *Refining of Synthetic Crudes*. Adv Chem Series 179, American Chem. Soc. p 13.

Adegoroye, A., Paterson, N., Morgan, T.J., Herod, A.A., Dugwell, D.R. and Kandiyoti, R. (2004) Fuel 2004; 83: 1949–1960.

Al-Muhareb, E.M., Karaca, F., Morgan, T.J., Behrouzi, M., Herod, A.A., Bull, I.D. and Kandiyoti, R. (2005) Submitted to Energy & Fuels.

ASTM. American Society for Testing Materials, Methods in Annual Year Books.

Avid, B., Purevsuren, B., Paterson, N., Zhuo, Y., Peralta, D., Herod, A.A., Dugwell, D.R. and Kandiyoti, R. (2004) Fuel 83, 1105–1111.

Bartle, K.D. (1985) in Handbook of Polycyclic Aromatic Hydrocarbons Vol 2. A Bjorseth and T Ramdahl eds. Marcel Dekker New York, Chapter 6.

Bartle, K.D. (1989) in Spectroscopic Analysis of Coal Liquids. JR Kershaw ed, Elsevier Amsterdam, Chapter 2.

Bartle, K. D. and Myers, P. (2002) Trends Anal. Chem. 21, 547.

Bartle, K.D., Collin, G., Stadelhofer, J.W. and Zander, M. (1979) J. Chem. Tech. Biotech. 29, 531.

Bartle, K.D., Lee, M.L. and Wise, S.A. (1981) Chem. Soc. Review 10, 113.

Bartle, K.D., Barker, I.K., Clifford, A.A., Kithinji, J.P., Raynor, M.W. and Shilstone, G.F. (1987) Analyt. Proc. 24, 299.

Bartle, K.D., Burke, M., Mills, D.G., Pape, S. and Lu, S.L. (1992) Fuel Sci. Tech. Int. 10, 1071.

Bartle, K.D., Bottrell, S., Burke, M.P., Jones, C., Louie, P.K., Lu, S.L., Salvado, J., Taylor, N. and Wallace, S. (1994) Int. J. Energy Res. 18, 299.

Bartle, K.D., Clifford, A.A., Myers, P., Robson, M.M., Sealey, K., Tong, D., Batchelder, D.N. and Cooper, S. (2000) in Parcher J.F. and Chester T.L. (eds) Unified Chromatography; ACS Symp. Series 748, Amer. Chem. Soc. Washington DC., p142.

Begon, V., Islas, C.A., Lazaro, M-J., Suelves, I., Herod, A.A., Dugwell, D.R. and Kandiyoti, R.(2000) Eur. J. Mass Spectrom. 6, 39–48.

Bemgard, A., Colmsjo, A. and Lundmark, B.O. (1993) J. Chromatogr. A 630, 287.

Bocchini, P., Galletti, G.C., Seraglia, R., Traldi, P., Camarero, S. and Martinez, A.T. (1996) Rapid Commun. Mass Spectrom. 10, 1144–1147.

Boden, A.R., Ladwig, G.E. and Reiner, E.J. (2002) Polycyclic Aromatic Compounds 22, 301.

Boduszynski, M.M., Hurtubise, R.J., Allen, T.W. and Silver, H.F. (1983) Anal. Chem. 55, 225.

Borwitzky, H., Schomburg, G., Sauerland, H.D. and Zander, M. (1978) Erdol, Kohle, Ergas, Petrochemie 31, 371.

Cagniant, D., Gruber, R., Lacordaire-Wilhelm, C. and Schulten, H-R. (1992) *Energy & Fuels, 6*, 694–701.

Campbell, R.M. and Lee, M.L. (1986) Anal. Chem. 58, 2247.

Carbognani, L. (1997) J. Chromatography A 788, 63–73.

Caslavsky, J. and Kotlarikova, P. (2003) Polycyclic Aromatic Compounds 23, 327.

Challinor, J.M. (1995) J Anal. Appl. Pyrol. 35, 93–107.

Chapman, J.R. (1993) Practical Organic Mass Spectrometry 2nd ed Wiley, London,

Chang, H.K.C., Nishioka, M., Bartle, K.D., Wise, S.A, Bayona, J.M., Markides, K.E. and Lee, M.L. (1988) Fuel 67, 45.

Chester, T.L. and Pinkston, J.D. (2004) Anal. Chem. 76, 4606–4613.

Cramers, C.A., Janssen, H.-G., van Deursen, M.M. and LeClerq, P.A. (1999) J. Chromatogr. A 856, 315.

Davies, I.D., Raynor, M.W., Williams, P.T., Andrews, G.E. and Bartle, K.D. (1987) Anal. Chem. 59, 2579.

Deelchand, J-P., Naqvi, Z., Dubau, C., Shearman, J., Lazaro, M-J., Herod, A.A., Read, H. and Kandiyoti, R. (1999) J. Chromatogr. A. 830, 397–414.

Eight Peak Index of Mass Spectra, (1992) 4th Ed, Royal Society of Chemistry, Cambridge UK.

Evans, R.J. and Milne, T.A. (1987a) Energy & Fuels 1, 123–137.

Evans, R.J. and Milne, T.A. (1987b) Energy & Fuels 1, 311–319.

Evershed, R.P., Jerman, K. and Eglinton, G. (1985) Nature 314 (6011) 528.

Fraga, A.R., Gaines, A.F. and Kandiyoti, R. (1991) Fuel 70, 803–809.

Genuit, W.J.L. (2005) Chapter 8 Determination of organic compounds by methods using mass spectrometry, in The Encyclopedia of Mass Spectrometry vol 4 Fundamentals of and Applications to organic (and organometallic) compounds Ed N.M.M. Nibbering, Elsevier, Oxford UK. Pages 781–790.

Guell, A.J., Li, C.-Z., Herod, A.A., Stokes, B.J., Hancock P. and Kandiyoti, R. (1993a) Proc. Advances in Thermochemical Biomass Conversion, Interlaken, Switzerland, 11–15 May 1992. Blackie, London, 2, 1053–1067.

Guell, A.J., Li, C.-Z., Herod, A.A., Stokes B.J., Hancock P. and Kandiyoti, R. (1993b) Biomass and Bioenergy, 5(2), 155–171.

Hamilton, R. and Hamilton, S. (1987) Thin Layer Chromatography, ACOL, John Wiley & Sons Chichester UK.

Hamilton, J.F., AC Lewis, M. Millan, KD Bartle, AA Herod, R. Kandiyoti (2005) paper in preparation.

Hamming, M.C. and Foster, N.G., *Interpretation of mass spectra of organic compounds*, Academic Press , NY, USA 1972.

Heaton, D.M., Bartle, K.D., Clifford, A.A., Myers, P. and King, B.W. (1994) Chromatographia 39, 607.

Herod, A.A. (1989) *Institute of Petroleum, Quarterly Journal of Technical Papers*, April–June, 15–53.

Herod, A.A. (1998) Chapter 7 in *PAHs and related compounds* part 3I Neilson A.H. ed, Springer-Verlag, Berlin. pages 271–323.

Herod, A.A. (2005). Chapter 8 Determination of organic compounds by methods using mass spectrometry, in The Encyclopedia of Mass Spectrometry vol 4 Fundamentals of and Applications to organic (and organometallic) compounds Ed N.M.M. Nibbering, Elsevier, Oxford UK. Pages 790–803.

Herod, A.A., Hodges, N.J., Pritchard, E. and Smith, C.A. (1983) Fuel, 62, 1331–1336.

Herod, A.A., Ladner, W.R., Stokes, B.J., Berry, A.J., Games, D.E. and Hohn, M. (1987) Fuel, 66, 935–946.

Herod, A.A., Ladner, W.R., Stokes, B.J., Major, H.J. and Fairbrother, A. (1988) The Analyst, 113, 797–804.

Herod, A.A., Stokes, B.J. and Radeck, D. (1991) Fuel, 70, 329–340.

Herod, A.A., Stokes, B.J. and Schulten, H.-.R. (1993) *Fuel, 72*, 31–43.

Herod, A.A., Stokes, B.J., Hancock, P., Kandiyoti, R., Parker, J.E., Johnson, C.A.F., John, P. and Smith, G.P. (1994) J. Chem. Soc. Perkin 2, 499–506.

Herod, A.A., Hellenbrand, R., Zhang, S.F., Xu, B. and Kandiyoti, R.,(1995a) Fuel 1995, 74, 1739–1752.

Herod, A.A.and Kandiyoti, R. (1995b) *J. Chromatogr*. A, *708*, 143–160.

Herod, A.A.; Islas, C.A.; Lazaro, M-J.; Dubau, C.; Carter, J.F.; Kandiyoti R.; (1999) *Rapid Commun. Mass Spectrom*. 1999, *13*, 201–210.

Herod, A.A.; Millan, M.; Karaca, F.; Behrouzi, M.; Kandiyoti, R. (2005) unpublished work.

Howsam, M. and Jones, K.C. (1998) Chapter 4 in *PAHs and related compounds* part 3I Neilson A.H. ed, Springer-Verlag, Berlin. pages 137–174.

ICS Reference Library, vol 33A, (1920) Properties of Gases, Mine Gases, Mine Ventilation, Geology of Coal, Rock Drilling, Explosives and Shot-Firing, Mine-Air Analysis, Geological Maps and Sections, section 31, International Correspondence Schools Ltd, London.

Islas, C.A.; Suelves, I.; Carter, J.F.; Herod, A.A.; Kandiyoti, R. (2000) *Rapid Commun. Mass Spectrom*. 2000, *14*, 1766–1782.

Islas, C.A. (2001) PhD Thesis University of London.

Islas, C.A., Suelves, I., Carter, J.F., Herod A.A. and Kandiyoti, R. (2001) Rapid Commun. Mass Spectrom. 15, 845–856

Islas, C.A.; Suelves, I.; Carter, J.F.; Li, W.; Morgan, T.J.; Herod, A.A.; Kandiyoti, R. (2002) *Rapid Commun. Mass Spectrom*. 16, 774–784

Islas, C.A.; Suelves, I.; Carter, J.F.; Apicella, B.; Herod, A.A.; Kandiyoti, R.; (2003a) *Comb. Sci. Technol*. 2003, *175*, 775–791.

Islas, C.A., Suelves, I., Li, W., Morgan, T.J., Herod, A.A. and Kandiyoti, R. (2003b) Fuel 2003; 82:1813–1823.

Kelly, G. and Bartle, K.D. (1994) J High Res. Chromatogr. 17, 390.

King, R.C., Lee, M.L, Tominaga, R., Pratap, R., Iwao, M. and Castle, R.N. (1982) Anal. Chem. 54, 1802.

Kithinji, J.P., Raynor, M.W., Egiu, B., Davies, I.L., Bartle, K.D. and Clifford, A.A. (1990) J. High Res. Chromatogr. 13, 27.

Klatt, L.N. (1979) J. Chromatogr. Sci. 17, 225.

Kühn, G., Weidner, St., Just, U., Hohner, G. (1996) J. Chromatogr. A 732 111–117.

La, i W.-C. and Song, C. (1995) Fuel 10 1436.

Lazaro, M.J., Islas, C.A., Herod, A.A. and Kandiyoti, R. (1999) Energy & Fuels, 1999, 13, 1212–1222.

Lederer M. (1994) Chomatography for Inorganic Chemistry, John Wiley & Sons Chichester UK.

Lee, A.L., Bartle, K.D. and Lewis, A.C. (2001) Anal. Chem. 73 1330.

Lee, M.L., Vassilaros, D.L., White, C.M. and Novotny, M. (1979) Anal. Chem. 51, 768.

Lee, M.L., Yang, F.J. and Bartle, K.D. (1984) Open Tubular Column Gas Chromatography, Wiley, New York.

Lipsky, S.R. and Duffy, M.L. (1986) J. High Res. Chromatogr. 9, 376.
Malhotra, R., McMillen, D.F., Watson, E.L. and Huestis, D.L. (1993) *Energy & Fuels, 7*, 1079–1087.
Marriott, P.J., Massil, T., Hugel, H. (2004) J. Sep. Sci. 27, 1273–1284.
McLafferty, F.W. and Turecek, F., (1993) *Interpretation of mass spectra*, 4th ed, University Science Books, Mill Valley, Ca, USA.
McRae, C., Sun, C.G., McMillan, C.F., Snape, C.E. and Fallick, A.E. (2000) *Polycycl. Aromat. Compd. 20*, 97–109.
Menéndez, R.; Blanco, C.; Santamaría, R.; Domínguez, A.; Blanco, C.G.; Suelves, I.; Herod, A.A.; Morgan, T.G.; Kandiyoti, R.(2002) *Energy & Fuels 16*, 1540–1549.
Millan, M., Morgan, T.J., Behrouzi, M., Karaca, F., Galmes, C., Herod, A.A. and Kandiyoti, R. (2005) Rapid Commun. Mass Spectrom. 19, 1867–1873.
Novotny, M., Hirose, A. and Wiesler, D. (1984) Anal. Chem. 56, 1243.
Owen, J. (1979) Chapter 9 in Coal and Modern Coal processing: an Introduction. Eds Pitt G.J. and Millward G.R.; Academic Press London
Parker, J.E.; Johnson, C.A.F.; John, P.; Smith, G.P.; Herod, A.A.; Stokes, B.J.; Kandiyoti, R. (1993) *Fuel 72*, 1381–1391.
Phillips, J.B. and Venkatramani, C. (1993) J Microcolumn Sep. 5, 511–516.
Pindoria, R.V., Megaritis, A., Chatzakis, I.N., Vasanthakumar, L.S., Lazaro, M.J., Herod, A.A., Garcia, X.A., Gordon, A. and Kandiyoti, R. (1997a) Fuel, 76, 101–113.
Pindoria, R.V., Lim, J.-Y., Hawkes, J.E., Lazaro, M.-J., Herod, A.A. and Kandiyoti, R. (1997b) Fuel 76, 1013–1023.
Pipatmanomai, S., Islas, C.A., Suelves, I., Herod A.A. and Kandiyoti, R. (2001) J. Anal. Appl. Pyrol. 58, 299–313.
Purevsuren, B., Herod, A.A., Kandiyoti, R., Morgan, T.J., Avid, B., Gerelmaa, T. and Davaajav Ya. (2004a) Fuel 83: 799–805.
Purevsuren, B., Herod, A.A., Kandiyoti, R., Morgan, T.J., Avid, B. and.Davaajav Ya. (2004b) Eur. J. Mass Spec. 10, 101–108.
Reddy, K.M., Wei, B. and Song, C. (1998) Catalysis Today 43, 187–202.
Robinson, N., Evershed, R.P., Higgs, J., Jerman, K. and Eglinton, G. (1987) The Analyst 112, 637.
Romanowski, T., Funcke, U., Grossman, I., Konig, J. and Balfanz, E. (1983) Anal. Chem. 55, 1030.
Saini, A.K. and Song, C. (1994) Amer. Chem. Soc. Div. Fuel Chem. 39(3) 796
Sharriff, S.M., Tong, D. and Bartle, K.D. (1994) J. Chromatogr. Sci. 32, 541.
Sharriff, S.M., Robson, M.M. and Bartle, K.D. (1997) Polycyclic Aromatic Compounds 12, 147.
Shariff, S.M., Robson, M.M., Myers, P., Bartle, K.D. and Clifford, A.A. (1998) Fuel 77 927.
Shellie, R., Marriott, P., Leus, M., Dufour, J-P., Sun, K., Winniford, B., Griffith, J., Luong, J., Mondello, L., Dugo, G. (2003) J. Chromatogr. A 1019, 273–278.
Scott, A.C. and Fleet, A.J., Eds (1994) Coal and Coal-bearing Strata as Oil-prone Source Rocks? Geological Society Special Publication no. 77, The Geological Society, London.
Snape, C.E., Ladner, W.R. and Bartle, K.D. (1985) Fuel 64, 1394–1400.
Song, C., Schobert, H.H. and Hatcher, P.G. (1992) *Amer. Chem Soc. Div. Fuel Chem., 37(2)*, 638–645
Suelves, I., Islas, C.A., Herod, A.A. and Kandiyoti, R. (2001a) Energy & Fuels 15 429–437.
Suelves, I., Lazaro, M-J., Begon, V., Morgan, T.J., Herod, A.A., Dugwell, D.R. and Kandiyoti, R. (2001b) Energy & Fuels 15, 1153–1165.
Takagama, Y., Takeichi, T. and Kawai, S. (1988) J. High Res. Chromatogr. 11 732.
Tong, D., Bartle, K.D., Robinson, R.E. and Altham, P. (1993) J. Chromatogr. Sci. 31, 77.
Touchstone JC. (1992) Practice of Thin Layer Chromatography, 3rd ed. John Wiley & Sons Inc. New York.

Trestianu, S., Zilioli, G., Scironi, A., Saravelle, C., Munari, F., Galli, M., Gaspar, G., Colin, J.M. and Jovelin, J.L. (1985) J. High Res. Chromatogr. 8, 771.

Tromp, P.J.J., Moulijn, J.A. and Boon J. (1988) In New Trends in Coal Science, ed. Y. Yurum, Kluwer Academic Publishers, 241–269.

Tromp, P.J.J., Moulijn, J.A. and Boon, J. (1989) *J. Anal. Appl. Pyrolysis, 15*, 319–331.

Vassilaros, D.L., King, R.C., Later, D.W. and Lee, M.L. (1982) J. Chromatogr. 252, 1.

Walton, S.T. (1993) Fuel 72, 687.

Watson, J.T. and Sparkman, O.D. (2005) Chapter 7, Rules for mass spectral interpretation: the standard interpretation procedure. The Encyclopedia of Mass Spectrometry vol 4 Fundamentals of and Applications to organic (and organometallic) compounds Ed N.M.M. Nibbering, Elsevier, Oxford UK. pages 719–731.

Wilson, R., Johnson, C.A.F., Parker, J., Herod, A.A. (1987) Organic Mass Spectrometry, 22, 115–116.

Wise, S.A. (1983) in Bjorseth A. and Ramdahl T. (eds) Handbook of Polycyclic Aromatic Hydrocarbons vol. 1 Ch 5, Marcel Dekker, New York.

Wise, S.A. (1985) in Bjorseth A. and Ramdahl T. (eds) Handbook of Polycyclic Aromatic Hydrocarbons, vol 2, Ch 5, Marcel Dekker, New York.

Wise, S.A. (2002) Polycyclic Aromatic Compounds 22, 197.

Wood et al 1985 Wood, K.V., Albright, L.F., Brodbelt, J.S. and Cooks, R.G.; (1985) *Analytica Chimica Acta, 173*, 117–127.

CHAPTER 8

Analytical Techniques for High Mass Materials: Method Development

8.1 Introduction

In Chapter 7, we examined some of the most powerful techniques available to analytical chemistry for examining coal and biomass derived tars and extracts. Typically, these tars and extracts are similar to materials recovered during pyrolysis, gasification and liquefaction experiments described in Chapters 3–6. Large proportions of such samples are of molecular masses above the ranges amenable to analysis by gas chromatography and associated methods such as GC-mass spectroscopy (GC-MS).

In this Chapter, we consider analytical methods able to characterize molecules, which lie beyond the range of such standard techniques. Mostly we will focus on materials with masses greater than 500 u, contained in tars and extracts from coal or biomass as well as crude oil and its fractions. Standard molecules are not normally available for such materials. We will first outline developments in bulk methods for mass estimation, starting with size exclusion chromatography. The technique has been used for examining mass distributions of predominantly aromatic materials and is calibrated by using standard compounds and polymers of many different types.

We will next describe mass spectrometric methods likely to ionize and detect materials with molecular masses above the m/z 500 range, normally beyond the working range of mass spectrometric methods introduced in the previous Chapter. Methods described in the present Chapter are usually coupled to chromatographic fractionation and involve several different combinations of sample introduction (e.g. liquid chromatography, supercritical fluid chromatography, thin layer chromatography) and ionization techniques. The latter include desorption chemical ionization (DCI), field ionization (FI), field desorption (FD), fast atom bombardment (FAB), plasma desorption (PD), matrix assisted laser desorption/ionization MALDI- and laser desorption (LD) mass spectrometry and electrospray ionization (ESI). Finally, we will consider the analysis of trace metal elements using inductively coupled plasma mass spectrometry (ICP-MS). Units of molecular mass based on the unified atomic mass scale (^{12}C) are shown as u [Mills *et al.*, 1993] rather than the alternative Dalton (Da), while masses determined by mass spectrometric methods are shown as mass to charge ratio (m/z).

Structural features of these larger molecular mass materials are examined with methods such as UV-fluorescence, NMR, FT-IR. Applications of vapor pressure osmometry (VPO) will be described for molecular masses greater than 500 u, although the technique is better suited to lower mass measurement. Examples have been included from the examination of coal liquids, soots, petroleum products and residues, biomass tars, and kerogens. Some aspects of this work have been reviewed previously [Zhuo *et al.*, 2003; Herod *et al.*, 2003b; Herod 2005].

8.2 SEC as Method to Examine Molecular Mass or Size Ranges of Complex Mixtures

Size exclusion chromatography (SEC) is the subject of regular reviews in Analytical Chemistry [Barth *et al.*, 1996, 1998]. Calibration methodologies have been reviewed [Kostanski *et al.* 2004]. Normal and reverse phase liquid chromatography methods depend on interactions between solute molecules and the surface of the column packing material with changes in solvent composition being used to elute components of different polarity from the surface of the packing. SEC on the other hand, depends for its action on the absence of interaction with the surface of the column packing material and only depends on the ability of different molecular sizes to penetrate the porosity of the column-packing medium. In normal and reverse phase liquid chromatography, the smallest molecules usually elute first, followed by the larger and more polar molecules. In SEC, the largest size molecules elute first because they cannot penetrate the porosity and elute through the inter-particulate voids of the column; the smallest molecules that can penetrate the smallest porosity of the packing elute last and define the permeation limit of the column. No part of any sample should elute after the permeation limit in true SEC. The measure of molecular size is the hydrodynamic volume in the solvent used to elute the column, which depends on solvation effects and this may not necessarily equate with molecular mass.

Parameters that define SEC include the void volume (the earliest possible time or elution volume that corresponds to the elution of the largest size molecules through the column void volume), the permeation limit (the latest time or greatest elution volume required for elution of small molecules), number average molecular weight (M_n), weight average molecular weight (M_w) and peak average molecular weight (M_p). Polydispersity is defined as M_w/M_n and serves as an index of the breadth of the molecular weight distribution. For a sample of narrow polydispersity ($M_w/M_n < 1.1$) it may be expected that $M_n \approx M_w \approx M_p$.

The following equations define M_n and M_w:

$$M_n = \Sigma_i n_i M_i/\Sigma_i n_i \text{ and } M_w = \Sigma_i n_i M_i^2/\Sigma_i n_i M_i,$$

where n_i is the number of molecules in the ith elution volume interval of mass M_i.

Size exclusion chromatography is the liquid chromatographic method most suitable for this purpose but has the drawback that calibration of the relation between molecular size and elution-time or elution-volume must be achieved by using standard molecules and molecular mass standards. The method also assumes that the calibrant and analyte molecules have the same relation between molecular size and molecular mass. For tar and asphaltene components of mass greater than 500 u, this presents difficulties because the molecular structures are not generally known. The extrapolation of molecular structures from aromatics known from gas chromatography (see Chapter 7), by addition of successive benzo groups [replacement of 2H by $-C_4H_4-$, net mass increase 50 u], a methene bridge [replacement of 2H by $-CH_2-$, net addition of 12 u], ethylene bridges [replacement of 2H by –CH=CH–, net addition of 24 u] or ethyl bridges [replacement of 2H by $-CH_2-CH_2-$, net addition 26 u], leads to structures too highly aromatic in relation to the bulk atomic composition of the sample. In the case of the coal tar pitch used for much of the work described below, the C/H ratio is 1.9 whereas for structures such as coronene, the C/H ratio is 2. For ovalene the ratio is 2.28 and increases for more condensed structures.

Distributions of molecular masses in coal- and petroleum-derived liquids are relevant to research in fields as diverse as catalytic hydrocracking and environmental pollution diagnostics. Recent work [Herod *et al.*, 2000a] has attempted to develop tools for the characterization of 'heavy' coal liquids through the parallel use of size exclusion chromatography (SEC) and laser-desorption mass spectrometry. Progress in the parallel use of these two independent techniques has provided evidence for the presence of high mass materials in coal-derived and other samples. The mass spectrometric work is described below in section 8.5, with evidence from other supporting techniques in section 8.6.

The use of SEC in the estimation of molecular mass ranges of hydrocarbon mixtures relies on the assumption that the sizes of standard molecules or polymers relate to the sizes of the sample molecules of similar molecular masses. The general principles of the method have been given [Wu, 1995; Kostanski *et al.*, 2004]. For a fraction of narrow molecular mass range, the peak mass (M_p) in SEC approximates to both M_n and M_w. Polymer standards of narrow polydispersity ($M_w/M_n < 1.1$) tend to show linear relations between $\log_{10}$ molecular mass and elution time (or elution volume) over wide ranges of mass. In principle, a universal calibration can be applied to SEC where a graph of log 10 ($[\eta']M_n$) against elution time or volume ($[\eta']$ is the intrinsic viscosity) should account for the elution behavior of all polymers in a given solvent. Malawer [1995] and Kostanski *et al.* [2004] have given detailed discussions of the method.

The major problem with the estimation of molecular mass ranges of coal liquids or petroleum distillation residues is that the upper mass limit is undefined and unknown. Accordingly, it is necessary to assemble evidence from as many sources as possible to enable comparisons to be made with the intent of reaching some indication of what the upper limits may be. Size exclusion chromatography is one method that is not limited by volatility considerations but is limited by solubility in the chosen solvent.

8.2.1 Summary of previous SEC work in solvents (eluents) other than NMP

Size exclusion chromatography relies on separating analytes in solution on the basis of their molecular sizes or hydrodynamic volumes in solution in the chosen solvent. Much early SEC work on coal-derived materials was carried out using tetrahydrofuran (THF) as eluent [Bartle *et al.*, 1983, 1984, 1986]. Solutions of coal extract subfractions in THF had intrinsic viscosities very close to that of the solvent and the procedure of universal calibration for SEC in that solvent was considered experimentally unreliable [Bartle *et al.*, 1984].

Characterization work on pyrolysis tars and liquefaction extracts [Bartle *et al.*, 1983, 1984, 1986] by SEC, using tetrahydrofuran (THF) as eluent indicated that the upper limit of molecular masses (MMs) was in the range 4,000–6,000 u [Li *et al.*, 1993a,

1993b, 1994a]. These results were comparable with those from similar research involving the use of SEC coupled to calibrations based on the vapor pressure osmometry (VPO) of coal derived liquid fractions. The use of pyridine as eluent [Boenigk *et al.*, 1990] led to average mass values up to 2,500 u, while Larsen and co-workers [Larsen & Wei, 1988; Nishioka & Larsen, 1990] reported average MM-values of 5050 u, with indications of masses up to possibly 15,000 u.

However, neither THF nor pyridine completely dissolves many common coal derived liquids. When using either THF or pyridine as eluent, the progressive increase in back pressure across SEC-columns has been widely recognized as a symptom of the precipitation of sample out of solution. Clearly, some of the higher molecular mass and/or most polar materials were not observed by SEC in THF and in pyridine. Furthermore, insufficient solvent power (notably of THF and pyridine) leads to adsorption effects between solute and column packing, known to distort chromatograms [Yau *et al.*, 1979] and lead to smaller apparent MM-distributions. The shortcomings of still weaker solvents, such as toluene, chloroform, dichloromethane, benzonitrile and dimethylformamide, have been reviewed by Johnson *et al.* [1998]. It does not appear possible to dissolve tar or extract samples in these solvents completely. Results from *any* solution-based technique, obtained with samples dissolved in the latter solvents, must therefore be regarded as reflecting only the more soluble parts of the samples. For asphalts however, THF solvent dissolves all the material and SEC in THF [Davison *et al.*, 1995] showed an excluded peak, which concentrated the material of the high molecular mass fractions. The association of the sample in solvents used for vapor pressure osmometry, was clearly demonstrated by an increase of apparent molecular mass in less polar solvents [Davison *et al.*, 1995]. The dissociation with time in solution in dichloromethane of asphaltene aggregates has been observed in SEC as the shifting of the relative intensities of peaks from early- to late-eluting [Strausz *et al.*, 2002]. Sato *et al.* [2005] have shown that chloroform in use as an SEC eluent and calibrated using polystyrene standards, significantly underestimated the molecular masses of asphaltene molecules below m/z 800.

These necessarily incomplete data from SEC suggested that coal tars or extracts contained significant proportions of material with MMs above 1,000 u and that smaller amounts of these materials could have molecular masses well above 5,000 u. However, results from SEC using THF or pyridine as eluent have often (and justly) been criticized for the structure-dependence of measured retention volumes, inaccuracies inherent in VPO-based calibrations at high MMs and the scarcity of model compounds above 300 u that could be used as calibrants. Confirmation of results from SEC by independent techniques was also clearly necessary, but much of the early work based on mass spectrometric methods did not show molecular masses much above 1,000–1,200 [e.g. cf. Larsen & Wei, 1988].

Problems encountered with THF included surface interactions between packing and sample molecules and partial precipitation of sample with attendant gradual increases in column back pressure [Herod *et al.*, 1995c, 1996b]. The next logical step in the development of size exclusion chromatography was the removal of solubility limitations imposed by the use of THF as solvent and as eluent. The inadequate solvent power of THF has eventually led to its abandonment in favor of NMP (1-methyl-2-pyrrolidinone) [Herod *et al.*, 1995d, 1996b].

8.2.2 SEC using NMP as solvent

NMP was first used as eluent in SEC by Lafleur and Nakagawa [1989] who attempted to establish a molecular mass calibration using a polydivinylbenzene column at ambient temperature. Their use of NMP allowed the detection of an additional range of materials in coal derived liquids with considerably shorter retention times (i.e. larger apparent MMs) than those observed in SEC using THF as eluent. In particular, a large peak was reported near the exclusion limit of the column [Lafleur & Nakagawa, 1989] in the pyridine extracts of coals. However, some polar model compounds eluted from their column at times that were far shorter than would have been expected from the column calibration against polystyrene MM-standards. In attempting to characterize the material eluting near the column exclusion limit, these researchers used heated probe-MS, which showed little ion abundance between m/z 250 and 500, the upper limit of the scan. These findings were interpreted in terms of a multimode mechanism for SEC, based on molecular size but with earlier-than-expected elution of polar molecules. The overall conclusion from the work was that materials observed at or near the exclusion limit of the column were aggregates of smaller, polar molecules, held together by ionic forces, and *not* of large molecular mass. On the other hand, if the molecules responsible for the early eluting peak were indeed of large mass, it would not have been possible to detect them by mass spectrometry in the m/z 250–500 range.

NMP appears to completely dissolve most coal derived 'liquids' investigated, including a coal tar pitch, which is normally a solid at room temperature. Extended periods of use have not led to the build up of backpressure in SEC columns. This contrasts with the progressive blockage of the SEC guard columns placed in the circuit when using THF as eluent. These solids were recovered by dissolving the deposit in NMP, which gave a black solution that showed the presence of large molecules in SEC, when using NMP as eluent. The SEC chromatogram on the Mixed-E column (not shown) of material corresponding to THF and pyridine insoluble materials contained a large proportion of excluded material [Herod *et al.*, 1995d]. On the other hand, polymerization by bubbling air through heated pitch was observed to lead to some insolubility in NMP [Menendez *et al.*, 2001, 2002]. Furthermore, NMP is not miscible with aliphatic species.

The SEC columns used in work described below include Polymer Laboratories (Church Stretton, UK) Mixed-Bed columns labeled as "Mixed-E", "Mixed-D" and "Mixed-A". The columns were packed with progressively larger (3, 5 and 20 μm diameter, respectively) particles of polystyrene/polydivinylbenzene copolymer with different ranges of porosity (not specified by the manufacturer). These porosity ranges give linear relations between $\log_{10}$ molecular mass of polystyrene and elution time (min) from low-mass to 30,000 u (Mixed-E), from 200 to 400,000 u (Mixed-D) and from 1000 to 40 million u (Mixed-A) when using THF as eluent. Meehan [1995] has described the behavior of these columns in use with THF. The Mixed-D column is claimed to be better able to resolve smaller molecules than the Mixed-A, which is better at resolving large polystyrene oligomers. The use of the Mixed-E column was discontinued in favor of the Mixed-D column. The Mixed-E column was converted to analyze aliphatics using heptane as eluent, as described in Chapter 7. Operating conditions in NMP were defined by the backpressure on the pump caused by the viscosity of eluent and the particle size of the column packing. The Mixed-E column operated at 85°C and a flow rate of 0.45 ml min^{-1}. The Mixed-D column operated at 80°C and a flow rate of 0.5 ml min^{-1}, while the Mixed-A column operated at ambient temperature and a flow rate of 0.5 ml min^{-1}. Detection was by UV-absorbance at 280, 300, 350, 370 and 450 nm. The evaporative light scattering (ELS) detector and other details of operation have been given elsewhere [Zhang *et al.*, 1997b, Lazaro *et al.*, 1999b, Herod *et al.*, 2000a, Islas *et al.*, 2001c, 2002b].

An important feature of SEC results described below [e.g. Islas *et al.*, 2003b] as well as in the original work [Lafleur & Nakagawa, 1989] was the appearance of an early peak eluting by exclusion from the porosity of the column, with a second peak which apparently consisted of material able to penetrate the porosity of the column packing. This bimodal distribution has subsequently been observed as a standard feature of SEC using NMP (see section 4 below) and it raises several important questions: (i) is the early peak formed by very large molecules or by aggregates of small polar molecules or is there another mechanism at play? and, (ii) what is the significance of the valley between the two peaks? These questions are addressed in this Chapter in sections on calibration and aggregation below. Polystyrene molecular mass standards up to 15×10^6 u, polymethylmethacrylate standards up to 1×10^6 u and polysaccharide standards up to 788,000 u have been eluted in NMP with apparently minimum interference from surface effects [Islas *et al.*, 2001b]. The latter may be diagnosed by delayed elution of samples, at times longer than the permeation limit of the column, i.e. longer than the time required for the elution of the smallest molecules such as benzene. The polysaccharide standards eluted at earlier times than polystyrene or polymethylmethacrylates, compared to polystyrene standards of similar mass, presumably due to differences in hydrodynamic volumes in solution. It would, in any case, have been possible to produce a universal calibration curve covering all three polymers through calculation of their Mark-Houwink constants. However, such a universal calibration would be of no purpose in the evaluation of molecules of unknown structure, found in coal, petroleum and biomass derived liquids. We noted above that the universal calibration method did not produce reliable results in THF solvent because the intrinsic viscosities of the solutions were very close to that of the solvent alone. In our work with NMP, we have attempted to use a triple detector (courtesy of Viscotek UK). The viscosity detector element of the system failed to produce any signal for coal tars or their high-mass fractions including pitch, coal extract and a low temperature tar, or for asphaltenes from petroleum residues, biomass tars, amber extracts or from soots although signal was detected from polystyrene standard polymers. In this unpublished work [Herod & Kandiyoti, 2000] one element of the detector system was a refractive index detector and because this is insensitive relative to UV-absorbance, highly concentrated solutions (5–20 mg ml^{-1}) were injected onto the SEC column to give a signal on the RI detector. We assume that the viscosities of the eluting sample peaks were insufficiently different from that of NMP alone to provide a useable signal, indicating that the universal calibration method would not be applicable in NMP to the types of samples studied. Further details of the universal calibration method and of the viscosity and light-scattering detectors used in the triple detector mass measurement in SEC can be found elsewhere [Malawer,1995; Kostanski *et al.*, 2004].

In this type of work, the initial aim is to obtain a calibration, which provides reasonable estimates of molecular mass distributions of heavy hydrocarbon liquids, irrespective of their origins and of the structural features of their components. To the extent that such samples contain multiplicities of molecules with different and largely ill-defined structural features, it is important to minimize the level of structure dependence of elution times. The aim of the work described below, was to obtain calibration curves for two SEC-columns, using NMP as eluent, and to test the level of dependence of the calibrations on chemical structures.

Calibration with polymers: in SEC, a calibration curve is prepared by measuring the elution times of a set of polymer standards of known molecular mass and of low polydispersity. The curve is obtained by plotting the logarithms of the known molecular masses ($\log_{10}$MM) against their elution times. Most commonly, these standards are synthetic materials with narrow molecular mass distributions, such as a set of polystyrenes [Vander Hayden *et al.*, 2002]. Figures 8.1 and 8.2 present calibration data [$\log_{10}$MM vs. elution time] for the Mixed-A and Mixed-D columns, respectively; the Mixed-E calibration is not shown. Three sets of polymer standards with known MMs were used to calibrate the Mixed-D and -A columns: polystyrenes (PS), polymethylmethacrylates (PMMA) and polysaccharides (PSAC), the latter with masses up to 788,000 u. Only polystyrenes were used to calibrate the Mixed-E column. The ELS detector was used when working with the PMMA and PSAC samples, which do not contain UV-absorbing chromophores. The linear part of the calibration plot corresponds to the molecular size range resolved by a particular SEC-column. At the high mass end, the void volume represents the shortest possible elution time. Void volumes for both the polystyrene and the polymethylmethacrylate standards were found to be earlier than the elution time (volume) of the standard of mass 1.85 million u. Exclusion limits of the two columns are indicated by the onset of departure from linearity at the short elution time end of the straight-line. This occurred at ~14 min for the Mixed-A column, just short of the largest commercially available polystyrene standard of 15 million u (Figure 8.1) and ~11 min for the Mixed-D column at a mass between ~200,000–400,000 u in Figure 8.2 [Lazaro *et al.*, 1999b, Herod *et al.*, 2000a, Apicella *et al.*, 2002].

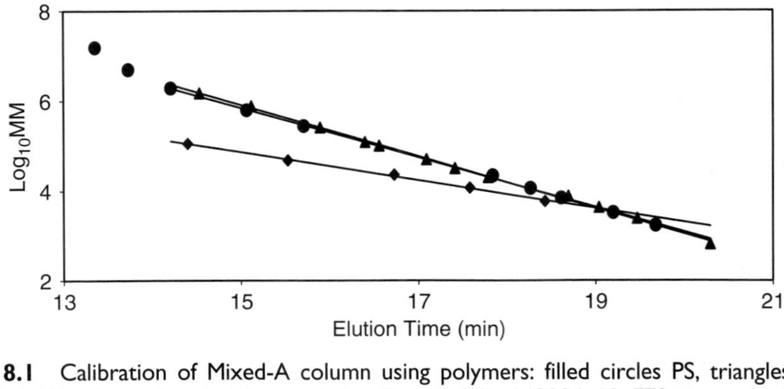

Figure 8.1 Calibration of Mixed-A column using polymers: filled circles PS, triangles PMMA, diamonds PSAC. Reproduced with permission: Energy & Fuels 2004, 18, 778; copyright American Chem. Soc.

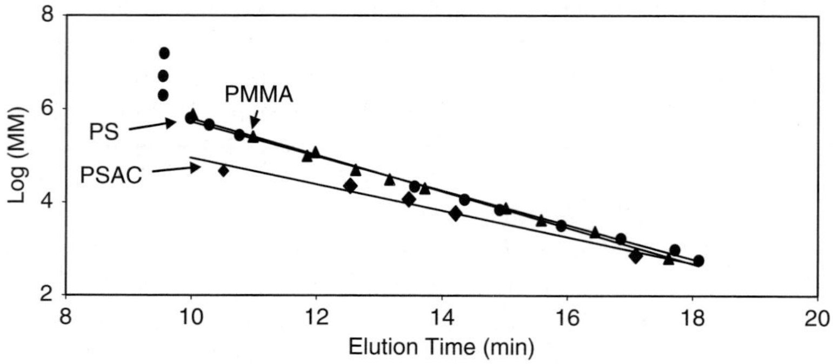

Figure 8.2 Calibration of Mixed-D column using polymer standards PS, PMMA and PSAC. Reproduced with permission: Energy & Fuels 2004, 18, 778; copyright American Chem. Soc.

Parameters were determined for these calibration graphs on Mixed-A and Mixed-D columns in the equation $\log_{10}$ molecular mass = A – B. elution time (in minutes) for the linear regions, as follows:

Mixed-A	PS	y = 14.15 – 0.5533x	R^2 = 0.9993;
	PMMA	y = 14.539 – 0.5747x	R^2 = 0.9984;
	PSAC	y = 9.5362 – 0.3116x	R^2 = 0.9976;
Mixed-D	PS	y = 9.4024 – 0.3677x	R^2 = 0.997;
	PMMA	y = 9.6756 – 0.3884x	R^2 = 0.9951;
	PSAC	y = 7.7816 – 0.2826x	R^2 = 0.9744.

Figures 8.1 and 8.2 also show that in both columns, the elution times of PS and PMMA standards up to masses of ~1.5 million u were statistically nearly indistinguishable [Islas *et al.*, 2001b]. In view of the structural differences between PS and PMMA, these findings are encouraging. However, the elution times of the highly oxygenated polysaccharide (PSAC) samples traced a different curve. To the extent that these differences are observed, the present column/eluent combination cannot be said to operate in a fashion independent of chemical structure. Considering Figure 8.1 in detail, closer agreement was observed between the three calibration lines at longer elution times (smaller MMs). The data were nearly indistinguishable at 19–20 min, but differed by a factor of ~1.8 at 18 min [~15.5 k u vs. 8.5 k u] and by a factor of between 12 and 15 at 14.5 min [~ 1.34 (for PS) or 1.61 (for PMMA) million u vs. 0.104 (for PSAC) million u]. To put these elution times into context, 18 min corresponded to the forward edge of the resolved peak in pitch (see Figure 8.8, Section 8.4.1 below for pitch fractions), while 14.5 min was on the low-mass tail of the excluded peak. The calibration equations provide estimates of mass for unknown materials, based on the molecular masses of the standard polymer eluting at the same time.

These results show the necessity for caution in the levels of confidence attached to *quantitative* estimates of molecular masses of structurally less well-defined samples, when using the PS-PMMA calibration line. However, the structural differences between hydrocarbon liquids (our main focus) and the PSAC-polymers are quite marked. The results also show that below 15k u, even the MMs of PSACs may be estimated to within a factor of ~2–2.5. Clearly, confidence in quantitative determinations appear subject to increasing uncertainty with increasing MM. These low levels of precision of mass measurement bear no relation to the mass measurement achieved by working with GC-MS and other high precision methods of analysis. However, many materials lie outside the reach of these well-known techniques and SEC is useful in attempting to assess the ranges of MMs of these samples.

Calibration with standard polymer molecules: Figure 8.3 presents plotted elution times for several different types of polymers of wider polydispersity than the calibration standards that have already been examined. They were: polyethylene oxide (M_p 58,400 polydispersity 1.03) and polyethylene glycol (M_p 4120, polydispersity 1.02) from Polymer Laboratories; poly-N-vinylcarbazole (M_p 90,000), two different samples of polyvinylpyrrolidinone (M_p 3,500 and 58,000), polyvinylacetate (M_p 170,000), and polyethylene adipate (M_p 10,000) from Acros Organics, supplied by Fisher Scientific of Loughborough, UK. Polydispersities of the latter were not supplied but were expected to be greater than the polydispersity (1.04) of the PS and PMMA standards. Figure 8.3 shows the elution times of these seven polymeric MM-standards plotted alongside the PS-PMMA-calibration line of the Mixed-A column. The interest in these samples arises because of their considerable structural differences from PSs, PMMAs and PSACs. Of the seven polymer standards, four eluted within half a minute of the calibration line, despite polydispersities greater than those of the PSs and PMMAs. Only poly-N-vinylcarbazole (90,000 u) eluted ~1.2 minutes early, indicating nearly an order of magnitude difference with the calibration line, between ~100,000 and 1 million u. The two oxygenates, polyethylene-oxide and polyethylene–glycol also eluted early compared to the straight line, the first by nearly a minute, and the second by less than half a minute. Comparing Figures 8.1 and 8.2 to results for these polymers of different structures helps put differences between the PS-PMMA and the PSAC lines into a broader context. The

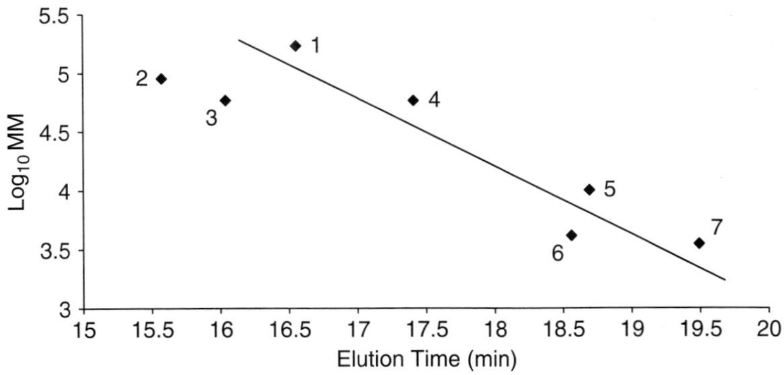

Figure 8.3 Miscellaneous polymers and polystyrene calibration line; 1-polyvinylacetate (Mp 170,000), 2-poly(N-vinylcarbazole), Mp 90,000; 3-poly(ethylene oxide) Mp 58400; 4-poly(vinylpyrrolidinone) Mp 58,000; 5-polyethylene adipate (Mp 10,000); 6-poly(ethylene glycol) 4120; 7-poly(vinylpyrrolidinone) Mp 3,500. Reproduced with permission: Energy & Fuels 2004, 18, 778; copyright American Chem. Soc.

smaller deviations observed in Figure 8.3 suggest that to a first approximation, we might consider deviations between the PSAC and PS-PMMA lines in Figures 8.1 and 8.2 as an upper limit to levels of error arising from the structure dependent variations in the elution behavior of these samples. In other work [Apicella *et al.*, 2003b] using a Mixed-D column, NMP eluent and polystyrene calibration, showed that polyacenaphthylene of mass 5,000–10,000 eluted very close to the polystyrene calibration line.

Calibration with small molecules: The linear mass analysis range quoted for the Mixed-A column by the manufacturers is between 1,000 and 10 million u. This column would not normally be selected for resolving mixtures of relatively small molecules. It is necessary, however, to understand the elution behavior of small molecules, in order to interpret signal observed at long elution times during the characterization of complex mixtures.

Figure 8.4a shows the elution times of eleven model compounds plotted alongside the polystyrene calibration line. In parallel with the behavior of the highly oxygenated PSAC samples, pyrogallol eluted earlier than expected. Many of the other samples eluted at about 0.5 min longer than predicted by extrapolating the PS-calibration line. Acetone and anthracene eluted with slightly less than ~1 min delay. The largest deviations of nearly 2 minutes delay were observed for the smallest two aromatic compounds, benzene and toluene. The latter effect appears related to greater diffusivities of smaller polycyclic aromatic (PAC) compounds within the column packing, due to their compact shapes. These differences for small molecules compared with polymers indicate that the relation between hydrodynamic volume and molecular size is not understood for small molecules – in this particular column. Solvation effects might become more important because small molecules are unable to change shape whereas large molecules may

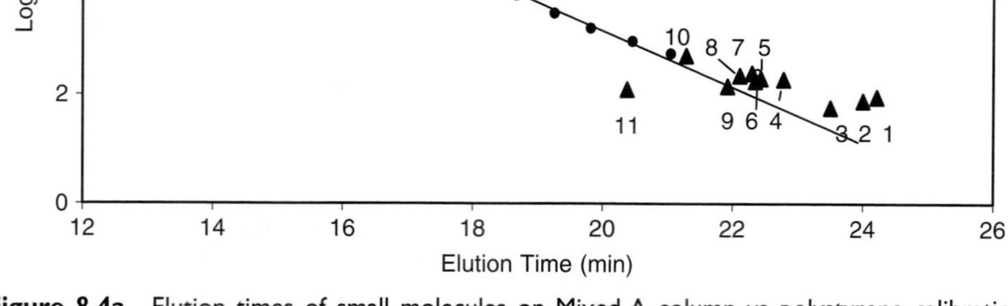

Figure 8.4a Elution times of small molecules on Mixed-A column vs polystyrene calibration; compounds are: 1 toluene (92); 2 benzene (78); 3 acetone (58); 4 9-methyl anthracene (192; 5 fluoranthene (202); 6 5,6 benzoquinoline (179); 7 perylene (252); 8 chrysene (228); 9 2,3 dimethylindole (145); 10 rubrene (532); 11 pyrogallol (126); molecular masses in parentheses. Reproduced with permission: Energy & Fuels 2004, 18, 778; copyright American Chem. Soc.

change conformation in solution. We note that the shape of the calibration curves determined using small molecules do not follow the expected S-shaped curve at the small molecule end of the elution range to define the permeation limit [Kostanski *et al.*, 2004]. The effect is apparent on both columns described here (Figures 8.4a and 8.4b) when using the series of small molecules as sample, with narrow-time fractions from the SEC separation of pitch (Figure 8.5) and with three volatile essential oils discussed below.

Figure 8.4b presents analogous data for three sets of model compounds plotted alongside the PS-PMMA calibration line of the Mixed-D column. The list of PAC compounds (squares), oxygenated compounds (cross) and nitrogen bearing compounds (triangles) is reproduced here as Table 8.1, arranged by increasing molecular mass [Lazaro *et al.*, 1999b]. The PAC standards (including cataannelated, pericondensed, non-planar and some alkyl substituted species) were found to behave much as they did in the Mixed-A column, systematically eluting at slightly longer times than expected from their molecular masses. As above, the effect appears related to greater diffusivities due to the compact shapes of these molecules. In calculating average MMs of complex mixtures, the deviation from linearity of smaller PAC species needs to be taken into account. The nitrogen bearing compounds (ranging from pyridine to the dye, alcian blue, of MM 1086 u) showed more symmetric scatter about the PS-PMMA calibration line than the PAC-group, while elution times of oxygenates, ranging from acetone to stearyl alcohol, MM 270 u were more heavily weighted toward shorter elution times, reflecting the behavior of the PSAC standards and other oxygenates. Only the mixtures of C60 and C70 fullerenes with their rigid three-dimensional structures showed anomalous behavior, eluting near the exclusion limit of both the Mixed-A and the Mixed-D columns [Lazaro *et al.*, 1999b, Karaca *et al.*, 2004]. This constituted the greatest single departure from the PS-PMMA line and will be discussed below. Apicella *et al.* [2003b] using a Mixed-D column and NMP eluent, showed that dicoronylene (596 u), a mixture of PACs of average molecular mass 200 and toluene all eluted close to the polystyrene calibration line.

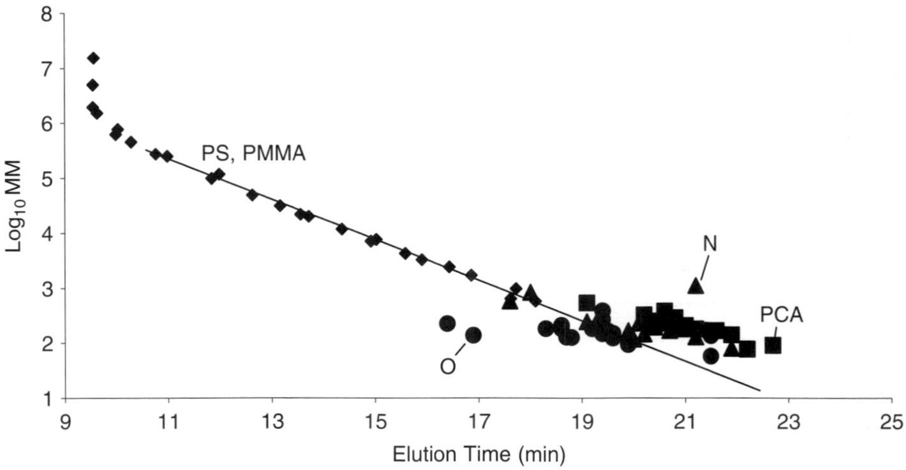

Figure 8.4b Elution times of small molecules of Table 8.1 on Mixed-D column vs polymers. Reproduced with permission: Energy & Fuels 2004, 18, 778; copyright American Chem. Soc.

Table 8.1 Calibration of Mixed-D column using standard compounds

Compound name	*MW*	t_{exp}	t_{calc}	Δt_{calc}
Acetone	58	21.5	21.4	+0.1
Benzene	78	22.2	21.0	+1.2
Pyridine	79	21.9	21.0	+0.9
Toluene	92	22.7	20.8	+1.8
Phenol	94	19.9	20.8	−0.9
Indole	117	20.2	20.5	−0.5
Benzoic acid	122	19.6	20.4	−0.8
Pyrogallol	126	18.8	20.4	−1.5
Quinoline	129	21.2	20.4	+0.9
Cyanuric acid	129	18.7	20.4	−1.7
Benzo[*b*]thiophene	134	21.5	20.3	+1.2
Salicylic acid	138	16.9	20.3	−3.4
1-methyl naphthalene	142	21.9	20.2	+1.6
2,3-dimethylindole	145	20.2	20.2	0.0
3-phenylpropenoic acid	148	19.4	20.2	−0.8
3-nitrobenzyl alcohol	153	19.6	20.1	−0.5
3-dimethylaminobenzoic acid	165	19.5	20.0	−0.5
Fluorene	166	21.6	20.0	+1.6
Carbazole	167	19.9	20.0	−0.1
2-mercaptobenzothiazole	167	20.7	20.0	+0.7
Dibenzofuran	168	21.4	20.0	+1.4
Anthracene	178	21.2	19.9	+1.3
Phenanthrene	178	21.2	19.9	+1.3
5,6-benzoquinoline	179	20.9	19.9	+1.0
Phenanthridine	179	20.9	19.9	+1.0
Coniferyl alcohol	180	18.6	19.3	−1.3
2-aminofluorene	181	19.4	19.9	−0.5
Hydrocaffeic acid	182	18.3	19.9	−1.6
3,5-dimethoxybenzoic acid	182	19.2	19.9	−0.7
3,5-dimethoxy-4-hydroxycinnamic acid	182	18.3	19.9	−1.6
4,4-dimethyl 2,2 bipyridyl	184	21.3	19.9	+1.4
9-methylanthracene	192	20.4	19.8	+0.6
Fluoranthene	202	21.0	19.8	+1.2
Anthraquinone	208	20.7	19.7	+1.0
Sinapyl alcohol	210	18.6	19.7	−1.1
9-anthracene carboxylic acid	222	16.4	19.7	−3.2
Chrysene	228	20.8	19.6	+1.2
Triphenylene	228	20.7	19.6	+1.1

Table 8.1 *Continued*

2,6-diphenylpyridine	231	20.1	19.6	+0.5
n-phenylcarbazole	243	20.6	19.5	+1.1
2-aminochrysene	243	19.1	19.5	−0.4
9-aminochrysene	243	19.1	19.5	−0.4
Perylene	252	20.8	19.5	+1.3
Benzo[*a*]pyrene	252	20.8	19.5	+1.3
1,1-binaphthyl	254	20.4	19.5	+1.0
9-phenylanthracene	254	20.6	19.5	+1.1
1,2-benzanthraquinone	258	20.5	19.4	+1.1
1-acetaminopyrene	259	19.3	19.4	−0.1
Stearyl alcohol	270	19.4	19.4	0.0
Benzo[*ghi*]perylene	276	20.8	19.3	+1.5
Picene	278	20.2	19.3	+0.8
20-methylcholanthrene	286	20.8	19.3	+1.5
Rubicene	326	20.2	19.1	+1.1
dibenzopentacene	378	20.6	18.9	+1.7
3,4,9,10-perylen tetracarboxylic dianhydride	392	19.4	18.9	+0.5
Rubrene	532	19.1	18.5	+0.6
N, N'-bis(3-aminophenyl)-3,4,9,10-perylenetetracarboxyldiimide	573	17.6	18.4	−0.8
Fullerene	720–840	9.0	17.5–17.8	−8.5–8.8
Alcian yellow	838	18.0	17.9	+0.1
Alcian blue (pyridine variant)	1086	21.2	17.5	+3.7

t_{exp} is the measured elution time (min); tcalc is the elution time calculated from the polystyrene calibration; Δt_{calc} is the difference between texp and t_{calc}.

Calibration with pitch fractions of narrow polydispersity: MALDI-mass spectrometry of successively eluting SEC fractions of a coal tar pitch: Two sets of samples have been prepared using the same coal tar pitch, by collecting fractions exiting from a preparative scale SEC column (details below). The first set consisted of successively eluting 3-minute fractions. The second set was made up of 10 s fractions collected every 3 minutes. The elution times of these fractions were determined in the analytical Mixed-D column. A third set of fractions was prepared using 30 s fractions eluting from the same analytical Mixed-D column. MMs of all three sets of fractions were evaluated by their masses at peak intensity (M_p) in MALDI and LD-mass spectra. For fractions of narrow molecular dispersity, the mass at peak intensity, M_p, is expected to be closer to the number average molecular mass, Mn. Mercaptobenzothiazole was used as matrix for the 3 min samples, and sinapinic acid for the 10 s fractions. The third set was examined by LD-MS (i.e. no matrix).

Figure 8.5 presents plots of $\log_{10}$ (M_p) vs. elution time in the Mixed-D column, for all three sets of fractions. The data were internally consistent and reproducible, despite

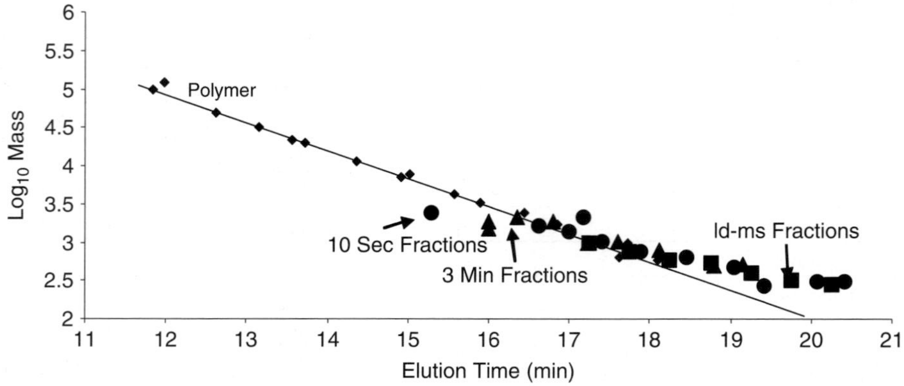

Figure 8.5 Plots of $\log_{10}$ Mp from MALDI-MS and LD-MS of narrow-time fractions of pitch from SEC vs elution times measured on Mixed-D column. Reproduced with permission: Energy & Fuels 2004, 18, 778; copyright American Chem. Soc.

accumulation of results at different times, from three separate SEC-experiments, using two different matrices, two different mass spectrometers, and several spectrometer operators. Later-than-expected elution times were observed for smaller molecular mass samples in all three sets, in parallel with shifts observed for known model compounds in Figures 8.4a and 8.4b. Above 3,000 u, the MALDI derived M_p values were observed to flatten out for samples eluting at shorter elution times, i.e. larger MMs. For molecular masses of pitch fractions up to about m/z 3,000, reasonably good agreement was found between the PS-PMMA calibration of the Mixed-D column and M_p values arrived at by MALDI and LD-MS.

We have thus noted that (i) the largest deviations from the PS-PMMA line were found in the case of oxygenated samples, (ii) that even for these samples, MMs may be estimated to within a factor of ~2–2.5 below about 15 k u, and that (iii) other structural features (e.g. ring-embedded nitrogen) are likely to give rise to smaller departures from the PS-PMMA calibration line. Put in this context, the agreement observed between findings from SEC and MALDI-MS up to about 3,000 u in Figure 8.5 suggests that up to this limit and perhaps a little beyond, SEC as defined in the sections above may be considered as a quantitative tool, the accuracy of the measurement being subject to greater uncertainty with increasing molecular mass.

Comparison of calibration vs other polymers: A comparison of SEC techniques was arranged [IUPAC Round Robin Test on the Molecular Characterisation of Epoxy Resins by Liquid Chromatography; Podzimek 2004] using two epoxy resins based on bisphenol A. The two resins were examined by vapor pressure osmometry (VPO), method of end groups (EG), static multi-angle light scattering (MALS), and mass spectrometry (MALDI-MS). The data from VPO for M_n and from MALS for M_w, were considered the best values for the two epoxy resins. These data indicate that the 'official' values from methods not using SEC, do not all agree, possibly pointing to the difficulties of using MALDI-MS in a relatively highly polydispersed sample. The results calculated from our work together with the 'official' results and some of the results from other

Table 8.2 Epoxy resin data from various sources [Podzimek, 2004] (g mol^{-1})

Measurement	*Epoxy 1*		*Epoxy 2*	
	M_n	M_w	M_n	M_w
VPO	1490		2780	
End Group	1500		4370	
MALDI-MS	1820	2320	2530	2950
MALS		3480		8330
SEC-D-PS	6920	12340	14210	26330
SEC-A-PS	5780	9770	12840	27000
SEC-D-PSAC*	1490/3250	3250/5200	690/6280	2860/10780
SEC-A-PSAC*	3940/4580	4580/5290	2590/6900	4430/8590
Mean of 21**	1780	5000	3960	12440
Range of 21**	1240–2380	3580–5590	2720–4710	9790–14190
Mean of 9***	1860	3590	3490	8240
Range of 9***	1120–2770	2900–4090	2300–4700	7400–9120
SEC PS calib.†	2180	5480	4350	13220
SEC Universal†	1570	3760	3090	8840

* Values shown in pairs are first, assuming signal proportional to number of molecules; second assuming signal proportional to number of molecules × mass.
** Various columns, with THF and mainly RI detection.
*** Results from SEC with molar mass sensitive detectors (6 light scattering, 2 viscometers).
† Results from two Styragel columns with calibration by polystyrenes and (bottom line) recalculated using universal calibration constants.

participants, are listed in Table 8.2. The mass estimates shown in Table 8.2 for the chromatograms were made initially using the polystyrene calibrations (SEC-D-PS and SEC-A-PS) and were recalculated later using the polysaccharide calibration (SEC-D-PSAC and SEC-A-PSAC). The polysaccharide-calculated results fall within the main spread of values for the resins. UV-absorbance detection was used for the Mixed-D column work and the ELS detector for Mixed-A results. The two integration methods used differed in terms of dependence on molecular mass. Method 1 assumed that the intensity of signal was proportional only to the number of molecules and that the detector response was independent of molecular mass [Kostanski *et al.* 2004]. Method 2 assumed the intensity of signal to be proportional to the product of number of molecules and the molecular mass.

The results obtained by conventional SEC, using a variety of columns, often in THF solvent, with detection using either refractive index or UV-absorbance and calibration with polystyrene standards, gave a wide range of molecular weights, significantly greater than the preferred values arrived at by VPO and MALS. However, the application of the universal calibration constants for polystyrenes and for epoxy resins to one set of results in THF solution, did bring the SEC values close to the preferred values. These results are shown in Table 8.2 as the last two rows. This illustrates the fact that the common practice of calibration with polystyrenes can be useful if used with the universal calibration method, but otherwise, the results can be deviate by a factor of two or more. In our samples where the structures of the asphaltenes or coal-derived liquids have not been described in detail, the universal calibration constants to apply in the Mark-Houwink equation are not determined.

At first sight, the experimental observation that data for both coal derived and polystyrene fractions in NMP fall on the same line on a graph of log M versus retention volume in might seem to conflict with the universal calibration procedure for SEC [see Kostanski *et al.*, 2004]. The procedure predicts that a single graph of log [intrinsic viscosity] M versus retention volume should account for the retention behavior of all polymers in a given solvent. The assumption underlying this treatment is that a flexibly coiled polymer behaves as a hydrodynamic sphere equivalent, the volume of which governs retention in SEC. The universal calibration procedure has been shown not to apply to petroleum asphaltenes [Reerink & Lijzenga, 1975], the hydrodynamic behavior of which may be more realistically modeled by ellipsoidal discs rather than flexible coils [also confirmed by Acevedo *et al.*, 1994]. Previous attempts by Bartle *et al.* [1984] to assess the relevance of universal calibration in the SEC of coal derived fractions in THF were unsuccessful, although results for the highest MW fractions suggested similarities to the behavior of petroleum asphaltenes.

Three-dimensional standards: *solid or rigid samples with known diameters*: In discussing Figure 8.4b and the data of Table 8.1, we noted that a sample of mixed C60 and C70 fullerenes eluted near the exclusion limit of the column. In view of their MMs of 720 and 840 u respectively, these materials could be classed as the two major outliers with respect to *all* the calibration curves described above. Clearly, much of size exclusion chromatography depends on a broad correlation between molecular mass and molecular size. With diameters of ~1 nm, the shapes of the fullerenes make them the only species in the range of compounds tested, with as great a discrepancy between molecular *mass* and molecular *size*.

Previous reports [Herod *et al.*, 2000b, Apicella *et al.*, 2002; also see section 4.2 below] have shown that soot samples could be fractionated by filtering out material caught by a 20 nm filter. When the material retained on the filter was re-dissolved or re-dispersed in NMP, the earlier chromatograms could be reproduced. Clearly, the higher mass components of the samples we have been putting through these SEC-columns contain material with diameters in tens of nm. These findings suggested that there might be a direct relationship between elution times and the sizes of more fully 3-dimensional objects in these columns. A set of colloidal silica samples (Nissan Chemical Industries Ltd. of Houston, Texas, USA) of diameters 22, 12 and 9 nm have been used to test the relationship between actual particle diameters and elution time. Figures 8.6(a,b) show plots of $\log_{10}$ (particle diameter) vs. elution times, in the Mixed-A and Mixed-D columns respectively. Both plots appear to conform to linear behavior and are independent of particle density or molecular mass. The fourth data-point pairs in the two diagrams belong to the mixture of C60 and C70 fullerenes, eluting at ~13.5 min in the Mixed-A and ~10.5 min in the Mixed-D columns.

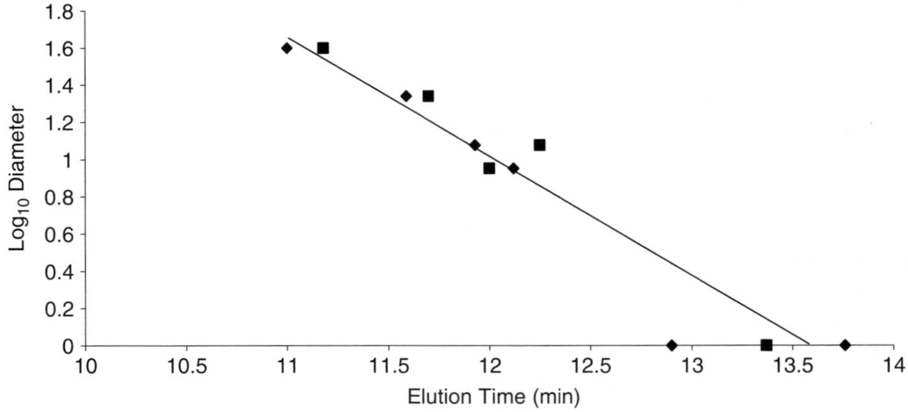

Figure 8.6a Three dimensional standards (soot, colloidal silica and fullerenes) on Mixed-A column. From the data of Karaca *et al.*, 2004. Diamonds and squares represent replicate determinations.

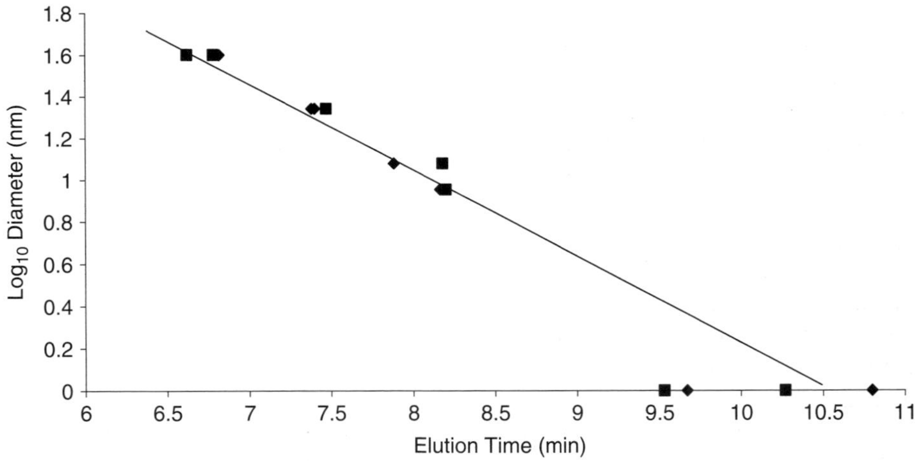

Figure 8.6b Three dimensional standards (soot, colloidal silica and fullerenes) on Mixed-D column. From the data of Karaca *et al.*, 2004. Diamonds and squares represent replicate determinations.

Figures 8.6(a,b) indicate that the sample of mixed fullerenes have eluted according to molecular *size* rather than molecular *mass*. Considering the fullerene data in a little more detail, in the Mixed-D column, the fullerene mixture gave two peaks, shown at ~1 nm diameter; it is likely that the later eluting peak corresponded to C60 and the earlier peak to C70, which in this calibration would have an indicated diameter of about 2.5 nm. On the Mixed-A column, two peaks were observed during initial runs but only one in the repeat run. All these data points have been shown in Figure 8.6a. We note that the possibility of reaction between NMP and fullerene in solution has been raised by

Yevlampieva *et al.* [2002] and it is not clear what effect this would have had on the elution time of fullerene.

Behavior of steam-distilled samples: the behavior of chemicals from natural sources and known to consist of small molecules was investigated [Morgan *et al.*, 2005b] to show that they behaved as small molecules and produced no excluded peaks. Three essential oils produced by steam distillation of plant materials were examined by the Mixed-A and Mixed-D columns using NMP as eluent. The results showed that these samples did indeed elute as small molecules and showed no evidence of aggregation. In addition, the three essential oils, Tea Tree Oil from leaves of the tree *Melaleuca alternifolia* and two lavender oils (Lavendula Angustifolia and Lavendula Intermedia Lavandin) produced by steam distillation were examined by GC-MS. The molecular components of these oils are thought to be diterpenes, sesquiterpenes and oxygenated derivatives of terpenes of relatively low mass. In SEC using NMP as eluent, all three oils were completely soluble. The chromatograms gave only late eluting peaks commensurate with the small molecules expected. No excluded 'aggregate' peaks were detected; a chromatogram of one of the oils is presented in Figure 8.7.

Summary: The calibration data presented in this section cover a wider range of polystyrene (PS) standards compared to previous work and have been overlaid with elution times of a set of polymethylmethacrylate (PMMA) standards. The resulting PS-PMMA calibration has been compared against elution times of several sets of standards: (i) a set of polysaccharides (PSAC), (ii) known high-mass polymers of various origins and narrow polydispersity, (iii) model compounds, (iv) colloidal silica spheres with graded particle sizes and (v) samples known to consist of only small molecules. The level of agreement between SEC and MALDI mass spectrometry has also been examined, using narrow SEC elution fractions of a coal tar pitch. Recorded elution times have been matched against molecular mass values observed in MALDI-MS. The two techniques are independent and the level of agreement between them provides a yardstick with which to evaluate the SEC results.

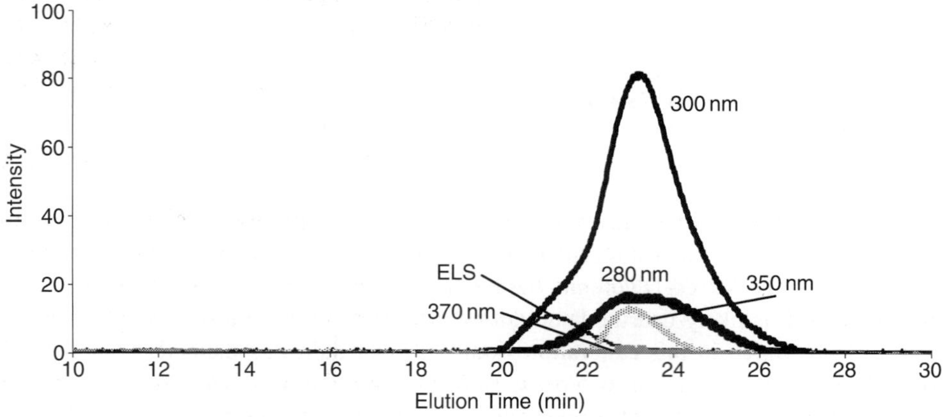

Figure 8.7 SEC chromatogram on Mixed-A column of ***Lavandula intermedia lavandin*** oil.

We note that in SEC, problems may arise from two sources. First, solvent flow rates should be low to avoid shear degradation of polymers in the porosity of the column packing or in passage through the porous plugs placed at either end of the column to retain the packing. Second, there is the possibility of breaking covalent bonds of polymers during extensive ultrasonication, often used to get the polymers into solution. These problems have been described by several authors in relation to polymers of high molar mass, such as those used in this work [Aust, 2003; Strlic & Kolar 2003]. The largest materials found in coal tars, biomass tars and petroleum asphaltenes elute at similar times as, or earlier than, the polystyrene standard of mass 20 million u and could possibly suffer from degradation of molar mass in the columns used. However, where early-eluting material has been collected at the exit of the UV-absorbance detector and re-injected, the elution times have been observed to remain unchanged. In one case, shorter times were observed, probably because in the initial injection, the capacity of the excluded region of the column was exceeded (i.e. overloaded).

Another problem seems to arise when the oxygen content of the analyte is high, as in humic and fulvic acids. These materials are polyelectrolytes, soluble in NMP in the protonated form but insoluble in NMP when in the neutral state in soil or peat.

8.3 Fractionation Methods to Isolate Molecules of Large Mass or Size

In view of their high polydispersity, it was found necessary to fractionate many of the fuel-derived samples [Herod & Kandiyoti, 1995e; Herod *et al.*, 1995f]. The results obtained from the fractions recovered are in part defined by the fractionation method used, particularly for the larger mass or more polar fractions. Useful fractionation methods in this type of work include planar chromatography, column chromatography, preparative size exclusion chromatography and ordinary solvent solubility separations.

8.3.1 Planar chromatography (Thin layer chromatography)

Planar chromatography has been the subject of many reference and teaching books [Touchstone, 1992; Hamilton & Hamilton, 1987] as well as regular reviews in Analytical Chemistry [Sherma, 1996, 1998, 2000]. Viewed as outmoded by some, it has proved a valuable technique for fractionating heavy hydrocarbon liquids. One advantage of the technique is the ability to recover the heavier fractions that are immobile in the set of solvents used. By contrast, other chromatographic methods lose the heavier components either as involatiles sticking to the injection system in gas chromatography or remaining immobile on the column in HPLC, when using inadequate solvents. Fraction recovery is usually not quantitative but the method is rapid and cheap to operate; any combination of volatile solvents can be used to develop the chromatogram. Acetonitrile and pyridine have been used in planar chromatography on silica, to generate fractions equivalent to those from the solvent solubility method. After drying, the separated fractions may be recovered by dissolving in NMP [Herod *et al.*, 1996b,d, 1999; Lazaro *et al.*, 1999d; Deelchand *et al.*, 1999; Herod & Kandiyoti, 1996c; Herod & Lazaro, 2000; Suelves *et al.*, 2001a]. However, fraction sizes are usually small for analytical methods such as NMR. Some of the heavier material could not be recovered from the silica into NMP.

The common types of Whatman chromatographic plates used in the work were the 20 × 20 cm K6 analytical and preparative plates. 10 × 20 cm plates have been used for comparison purposes, with 5 × 5 cm plates for rapid investigations. For all developments other than those of the 5 × 5 cm plates, the development tanks were lined with filter paper, solvent added and allowed to stand for at least half an hour to equilibrate liquid and vapor phases. Plates were first washed using the most polar solvent of the sequence to be used, to sweep contaminants to a band along the top of the plate. After drying, sample was added to the plates often as a slurry in pyridine either as a band of material at the origin, or as separated spots for comparison of several samples. After drying the slurry solvent, the plate was developed using the most polar solvent first. A subsequent development using the second, less polar solvent followed after drying of the first solvent, using a separate, equilibrated tank. The third, less polar solvent was used to develop the plate subsequently after drying the second solvent. Development distances were calculated from the available development distance and the number of solvents intended for use, such that each development covered a similar distance from the previous solvent front.

Solvents were used with and without added salts to investigate claims that salts added to the eluent in SEC could disaggregate the material excluded in SEC using NMP as eluent. Details of that work will be given in Section 8. In some cases where preparative fractions were collected, multiple solvent development was used to improve the fractionation. Multiple solvent development is the repeated use of one solvent to wash completely all material mobile in that solvent up to the solvent front. The method sharpens the band of mobile material up to the solvent front. Fractions were recovered from the plates by scraping the appropriate area of silica from the plate into NMP, exposing the mixture in an ultrasonic bath for a short time and filtering the liquid to remove silica particles. The method is rapid, low cost; the heaviest fraction remains on the plate and can be examined in situ if necessary. Almost any analytical method can be applied to the separated fractions. Somsen *et al.* [1995] have reviewed methods for combining planar chromatography with other spectroscopic methods. The method has been used to prepare fractions of a coal digest for examination of the iron content by Mössbauer spectroscopy [Herod *et al.*, 1995g, 1996d; Richaud *et al.*, 2000b].

8.3.2 Column chromatography

A column chromatography method was developed, eluting fractions of the sample successively with acetonitrile and pyridine. NMP was used to sweep the remaining material immobile in pyridine [Islas, 2001; Islas *et al.*, 2003b; Suelves *et al.*, 2001a]. This method allows a quantitative estimation of fractions and provides sufficient material for further analyses by a variety of methods. Some material could not be recovered and the silica remained colored; sample losses were about 10% overall as volatiles and insolubles.

The method was developed to produce up to 1 gram of sample fractions; it would have been difficult to produce comparable amounts of sample by planar chromatography. It was based initially on the use of silica gel with sequential elution using acetonitrile, pyridine and NMP. In subsequent work, pentane and toluene were added to the sequence.

A total volume of 100 ml of each solvent was used, 50 ml with gravity elution and the 50 ml under vacuum and then NMP (100 ml) with vacuum elution. Water (100 ml) was added to wash out the NMP. Some loss of material in each fractionation was through a combination of high mass material retained on the silica and low mass material lost with evaporation of solvent during distillation and in the vacuum oven. However, unlike during planar chromatography, an approximate mass balance could be achieved.

8.3.3 Solvent solubility

The traditional method consists of separation by solvent solubility into oils (pentane solubles), asphaltenes (pentane insoluble, benzene solubles) and preasphaltenes (benzene insolubles) [e.g. see Suelves *et al.*, 2001b]. In much of the work described in this Chapter, acetonitrile (or acetone) and pyridine were used to produce analogous solubility fractions, to concentrate the larger molecules in the acetonitrile insoluble-pyridine soluble and the pyridine insoluble fractions [Lazaro *et al.*, 1999d; Herod & Lazaro, 2000]. Solubility in heptane was used to isolate heptane insoluble materials from several heavy petroleum fractions [Suelves *et al.*, 2001a, 2001b].

This method was used as an alternative that avoids losses of the heaviest material on the silica of plates (in planar chromatography) or columns (in column chromatography). Sample was mixed with the least polar solvent, acetonitrile (or acetone) in excess, shaken and stood in the dark. The solution was then decanted off the residual solid. The extraction was repeated several times, with collection of the acetonitrile (or acetone) solubles. The residual solid was dried and extracted using pyridine in the same way. The solvent-soluble fractions and the pyridine-insolubles were dried to give a mass balance. The method uses large quantities of solvent and gives a less clear separation than either planar chromatography or column chromatography, but has the advantage that no large mass material is lost on the filtration or separation media. Mass closures were usually slightly greater than 100% because of the difficulty of completely removing the solvents. NMP was not used in the fractionation, an important point given the difficulty of drying this solvent without polymerization.

8.3.4 Preparative size exclusion chromatography

A Perkin-Elmer LC 250 isocratic pump was used to maintain an NMP flow rate of 2 ml min^{-1}. The preparative SEC column (600 mm; 25 mm ID packed with polystyrene/polydivinylbenzene beads) used to fractionate the sample was maintained at 85°C and connected to a Perkin-Elmer LC 290 UV-absorbance detector at 450 nm. A single run of 100 minutes gave enough sample for subsequent analyses. Two sequences of collection have been used. In the first, fractions were collected at consecutive 3 min intervals. In the second sequence, narrow fractions corresponding to about ten second elution periods were obtained with three minutes intervals between the fractions. This method was used to minimize overlapping between different fractions and to restrict the total number of samples collected [Islas 2001; Islas *et al.*, 2001a, 2002a]. A Mixed-D analytical column has also been used to prepare fractions, with collection of eluting material at consecutive 0.5 min intervals [Karaca *et al.*, 2004; Millan 2005].

8.4 Application of SEC and Fractionation Methods to Samples

8.4.1 Coal-derived materials

Samples which have been characterized by these methods include a coal tar pitch [Parker *et al.*, 1993; Herod *et al.*, 1999, 2000a; Lazaro *et al.*, 1999d; Menendez *et al.*, 2001; Begon *et al.*, 2000; Islas *et al.*, 2001a, b, 2002a, 2003b], coal extracts prepared using different solvents and thermal treatments [Islas *et al.*, 2000; Suelves *et al.*, 2001b; Begon *et al.*, 2002; Bodman *et al.*, 2002, 2003], a low temperature coal tar [Islas, 2001; Islas *et al.*, 2002b], direct solvent extracts of coal [Richaud *et al.*, 2000a], pyrolysis tars [Avid *et al.*, 2004] and tars from the co-pyrolysis of coal and waste lubricating oil [Lazaro *et al.*, 2001]. Much of the method development work has been done using a sample of coal tar pitch, because of its chemical stability and availability in quantity. Analytical SEC profiles of fractions [Islas *et al.*, 2001b] obtained by sample collection from a preparative SEC column showed regular small shifts to smaller size molecules with increasing elution time [Islas 2001; Islas *et al.*, 2002a].

Discussion of the shape of the SEC chromatogram: Figure 8.8 shows SEC profiles of three coal-derived samples, a pitch, a coal extract and a low temperature tar eluted using the Mixed-D column. The SEC chromatograms of fractions from these samples have been shown elsewhere [Herod *et al.*, 1999; Islas, 2001; Islas *et al.*, 2000, 2002b, 2002c]. However, in common, they all exhibit a gap in elution time between about 10 min and 13 min during which no sample is detected. We have noted that the exclusion region of the column, before 10 min, tends to overload before the porosity (retained region) becomes overloaded [Lazaro *et al.*, 1999b] and therefore it is essential to operate the columns with relatively dilute solutions of sample, 1 or 2% by weight at most. When higher concentrations are used, the excluded material cannot pass through the column in the correct time band and elutes at later times, gradually filling the gap. Much of the

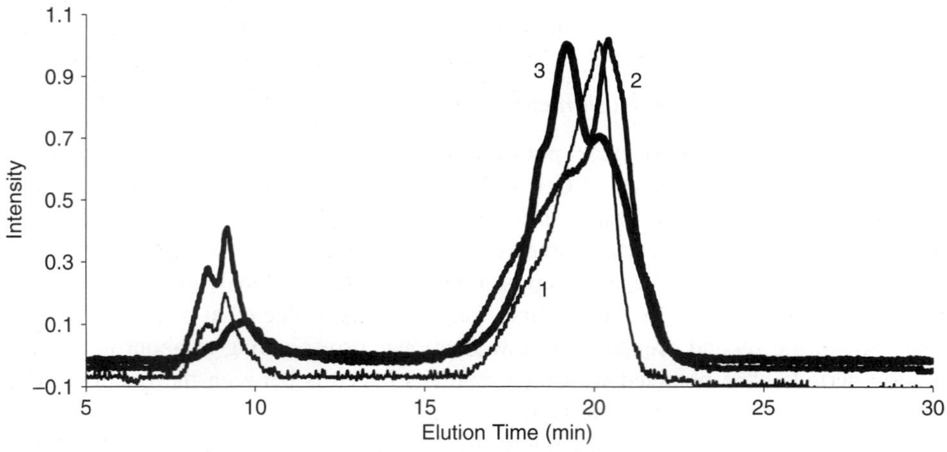

Figure 8.8 SEC chromatograms on Mixed-D of 1) pitch, 2) a coal digest and 3) a low temperature tar; zeros offset slightly; UV-absorbance detection at 350 nm.

data generated using the Mixed-E column showed overloading with a non-zero intensity region between material excluded and retained by the porosity [Zhang *et al.*, 1996a, 1996b, 1996c, 1997a, 1997b; Johnson *et al.*, 1997]. A similar overloading of the excluded region appears to have occurred in other work, where refractive index detection was used [Chen & Iino, 2001]. The RI detector is less sensitive than either UV or ELS detectors and requires greater amounts of sample to be injected.

The excluded peak would be expected to diminish in height or completely disappear on dilution if it were made up of smaller, aggregated molecules. In fact, dilution of these samples leads to the sharpening of both peaks. As dilution normally favors disaggregation, these findings tend to suggest that the excluded material is of high mass rather than being aggregates of small molecules. Furthermore, the relative intensity of UV absorbance over the range of wavelengths used shifts toward 450 nm for the excluded material rather than the 280–300 nm maximum for the retained material – indicating that the two sets of material are structurally different. In another set of experiments, coal samples were extracted in parallel, with nearly boiling NMP and with cold NMP. All SEC chromatograms showed two peaks, the one showing signal for material resolved by column porosity and the other excluded, with a zero-intensity gap. However, the hot extraction gave a more intense excluded peak than the cold extraction, an expected result if dissolving in the hot solvent increases the amounts of larger molecules extracted [Richaud *et al.*, 2000a].

Supporting evidence for this interpretation may be found in work involving the treatment of pitch for the production of carbon electrodes. Pitch polymerization by several different heat treatment methods showed that the SEC profile shifted to shorter elution times (larger MMs). Extended treatment gradually caused the retained peak (small MMs) to be reduced. The mechanisms of polymerization varied according to treatment; however, the *initial* pitch or anthracene oil samples were almost entirely soluble in NMP and observable by SEC [Menendez *et al.*, 2001, 2002; Bermejo *et al.*, 2001a,b].

Correlation of structural features with changing molecular mass: a sample of coal tar pitch, a coal extract and a low temperature coal tar have been fractionated by column chromatography and examined by SEC as well as by pyrolysis-GC-MS [Herod *et al.*, 1999, Islas 2001, Islas *et al.*, 2000, 2002b]. SEC chromatograms of the three coal extract fractions using the Mixed-D column are shown in Figure 8.9. A common feature of the three column-chromatography separations was the parallel shifts in molecular size distribution to shorter elution times in increasing order of solvent solubility from acetonitrile to pyridine and NMP solubles. The pyridine and NMP soluble fractions gave very little signal by py-GC-MS, (mostly residual solvent), indicating that sample either charred during pyrolysis and that any resulting volatile pyrolytic fragments were too large to pass through the chromatographic column. Comparing with the quality of signal from the pyrolysis-GC-MS of the acetonitrile solubles suggests that the fragmentation of the larger molecules displayed a very different pattern from those of the lighter fraction [Herod *et al.*, 1999; Islas, 2001; Islas *et al.*, 2000, 2002b].

Pitch fractions from preparative SEC using NMP as eluent [Islas *et al.*, 2002a] showed that even those molecules that eluted within the retained region of the Mixed-D column but were too large for GC-MS themselves, produced a wide range of fragments

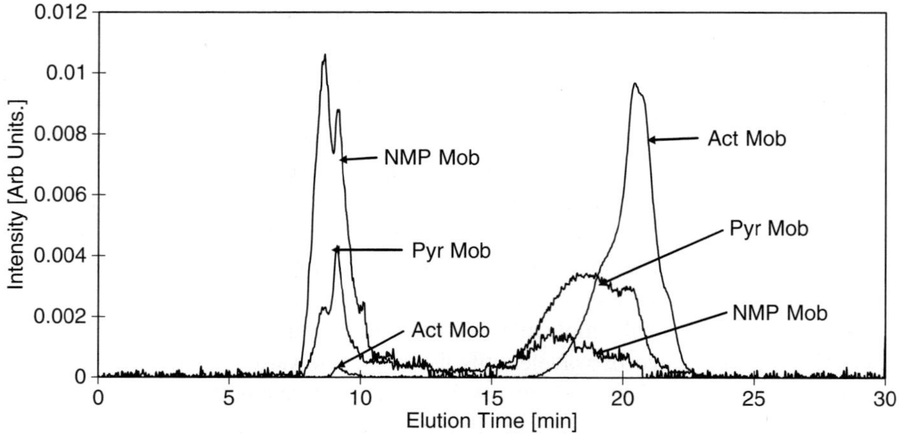

Figure 8.9 SEC on Mixed-D of column chromatography fractions of coal digest. Reproduced with permission from Fuel 2003, 82, 1813; copyright Elsevier.

and structural isomers when analyzed by pyrolysis-GC-MS. These fragments were mostly the usual array of small polycyclic aromatics, readily observed by GC of the whole pitch.

Application of SEC to liquefaction and hydrocracking: a set of coal macerals, including a vitrinite, a liptinite and two inertinites was liquefied in tetralin at 450°C and the coal extracts examined by SEC in NMP, using the Mixed-E column [Begon *et al.*, 2002]. Comparisons of MM-distributions of coal-derived materials at different stages of the liquefaction process showed the expected trends. The subsequent catalytic hydrocracking in tetralin of these [Begon *et al.*, 2002] and other coal extracts [Herod *et al.*, 1996a; Zhang *et al.*, 1996a, 1996b, 1996c, 1997a, 1997b; Bodman *et al.*, 2002, 2003], reduced the proportion of excluded material and produced a shift of the retained material to longer elution times.

In one exploratory experiment, the pyridine insoluble fraction of pitch was catalytically hydrocracked [Begon *et al.*, 2000]. This corresponds to a sample where material appearing under the excluded peak of the SEC chromatogram would have been concentrated. The Mixed-D column was used to characterize the product distributions. This experiment showed that the excluded material reacted rapidly under catalytic hydrocracking conditions with loss of the excluded SEC peak and generation of a retained peak at a longer elution time than observed in the starting material. This new peak contained material of molecular masses less than m/z 600 that were detectable by probe mass spectrometry. Part of the material adhering to the catalysts could be extracted into NMP. It was found to contain relatively high proportions of excluded material by SEC. Figure 8.10a shows the SEC chromatograms of the pitch pyridine insoluble fraction and the product after 30 min hydrocracking, while Figure 8.10b shows the synchronous UV-fluorescence spectra of the pyridine insoluble fraction and the hydrocracked product. The shift to smaller molecules and smaller aromatic clusters on hydrocracking is clear using both methods.

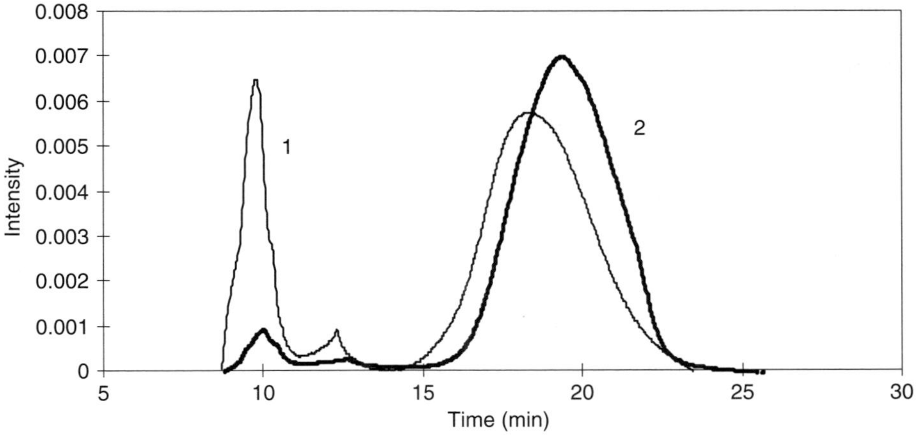

Figure 8.10a SEC chromatograms of the pitch pyridine insoluble fraction (curve 1) and the hydrocracked product (curve 2); data from Begon *et al.* 2000. Mixed-D column at 350 nm UV-A.

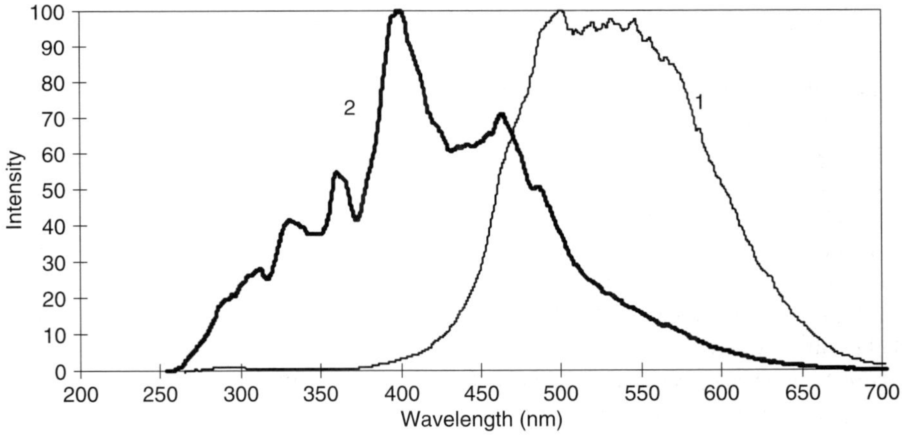

Figure 8.10b Synchronous UV-fluorescence spectra of the pitch pyridine insoluble fraction (curve 1) and the hydrocracked product (curve 2) From the data of Begon *et al.*, 2000.

8.4.2 Soots

Several soot samples have been examined including coal and wood soots from domestic open fires, soots or tars from the combustion of coal in a Bunsen flame and a candle soot from condensation from a burning candle onto a cool glass tube [Herod *et al.*, 2000b, 2003b, 2004; Apicella *et al.*, 2002, 2003a,b]. A series of soots produced from two fuel-rich burner systems fed by ethylene gas, rape seed oil, diesel oil and heavy petroleum oil have also been examined. In all cases, a distinguishing feature of the SEC profiles was a very early eluting peak at 6–7 min from the Mixed-D column and a peak in the Mixed-A column eluting well before the elution time of polystyrene of mass 15 million. When the candle soot was examined by transmission electron microscopy (TEM), it showed some material with approximately 50 nm diameter [Herod *et al.*, 2000b]. This material could be filtered out of NMP solution using a 20 nm porosity filter, re-dissolved by ultrasonication into NMP, re-injected into SEC, with no change of elution time and no generation of smaller molecules. Other soots showed the same behavior after filtration [Apicella *et al.*, 2002, 2003a,b]. The soots were assumed to be in colloidal solution, but again, the existence of gaps in the elution profile between dichloromethane soluble small molecules and the soot peak at 6–7 min points to abrupt changes of molecular size and shape with increasing mass. These findings are in agreement with evidence from other techniques applied to soot formation in flames where diameters of primary soot particles ranged from 20 to 60 nm [Hessler *et al.*, 2002; Choi & Mochida, 2002; Blevins *et al.*, 2002; Pacey & Glasier, 2002]. Soot of diameter about 40–50 nm has been detected in the dust from blast furnace operation where coal injection was used to reduce the quantity of high-strength coke needed to support the blast furnace burden [Pipatmanomai *et al.*, 2004]. Soot particles from ethylene combustion were reported to have diameters of about 60 nm [Kim *et al.*, 2005], of similar size to those of the other flames while primary soots from diesel engines ranged from 18 to 33 nm [Mathis *et al.*, 2005]. Apicella *et al.* [2003b] have confirmed these observations on soots, using samples filtered above and below diameters of 20 nm, and a carbon black sample. Carbon black and the >20 nm fraction of soot were excluded from the column porosity and were assumed to be of very large size and similar to the observations described above.

8.4.3 Petroleum residues, oil shale and bitumen

A major discussion has centered on the molecular mass ranges of petroleum vacuum residues and asphaltenes because of the importance of this property for the conversion and use of the heavy parts of crude petroleum feedstocks. Strausz *et al.* [2002] have suggested that asphaltenes in solution exhibit colloidal behavior and associate into micelles with apparently very large molecular masses but that, nevertheless, the system contains some molecules with relatively large masses. The evidence for micelle formation came from the SEC of asphaltene solutions in dichloromethane and showed that the relative intensities of the three peaks detected shifted with time of standing of the solution. The evidence for the formation of asphaltenes into colloidal particles as determined by small-angle neutron and X-ray scattering techniques has been challenged as being caused by 'ephemeral dynamically fluctuating inhomogenieties' and not a valid diagnostic of aggregation [Sirota, 2004]. The use of SEC with asphaltenes using various solvents (but not NMP) has been discussed using refractive index detection [Merdrignac *et al.*, 2004]. Weight average masses for a Safaniya asphaltene were around 8,000 u and the molecular weight profile varied with eluent. In this work, the concentrations of solutions on the SEC column were much higher than those in the work described below, because the refractive index detector is much less sensitive than UV absorbance detectors. The discussion above on calibration has indicated that for aromatic molecules, the use of solvents other than NMP is liable to produce interactions with the packing surfaces and destroy the size separation mechanism. Only NMP appears to produce SEC chromatograms from vacuum residues and asphaltenes with two peaks, one excluded and one within the porosity range of the column [Pindoria *et al.*, 1997a; Deelchand *et al.*, 1999; Suelves *et al.*, 2000, 2001a,b, 2003; Ascanius *et al.*, 2004; Morgan *et al.*, 2005a], apart from the work of Strausz *et al.* [2002]. Figure 8.11a shows size exclusion chromatograms of three asphaltenes, insoluble in heptane but soluble in NMP, isolated from the three vacuum bottoms samples shown in the chromatogram of Figure 8.11b. The asphaltenes are only minor components of the whole residues but show a significant shift to larger size molecules for the retained peaks. In addition, the excluded peaks

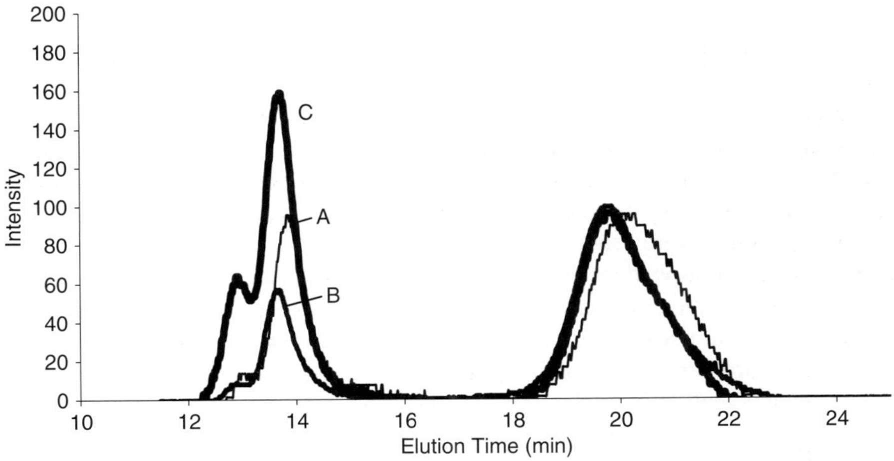

Figure 8.11a SEC in NMP on Mixed-A column of asphaltenes from three vacuum bottoms, detection at 300 nm UV-A. Reproduced from Morgan *et al.* 2005a, with permission. Copyright Elsevier 2005.

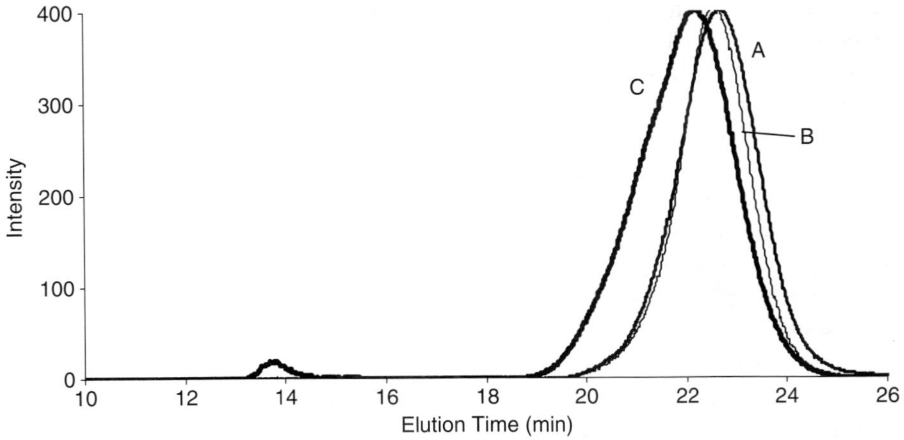

Figure 8.11b SEC in NMP on Mixed-A column of three vacuum bottoms, detection by UV-A at 300 nm. Reproduced from Morgan *et al.* 2005a, with permission. Copyright Elsevier 2005.

become a much more significant proportion of the fraction than in the whole sample. These data indicate that the excluded material concentrates into the heptane insoluble material along with some of the larger size molecules of the material in the retained region. A non-fluorescent fraction of asphaltenes was not soluble in NMP [Ascanius *et al.*, 2004] and probably consists of waxes.

Another view of asphaltene structures is that there are only small molecules present [Groenzin & Mullins, 2000; Badre *et al.*, 2006] as shown by fluorescence depolarization techniques and that apparently high values are caused by aggregation. However, three independent studies have suggested that the technique cannot detect those components of the asphaltenes that do not fluoresce [Strausz *et al.*, 2002; Ascanius *et al.*, 2004; Morgan *et al.*, 2005a]. An example of a vacuum bottoms sample examined by SEC with UV-absorbance and-fluorescence detectors in series is shown in Figure 8.12, showing poor sensitivity by the fluorescence detector at shorter elution times. The mass ranges indicated by fluorescence depolarization therefore constitute an under-estimate of sample molecular masses. The small molecule nature of asphaltenes has, nonetheless, found support from molecular weight determinations using surface tension measurement to establish a critical concentration for micelle formation [Monte *et al.*, 2004]. Values of 500–1,000 g mol^{-1} were reported. These average values were considered to represent the non-aggregated asphaltenes and point to larger mass values for the upper end of the range. An investigation of asphaltene molecular weights using atmospheric pressure chemical ionization mass spectrometry [Cunico *et al.*, 2004] has also indicated a small range of mass around m/z 500–600, although the mass spectra were of very low intensity for such a potentially complex mixture of components.

Petroleum vacuum residues have been examined by SEC using NMP as eluent with the Mixed-E [Pindoria *et al.*, 1997a] and the Mixed-D columns [Deelchand *et al.*, 1999; Suelves *et al.*, 2000, 2001a,b, 2003]. The vacuum residues examined on the Mixed-D column showed two peaks: a small excluded peak and a large retained peak that eluted

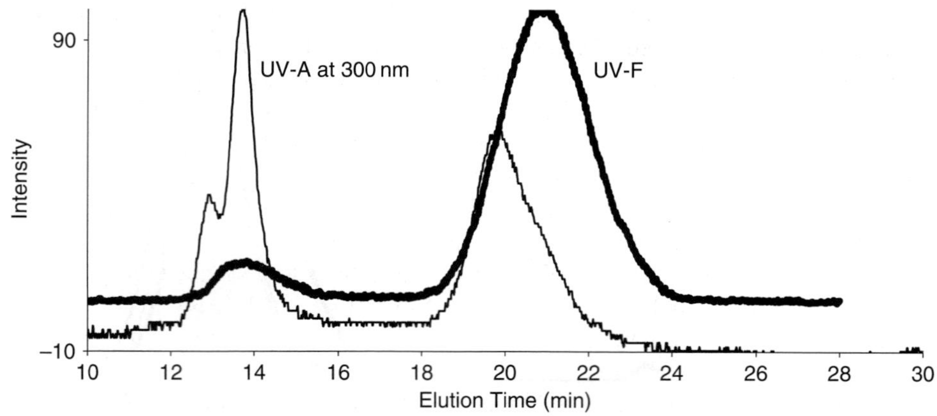

Figure 8.12 SEC of heptane insolubles of vacuum bottoms C. Chromatograms were obtained using UV-absorbance and UV-fluorescence detectors in series. Mixed-A column, both chromatograms show two separated peaks; from Millan *et al.*, 2005b, copyright Elsevier.

before the retained peaks observed for coal-derived materials [Deelchand *et al.*, 1999]. In considering the molecular masses of the petroleum residues, the range of masses from the SEC calibration covering the main, retained peak, corresponded to between 170 and 3,000 u with the peak maximum at about 400 u. These values lie within a factor 2 of equivalent values from fluorescence depolarization studies, [Groenzin & Mullins, 2000]. This level of agreement may be considered reasonable. The SEC calibration with polystyrene then appears to predict the masses of petroleum residues within a factor 2 (about 1 min difference in elution time). The question of what mass or structures the excluded material has, remains to be answered but in the samples examined, the excluded material constituted not more than 1–2% of the whole.

Athabasca bitumen was fractionated by preparative SEC using THF eluent [Domin *et al.*, 1998]. The fractions were examined by analytical SEC using NMP. The THF eluent was found to have produced a very poor fractionation but estimates of peak masses from analytical SEC in NMP agreed reasonably well with estimates from VPO and MALDI-ms. ^{252}Cf plasma desorption mass spectrometry showed similar trends to SEC, MALDI and VPO but underestimated the masses. The masses of polystyrene standards having similar elution times to the peak intensities of the retained peaks of the fractions were 3200, 1005, 480, 390 and 370. The comparison of SEC in THF and NMP indicated that the material excluded in the NMP system had been spread throughout the material retained in THF within the column porosity. This evidence provides a clear indication that THF is unsuitable for this application of SEC. There is further discussion of this experimental result in section 8.5.8.

A waste motor oil has also been examined by the Mixed-E column [Lazaro *et al.*, 2001] and shown to contain mainly small molecules, with the peak intensity at about 22.5 min, equivalent to polystyrene of mass about 200 u. Some excluded material was observed and would correspond to polystyrene of mass greater than 20,000 u. This suggests that even after use in an engine, the large excluded molecules are not completely degraded.

Asphaltenes sub-fractionated by NMP have been examined using a Mixed-D column with NMP as eluent [Ascanius *et al.*, 2004]. Two peaks were observed, one excluded and one within the resolution of the column, showing that the excluded peak is a feature of the samples and only detected when using SEC with NMP as eluent.

Fractions of a vacuum residue separated by planar chromatography have been examined by SEC with NMP as eluent [Li *et al.*, 2004]. It was found that that the proportion of material excluded from the column porosity increased with increasing polarity of the solvent used to develop the planar chromatographic plates. The retained SEC peak (within the range of column porosity) shifted to slightly earlier elution times (higher mass) with increasing polarity of the solvent used in planar chromatography but was still of relatively small mass. An oil shale from Mongolia gave an oil by pyrolysis showing the characteristic double peaks observed in SEC for other oil-derived products, with insignificant signal between the peaks; the yield of oil was low (7% wt) following pyrolysis at 800°C [Avid *et al.*, 2004].

The molecular mass ranges of petroleum vacuum residues and asphaltenes were found to be low compared to the heavier fractions of coal tar pitch. Preliminary results from a comparison of SEC with NMP as eluent and MALDI-mass spectrometry have indicated the main range of masses of an asphaltene at up to 10,000 u with low intensity masses continuous up to about mass 40,000 u. It is thought likely that the material excluded in SEC broadly corresponded to material observed in MALDI in the mass range from 10,000 to 40,000 u [Millan *et al.*, 2005b].

In contrast, the coal tar pitch sample used extensively in the work described above has proportionally more excluded material, but the retained peak is shifted to longer elution times (smaller masses) than found in vacuum residues. Figure 8.13 compares "Vacuum Bottoms-C" with the pitch acetone-soluble and pyridine-soluble fractions. The vacuum bottoms showed somewhat larger masses than the acetone solubles and was more similar to the pitch pyridine-solubles.

Hydrocracking of a petroleum residue using tetralin as solvent and with different catalysts has shown that the SEC chromatogram of the residue shifted to smaller molecular sizes after reaction, with the greatest shift to smaller molecules shown using pillared clay catalysts [Bodman *et al.*, 2002, 2003]. In the absence of catalyst, polymerization of the residue was observed with a marked shift of the peak intensity of the chromatogram to shorter times than observed for the initial residue sample.

8.4.4 SEC of Kerogen extracts

Kerogen samples have been pyrolyzed using the wire-mesh reactor described in Chapter 3 [Madrali *et al.*, 1994, Rahman *et al.* 2000]. Similar samples have also been extracted in NMP and liquefied in the flowing solvent reactor described in Chapter 5 [Li *et al.*, 2002]. The aim of the work was to correlate tar yields with oil production and the structure and maturity of the kerogen.

Extracts of immature kerogen samples in NMP showed behavior that could be considered as anomalous and characteristic of soots. In SEC, peaks were observed at the

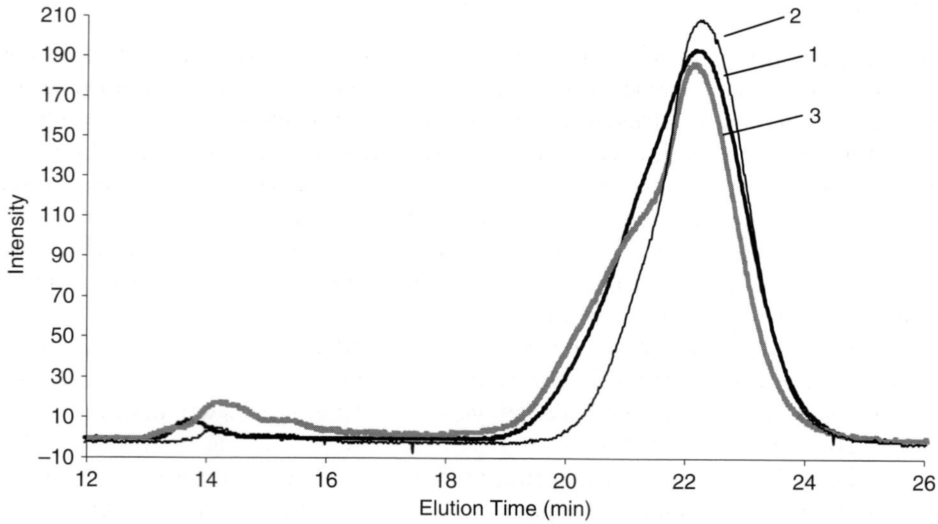

Figure 8.13 SEC of Vacuum Bottoms-C (curve 1) compared with pitch acetone-soluble (curve 2) and pyridine-soluble (curve 3) fractions, Mixed-A column, 300 nm UV-A.

unexpectedly early elution times of 6–7 min, as well as at the 9–10 min and 14–20 min ranges [Herod *et al.*, 1997]. The masses of the early eluting 6–7 min peaks cannot be estimated, but may be equivalent to spherical particles of diameter up to 40 nm and similar to the spherical particles found in soots. This peak was more prominent for the younger kerogens than for the older samples. The excluded peaks at 9–10 min eluted as if similar to the largest polystyrene standard available of 15 million u, down to the exclusion limit of around 200,000 u; the material corresponds to very large molecules of unknown mass. The retained peaks (at maximum intensity) were equivalent in mass to polystyrene standards of mass between 1600 and 760 u; signal in the retained region extended up to 15 min, equivalent to polystyrene of mass about 7,000 u. Extracts in NMP of the older immature kerogens had proportionally more signal in the retained region of the Mixed-D column than was found for younger but also immature kerogens. However, the signal in the retained region from extracts from younger kerogens (Types I and III) was shifted to slightly longer elution times (smaller molecular size) than for the older kerogens.

Three kerogens of type I and different maturities (immature, in the oil-producing window and mature) have been extracted [Li *et al.*, 2002] using the flowing solvent liquefaction system using NMP. The mass of kerogen extracted in the flowing solvent method using NMP at 300°C varied, with the least in the mature sample (3.4%), the most in the sample within the oil production window (51.2%) and less extract from the immature sample (37.6%). The proportions of material (in direct solvent extracts in NMP) in the excluded and retained regions changed with the most excluded signal from the kerogen within the oil-producing window and the least in the extract from the mature sample. The possibilities of the product characterization methods described remain to be fully explored. However, these initial results suggest far greater possibilities for examining the fundamental properties of kerogens than studies using Rock-Eval tests.

8.4.5 Biomass and amber extracts and tars

Extracts have been obtained using NMP to partially dissolve wood fragments known as 'forest residue' over a period of weeks at ambient temperature and by heating to the boiling point of NMP [Richaud *et al.*, 2000a]. The higher temperature extraction showed an increase of excluded material on the Mixed-D column, an expected result if dissolving macromolecules. The excluded material had relatively greater absorbance at 370 and 450 nm than the retained material, particularly when examining the sample recovered after the hot extraction. Peat from different sources has also been extracted using NMP to show that the extracted material was both unlike humic substances and unlike coal-oil contamination of soil [Morgan *et al.*, 2005c].

Samples of amber were pyrolyzed using the atmospheric pressure wire mesh pyrolysis reactor described in Chapter 3 [Pipatmanomai *et al.*, 2001]. Gases and tars evolved and there was little char residue left on the mesh, suggesting the absence of large polycyclic aromatic systems in amber. The tar recovered from the tar trap was examined by SEC and by GC-MS. Significantly, the amber pyrolysis tar was similar in GC-MS analysis to the solvent extracts of amber, showing mainly a distribution of small aromatic groups. The amber has also been extracted using pentane, toluene and NMP with the extracts and the insoluble residue being examined by pyrolysis-GC-MS [Islas *et al.*, 2001d,e]. In

SEC, the proportion of excluded material increased from pentane to NMP extracts. However, unlike findings for coal liquids, pyrolysis-GC-MS showed that structures observed for each fraction *and* the insoluble residue were practically identical.

Taken together, these data showed that even when molecular masses increased, the structural units (including PAC groups) making up the amber matrix were relatively similar. The absence of large polycyclic aromatic systems in amber is interesting. Whether the maturation process of resin-derived biomass differs from the evolution of, say, coal vitrinites or whether further maturation would have led the amber toward the larger aromatics observed in coal-derived materials would be worth investigating.

Spruce wood was also pyrolyzed using the wire-mesh technique at 1,000°C in a stream of helium using different heating rates (1, 50, 200 and 1,000°C sec^{-1}) and the tars analyzed on the Mixed-E column [Messenbock, 1998]. The excluded material was more intense than the retained material in each case, but greatest in proportion for the fastest temperature rise.

The pyrolysis products of eucalyptus wood in the wire-mesh and a fixed-bed ("hot-rod") reactor (Chapter 3) have been compared by Pindoria *et al.* [1997b]. The hot-rod reactor was heated at slower heating rates, ranging from 10°C s^{-1} to 10°C min^{-1}. The wire mesh reactor with a rapid temperature rise and a sample bed of dispersed particles would be expected to generate tars with larger molecular sizes than those from a packed bed, where tar cracking through contact with heated solid particles would tend to narrow molecular mass distributions. As expected, the SEC chromatograms indicated that tars from the "hot-rod" reactor were of smaller molecular size than the tar from the wire-mesh reactor and that molecular size decreased with increasing bed thickness (increased chance of cracking).

In subsequent work the effect of hydrocracking on the eucalyptus tar was to sharply reduce the yield of tar while increasing the gas and char yields [Pindoria *et al.*, 1998]. The effect on the recovered tar was to produce some smaller molecular mass material as well as increasing the proportion of excluded material. A sugar cane bagasse sample was also pyrolyzed using both wire-mesh and hot-rod reactors [Pindoria *et al.*, 1999] at different pressures. The hot rod reactor tar contained smaller molecules than the wire mesh tar, but tars from both reactors contained large proportions of excluded material. It was concluded that the hydropyrolysis route was not a good method of producing bio-oils for potential use as transport fluids. The reason was that the catalytic hydrocracking route to deoxygenation of the tars produced increasing quantities of volatiles and gas and little distillate-grade fuels with the prolonged us of the catalyst.

Other materials pyrolyzed and examined by SEC include silver birch wood alone or in mixtures with coal [Collot, 1999], wood, rice husk, coconut coir, agricultural waste [Vasanthakumar *et al.*, 1998], and various wood tar samples [Tei *et al.*, 1997]. A sample of Stockholm tar was examined following fractionation by planar chromatography [Lazaro *et al.*, 1999c]. This material formed by destructive distillation of pine wood has been produced for many centuries and used as tar in wooden ships [Robinson *et al.*, 1987]. It proved to contain material excluded from the Mixed-E column, in all fractions from planar chromatography (Figure 8.14). Because the solutions injected into the SEC column were dilute, the valley between excluded and retained peaks approached zero intensity for the acetonitrile-mobile and pyridine-mobile fractions as with other samples

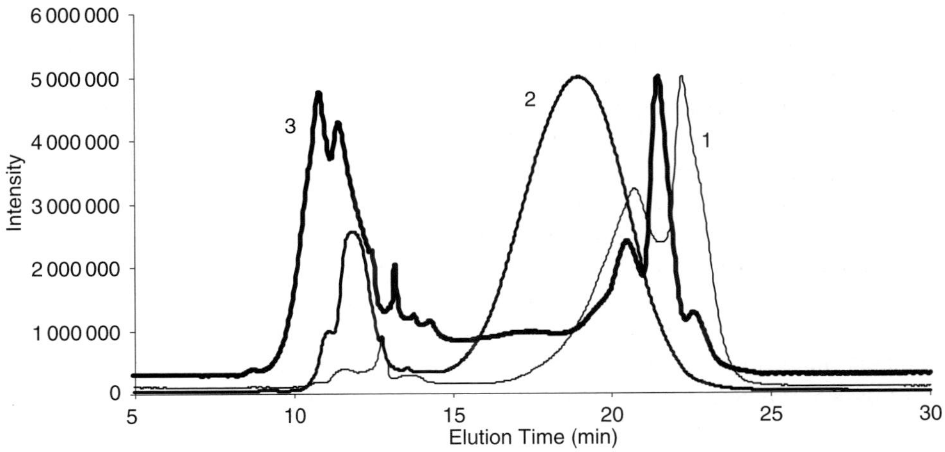

Figure 8.14 Stockholm tar fractions from TLC 1) acetonitrile mobile, 2) pyridine mobile and 3) pyridine immobile, detection by ELS, Mixed-E column; from data of Lazaro *et al.*, 1999.

described above. The pyridine-immobile fraction in Figure 8.14 shows evidence of overloading of the excluded region during SEC, as the signal in the valley between the two peaks was greater than zero.

The pyrolysis of sewage sludge produced a tar that in SEC showed a bimodal chromatogram [Adegoroye *et al.*, 2004]. Tars produced by the pyrolysis of animal bones and casein from milk production [Purevsuren *et al.*, 2004 a,b] also showed the bimodal pattern in SEC. It seems likely therefore that the pyrolysis of any biomass material might produce some material eluting in the excluded region as well as the range of smaller molecules eluting within the range of porosity of the columns used. Although these molecules have not been defined except in terms of their probable three-dimensional shapes, they are clearly not detectable by GC-MS methods or indeed, by mass spectrometric methods other than MALDI-MS or LD-MS (see below).

8.4.6 Humic and fulvic acids

Much of the organic material of soils, sediments and natural waters remains beyond the molecular range of characterization methods normally applied to such samples such as gas chromatography [Hedges *et al.*, 2000; Poirier *et al.*, 2000]. The behavior of humic and fulvic acids in the SEC system was studied in relation to determining the sources of soil contamination and distinguishing between residues from old coal processing plant and from naturally occurring humic substances.

Humic and fulvic acids obtained from the International Humic Substance Society, have been examined using the Mixed-D and Mixed-A columns [Herod *et al.*, 2003b; Morgan *et al.*, 2005c]. The normal characterization method by SEC for these materials uses aqueous solvents. They are however, very soluble in NMP and give clear, dark-colored solutions after ultrasonic treatment. Their SEC chromatograms have shown relatively simple profiles, with a single peak of relatively narrow spread. Using the Mixed-D column indicated that the humic acids were excluded from the porosity while the fulvic acids, eluting slightly later than humic acids, were of slightly smaller size. On the Mixed-A column, both humic and fulvic acids eluted within the linear region of the column and approximated to polystyrene masses of around 1 million u (humic acids) and 0.5 million u (fulvic acids).

Figure 8.15 presents the SEC chromatograms of humic and fulvic acids. NMP extracts of peat gave solutions quite unlike those of standard fulvic or humic acids prepared from peat by extraction using NaOH solution. The humic materials purchased as standards were in the protonated polyelectrolyte form whereas the humic substances in the peat were in the native form as polyelectrolyte salts that would not be expected to be soluble in NMP. Estimates of mass using the polysaccharide calibration suggest average masses of humic and fulvic acids to be 124,000 and 66,000 respectively.

A further problem with defining the molecular masses by the calibration of the columns as above is that the humic substances are highly oxygenated and are polyelectrolytes. The polysaccharide calibration might overestimate the masses since the standard molecule pyrogallol (trihydroxybenzene) of mass 126 eluted 1 minute before rubrene of mass 532 [Karaca *et al.*, 2004] and close to the standard polysaccharide of mass 738 u. Hydrodynamic volumes of polyelectrolytes and polysaccharides may differ despite the oxygen contents of humic substances and PSAC standards being high (32–49%).

A summary of findings from studies using organic eluents other than NMP has been given by Johnson *et al.* [1998]. In aqueous phase SEC, high molecular masses up to 10^5 u have been observed in alkaline solutions of lignin [Wong & de Jong, 1996]. Molecular weights of alkali-solubilized coal were estimated by Polman and Quigley [1991] at about 26,000 u. Alkali solubles of a THF-insoluble brown coal extract gave molecular masses ranging from 355,000 to 33,000 u, in agreement with values determined by ultrafiltration techniques [Ralph & Catcheside, 1996].

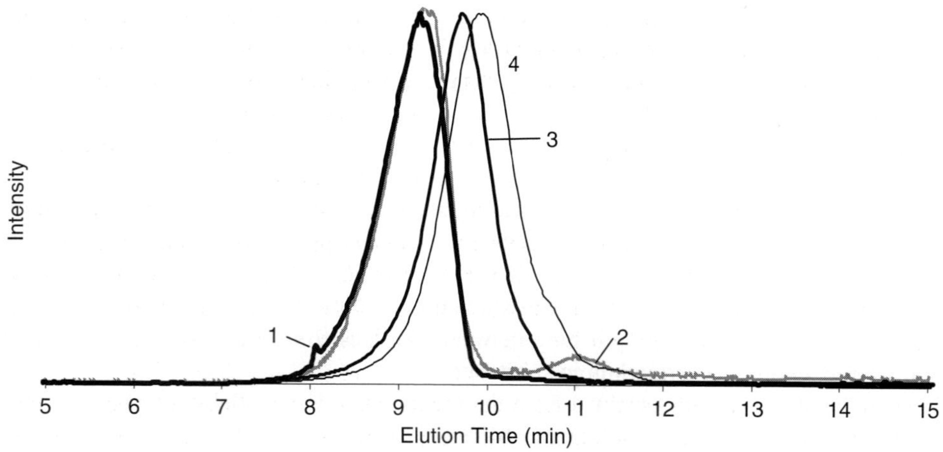

Figure 8.15 Standard humic and fulvic acids on Mixed-D column; curves are 1-peat standard humic, 2-peat reference humic, 3-peat fulvic, 4-soil fulvic, by ELS detection. Reproduced from Morgan *et al.*, 2005c copyright Elsevier.

More recent studies of humic and fulvic acids by mass spectrometry have indicated much smaller molecular mass ranges for these materials. Literature reports of attempts to examine humic and fulvic acids by MALDI-MS and LD-MS have given weak spectra consistent with the detection of only very minor quantities of the sample [Pokorna *et al.*, 1999; Haberhauer *et al.*, 1999, 2000; Ziechmann *et al.*, 2000]. Electrospray ionization with FT-ICP-MS has indicated three possible scenarios concerning the molecular mass of Suwannee river humic acid [Stenson *et al.*, 2002]. (i) The material is composed of low molecular mass materials, which appear larger in other techniques. (ii) The humic acid is made up of large molecules, which fragment in the ion source. (iii) The acid consists of large molecules (5000–10,000 u) that are incapable of acquiring sufficient charge to be detected.

Options 1 and 3 appeared to be the most likely and were investigated further. Aqueous SEC has been coupled to electrospray MS for the analysis of Suwannee River humic and fulvic acids [Reemstma and These 2003]. Molecular ions up to m/z 2,000 were the highest mass ions detected. Suwannee River fulvic acid examined by electrospray ionization [Leenheer *et al.*, 2001] detected ions up to m/z 2,500 and by tandem mass spectrometry showed the presence of carboxylic acid groups. None of these mass spectrometric examinations of humic substances has defined the upper mass limits of these materials, because of the limitations of the electrospray process that has not been shown to ionize the largest molecules of complex mixtures [These & Reemstma, 2003; These *et al.*, 2004; also see section 5.10]. Therefore, the masses of these materials (peak and average masses) remain a source of conjecture.

8.5 Aggregation of Small Polar Molecules to Appear as Large Molecules – in NMP?

Aggregation of small molecules in solution in NMP has been held as the cause of the excluded peak in SEC when using NMP as eluent. Initially, LiBr was held to dissociate these aggregates when added to the eluent NMP [Takanohashi *et al.*, 1994; Masuda *et al.*, 1996; Mori, 1983; Chen & Iino, 2001] by dissipating the ionic forces holding the aggregates together. The observed shift of chromatograms to longer elution times (i.e. smaller apparent MMs) upon LiBr addition to the NMP eluent was thus explained in terms of dis-aggregating large polar molecular aggregates [Masuda *et al.*, 1996].

However, upon examination, it was possible to show that LiBr ruined the size exclusion mechanism by promoting surface effects and causing even non-polar materials to elute after the permeation limit of the column [Herod *et al.*, 1998]. The addition of LiBr to the eluent NMP was found to shift chromatograms of two non-polar samples (a naphthalene mesophase pitch and a mixture of C60 and C70 fullerenes) to longer retention times, i.e. to smaller apparent molecular masses [Herod *et al.*, 1998]. These findings were interpreted to mean that the LiBr was altering column performance, irrespective of sample polarity. More to the point, LiBr addition caused SEC chromatograms to shift to retention times longer than the permeation limits of the analytical columns. LiBr addition appears to affect adversely the solvent properties of NMP and to promote surface sorption mechanisms rather than size exclusion. In the same study, excluded peaks were observed in the ordinary SEC-chromatograms of the naphthalene mesophase pitch, an observation clearly unlikely to be related to 'aggregates of smaller, polar molecules, held together by ionic forces'. MALDI-mass spectra of the same naphthalene pitch have shown molecular mass distributions with large high-mass limits [Herod *et al.*, 1998].

In a separate experiment, LiBr was added to the solution of sample in NMP, which was then injected into the SEC column using pure NMP as eluent. The profiles of the sample with and without LiBr were identical [Herod *et al.*, 1996b]. It is difficult to envisage a disaggregation in the sample solution with added LiBr, which could reconstitute the 'aggregates' exactly as before the addition of salt, upon injection and subsequent dilution in the eluent NMP. In view of the outline presented above, the perceived absence of high-MM materials in the heated probe-MS analysis performed by Lafleur and Nakagawa [1989] – attributed to aggregate formation – may be reasonably interpreted in terms of inherent limitations of the heated probe-MS method.

Based on findings described above, LiBr was withdrawn as a candidate. Subsequently, however, other salts were claimed to perform a similar disaggregation function [Chen & Iino, 2001; Shui *et al.*, 2002]. To test the new hypothesis, it was decided to separate the action of the salt from the SEC mechanism [Karaca *et al.*, 2005a,b]. The fractionation of pitch using thin-layer chromatography was performed with and without salt additions (LiBr, TBAA, TCAA) to the solvents. Bands of material mobile in pure and salt-added solvents (acetonitrile and pyridine) were examined by SEC. In all cases tested, including some experiments using column chromatographic separation rather than TLC, the salt-added solvent mobilized some sample in addition to that mobilized by the pure solvent and the extra sample was of larger molecular size than that mobilized by the pure solvent. In terms of action at the silica surfaces, the solvent plus salt has greater polarity than the pure solvent and can displace more material from the polar sites of silica than the pure solvent. In SEC, the salt added to the eluent promotes surface interactions causing the samples to elute by a mixture of size and surface effects and delaying exit from the column. Despite the diversion of the salt-addition studies, it seems reasonable to consider that aggregates do not form in NMP under the dilute conditions used [Thiyagarajan & Cody, 1997], but that they can and do form in other, less powerful solvents, or at higher concentrations than those prevailing in the detection stage of the SEC systems. It has been estimated that sample concentrations (in NMP) in the detector stage of an SEC device are lower by more than an order of magnitude from those likely to cause micelle aggregation in nitrobenzene, by petroleum asphaltene molecules [Karaca *et al.*, 2004].

Do they aggregate or do they not?

Aggregation at low concentrations in NMP solution appears to be a difficult phenomenon to observe. None of the polycyclic aromatic hydrocarbon standards used in tests or calibrations of the set of columns used (Mixed-E, Mixed-D and Mixed-A), have shown any tendency to form an early eluting peak indicative of the presumed π-π bonding assumed to cause such molecules to aggregate. We note however that the standard polycyclic aromatics available are all relatively small and few are outside the GC-MS range. Three essential oil samples consisting of diterpenes, sequiterpenes and oxygenated terpenes produced by steam distillation (and therefore containing small volatile molecules) gave no excluded peaks in SEC [Morgan *et al.*, 2005b]. Of the range of standard materials other than aromatics used to test the SEC calibrations, none gave rise to an early eluting peak except the mixture of fullerenes. As already explained, this exception is believed to be due to its spherical structure giving a far larger profile than would be warranted by its molecular mass.

The early-eluting peak of coal tar pitch was fractionated using analytical SEC [Lazaro *et al.*, 1999b]. On re-injection, the first two fractions eluted at the times of collection while the third fraction eluted earlier, showing that far from dissociating on dilution, the material had been delayed in elution by overloading the column. None of these fractions gave rise to a late-eluting peak that might have been expected if the 'aggregating' material had been diluted and disaggregated.

Fluorescence spectroscopy has been used on samples and fractions in both static solutions and in SEC effluents as detectors in series with UV-absorbance. The synchronous fluorescence spectra of late-eluting peaks, of presumably small molecular mass material, tend to be rather intense. Sample dilution initially increases signal intensity; the sample is diluted until the (increasing intensity) effect due to reduction in self absorption gives way to weakening of signal due to dilution. When the fractions of tars insoluble in acetonitrile but soluble in pyridine, or insoluble in pyridine but soluble in NMP are added to the fluorescence cell at dilutions used for strongly fluorescing molecules, no signal can be detected. It is necessary to increase the concentration significantly before any fluorescence can be detected from the sample and the fluorescence maximum intensity is shifted toward the red end of the spectrum compared with strongly fluorescing molecules.

When the fluorescence instrument is used as a detector in the SEC flow mode, it is necessary to inject very dilute solutions to avoid overloading this detector with the fluorescence of the small, strongly-fluorescing molecules. This requirement means that the UV-absorbance signal is very weak (cf Figure 8.10 above). Under these conditions, the early-eluting materials of the pyridine-solubles and the pyridine-insolubles show no significant fluorescence, and the fluorescence that is detected comes from the smallest molecules of these fractions [Morgan *et al.*, 2005a]. Even when these fractions that show no significant fluorescence in the SEC flow mode are examined in the static fluorescence cell, the fluorescence remains insignificant. It is noted that the early eluting material (still showing intense UV absorption) does not change nature by the dilution in the SEC detector systems. If these materials were composed of molecular aggregates, certainly, they appear to show no indications of disaggregating at these very low concentrations.

Sheu *et al.* [2005] have published data on surface tension of solutions of heavy hydrocarbon liquids that indicates that the surface tension of the solution changes at a critical concentration where aggregates form. The concentrations of samples in the SEC flow [Karaca *et al.*, 2004; Morgan *et al.*, 2005a] are lower than those indicated by Sheu *et al.* [2005] by a factor of at least 10, with even lower concentrations in the fluorescence work.

Pyrolysis of the pyridine-soluble or pyridine-insoluble fractions of pitch, coal extract or low temperature tar using the pyroprobe of the GC-MS system, showed that these fractions gave no significant components able to pass through the GC column. When a sample of the pitch pyridine-insoluble fraction was pyrolyzed in the atmospheric pressure wire-mesh reactor described in Chapter 3, little tar was produced (~13% by weight), none of which could be examined by GC-MS. Furthermore, pyrolysis-GC-MS of a series of fractions of low temperature tar collected from the SEC effluent (equivalent to acetonitrile solubles) indicated that only the last few fractions (corresponding to the smallest molecules of the sample) were able to give any signal in the MS detector. These experiments provide strong evidence that the largest molecules in these coal-derived samples are not aggregates of small molecules and cannot be simply disaggregated or cracked easily into smaller molecular fragments. The inability of elemental analyzers to give a reasonable carbon content when using a slug of oxygen to meet the sample in the combustion furnace also points to the char-forming tendency of the large molecular size fractions.

An additional consideration comes from the study of SEC using THF as eluent. No examples of chromatograms of coal-derived liquids using THF have showed any material excluded from the column porosity except for material dissolved from coal at 450°C [Gibbins *et al.*, 1991; Li *et al.*, 1995] where an extra peak was observed at the earliest elution times. However, chromatograms of masticated natural rubber using THF and refractive index detection [Homma & Tazaki, 1995] have shown excluded material of molecular mass from 1 to 10 million. The excluded material is likely to be particulates produced by milling from previously insoluble rubber gels and similar to the colloidal silicas used to calibrate the excluded region (above) in SEC. Strausz *et al.* [2002] observed three peaks in the SEC of asphaltenes using dichloromethane as eluent and at least a part of the excluded material corresponded to aggregates in the initial solution that disaggregated on standing; the remaining excluded material did not disaggregate and may be of large molecular size. In recent work with two SEC columns in NMP and in heptane, respectively, the colloidal silica samples eluted from each one at the exclusion limit, indicating that colloidal particles could be swept through the columns in the interparticulate voids [Karaca *et al.*, 2004, 2005c]. This feature applied also to the column used for the SEC of aliphatics, described in Chapter 7 section 7.7.2, where the colloidal silica particles eluted at the void volume using heptane as eluent. SEC of asphalts using THF and other solvents [Davison *et al.*, 1995] showed no excluded peaks although excluded peaks were observed in work with NMP eluent using vacuum residues of petroleum and asphaltene fractions [Suelves *et al.*, 2000a, 2003; Li *et al.*, 2004; Millan *et al.*, 2005b]. Because SEC using solvents other than NMP does appear to be able to elute colloidal particles in the void volume of the column, but show no evidence of separated excluded peaks (as in NMP) we consider that there are no aggregates in other solvents, since the micelles or nano-aggregates might be expected to elute in the void volume – none are detected. The evidence from the study of Athabasca bitumen discussed next, indicates that large molecules are present in THF solution but are not separated from small molecules until examined in SEC with NMP.

The fractionation of Athabasca bitumen (into five equal-time fractions) using preparative SEC in THF eluent, followed by analytical SEC in NMP eluent showed that although the THF fractionation gave no evidence of excluded material, all the fractions did show some excluded material when analyzed subsequently in NMP solution by SEC [Domin *et al.*, 1999]. The proportion of excluded material decreased with elution time on the preparative column. This change was interpreted to indicate that the largest (excluded) molecules observed in SEC with NMP were not separated sufficiently by THF solvent and eluted from the SEC column in THF in part by a surface interaction mechanism which delayed their elution well beyond the excluded region. Thus, these materials appeared to have molecular size or mass much smaller in THF than observed in NMP solution. The SEC chromatograms of the whole sample in THF and NMP and of the fractions in NMP are shown in Figure 8.16 as: a) the THF SEC repeated several times and b) the NMP SEC of the fractions, 1 being the earliest and 0 being the whole sample.

The chromatograms of Figure 8.14 show quite definitely that the fractions recovered from SEC in THF did contain material that eluted at the exclusion limit when examined in SEC using NMP as eluent. Therefore, the use of SEC with THF as eluent for similar

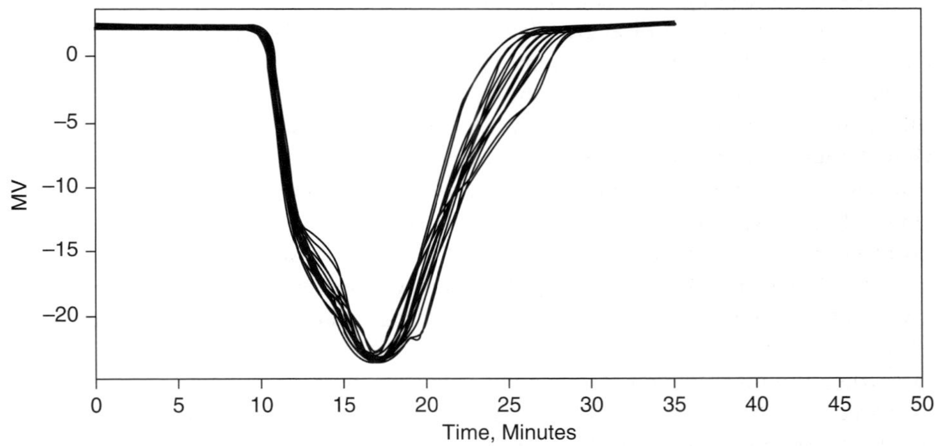

Figure 8.16a Athabasca bitumen, preparative SEC in THF, repeated runs. From Domin *et al.*, 1999, Energy & Fuels 13, 552; copyright Amer. Chem. Soc., with permission.

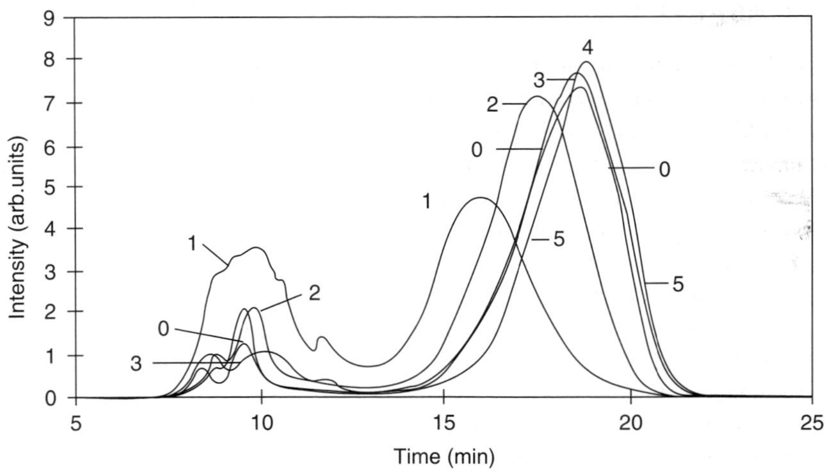

Figure 8.16b SEC in NMP of Athabasca bitumen (0) and of five equal time fractions from preparative SEC in THF, 1 being the earliest and 5 the last. From Domin *et al.*, 1999, Energy & Fuels 13, 552; copyright Amer. Chem. Soc., with permission.

samples will avoid seeing the excluded aromatic material by losing it amongst the smaller molecular size components. The preparative SEC did produce some fractionation of the small molecules as detected by the analytical SEC, with fractions 1 and 2 showing the largest effect. The remaining three fractions however indicate that no significant fractionation took place later in the preparative column. Hence, we consider that SEC using THF is pointless for aromatic samples.

Summary of evidence against aggregation in NMP: the various attempts to use salts in the SEC solvent to disaggregate the excluded material, described above, have shown only that salts have an effect on the mechanism of SEC, resulting in excluded and retained material eluting later than in the absence of salts. No-one has demonstrated that the excluded material has disaggregated; an alternative method of investigation would be necessary to achieve this result. In the work with Athabasca bitumen, described in the previous paragraph, we showed that the THF soluble material recovered from SEC in THF eluent (where no excluded peak was observed) did produce an excluded peak when examined by analytical SEC in solution in NMP, a stronger solvent than THF. This evidence indicates that in SEC using THF, the very large material elutes later than expected from SEC in NMP and therefore, the THF solvent is unable to prevent interaction between the large molecules and the surface of the column packing material.

The other evidence presented for SEC systems with solvent other than NMP, indicates that excluded material may have been detected occasionally but probably as particulate or colloidal matter rather than as large molecules. Such particulate matter as colloidal silicas elute from our columns using the two solvents, heptane or NMP; we regard these two as the extremes of poor solvent (heptane) and good solvent (NMP) for minimising the interactions between column surface and aromatic materials in SEC.

8.6 Molecular Mass Methods – Mass Spectrometry of High Mass Materials >500 u

There are many different methods for the introduction of high-mass materials into mass spectrometers, ranging from liquid chromatography, through desorption chemical ionization, field ionization, field desorption, fast atom bombardment, plasma desorption, laser desorption, matrix assisted laser desorption and electrospray ionization. Burlingame *et al.* [1996] have reviewed many of these applications. All of the sample introduction methods work well for specific applications, often to biological molecules, but their performance when applied to material of relatively wide polydispersity and of unknown upper mass, leaves much to future development. None of the available methods can define an upper mass limit independent of other measurements and it is necessary to assemble evidence from several techniques to compare data from individual samples or fractions of samples. Experience suggests that an absence of signal from an experimental method cannot necessarily be taken to indicate an absence of material, merely an indication that some method development may be needed to investigate the lack of signal.

The sample introduction (gas chromatography and complete thermal evaporation) and ionization methods described in the previous Chapter are only able to deal with relatively small molecules. In the present Chapter, sharp differences are shown between results from the low-mass techniques and those described below, using MS techniques where ionization is thought to take place before or during evaporation (*e.g.* FD, FAB, laser-desorption), where the molecular ions are released into the gas phase by processes other than thermal evaporation [Herod *et al.*, 1993a]. As in Chapter 7, however, the analyses that can be attempted will depend on the available mass spectrometers, the types of ionization methods and sample inlet systems they are able to accommodate and the type of mass analysis systems (quadrupole, time of flight, synchrotron etc.) they are connected to.

The current trend in biological mass spectrometry to use electrospray ionization has superseded the use of fast atom-bombardment (FAB) methods in that field, to the extent that FAB may not now be available without searching store-cupboards for old equipment, unused but as yet not discarded. Furthermore, the problems faced in the analysis of complex fuel derived liquids are different from those encountered in biomedical or biochemical work. In the former, we search for methods able to detect large molecules in the presence of an excess of small ones. Meanwhile, new developments in mass spectrometric ionization methods are generally in the biological chemicals field, where the requirement is that of confirming the mass of a molecule of known or suspected structure.

General texts on mass spectrometry include Chapman [1993], Hamming and Foster [1972], McLafferty and Turecek [1993] while papers covering specific topics such as the invention of the MALDI techniques [Hillenkamp *et al.*, 1991] are in the voluminous mass spectrometry literature. The background and evidence of large molecules in coal and petroleum derived liquids by mass spectrometry has been reviewed by Herod [2005]. The work described below has shown that material appearing under the 'excluded' peaks in SEC [Herod *et al.*, 1996b] correlate with MALDI mass-spectra showing, for some samples, material extending to molecular masses as high as 40,000 u [Herod *et al.*, 1995d; Lazaro *et al.*, 1997; Domin *et al.*, 1997a; Millan *et al.*, 2005a]. We will next present an outline of sample introduction methods into mass spectrometers, with reference to samples likely to contain high mass materials.

8.6.1 Liquid chromatography

The requirement for analytes to be volatile in gas chromatography is overcome in liquid chromatography, but sample introduction can still be a significant problem at the mass spectrometer interface. Many different interfaces have been devised to avoid thermal degradation during transfer to the mass spectrometer, including the moving belt, thermospray, direct liquid introduction and the particle beam [Chapman, 1993]. The moving belt and the direct liquid introduction methods rely on evaporation of analyte in the ion-source vacuum chamber, which is less likely to produce pyrolysis than evaporation at atmospheric pressure. Thermospray and particle beam interfaces create droplets in vacuum from which solvent can evaporate, leaving the analyte in the vapor phase.

Liquid chromatography has not been used extensively for the analysis of coal or petroleum-derived materials. The reasons are twofold: (i) the HPLC methods have not been worked up to apply to whole samples, but only to the PAC fractions of relatively low mass and (ii) the high cost of mass spectrometers able to accept the input from such columns. The general principles and instrumentation of LC-MS [Niessen & Tinke,

1995] and applications to the detection of sulfur and nitrogen compounds in coal liquids have been reviewed by Neal [1995] and Herod [1998].

Moving belt interfaces have been used to examine pyrolysis and hydropyrolysis tars. Aliphatics including the biomarker pentacyclic triterpanes and aromatics from alkyl benzenes to alkyl dibenzopyrenes were detected, up to molecular masses of about m/z 550 [Herod *et al.*, 1987] with high molecular mass alkanes and cycloalkanes up to m/z 900 (C_{64}) as well as aromatic homologous series (up to 40 members) to similar high masses [Herod *et al.*, 1988]. Polar components of the aromatic fraction were not identified but the evidence from electron impact and chemical ionization pointed to the presence of nitrogen containing compounds. Similar data have been obtained using a particle beam interface to a magnetic sector mass spectrometer using similar samples [Herod *et al.*, 1993a]. The LC separation of the aromatic fraction was by compound type such as alkyl benzenes and alkyl benzopyrenes. Metalloporphyrins extracted from coals have been examined by Bonnett *et al.* [1991] using a moving belt interface, reverse phase chromatography and fast atom bombardment ionization, giving mass spectra up to m/z 600. The crude porphyrin extract was isolated by TLC and demetallated before examination by mass spectrometry. However, it was considered that a direct analysis might be possible to avoid the demetallation step. A reversed phase liquid chromatographic separation of nitrogen-containing polynuclear aromatics from a solvent refined coal liquid was used to generate fractions for off-line mass spectrometry [Borra *et al.*, 1987]. Particle beam spectra of a hydropyrolysis tar have been shown with components to m/z 400 [Herod *et al.*, 1993a].

LC-mass spectra are more complex than GC-mass spectra but LC-MS has not been used to examine the largest molecular mass components of coal liquids such as pyridine insolubles. Further work is needed to extend the range of molecular masses and types eluted from HPLC columns, probably by using more polar solvents, such as pyridine and NMP, than have been used hitherto. The capital cost of HPLC columns has tended to argue in favor of caution when investigating new methods and solvents. The moving-belt interface is no longer available but thermospray methods may be able to overcome the problems of introducing into a mass spectrometer samples dissolved in relatively involatile solvents such as NMP. Size exclusion chromatography has been used off-line to produce fractions of pitch for examination by MALDI-MS [Islas *et al.*, 2003c]. Mass spectra showed envelopes of ion intensity over ranges of mass for fractions of relatively narrow polydispersity.

8.6.2 Supercritical fluid chromatography

This chromatographic method exploits the enhanced solvent properties of fluids under conditions of temperature and pressure in excess of their critical points. The method should have many advantages over gas or liquid chromatography but the promise has not yet been fulfilled. The carrier fluid normally used is carbon dioxide with modifiers to improve the solubility of coal-derived samples, such as methanol or ethanol. To extend the range of this work, solvents are needed that are more aggressive, in terms of dissolving large aromatics. The theory of the chromatographic method has been described in Chapter 7 [Bartle, 1988]. In past work, the range of aromatics examined by

SFC did not extend much beyond the capability of GC. Off-line extraction of coal tar pitch combined with heated probe mass spectrometry extended the mass range to m/z 500 [Kershaw & Smart, 1994].

8.6.3 Thin layer or planar chromatography

The technique is useful for fractionating samples prior to mass-spectrometric or other analyses. Its main advantage stems from the ability to recover and examine all fractions, including the heaviest, without significant loss of sample. The technique is non-destructive and many samples can be run together (i.e. on the same plate) together with standards to monitor the separation. Direct interfacing with mass spectrometry is possible but can be a disadvantage. In off-line operation, different mass spectrometric methods can be used to analyze different parts of one sample development if required. Reviews have covered the combination of planar chromatography with mass spectrometry [Somsen *et al.*, 1995] and the application to coal derived materials [Herod, 1994], as well as nitrogen and sulfur compounds [Herod, 1998].

Pitch was fractionated on silica gel plates with development in solvent series from polar to non-polar [Parker *et al.*, 1993; Herod *et al.*, 1995a,b, 1996a; Domin *et al.*, 1998]. The fractions were examined directly by probe mass spectrometry showing PACs, neutral and basic nitrogen compounds with molecular mass increasing as mobility on the plate decreased. Subsequently, fractions were analyzed by MALDI-MS or plasma desorption mass spectrometry (PDMS). Observed increases in molecular mass with mobility in more polar solvents were found to correlate with decreasing elution times (i.e. increasing apparent molecular size) in SEC with NMP as eluent. This evidence tends to indicate that the planar chromatographic separation took place largely on the basis of molecular size [Li *et al.*, 2004]. This work has involved the fractionation of coal tar pitch and a petroleum distillation residue (Petrox from Concepcion, Chile) using pentane, toluene, acetonitrile and pyridine. Fractions were collected from each migration zone and examined by SEC. For both samples, the elution time of the retained material in SEC shifted to earlier times (larger molecules) with decreasing mobility in TLC. Meanwhile, the proportion of excluded material increased with decreasing mobility in TLC. The fractions were not examined by MS methods but both SEC and TLC indicated that the molecular size of the mobile material increased with decreasing mobility. We believe this may be the first indication of such an agreement between the two methods when applied to coal and petroleum liquids, with potential for examination by mass spectrometric techniques.

8.6.4 Field ionization

FIMS and pyrolysis FIMS have been used to analyze coal and coal liquids [Schulten 1982; Malhotra *et al.*, 1993; Cagniant *et al.*, 1992]. Field ionization mass spectrometry can detect those molecules, which are volatile in vacuum, or which form during pyrolysis in vacuum. The method works by passing volatile molecules through an electric field between a sharp point or edge (e.g. a razor blade) and an earthed electrode. An electron can tunnel out of the molecule to the electrode leaving a positively charged molecular ion, with no significant transfer of vibrational energy to the molecular ion that could lead to fragmentation. High resolution FIMS may allow determination of atomic compositions of peaks, but the complexity of mass spectra observed for asphaltenes and preasphaltenes tends to limit the unambiguous use of high resolution. The problem is that the number of components with the same nominal mass (integer mass) increases as the number of atoms increases and as the mass number increases and mass resolution quickly becomes unable to deal with the complexity of incoming signal. Furthermore, the relative intensity of odd-mass peaks to even-mass peaks in spectra of asphaltenes tends to be greater than in spectra of aromatics or oil fractions, indicating that the molecular complexity has increased by the introduction of heteroatoms, further reducing the ability of high resolution mass measurement to give unambiguous results. The upper mass limit of the technique, defined by the requirement of volatility under the operating conditions, appears to be about m/z 1200, but that need not correspond to the largest mass present in any sample. Mass spectra of aromatics extended to m/z 700 [Herod *et al.*, 1993a].

8.6.5 Field desorption

Field desorption mass spectrometry should be able to ionize molecules which are thermally labile or involatile in FIMS, because ionization is from the solid phase rather than the vapor phase. The mechanism of ionization is similar to that of FIMS. The sample is deposited on a dendritic wire carrying multipoint micro-needles with tips of small radius. The intense electric fields at the tips (10^7–10^8 V cm^{-1}) strip one electron from the analyte molecules and force them into the gas phase. The emitter wire is normally heated to encourage desorption of the analyte and some thermal decomposition can occur at high emitter wire currents. Little internal energy is transferred in the ionization process that produces molecular ions from the solid phase. Mass ranges observed using this technique appear to exceed the range observed for FIMS [Herod *et al.*, 1993a; Herod, 2005]. The difference between the FIMS spectrum and the FD spectrum of the same asphaltene fraction, suggests the two techniques may be detecting different materials found within the samples. The maximum intensities appear at different masses with more intensity by FD above m/z 500. This difference may result from decreasing thermal volatility of the PAC with increasing molecular mass, a situation that should not affect FD-MS.

Field desorption MS of an asphaltene fraction from Maya crude gave ions up to m/z 800 [Douda *et al.*, 2004]. However, the proportion of the asphaltene ionized was not apparent. This exemplifies the problem with the FD method applied to mixtures of unknown mass range. The upper mass limit is not apparent and may not be reached by the FD method at all. Experimental details of a Maya asphaltene examined by SEC indicate the presence of material excluded from the column porosity that might be of relatively high mass, as in the pyridine insoluble fraction of pitch, possibly up to m/z 100,000. In any case, the sample almost certainly contained material of larger masses than those indicated by FD-MS [Millan *et al.*, 2005b; Morgan *et al.*, 2005a].

8.6.6 Fast atom bombardment

Fast atom bombardment mass spectrometry (FAB-MS) relies on ionization from the liquid phase, with the probable requirement that the sample molecules congregate at the surface of the liquid matrix in the vacuum. Pyrolysis tars from the Argonne set of coals [Vorres, 1990] gave FAB spectra up to m/z 1200 [Winans, 1991]. FAB spectra of pentane insoluble material from coal liquefaction extracts showed signal up to about m/z 4000 [Herod *et al.*, 1993d; Herod, 2005]. Spectra of asphaltene and pre-asphaltene (benzene-insoluble) fractions of a hydropyrolysis tar were found to be similar, suggesting the technique may not have ionized the entire sample. Fractions collected from SEC of a hydropyrolysis tar asphaltene fraction also reached m/z 2,000 [Herod *et al.*, 1993a]. The tar had been prepared in a hot-rod reactor. The calibration of the FAB source using CsI salt indicated that m/z values up to about 5,000 were detected, but such high mass ions were not observed from coal-derived liquids, where the upper mass limit was around m/z 3000–4000 [Herod *et al.*, 1993d]. The method does ionize quite large biomolecules, however, and the reasons for the lack of high mass ions from coal liquids may be associated with the liquid matrix and the addition of additives such as acids, to the matrix. Further investigative work is needed to extend the mass ranges observed by FAB-MS.

8.6.7 ^{252}Cf Plasma desorption-mass spectrometry (PDMS)

Plasma desorption mass spectrometry produces ions from solid samples. Many requirements of this technique are similar to those of MALDI-MS. In particular, a sufficiently large ion extraction voltage is required to accelerate ions through the time-of-flight analyzer and allow detection of slow (large) as well as fast (small) ions from an unfractionated sample or one of relatively wide polydispersity. Plasma desorption mass spectrometry has been used to analyze heavy distillation residues from direct coal liquefaction processes [Larsen *et al.*, 1994]. Number average molecular masses derived from plasma desorption were compared with those from gel permeation chromatography. Mass ranges extended to about 2,000 u for the plasma desorption results using oils, asphaltenes and preasphaltenes. A comparison of this method with MALDI-MS showed that although PDMS showed similar trends as MALDI, the maximum ion accelerating voltage of the plasma instrument was insufficient to allow detection of high mass ions [Domin *et al.*, 1998; Johnson *et al.*, 1999].

An alternative view has been expressed favoring PDMS over MALDI-MS on the basis that VPO methods overestimate molecular masses because the sample in solution must consist of aggregates at the solution concentrations necessary for the method to work [Strausz *et al.*, 2002]. Therefore, if the MALDI-MS results were comparable with VPO results, they too must be an overestimate. However, our work indicates that aggregation does not become important in NMP solutions at the concentrations used (section 8.5). Furthermore, MALDI is generally thought not to ionize/desorb aggregates but only molecular species; it seems more necessary than ever to look for experimental confirmation (or otherwise) for this last statement.

8.6.8 Laser desorption-mass spectrometry (LD-MS) and matrix assisted laser desorption/ionization mass spectrometry (MALDI-MS)

Lasers and mass spectrometry have been combined to study fossil fuel derived materials for over 30 years. In early work, rapid laser pyrolysis of coal was observed in the mass spectrometer vacuum [Joy *et al.*, 1970]. The spatial variation of the elemental composition of coal macerals in UK and US coals was studied by Lyons *et al.* [1987, 1990]. The laser ablation of coals, coal derived materials and polynuclear aromatics was examined by Greenwood *et al.* [1990]. One salient feature of laser desorption of UV-adsorbing carbon based materials is the formation of positive and negative ion carbon clusters at relatively low masses, from C_1 up to about C_{29} [Herod *et al.*, 1993c; Jiao *et al.*, 1993; Burroughs *et al.*, 1993]. These clusters probably form the effective matrix during the desorption step, in terms of energy transfer from absorbance of photons at the laser wavelength as electronic excitation in the solid state, to the translational energy necessary to generate molecules and ions in the vapor phase. Greenwood [1994] reviewed the relation between mass spectrometry, coal and carbon clusters. The cluster formation increases as the laser power increases and the molecular ion intensity of pure compounds decreases [Herod *et al.*, 1994a]. As the laser power increases, the carbon clusters form the fullerene carbon compounds, from C_{60} (m/z 720) to higher masses. The presence of fullerenes in the spectra of coal-derived materials indicates that the laser power is considerably higher than the minimum fluence necessary for desorption, since fullerenes are easily desorbed by laser energy [Herod *et al.*, 1994a]. Fullerenes are much more readily formed from coal-derived materials than from graphite [Greenwood, 1994].

A number of laser desorption applications have avoided fullerene formation. Winans *et al.* [1991] examined vacuum pyrolysis tars from the Argonne series coals by laser desorption as well as by FAB and by desorption chemical ionization (DCI). Molecular mass ranges were only to about m/z 800 although the FAB spectrum was limited by the mass range of the instrument. LD-MS of high-temperature coal tar and supercritical-water extracts of coal gave major intensity peaks in the range m/z 200–400 with low-intensity ions up to m/z 15,000 [Wang & Takarada, 2003]. Laser desorption work on coal tars and pitch from high temperature coke ovens extended the range of material observed up to m/z 12,000 [Herod *et al.*, 1993b, 1999] and subsequently to m/z 200,000 by MALDI [John *et al.*, 1993, Parker *et al.*, 1993; Islas *et al.*, 2003a]. Molecular mass distributions in coal-derived tars and liquefaction extracts have similarly been shown by laser desorption to extend up to m/z 12,000 [John *et al.*, 1991; Parker *et al.*, 1993; Herod *et al.*, 1993b] and by MALDI to m/z 20,000 – 30,000 [John *et al.*, 1993; Parker *et al.*, 1993; Islas *et al.*, 2003a].

Coal samples directly exposed to MALDI-MS showed signal that extended to the limit of the same instrument, m/z 270,000 [John *et al.*, 1993; Parker *et al.*, 1993; Islas *et al.*, 2003a]. In these studies, the mass spectra showed a peak of intensity that was not instrument dependent, in the approximate mass range m/z 1,000 to 5,000 [John *et al.*, 1991, 1993; Parker *et al.*, 1993; Herod *et al.*, 1993b; Islas *et al.*, 2003a]. Similarly, MALDI-MS spectra were obtained from a series of kerogens mixed as fine powders with matrix by [Li *et al.*, 1994b]. In pyrolysis tars from maceral concentrates of a UK coal, this peak of intensity shifted significantly to higher masses in going from a coal vitrinite, to a coal liptinite and an inertinite [John *et al.*, 1994]. Similar work using the Argonne coals that range from lignite to semi-anthracite showed a range of intense peaks in the mass range m/z 1000–5000, extending at lower intensities up to m/z 270,000 [Herod *et al.*, 1994b]. The reproducibility of the ions of relatively low intensity at the upper mass range was not investigated but the MALDI method does not appear to generate ions corresponding to clusters of molecules when used in samples of wide molecular mass range.

Pyrolysis tars and liquefaction extracts of the Argonne coals showed intense ions over the same mass range, m/z 1000–5000 which was sample dependent [Herod *et al.*, 1994c]. As expected, liquefaction extracts extended to rather larger molecular masses. Liquefaction extracts required lower laser fluences to desorb compared to pyrolysis tars. Oxidation and decarboxylated products from Pocahontas coal were examined by laser desorption MS and the mass range of products exceeded m/z 1200 [Stock & Obeng, 1997]. Some problems associated with LD-MS and MALDI-MS of coal extracts have been discussed [Hunt & Winans, 1995]. These were the lack of high mass components in the mass spectra, the need to identify an effective matrix and the difficulty of detecting high mass ions in the presence of low mass ions generated from a sample of wide polydispersity. The problems of mass discrimination against the high mass components of mixtures with high polydispersity (>1.1) have been discussed by several other groups [Lloyd *et al.*, 1995, Montaudo *et al.*, 1994, Shimada *et al.*, 2001]. It is relatively easy to detect the smallest molecular ions of a complex mixture but more difficult to observe the highest mass ions at the same time, either quantitatively or qualitatively. 'Recipes' for the successful ionization of various polymers by methods of combining sample and matrix have been given by several researchers [Danis & Carr 1993; Garrozo *et al.*, 1995; Montaudo *et al.*, 1995].

Comparisons of MALDI-mass determinations with results from SEC determinations have been made [Lehrle & Sarson 1995; Sheng *et al.*, 1994]. Calculations on the MALDI mass spectrum of pitch pyridine insolubles to evaluate an upper mass limit for the spectrum suggested that m/z 95,000 could be the limit of that particular spectrum [Lazaro *et al.*, 1999a]. Figure 8.17a shows a laser desorption spectrum and Figure 8.17b shows the MALDI mass spectrum with MBTA as matrix, of the pyridine insoluble fraction of coal tar pitch extending to about m/z 9000 [Herod 2005; Millan *et al.*, 2005a]. The peak of intensity has shifted in the MALDI spectrum compared with the LD spectrum.

Evidence of the underestimation of high mass components of coal-liquids by MALDI-MS has come from the study of polymer standards of low polydispersity, where the relatively high mass molecular ions were less intense than expected when compared with the low mass molecular ions on a molar ratio basis [Lloyd *et al.*, 1995; Domin *et al.*, 1997b]. Fractionation of coal-derived samples to reduce the polydispersity of the fractions is seen as the way to overcome the mass discrimination caused by too wide polydispersities. Peak masses from MALDI-MS, LD-MS and masses equivalent to elution times in SEC calculated from the polystyrene calibration for three series of narrow time-fractions (collected from SEC) were found to be in quantitative agreement up to m/z 3000 [Islas *et al.*, 2003c]. However, reproducible methods for preparing targets to allow examination of higher mass material have not been defined.

The complementary nature of LD and MALDI has been shown in a study of dichloromethane solubles from soots produced from fuel rich flames [Herod *et al.*, 2004]. An oligomeric series was detected by LD-MS but not by MALDI-MS and proved to be siloxane polymers, introduced accidentally through the use of a silicone rubber tube for washing the burner assembly by dichloromethane to remove deposited soot. On further examination of a fraction of the dichloromethane solubles collected from preparative

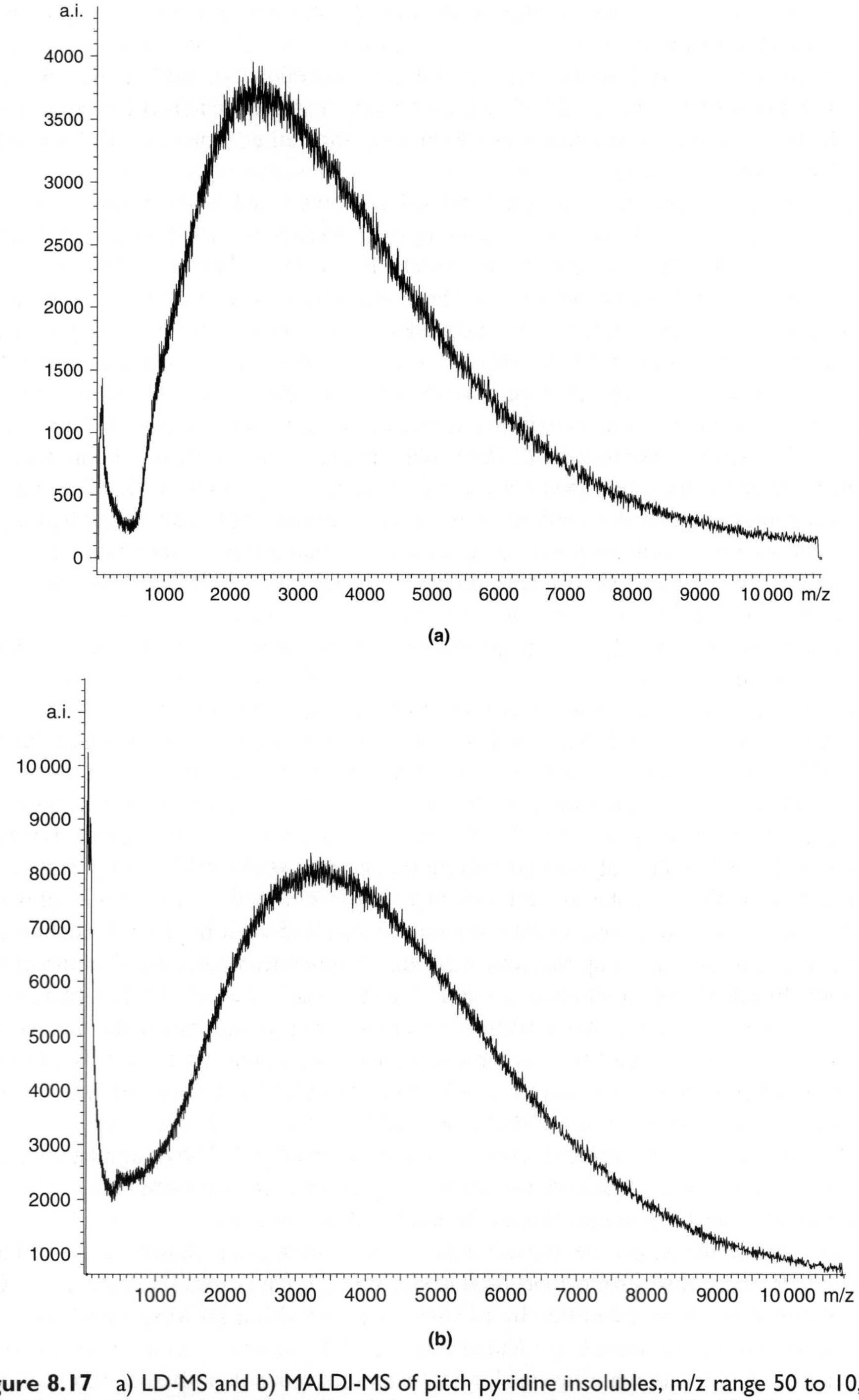

Figure 8.17 a) LD-MS and b) MALDI-MS of pitch pyridine insolubles, m/z range 50 to 10,800. Reproduced with permission from Rapid Commun. Mass Spectrom 2005, 19, 1867 copyright John Wiley & Sons.

SEC to concentrate the siloxane oligomers, another oligomeric series of carbon compounds was detected in addition to siloxanes, again only by LD-MS. This second series appeared to be involved in soot formation with the separation of groups of oligomers by 24 mass units over a wide range of masses. Clearly, the use of LD-MS in addition to MALDI-MS is able to give additional information that would otherwise be lost in the less-selective MALDI mass spectrum.

Recent work on coal tar pitch pyridine insolubles has involved changing the experimental parameters of the MALDI-mass spectrometer to find evidence of ions at m/z values in excess of 10,000 [Millan *et al.*, 2005a]. The starting point for this work was the comparison of LD-MS and MALDI-MS of this sample. Spectra are compared in Figure 8.17, which indicates that the action of the matrix shifts the peak of intensity to higher mass values than found by LD-MS. This range of relatively small ions (m/z range up to 6,000 or so) was extended by using an extraction voltage of 15 kV, reducing the amplifier gain voltage from 10 kV to 5 kV to avoid overloading the detector while at the same time increasing the laser power to 90% of the maximum available. Under these conditions, a second broad ion peak with a range of mass from about m/z 10,000 to 100,000 could be observed. Different operators could reproduce the spectra at different times. The magnitude of the second (broader) peak appears to give reliable trends, showing significant intensity for samples with visibly large peaks of material excluded from column porosity in SEC. Conversely, the second peak has been found to diminish to near zero intensity when samples with relatively small peaks of material excluded from SEC column porosity were tested.

The spectrum in Figure 8.18 shows the bimodal mass distribution observed using these parameters. The signal does not reach zero intensity in the valley between the two peaks of intensity. It is tempting to assign the two mass spectral peaks to the two peaks detected by SEC of the same sample, shown in Figure 8.19. In that case, the low mass MALDI peak of range up to m/z 6,000 would equate with the material retained in SEC (of mass indicated by the polystyrene calibration up to 8000 u) while the high mass peak would equate with the material excluded in SEC. The MALDI mass range from m/z 10,000 to m/z 100,000 is smaller than the mass range indicated by the SEC calibration (greater than 200,000 u and up to several million u). It would provide the first direct evidence for the change of structure (relation between molecular mass and effective size in SEC) in molecules of mass above 6,000 u or so indicated by the close approach to zero intensity of the SEC signal between the two peaks. Similar spectra have been obtained for petroleum asphaltenes [Millan *et al.*, 2005b] and a coal digest but extending to lower upper mass limits than m/z 100,000 [Millan, 2005].

Petroleum distillation residues have been examined by MALDI-MS and LD-MS and shown to have mass range distributions extending up to about m/z 6,000 depending on the origin of the residue or asphaltene fraction [Suelves *et al.*, 2001a, 2003]. Peak intensity masses were generally below m/z 1000 and in reasonable agreement with estimates from SEC. Recent work has extended the mass range for these samples to m/z 40,000 with relatively low intensity ion strengths above m/z 10,000 but showing a peak of maximum intensity around m/z 2000 [Millan *et al.*, 2005b]. It seems likely that the lower intensity ion abundances above m/z 10,000 represent the material excluded from the SEC column. Others have shown LD-MS spectra for asphaltenes with a peak of intensity

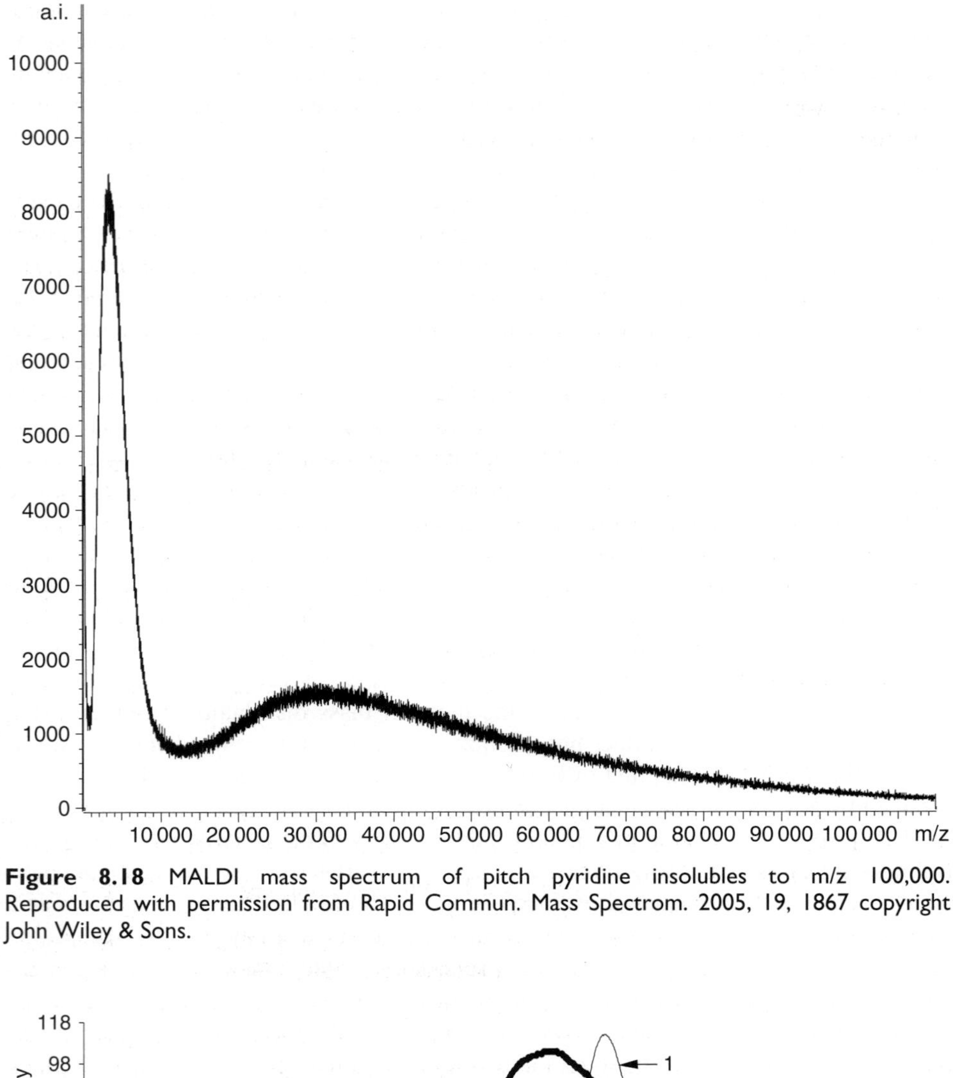

Figure 8.18 MALDI mass spectrum of pitch pyridine insolubles to m/z 100,000. Reproduced with permission from Rapid Commun. Mass Spectrom. 2005, 19, 1867 copyright John Wiley & Sons.

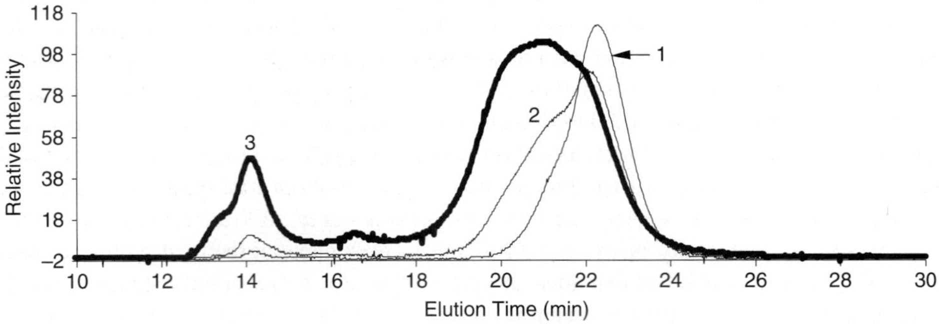

Figure 8.19 SEC chromatograms of pitch solubility fractions 1) acetone soluble, 2) pyridine soluble, 3) pyridine insoluble, Mixed-A column at 350 nm. From the data of Millan *et al.*, 2005a.

between m/z 1000 and 2000 and the upper limits of the spectra extending to m/z 10,000 [Tanaka *et al.*, 2004; Acevedo *et al.*, 2005]. These different works are in broad agreement in terms of both the peak of intensity and the upper mass limits. They also agree in that the laser power required to ionize and volatilize the larger components is well above that required to ionize the smaller mass components of the asphaltenes. Early work with fractions of an Athabasca bitumen showed that many of the fractions had a peak of maximum intensity below m/z 1000, lost in the matrix peaks, but that the fraction containing the largest molecular masses had a peak of maximum intensity around m/z 2,800 [Domin *et al.*, 1999]. In the whole bitumen sample and the two fractions with the largest molecular masses, the upper mass limits of the spectra exceeded m/z 20,000. The measured masses were in reasonable agreement with VPO and SEC data and the ^{252}Cf plasma spectra followed similar trends but gave lower masses.

Claims made in the past that would limit the masses of heavy petroleum and coal derived fractions to below 1,000 u appear unrealistic. There is mounting evidence that findings showing such low ceilings are limited by the capabilities of the techniques that have been used. Nevertheless, structures and upper mass limits of the high mass components of heavy fuel liquids remain to be investigated. It seems unlikely that they could be determined by mass spectrometry alone. The sheer complexity of the mass spectra coupled with the range of heteroatomic and structural isomeric types at any one nominal mass number insure that only some of the main structural features will be determined. Some information on the structures of these materials comes from techniques such as infrared spectra and X-ray diffraction and will be considered later in this Chapter.

8.6.9 Electrospray ionization

This method gives multiply charged ions for proteins but only singly charged ions for petroleum derived fractions. Petroleum-derived samples that have been studied include heavy crude oils by positive ion ESI [Qian *et al.*, 2001a, Hughey 2002b] and negative ion ESI [Hughey *et al.*, 2002a], acids in a heavy crude by negative ion ESI [Qian, 2001b], naphthenic acids by negative ion ESI [Barrow *et al.*, 2003), and Smackover oils [Hughey *et al.*, 2004]. These studies using high-resolution Fourier Transform Ion Cyclotron Resonance (FT-ICR) mass spectrometric techniques, produce a wealth of information on components of different atomic compositions but the same nominal molecular masses, information that is not available from other sources. A drawback is that the ion intensities are not quantitative so that relative abundances of the different heteroatomic types cannot be estimated. Furthermore, only low-intensity heteroatom containing components are ionized with this technique, while the major non-heteroatom-containing molecular types are not. Maximum masses detected were of the order of m/z 1200 and about 5,000 separate ion peaks were detected [Qian *et al.*, 2001]. The upper mass limit detected in this work is rather smaller by a factor 4 or more than the upper masses detected by the first peak in MALDI-MS of similar samples and well below indications from SEC that were discussed above.

Pyridine extracts of coals have been examined by high resolution FT-ICR-MS and impressive numbers of ion peaks were detected, as with petroleum samples above [Wu *et al.*, 2003, 2004]. However, the mass ranges did not extend to the same upper limits as observed by MALDI-MS or as indicated by SEC. The reasons are not yet evident. Negative ion electrospray ionization has been applied to a coal liquefaction residue and a hydro-processed product from it [Wu *et al.*, 2005]. The range of molecular types detected included oxygenates and nitrogen containing species. Fewer different types were found in the processed liquid. Although the sheer number of different molecules observed at the high resolution used is impressive, there remains the point that with increasing molecular mass, the probability of including several heteroatoms in each molecule becomes greater (the atomic compositions of the samples were not given) and therefore, the ion intensity should increase with increasing mass. The limit of the mass spectrum of the whole sample appeared to be about m/z 800 rather than the several thousand mass units that might be expected. The major polycyclic aromatics were not ionized by ESI and the authors make the point that the ion intensities are not quantitative.

Although not much work has been done on coal-derived liquids by electrospray ionization, the acetone-soluble fraction of a coal tar pitch has been examined. The sample contained mainly PACs as indicated by GC-MS, with minor concentrations of azaarenes such as acridine. The mass spectrum obtained by infusion of this acetone solution into the electrospray flow under conditions giving positive ionization gave a spectrum of masses 2 units larger than the main PACs [Herod *et al.*, 2005b]. To form these ions from PACs would require addition of $(2H)^+$ whereas addition of H^+ to the low-intensity azaarenes is the more likely ionization method.

This ionization route was confirmed using a mixture of PACs and azaarenes of different types. Further work using reversed phase LC-MS of acetonitrile-soluble fractions of coal tar pitch, a coal digest and a low temperature coal tar, have shown that the maximum mass detected by positive ion ESI was less than m/z 600 and included the ion m/z 414 which was tentatively identified as the protonated molecule methyl aza ovalene [Herod *et al.*, 2005b]. Given the nitrogen contents of these materials, it is expected that any molecule of mass greater than m/z 1200 would contain at least one N atom and should therefore ionize in electrospray. The absence of such large mass ions may indicate that the single-charged ions of high mass are not easily extracted into the mass spectrometer. Electrospray ionization giving negative ions would tend to reveal those components containing heteroatoms in acidic groups. Further work is necessary with standard materials and chromatographic separation but this method can only give ions for heteroatom-containing molecules. Addition of acid for positive ions and base for negative ions would extend the range of molecular masses capable of examination by electrospray-MS.

8.7 ICP-MS for Metallic Trace Elements

Many trace elements are known as components of coal and biomass and have been investigated by Mössbauer spectroscopy, ICP-MS, ICP-AES and atomic emission spectroscopy [Herod 1995; Herod *et al.*, 1995g, 1996d; Islas 2001; Kendrick *et al.*, 1992; Lachas *et al.*, 1999, 2000; Richaud *et al.*, 1998, 2000a,b,c, 2004].

Inductively coupled plasma (ICP) with mass spectrometry is a very sensitive technique for trace element analysis. The method, its strengths and weaknesses and applications have been described by several authors [Jarvis *et al.*, 1992; Vandecasteele & Block 1993; Holland & Tanner 1997; Beauchemin 2002]. A problem with ICP-MS is that the plasma is formed from nitric acid solutions dispersed as an aerosol into an argon gas stream and combinations of Argon atoms with other atoms in the solution or sample matrix can form polyatomic ions that might interfere with the measurement of the abundance of an ion specific to the trace element of interest. Another source of interference is spectroscopic such as overlap of isobars eg ^{58}Fe and ^{58}Ni although isobaric interference might be avoided by using high-resolution mass spectrometry. These sources of error have been described in detail [Jarvis *et al.*, 1992; Vandecasteele & Block 1993; Holland & Tanner 1997]. Where high-resolution mass spectrometry is not an option, other methods might be used to resolve the interferences, such as ICP-AES (ICP-atomic emission spectroscopy) for As and Se, or Mössbauer spectroscopy [Mössbauer, 1957] for ^{57}Fe. Whereas ICP-MS and ICP-AES give no information on the chemical states of binding to organic molecules unless combined with chromatography, Mössbauer spectroscopy can indicate chemical information.

Sample preparation for introduction into a plasma requires to prepare a solution in dilute nitric acid. This involves the digestion of the samples using acids to destroy organic or inorganic sample matrices and silicates. The method involves the use of concentrated acids in platinum crucibles to destroy silicates in minerals or the microwave digestion using nitric acid in bombs made of fluorinated plastics [Jarvis *et al.*, 1992]. These methods tend to lose volatile elements such as mercury and a separate method based on atomic emission spectroscopy has been used for quantifying mercury contents [Richaud *et al.*, 1998].

ICP-MS has been used to analyze coal-liquids to determine trace element loadings [Herod *et al.*, 2003a; Richaud *et al.*, 2000a,b] using small sample sizes, and to determine gallium porphyrins in coal extracts [Pretorius *et al.*, 1993]. The majority of trace elements in coal are associated with mineral matter and only a minority are in organic association [Richaud *et al.*, 2000b]. Solvent extracts of coal and of fractions of liquids prepared in different ways have been analyzed. Fractionation on silica lost most of the trace elements with the residue of sample (~15%) left on the column, but solvent separation without filtration media gave a reasonable mass balance [Herod *et al.*, 2003a]. In organic fractions, the trace elements are associated with the largest size molecules as indicated by SEC and insolubility in polar solvents. A coal tar pitch was separated into three fractions of approximately equal weight (soluble in acetone, soluble in pyridine and insoluble in pyridine) that contained approximately 10, 20 and 70% of the trace metal elements detected, respectively.

Trace elements can be found in fractions extracted from coal and biomass using NMP [Richaud *et al.*, 2000a] or tetralin [Herod *et al.*, 1995g, Richaud *et al.*, 2000b] and in a coal-derived liquid [Herod *et al.*, 1995g], in forms associated with the largest molecular size fractions. Fractionation of a coal digest by planar chromatography [Herod *et al.*, 1996d] with analysis by Mössbauer spectroscopy showed similar spectra for iron in each fraction despite the evidence from SEC showing the expected shifts of molecular size with solvent polarity. This indicates that the iron could form organometallic bonds to molecules of different sizes in the different fractions in the same manner. The closest identification of the iron-containing compound type involved in a coal extract was to ferritin, an iron storage protein [Richaud *et al.*, 2000b]. Liquefaction extracts from a rank series of coals in tetralin [Richaud *et al.*, 2000b] gave trace element distributions which were not related to the oxygen contents of the original coals. This tends to suggest that the carboxyl group concentration, expected to be greatest in low rank coals and possibly the site of attachment for trace element cations, was not of particular importance in trace element associations of these extracts.

A set of trace elements was measured in three coal-derived samples and their fractions from column chromatography using acetonitrile, pyridine and NMP [Islas 2001, Herod *et al.*, 2003a] and concentrations (ppm or mg/kg of sample) determined. Mass balances for each element were calculated using the fraction weights from column chromatography. In some cases, particular elements not detected in the original sample, notably aluminum and titanium were detected at low concentrations in one or more of the fractions. Where the element was found in the original sample at measurable concentration levels and not detected in any fractions (see below), it was concluded that sample material carrying the element had not been removed from the silica by the sequence of solvent washing. It also seems probable that the excessive quantities of sodium found in the fractions came from surface contamination or from contamination during the processing of solutions, although every attempt was made to avoid contamination.

In the three samples (pitch, PoA coal digest and low temperature tar), the oxygen content (by difference) decreased with increasing intensity of the thermal treatment and could be ranked as follows: LTT > PoA > pitch. It might also be expected that the concentration of carboxyl groups in the samples would decrease with increasing thermal treatment. Thus if oxygen contents correlated with trace element content, their concentrations would be expected to be greatest in the low temperature tar. The data showed that this was not the case but that the bulk of the trace elements were concentrated in the least soluble fractions and it was inferred that the trace elements were associated with the largest aromatic ring systems. In subsequent work, an approximate mass balance was achieved in fractions from the solvent separation of coal tar pitch [Herod *et al.*, 2003a], indicating quite definitely that the molecules containing the largest aromatic systems were associated with the highest concentration of trace elements, with approximately 30% of the sample containing 70% of trace elements.

8.8 Summary

The evidence presented in this Chapter indicates that there are molecules far larger than 1,000 u in coal and petroleum liquids and liquids produced from biomass materials by pyrolysis or simple solvent extraction. The largest molecules found to be excluded from the porosity of SEC columns probably have molecular shapes that are probably three dimensional and different from the shapes of smaller molecules. If this explanation holds, these materials would appear have different relations between molecular mass and molecular shape or elution time in SEC. Fractionation using solvents of different polarities produces materials with different size ranges as detected by SEC, with the smallest size molecules in the least polar solvents.

The excluded material of coal and petroleum samples is not likely to be polyelectrolytic or of high oxygen content, but the behavior of polyelectrolytes has not been examined apart from the protonated humic substances.

More recent MALDI and LD mass spectrometric data have shown agreement with mass ranges indicated by SEC by use of the polystyrene calibration curve. What makes the agreement powerful is that the two methods are entirely independent. The present view must be that we are seeing broadly realistic molecular mass distributions with MALDI and LD mass spectrometry, with the two peaks observed in either method showing broad agreement. It is thought that the higher masses indicated by signal under the excluded peaks are probably part amplified by three-dimensional conformations adopted by molecules larger than those indicated under the resolved peak in SEC.

Finally, the use of MALDI-MS, discussed above, has shown that for very polar molecules such as proteins and peptides and other biological molecules, the major ion formed is that of the monomer and if any dimer ion is detected, it tends to be at relatively low intensity only. There is no evidence to suggest that the complex materials of coal liquids are more likely to form cluster ions than proteins. Furthermore, while it is known that using laser power in excess of that required for ionization can produce multimer ions from a pure compound, as shown by Tanaka *et al.* [2004], there is a requirement that the laser power to ionize the largest molecules of a complex mixture should be in excess of that for the smallest molecules of the mixture. In such cases, the small molecules can act as matrix for the largest molecules even in the presence of a MALDI-matrix, since the samples themselves are black and aromatic and absorb at the laser wavelength, normally at 337 nm.

8.9 Structural Features of Large Molecular Mass Material Identified by SEC and MALDI-MS

So far in this Chapter, we have attempted to evaluate the molecular mass distributions of fossil fuel and biomass derived hydrocarbon liquids. Experimental methods described in this section aim to provide structural information on some of these large molecular mass species.

The exception is VPO, which measures an average mass for a sample. Results from this method have often been assumed to represent molecular masses of dimers or aggregates of molecules rather than an average mass representative of unassociated molecules. Examples will be presented where agreement between VPO and other measurements suggest that some VPO results are not caused by aggregation.

8.9.1 UV-fluorescence spectroscopy (UV-F)

The method has been reviewed [Lakowicz, 1986; Rendall, 1987; Oldham *et al.*, 2000] and examples of its use given [Wolfbeis, 1993]. Fluorescence spectra of complex aromatic mixtures have been discussed by Apicella *et al.* [2004]. UV-fluorescence spectroscopy (UV-F) cannot be used to identify particular structural features or precise sizes of polycyclic aromatics within complex mixtures. This is because of the multiplicity of possible types of structural changes that might contribute to shifts in the position and shape of the spectra. This is a common feature of results from analytical techniques when applied to coal-derived materials and other complex mixtures of hydrocarbons, where changes in a large number of structural features might contribute simultaneously to changes in the spectrum.

Fluorescence spectra may be generated in three ways from static solutions

(i) Emission spectra in which the excitation wavelength is varied while monitoring the emission at a certain wavelength.
(ii) Excitation spectra where the excitation wavelength is fixed while monitoring the emission over a range of wavelengths.
(iii) Synchronous spectra where the excitation and emission wavelengths are changed simultaneously with a fixed difference in wavelength (e.g. 20 nm) or a fixed difference in energy.

The synchronous spectrum is of more use for complex mixtures such as samples of coal, petroleum and biomass derived liquids. A single response is expected for each type of aromatic system, corresponding to excitation by the shortest wavelength for that system and emission by the longest wavelength for the system. The 20 nm difference corresponds approximately to the Stokes shift in wavelength between the lowest energy level of the excited state in excitation and the lowest energy level of the excited state after relaxation, but before emission. Although this form of operation simplifies the spectra, they remain very complex for pyrolysis and liquefaction tars because of the wide range of structures in the samples.

A simple classification of structural changes likely to lead to shifts of UV-F spectra of coal-derived materials to shorter wavelengths include: (i) loss of alkyl and alkoxy substituents, (ii) loss of aromatic substituent groups containing heteroatoms, (iii) a general reduction in size of the conjugated aromatic system, (iv) the loss of formally fixed double bonds e.g. in zethrenes (for instance the central double bond in 1,2 diphenyl ethene) and finally (v) the loss of structures with five-membered ring systems. Furthermore, the fluorescence spectra of large complex molecules is much weaker than that of small aromatic systems because of the greater ability of such molecules to lose the energy gained by absorbance of photons, that would normally be given out as fluorescence, by other pathways through non-fluorescence methods and degradation to vibrational energy and heat. The effect has been described in detail by Li *et al.* [1994c] and gives rise to low quantum yields from molecules in which one or more large PAC groups are embedded.

The difference in the fluorescence characteristics between ordinary aryl-alkyl compounds and molecules with large embedded PAC groups has give rise to a certain amount of confusion. Groenzin and Mullins [2000] have reported not observing material with molecular masses above about 1,000 u in petroleum asphaltenes. On the other hand, the tandem use of UV-fluorescence and UV-absorbance detectors in SEC has repeatedly shown that UV-fluorescence is unable to give any signal for material above about 3,000 u and is quite insensitive to material above 1,000 u. These findings were first reported by Li *et al.* in 1995. Since then, several other laboratories have independently reported similar findings. It has been shown that not all the material of coal-liquids and petroleum asphaltenes can fluoresce even though they strongly absorb UV light [Herod *et al.*, 1995c; Morgan *et al.*, 2005a; Ascanius *et al.* 2004; Strausz *et al.* 2002].

More generally, UV-fluorescence spectroscopy has been used in two distinct ways. One method uses a static solution at high dilution to avoid self-adsorption effects while the second uses a flow cell with solution flowing from the SEC column in real time with (widely variable) excitation and emission wavelengths set for the duration of the SEC run [Herod *et al.*, 1996b]. In static mode, the results show that the synchronous spectra of complex mixtures most closely resemble the spectrum of the smallest molecular mass fraction, most abundant and least polar sample fraction (in column chromatography) or in the most mobile material (in thin layer chromatography). Synchronous spectra of heavier fractions such as pyridine-solubles and pyridine-insolubles tend to shift to longer wavelengths and show much weaker fluorescence intensity than the smaller, less polar molecules.

Examples of synchronous fluorescence of fractions shown in Figure 8.20 have been height normalized since otherwise the spectra for the heavier fractions would be indistinguishable from the baseline of the spectrum of the lightest fraction because of their low quantum yields. Another example showing the effect of hydrocracking on the synchronous UV-fluorescence spectrum of the pitch pyridine insolubles has been shown in Figure 8.10b.

When used in the flow cell mode, the object has been to compare the signal from UV-absorbance with the fluorescence signal. For the lightest fractions, UV-A and UV-F chromatograms are similar but chromatograms of heavier fractions show that the excluded material and some of the earliest parts of the retained peaks of SEC give no fluorescence because of the low quantum yield for fluorescence. This lack of fluorescence is a sign that the molecular structures are very complex and the vibrational excitation induced by absorbing photons can be lost by methods other than fluorescence through forbidden transitions. An example of UV-F and UV-A detection in SEC is shown in Figure 8.12; other examples have been shown previously [Herod *et al.*, 1996b; Morgan *et al.*, 2005a] and similar effects have been observed in petroleum asphaltenes [Morgan *et al.*, 2005a, Millan *et al.*, 2005b]. Coal-derived molecules of mass greater than 3,000 and petroleum molecules of mass greater than 4,000 do not fluoresce significantly, as measured by the polystyrene calibration of SEC.

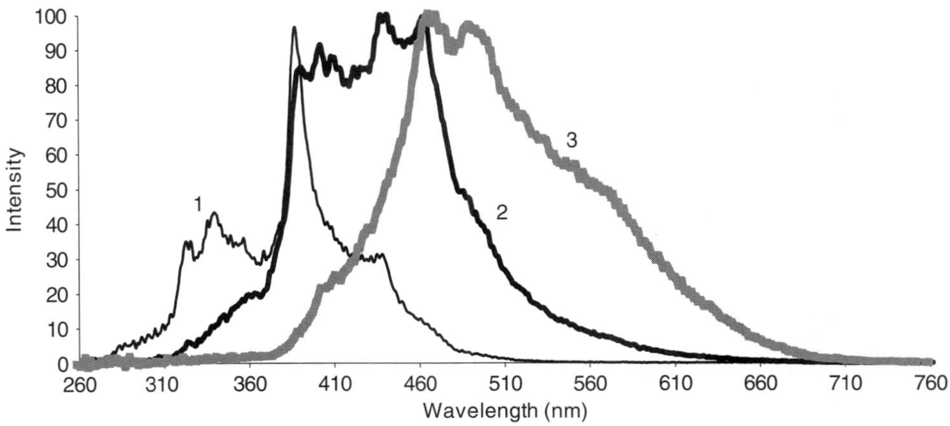

Figure 8.20 Synchronous UV-fluorescence spectra of 1) acetone solubles, 2) pyridine solubles and 3) pyridine insolubles of pitch; height normalized. Reproduced with permission from Rapid Commun. Mass Spectrom. 2005, 19, 1867; copyright John Wiley & Sons.

It is also useful to consider briefly how the fluorescence of polymers changes with mass. For a polymer such as polystyrene, where the benzene rings are not conjugated, the UV fluorescence of any polymer will remain that of the benzene group and there is no change of fluorescence with mass. For a polymer in which the conjugated system increases with added monomers, then the fluorescence will shift to longer wavelengths as the molecular mass increases. An example of this latter type is the polycyclic aromatics themselves, where the addition of successive fused benzo groups running for example, from naphthalene up to chrysene, shifts the fluorescence from the UV into the visible region. Müller and Müllen [2005] have described ways in which to tailor the shape and size of the aromatic core of perylene by enlargement of the aromatic system to produce both bathochromic and hypsochromic shifts in absorbance. For carotenes with increasingly conjugated polyene systems, up to carotene itself ($C_{40}H_{56}$), colors change from white to red and yellow, with (presumably) fluorescence shifting in a similar manner. Apicella *et al.* [2004] have drawn attention to heavy atom quenching of synchronous fluorescence spectra where chlorine in solvents could promote the loss of fluorescence in soluble material from soots.

8.9.2 Infra red spectroscopy

Reviews of the instrumentation and applications have been given in Analytical Chemistry [McKelvy *et al.*, 1996, 1998; Gillie *et al.*, 2000] and tables of IR frequencies are available [Socrates, 2001].

Spectra have been obtained for series of fractions from column chromatography (pitch, a coal digest, a low temperature tar) and fractions from solvent solubility (pitch) using both the KBr disk method and attenuated total reflection (ATR) using a diamond cell. The results from both methods agree in showing similarities between spectra of the whole samples and the smallest-molecule fractions (acetonitrile- or acetone-solubles). The main change in the spectra was evident in the NMP-solubles from column chromatography and pyridine-insolubles, where the absence of out-of-plane bending vibrations of substituted benzenoids at frequencies from 900–700 cm^{-1} was obvious. The loss of these frequencies points to the complexity of the structures of the largest molecules such that either there are very few of such sites left in the molecules or they are sterically hindered and prevented from vibrating. Both options argue for large, flat aromatic plates and probably three-dimensional structures in the largest (excluded in SEC) molecules. Large aromatic systems resembling graphitic structures have no infrared spectra because there is no dipole change in the available vibrational modes. Figure 8.21 shows the ATR spectra of the coal digest and fractions from column chromatography [Islas, 2001].

8.9.3 NMR methods

NMR methods for solid and liquid phase studies of coal, coal liquids and polymers have been described [Botto & Sanada 1993; Bovey & Mirau, 1996] for ^{13}C and ^{1}H. Periodic reviews of the techniques and applications of solid state NMR have been given in Analytical Chemistry [Dybowski & Bruch, 1996; Dybowski, 1998; Dybowski & Bai, 2000; Dybowski *et al.*, 2002].

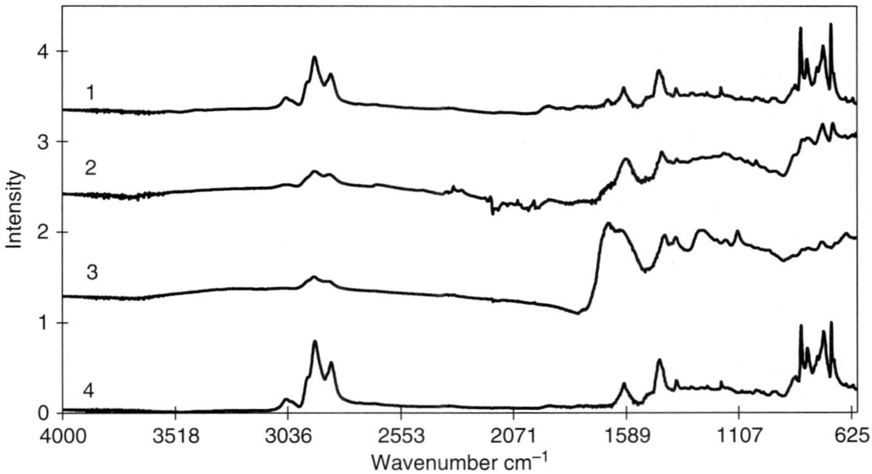

Figure 8.21 FT-ir spectra of coal digest and fractions, ATR cell; 1-acetonitrile solubles, 2-pyridine solubles, 3-NMP solubles, 4-coal digest. From Islas, 2001.

Solid state ^{13}C work on coals and coal tar pitch has been involved in part in attempting to make the technique quantitative in terms of detecting carbon atoms. A debate on the method [Snape *et al.* 1989a] came to the conclusion that the single pulse excitation method was likely to give the closest approach to quantitative analysis. A long relaxation time, in excess of 1 min, was considered necessary, making the acquisition of an adequate number of scans a time-consuming exercise. Following this report and the reviews above, our work has used the single pulse excitation method with a delay between scans of 120 sec. The relation between carbon aromaticities and hydrogen/carbon ratios in a series of coals has been observed as a linear function using the single pulse excitation method [Maroto-Valer *et al.*, 1998]. Aromaticity is defined here as the percentage of total carbon that is aromatic, that is the relative integrated areas of the spectrum as $100 \times (C_{aromatic}/C_{aromatic} + C_{aliphatic})$.

The work described above to define the molecular mass range of coal liquids has used the coal tar pitch sample as a standard to investigate the different analytical methods we have developed. We used the coal tar pitch as well, in investigations of NMR (solution and solid state) and it is appropriate that we consider next what others have found with similar materials. Snape *et al.* [1989b] recognized that most coal tar pitches contain a significant proportion of chloroform-insolubles that could not be analyzed by solution NMR techniques. Solid state methods were applied instead to the insoluble fraction. The aim of the investigation was to compare the concentrations of bridgehead carbons in the whole pitch and the toluene-insoluble fraction (comprising 44% of the whole). Using a combination of solution and solid state methods, the concentration of bridgehead carbons was found to increase from the toluene-solubles, to the whole pitch, to the toluene-insolubles (33:42:46% of total carbon respectively), as expected if the aromatic ring systems were becoming larger with increasing molecular mass. Bridgehead carbons are defined as those at the interface between adjoining aromatic groups; they carry no hydrogen atoms and only join other aromatic carbons in conjugated aromatic systems. The molecular weight (Mn) of the pyridine-soluble part of the toluene insolubles (20% by weight, equivalent to 8.8% of the pitch sample) determined by VPO was 880 u, suggesting that the pyridine-insoluble part of the fraction had molecular weights considerably in excess of this value. This was earlier than the work with SEC and NMP solvent described above (section 8.2.2.d, Figure 8.5) where the molecular mass up to 3000 u, as indicated by both polystyrene standards and MALDI-MS, was related to elution times.

A more recent paper [Andrésen *et al.* 1998] has examined toluene-insoluble fractions of coal tar pitch samples before and after heat treatment at temperatures much lower than the temperature of formation of the coal tar (>1000°C in a coke oven); the pitch was distilled from the coal tar at 350°C and heat treated at temperatures from 250 to 380°C. The determination of bridgehead carbons revealed that the number of these carbons in the initial pitch remained constant with the heat treatment, even though the proportion of toluene-insoluble material increased with treatment. This behavior was considered to reflect the polymerization of pitch components through single-bond links such as in diphenyl. Molecular masses were considered to be very low, with number averages less than 500 u. Subsequent work on heat treatment of pitches [Menéndez *et al.*, 2001, 2002] showed that in SEC using NMP as eluent, molecular mass increased significantly with increasing temperature of treatment or with air-blowing. Air-blowing polymerization did not increase the oxygen content of the sample, indicating that the mode of reaction was by formation of water and the joining of aromatic systems through covalent C-C bonds as in diphenyl.

Initial work with solid state ^{13}C NMR appeared to indicate that the largest molecules of the coal-derived materials (pyridine-solubles and -insolubles from thin layer chromatography [Herod *et al.*, 1999] and NMP-solubles from column chromatography [Islas *et al.*, 2002b] contained increasing quantities of carbonyl and aliphatic carbon with increasing molecular size. The molecules appeared to resemble humic acids in terms of their spectra. However, this was caused by a polymerization or reaction of NMP with the coal-liquid fraction. Subsequent work [Millan *et al.*, 2005c] showed that if NMP was not used in the fractionation, then no significant carbonyl signal was observed and the heavier, larger molecular fractions showed slightly increased aromaticity. A separate study of solution and solid state ^{13}C NMR of a coal digest and hydrocracked products showed that the common solvents used in NMR were insufficient to allow detection of the largest molecules as defined by the quaternary carbon content [Begon *et al.*, 2003].

The situation was improved using NMP as solvent but the aliphatic carbon signal of the solvent obscured the aliphatic carbon region of the samples. An additional problem arose when attempting to record the ^{13}C NMR spectra of pyridine-soluble and pyridine-insoluble fractions of pitch. These fractions are insoluble in the solvents often used for solution state NMR, such as chloroform as noted by Snape *et al.* [1989b], but are soluble in NMP. The normal parameters used for solution state NMR were applied to these fractions and no significant aromatic signal was detected. It appears that the instrumental parameters must be tuned to suit the aromatic carbon environments of the larger molecular mass fractions if adequate spectra are to be obtained, again indicating that these materials may have unusual structures. Another problem is that the heavy fractions of pitch (acetone-insoluble but pyridine-soluble and pyridine-insoluble) are insufficiently soluble in NMP for the normal procedures to be applied. Solution concentrations of 20–30% by weight are often cited for NMR measurements in solution, to give an adequate signal in a short analysis time of several minutes. Our experiments suggested that at concentrations much above a few % by weight, the sample was not in solution and no signal could be obtained in the normal few minutes of scanning for such a diluted solution. Solutions of these materials in NMP at these low concentrations appeared to be very black, hiding the material that was potentially soluble but remained insoluble because of an insufficient volume of solvent. To summarize the results, the acetone-soluble fraction gave a spectrum in solution that was very similar to that of the whole sample; the low-mass molecules would be soluble in most of the solvents commonly used in NMR, including NMP. The pyridine-soluble fraction in solution in NMP gave a weak spectrum in which the aromatic signal was detectable but without significant detail, while no aromatic signal was detected in the spectrum of the pyridine-insoluble fraction in solution in NMP. In the work described above on SEC and UV-fluorescence, the solution strengths involved were much less than a few percent by weight. A suitable test of the NMR method should be the determination of the proportion of carbon existing as bridgehead carbon, since this measure should increase with increasing molecular mass and complexity, as with the cases cited above. The future development of solution state ^{13}C NMR will include the extension of scanning time to take account of the low concentrations of the large-mass fractions of the samples.

Figure 8.22 shows solid state ^{13}C CPMAS TOSS spectra of the solvent separated pitch fractions and the whole pitch. The shape of the aromatic peak does not change between fractions while the aliphatic peak is most prominent in the small-molecule fraction (acetone soluble- about 4% aliphatic) and least prominent in the largest-molecule fraction (pyridine insoluble- about 2% aliphatic).

An alternative approach to the investigation of structure in coal tar pitch has involved solution in carbon disulfide with the identification of carbon atoms in aromatic rings attached to two or three others in condensed ring systems [Diaz & Blanco 2003]. This approach enables the calculation of bridgehead carbon atoms as a proportion of the total carbon for the sample in CS_2. However, it seems likely that not all of the coal tar pitch would be soluble in this solvent.

From the above discussion, it becomes apparent that the new approach to molecular weight determination of coal- and petroleum-derived liquids will have an impact on the interpretation of NMR results. The solution state work on the relatively insoluble fractions of coal liquids has not been done satisfactorily because of the limitations of the solvents able to dissolve the fractions. The work to date has defined the mode of operation in solid state scans – single pulse excitation with long relaxation times between scans, as well as the parameter to measure – the bridgehead carbon content. In addition, the atomic compositions of the heavy fractions are needed to allow for oxygen and nitrogen functions. While this measurement is routine on the micro scale for volatile molecules, we have found some difficulty in obtaining reliable measurements for the pyridine insolubles of pitch, for instance. The problem arises from the mode of combustion in chromatographic methods for elemental analysis. A plug of oxygen arrives at the furnace just after injection of the sample in a tin capsule and the sample is intended to combust completely

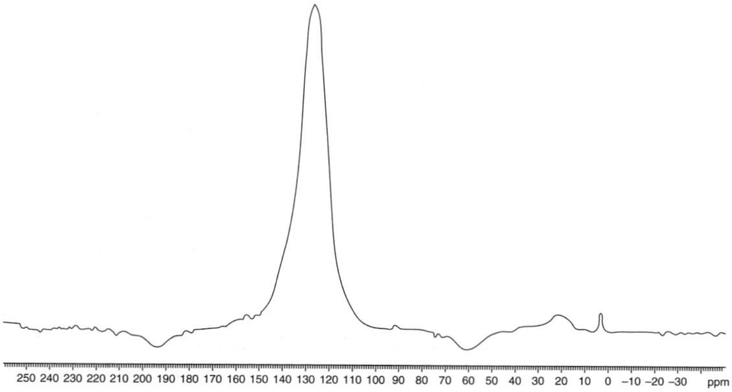

(a) Pitch

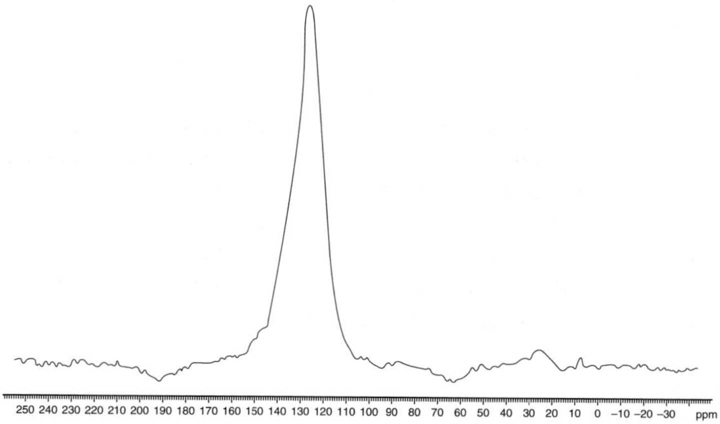

(b) Pyridine Insoluble

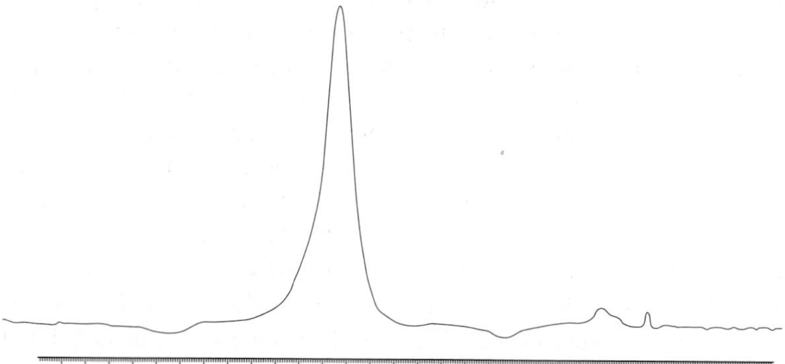

(c) Pyridine Soluble

Figure 8.22 Solid state ^{13}C NMR spectra of pitch and fractions; the peak at 3 ppm shift is the internal standard, tetrakis. The spectrometer was a Bruker MSL300 at 75.5 MHz (7.05 T); MAS frequencies were in the range 10.5 to 11.2 kHz with Bloch-decay single pulse excitation. The acetone soluble fraction was mixed with silica to provide a dry solid.

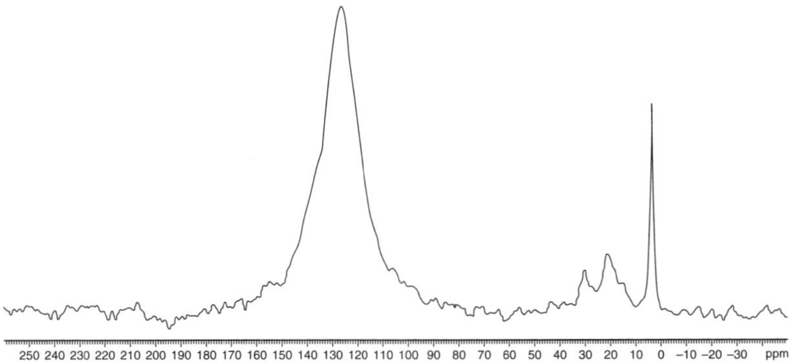

Figure 8.22 *Continued.*

before the oxygen plug is swept through the combustion zone. In our experience, the pyridine-insolubles of pitch do not combust completely under these conditions, probably because the pitch material forms a char rather than volatilizing in the furnace. Therefore the oxygen plug is not seen by most of the sample which fails to burn completely, giving low carbon contents even though there is no significant inorganic content to the sample. A modification of the experimental method is required and our future work will attempt to resolve these problems associated with solution state NMR.

8.9.4 VPO

Vapor pressure osmometry is a method based on the colligative property of the elevation of boiling point by reduction of the vapor pressure of a solvent by addition of an involatile solute [Atkins 1986]. The instrumentation normally used compares static drops of pure solvent and solution at a controlled temperature and in a vapor saturated with solvent. Under these conditions, the solute depresses the vapor pressure of the solution drop with the effect that solvent evaporates from the pure solvent drop and condenses onto the solution drop. The instrumentation detects the difference in temperature of the drops. Because colligative properties depend on the number of solute particles or the mole fraction of solute in solution, the VPO method is best suited to the measurement of the molar masses of relatively low mass materials since it becomes impossible to dissolve a high molar fraction of a polymer in a solvent. For a mixture, the determination gives a number average molar mass. Numerous studies have used the method to produce average masses of coal liquids and petroleum residues [Acevedo *et al.*, 1997; Domin *et al.*, 1999; Torregrosa-Rodriguez *et al.*, 2000; Ancheyta *et al.*, 2002, 2003] giving values up to several thousand molecular mass units for resins, asphaltenes. Because the solute must be present at relatively high concentration, the possibility that aggregation of solute in solution takes place cannot easily be avoided [Strausz *et al.*, 2002]. The VPO instrumental constant K has been shown to depend on the molecular weight of the solute, indicating an effect of solute adsorption [Wang & Cheng, 1998]. Results from VPO may indicate high number average molecular masses or the presence of aggregates of molecules and only comparison and agreement with determinations of molecular mass by other methods can reveal the truth (cf. Table 8.2 above).

8.10 Summary of Structural Features of the Largest Molecules

The work on SEC and different mass spectrometric methods has shown that the largest molecules from coal liquids, petroleum asphaltenes, and biomass tars are predominantly aromatic and apparently suffer a change of structure and/or conformation as molecular mass increases beyond masses of about 6,000 u. The structural change is thought to be one tending to three-dimensional conformations or folding in solution such that the relation between molecular size and mass observed for small molecules (<6,000 u) no longer applies to larger molecules. In support of this, the IR spectra of larger molecular mass fractions fail to show the benzenoid substitution patterns from 600–900 cm^{-1}. The solution state ^{13}C NMR spectra of the large aromatic systems have proved difficult to obtain without retuning the instrumental parameters and allowing for solutions of low concentration. However, this finding indicates that the largest molecules are structurally different from the smaller molecules.

The data presented have indicated that large molecules of coal liquids and petroleum asphaltenes do not fluoresce significantly when the mass rises above 3,000 u for coal liquids or 4,000 u for asphaltenes. The trace elements, many of them metals, are concentrated in the large molecular mass fractions of coal liquids and do not appear to be held there by carboxyl groups, since the concentrations in liquefaction extracts from a rank series of coals did not correlate with the oxygen content of the original coals. The molecular structures of petroleum asphaltenes and coal liquids are different, with the asphaltene molecules that elute within the porosity of the SEC columns being of larger size than those from coal liquids. However, the proportions of excluded molecules tend to be larger for coal derived liquids than for asphaltenes and their molecular masses as determined by MALDI-MS tend to extend to higher values than for asphaltenes. Because the solvent used for dissolving the largest molecules, NMP, can easily dissolve the coal-derived materials but does not completely dissolve the petroleum asphaltenes, the structures of the two kinds of materials must differ in some important manner. This may be the extent of alkylation or the degree of condensation of the aromatic systems.

The variety of analytical methods described in Chapters 7 and 8 all have their own drawbacks that we have attempted to highlight. It is important to understand these drawbacks in assessing the experimental results. Equally, it is desirable that the analytical methods are extended and developed to understand the large molecular materials that evidently exist in the coal-derived liquids, petroleum residues and biomass tars that have been described in this work. Part of this work is designed around the need to isolate the large molecular mass fractions of these samples, to improve the chances of understanding the changes in structure and size indicated by the SEC and MALDI-MS results. In particular, it is necessary to avoid fractionation methods liable to lose the largest molecules to the fractionation medium. The higher mass fractions have been observed to carry the major part of the trace metal element burden of these samples.

The various pyrolytic and liquefaction treatments described in Chapters 3 to 6 produce different proportions of the large-molecular mass material, excluded in SEC and presenting problems for further processing into useful products such as transport fuels. Although these large molecules can be hydrocracked to smaller molecules, as shown for the pitch pyridine insoluble fraction, the catalysts used for this purpose are those regularly used in the petroleum cracking processes. More effective catalysts, tailored to cracking these materials are clearly needed. We have already observed that the materials excluded from column porosity in SEC are not aggregates of small molecules, not least because they cannot be disaggregated by any of the methods attempted by us, or by others. In petroleum-derived asphaltenes, these large molecular mass materials present problems during further processing. The heavier crudes becoming available as lighter crudes are depleted tend to have higher levels of asphaltenes than previously. The large molecules might act as centers for precipitation when in liquid mixtures unable to dissolve them completely. In both coal liquids and petroleum asphaltenes, the largest molecules tend to form char rather than crack to small molecules, if treated by ordinary thermal methods.

The small molecules characterized by the methods of Chapter 7 are known in great detail, despite there being such an abundance of structural isomers and isosteres within that lower mass range. In particular, a wide variety of biological fossil markers (biomarkers) is known and their range continues to expand as science focuses on components of lower abundance in tars. It is not likely that such an intimate understanding of molecular structures will ever be available for the large mass molecules and the most that can be hoped, is that some general principles of structural change with increasing mass might emerge. The sheer number of molecular components and their possible isomers and isosteres that increases with increasing molecular mass, in the high mass material of tars (cf. Figure 8.18 showing continuous mass range to m/z 100,000) suggests that an intimate understanding of their structures is not within our immediate grasp.

References

Avecedo, S., Escobar, G., Renaudo, M.A. and Gutierrez, L.B. (1994) Fuel 73, 1807.

Acevedo, S., Escobar, G., Ranaudo, M.A., Pinate, J., Amorin, A., Diaz, M. and Silva, P. (1997) Energy & Fuels 11, 774.

Acevedo, S., Gutierrez, L.B., Negrin, G., Pereira, J.C., Mendez, B., Delolme, F., Dessalces, G. and Broseta, D. (2005) Energy & Fuels, 19, 1548–1560.

Adegoroye, A., Paterson, N., Morgan, T.J., Herod, A.A., Dugwell, D.R. and Kandiyoti, R. (2004) Fuel 83: 1949–1960.

Ancheyta, J., Centeno, G., Trejo, F., Marroquin, G., Garcia, J.A., Tenorio, E. and Torres, A. (2002) Energy & Fuels, 16, 1121–1127.

Ancheyta, J., Centeno, G., Trejo, F. and Marroquin, G. (2003) Energy & Fuels 17 1233.

Andrésen, J.M., Luengo, C.A., Moinelo, S.R., Garcia, R. and Snape, C.E. (1998) Energy & Fuels 12, 524–530.

Apicella, B., Ciajolo, A., Suelves, I., Morgan, T.J., Herod, A.A. and Kandiyoti, R. (2002) *Comb. Sci. Technol., 174* (11–12), 345–359.

Apicella, B., Ciajolo. A., Barbella, R., Tregrossi, A., Morgan, T.J., Herod, A.A. and Kandiyoti, R. (2003a) Energy & Fuels 17 565–570.

Apicella, B., Barbella, R., Ciajolo, A. and Tregrossi, A. (2003b) Chemosphere 51, 1063–1069.

Apicella, B, Ciajolo, A., and Tregrossi, A. (2004) Anal. Chem. 76, 2138–2143.

Ascanius, B.E., Merino-Garcia, D. and Andersen, S.I. (2004) *Energy & Fuels* 18 1827–1831.

Atkins, P.W. (1986) Physical Chemistry 3rd Ed., Oxford Univ. Press, UK.

Aust, N. (2003) J. Biochem. Biophys,. Methods, 56, 323–334.

Avid, B., Purevsen, B., Paterson, N., Zhuo, Y., Peralta, D., Herod, A.A., Dugwell, DR. and Kandiyoti, R. (2004) Fuel 83 1105–1111.

Badre, S., Goncalves, C.C., Norinaga, K., Gustavson, G. and Mullins, O.C. (2006) Fuel 85, 1–11.

Barrow, M.P., McDonnell, L.A., Feng, X., Walker, J., Derrick, P.J., (2003) *Anal. Chem*. 75, 860.

Barth, H.G., Boyes, B.E. and Jackson, C. (1996) Anal Chem 68 445R.

Barth, H.G., Boyes, B.E. and Jackson, C. (1998) Anal Chem 70 251R.

Bartle, K.D. (1988). Theory and principles of supercritical fluid chromatography. Chapter 1 in *Supercritical Fluid Chromatography*; R.M. Smith, ed., RSC Chromatography Monograms, Royal Soc. Chem. Cambridge.

Bartle, K.D., Taylor, N., Mulligan, M.J., Mills, D.G. and Gibson, C. (1983) Fuel 62 1181–1185.

Bartle, K.D., Mulligan, M.J., Taylor, N., Martin, T.G. and Snape, C.E., (1984) Fuel 63 1556–1560.

Bartle, K.D., Mills, D.G., Mulligan, M.J., Amaechina, I.O., and Taylor, N. (1986) Anal. Chem., 58 2403–2408.

Beauchemin, D. (2002) Anal. Chem. 74 2873.

Begon, V., Islas, C.A., Lazaro, M.-J., Suelves, I., Herod, A.A., Dugwell, D.R. and Kandiyoti, R. (2000) Eur. J. Mass Spectrom., 6 39–48.

Begon, V., Suelves, I., Li, W., Lazaro, M-J., Herod, A.A. and Kandiyoti, R. (2002) Fuel 81 185–202.

Begon, V., Suelves, I., Islas, C.A., Millan, M., Dubau, C., Lazaro, M-J., Law, R.V., Herod, A.A., Dugwell, D.R. and Kandiyoti, R. (2003) Energy & Fuels; 17: 1616–1629.

Bermejo, J., Menendez, R., Fernandez, A.L., Granda, M., Suelves, I., Herod, A.A. and Kandiyoti, R. (2001a) Fuel, 80 2155–2162.

Bermejo, J., Fernandez, A.L., Granda, M., Suelves, I., Herod, A.A., Kandiyoti R. and Menendez, R. (2001b) J. Chromatography A 919 255–266.

Blevins, L.G., Yang, N.Y.C., Mulholland, G.W., Davis, R.W. and Steel, E.B. (2002) Amer. Chem. Soc. Fuel Chem. Div. Preprints 47 no 2, 740–741.

Bodman, S.D., McWhinnie, W.R., Begon, V., Suelves, I., Lazaro, M-J., Morgan, T.J., Herod, A.A. and Kandiyoti, R. (2002) Fuel 81 449–459.

Boenigk, W., Haenel, M.W. and Zander, M. (1990) Fuel, 69, 1226–1232.

Bonnett, R., Burke, P.J., Dewey, C.R., Fairbrother, A.E., Major H.J. (1991) *Fuel, 70*, 1227–1229.

Borra, C., Wiesler, D. and Novotny, M. (1987) *Anal. Chem., 59*, 339–343.

Botto, R.E. and Sanada, Y. (1993) Magnetic Resonance of Carbonaceous Solids. Adv. Chem Series 229, American Chem. Soc. Washington DC, USA.

Bovey, F.A. and Mirau, P.E. (1996) NMR of Polymers Academic Press, San Diego USA.

Burlingame, A.L., Boyd, R.K. and Gaskell, S.J. (1996) Anal Chem 68 599R.

Burroughs, J.A., Cadre, B.M. and Hanley, L. (1993) *Amer. Chem. Soc. Div. Fuel Chem., 38(1)*, 318–323.

Cagniant, D., Gruber, R., Lacordaire-Wilhelm, C. and Schulten, H-R. (1992) *Energy & Fuels, 6*, 694–701.

Chapman, J.R. (1993) Practical Organic Mass Spectrometry 2nd ed Wiley, London.

Chen, C. and Iino, M. (2001) Fuel 80 929–936.

Choi, K. and Mochida, I. (2002) Amer. Chem. Soc. Fuel Chem. Div. 47 no 2 776–777.

Collot, A-G., (1999) PhD Thesis, University of London.
Cunico, R.L., Sheu, E.Y. and Mullins, O.C. (2004) Petroleum Sci Technol 22 (7&8) 787–798.
Danis, P.O. and Karr, D.E. (1993) Organic Mass Spectrom., 28 923–925.
Davison, R.R., Glover, C.J., Burr, B.L. and Bullin, J.A. (1995) Chapter 8 Size exclusion chromatography of asphalts. In Handbook of Size Exclusion Chromatography, ed Chi-san Wu, Marcel Dekker Inc, NY, pages 211–247.
Deelchand, J-P., Naqvi, Z., Dubau, C., Shearman, J., Lazaro, M-J., Herod, A.A., Read, H. and Kandiyoti, R. (1999) J. Chromatogr A. 830, 397–414.
Domin, M., Moreea, R., Lazaro, M.-J., Herod, A.A. and Kandiyoti, R. (1997a) Rapid Commun. Mass Spectrom. 11 638–645.
Domin, M., Moreea, R., Lazaro, M-J., Herod, A.A. and Kandiyoti, R. (1997b) Rapid Commun. Mass Spectrom. 11, 1845–1852.
Domin, M., Li, S., Lazaro, M.J., Herod, A.A., Larsen, J.W. and Kandiyoti, R. (1998) Energy & Fuels 12 485–492.
Domin, M., Herod, A.A., Kandiyoti, R., Larsen, J.W., Lazaro, M-J., Li, S. and Rahimi, P. (1999) Energy & Fuels, 13, 552–557.
Douda, J., Llanos, Ma. E., Alvarez, R., Navarette Bolanos J. (2004) Energy & Fuels 18, 736–742.
Dybowski C (1998) Anal Chem 70 1R.
Dybowski, C. and Bruch, M.D. (1996) Anal. Chem. 68 161R.
Dybowski, C. and Bai, S. (2000) Anal. Chem. 72 1R.
Dybowski, C., Bai, S. and Bramer SV. (2002) Anal. Chem. 74 2713.
Gallacher, J., Snape, C.E., Dennison, P.R., Bales, J.R. and Holder, K.A. (1991) Fuel 70, 1266–1270.
Garozzo, D., Impallomeni, G., Spina, E., Sturiale, L. and Zanetti, F. (1995) Rapid Commun. Mass Spectrom., 9 937–941.
Gibbins, J.R., Kimber, G.M., Gaines, A.F., Kandiyoti, R. (1991) Fuel 70, 380–385.
Gillie JK, Hochlowski J, Arbuckle-Keil GA (2000) Anal Chem 72, 71R.
Greenwood, P.F. (1994) *Organic Mass Spectrom., 29*, 61–77.
Greenwood, P.F., Strachan, M.G., Willett, G.D. and Wilson, M.A., (1990) *Org. Mass Spectrom., 25*, 353–362.
Groenzin, H. and Mullins, O.C. (2000) Energy & Fuels 14 677–684.
Haberhauer, G., Bednar, W., Gerzabek, M.H. and Rosenberg, E. (1999) MALDI-TOF-MS analysis of humic substances – a new approach to obtain additional structural information? In Ghabbour EA and Davies G (eds) Understanding humic substances: advanced methods, properties and applications. Royal Society of Chemistry, London, special publication no.247, (1999) pages 121–128.
Haberhauer, G., Bednar, W., Gerzabek, M.H. and Rosenberg, E. (2000) Application of MALDI-TOF-MS to the characterization of fulvic acids. In Ghabbour EA and Davies G (eds) Humic Substances, versatile components of plants, soils and water. Royal Society of Chemistry, London, special publication no. 259, (2000) pages 143–152.
Hamilton, R and Hamilton, S. (1987) Thin layer chromatography, ACOL, John Wiley London UK.
Hamming, M.C. and Foster, N.G. *Interpretation of mass spectra of organic compounds*, Academic Press, NY, USA 1972.
Hedges, J.I., Eglinton, G., Hatcher, P., Kirchman, D.L., Arnosti, C., Derenne, S., Evershed, R.P., Kögel-Knaber, I., de Leeuw, J.W., Littke, R., Michaelis, W. and Rullkötter, J. (2000) *Organic Geochemistry, 31*, 945.
Herod, A.A. (1994) *J. Planar Chromatography, 7*, 180–196.
Herod, A.A. (1998) Azaarenes and thiaarenes. Chapter 7 in *PAHs and related compounds* part 3I Neilson AH ed, Springer-Verlag, Berlin. pages 271–323.

Herod, A.A. (2005). Mass spectrometry of coal liquids; Chapter 8 Determination of organic compounds by methods using mass spectrometry, in The Encyclopedia of Mass Spectrometry vol 4 Fundamentals of and Applications to organic (and organometallic) compounds Ed N.M.M. Nibbering, Elsevier, Oxford UK. Pages 790–803.

Herod, A.J. (1995) PhD Thesis, University of Leeds, UK.

Herod, A.A., Ladner, W.R., Stokes, B.J., Berry, A.J., Games, D.E. and Hohn, M. (1987) Fuel, 66, 935–946.

Herod, A.A., Ladner, W.R., Stokes, B.J., Major, H.J. and Fairbrother, A. (1988) The Analyst, 113, 797–804.

Herod, A.A., Stokes, B.J. and Schulten, H.-.R. (1993a) *Fuel, 72*, 31–43.

Herod, A.A., Kandiyoti, R., Parker, J.E., Johnson, C.A.F., John, P., Smith, G.P. and Li, C-Z. (1993b) *J. Chem. Soc. Chem. Comm.* (9) 767–769.

Herod, A.A., Kandiyoti, R., Parker, J.E., Johnson, C.A.F., John, P., Smith, G.P. and Li, C-Z. (1993c) Rapid Commun. Mass Spectrom. 7 360–362.

Herod, A.A., Stokes, B.J., Tye, R.E., Kandiyoti, R., Gaines, A. and Li, C.-Z. (1993d) *Fuel, 72*, 1317–1325.

Herod, A.A., Stokes, B.J., Hancock, P., Kandiyoti, R., Parker, J.E., Johnson, C.A.F., John, P. and Smith, G.P. (1994a) J. Chem. Soc. Perkin 2 499–506.

Herod, A.A., Li, C.-Z., Parker, J.E., John, P., Johnson, C.A.F., Smith, G.P., Humphrey, P., Chapman, J.R. and Kandiyoti, R. (1994b) Rapid Commun Mass Spectrom, 8 808–814.

Herod, A.A., Li, C-Z., Xu, B., Parker, J.E., Johnson, C.A.F., John, P., Smith, G.P., Humphrey, P., Chapman, J.R. and Kandiyoti, R. (1994c) Rapid Commun. Mass Spectrom. 8 815–822.

Herod, A.A. and Kandiyoti, R. (1995a) Characterisation of large molecular mass materials in coal tar pitch fractions separated by planar chromatography. 8th ICCS, Oviedo, Spain 10–15 Sept.1995. in Coal Science vol I, eds J.A. Pajares and J.M.D. Tascon, Elsevier, Amsterdam, 949–952.

Herod, A.A. and Kandiyoti, R. (1995b) Journal of Chromatography A,708, 143–160.

Herod, A.A., Bartle, K.D., Carter, D.M., Cocksedge, M.J., Domin, M. and Kandiyoti, R. (1995c) *Rapid Commun. Mass Spectrom., 9*, 1446.

Herod, A.A., Johnson, B.R., Bartle, K.D., Carter, D.M., Cocksedge, M.J., Domin, M. and Kandiyoti, R. (1995d) Rapid Commun. Mass Spectrom., 9 1446–

Herod, A.A. and Kandiyoti, R. (1995e) *Fuel, 74*, 784–786.

Herod, A.A., Johnson, B.R., Bartle, K.D. and Kandiyoti, R. (1995f) Large molecules in coal, where do they hide? Paper to 8th ICCS, Oviedo, Spain, Sept 10–15 1995. in Coal Science vol I, eds J.A. Parajes and J.M.D. Tascón, Elsevier, Amsterdam 1995, 315–318.

Herod, A.J., Gibb, T.C., Herod, A.A., Zhang, S.F., Xu, B. and Kandiyoti, R.; (1995g) Fuel 1995, 75, 437.

Herod, A.A., Zhang, S-F., Carter, D.M., Domin, M., Cocksedge, M.J., Parker, J.E., Johnson, C.A.F., John, P., Smith, G.P., Johnson, B.R., Bartle, K.D. and Kandiyoti, R. (1996a) Rapid Commun. Mass Spectrom. 10 171–177.

Herod, A.A., Zhang, S-F., Johnson, B.R., Bartle, K.D. and Kandiyoti, R. (1996b) *Energy & Fuels*, 10, 743–750.

Herod, A.A. and Kandiyoti, R. (1996c) J. Planar Chromatography 9 16–24.

Herod, A.J., Gibb, T.C., Herod, A.A., Shearman, J., Dubau, C., Zhang, S-F. and Kandiyoti, R. (1996d) J Planar Chromatogr, 9 361–367.

Herod, A.A., Lazaro, M-J., Rahman, M., Domin, M., Cocksedge, M.J. and Kandiyoti, R. (1997) Rapid Commun. Mass Spectrom. 11 1627–1634.

Herod, A.A., Shearman, J., Lazaro, M-J., Johnson, B.R., Bartle, K.D. and Kandiyoti, R. (1998) Energy & Fuels 12 174–182.

Herod, A.A., Islas, C.A., Lazaro, M-J., Dubau, C., Carter, J.F. and Kandiyoti, R. (1999) Rapid Commun. Mass Spectrom. 13 201–210.

Herod, A.A. and Kandiyoti, R. (2000) unpublished work Oct.

Herod, A.A., Lazaro, M.-J., Domin, M., Islas, C.A. and Kandiyoti, R. (2000a) Fuel 79 323–337.

Herod, A.A. and Lazaro, M.J. (2000) Thin Layer (Planar) Chromatography. pp 3690–3701 in Encyclopedia of Separation Science vol III. Eds I. Wilson, E.R. Adlard, C.F. Poole and M. Cook, Academic Press.

Herod, A.A., Lazaro, M-J., Suelves, I., Dubau, C., Richaud, R., Shearman, J., Card, J., Jones, A.R., Domin, M. and Kandiyoti R. (2000b) Energy & Fuels, 14 1009–1020.

Herod, A.A., George, A., Islas, C.A., Suelves, I. and Kandiyoti, R. (2003a) *Energy & Fuels, 17*, 862–873.

Herod, A.A., Zhuo, Y. and Kandiyoti, R. (2003b) J Biochem Biophys Methods, 56, 335–361.

Herod, A.A., Apicella, B., Ciajolo, A., Millan, M., Galmes, C. and Kandiyoti, R. (2004) Rapid Commun. Mass Spectrom.; 18: 331–338.

Herod, A.A., Millan, M., Karaca, F., Behrouzi, M. and Kandiyoti, R. (2005a) unpublished work.

Herod, A.A., Millan, M., Morgan, T.J., Li, W., Feng, J. and Kandiyoti, R. (2005b) Eur. J. Mass Spectrom 11, 429–442.

Hessler, J.P., Seifert, S. and Winans, R.E. (2002) Amer. Chem. Soc. Fuel Chem. Div. Preprints 47 no 2, 736–737.

Hillenkamp, F., Keras, M., Beavis, R.C. and Chait, B.T. (1991) *Anal. Chem., 63*(24), 1193A.

Holland, G. and Tanner, S.D. (1997) Plasma source mass spectrometry: developments and applications.Special Publication No. 202, Royal Society of Chemistry, Cambridge UK.

Homma, T. and Tazaki, M. (1995) 'Size exclusion chromatography of natural and synthetic rubbers', Chapter 7 in Handbook of Size Exclusion Chromatography, Ed Chi-san Wu, Chromatographic Science Series vol 69, Marcel Dekker, New York, pages 185–210.

Hughey, C.A., Rodgers, R.P., Marshall, A.G., Qian, K.N. and Robbins, W.K., (2002a) *Org. Geochem*. 33, 743 (2002).

Hughey, C.A., Rodgers, R.P. and Marshall, A.G., (2002b) *Anal. Chem*. 74, 4145 (2002).

Hughey, C.A., Rodgers, R.P., Marshall, A.G., Walters C.C., Qian K.N. and Mankiewicz, P., (2004) *Org. Geochem*. 35, 863 (2004).

Hunt, J.E. and Winans, R.E. (1995) *Amer Chem Soc Div Fuel Chem, 40*(3) 449–452.

Islas, C.A., Suelves, I., Carter, J.F., Herod, A.A. and Kandiyoti, R. (2000) *Rapid Commun. Mass Spectrom., 14*, 1766–1782.

Islas, C.A., Suelves, I., Carter, J.F., Herod, A.A. and Kandiyoti, R. (2001a) Pyrolysis-gc-ms of pitch fractions from preparative size exclusion chromatography. In Advances in Mass Spectrometry vol 15, ed Emilio Gelpi, Wiley, Chichester, pages 923–924.

Islas, C.A., Suelves, I., Herod, A.A. and Kandiyoti, R. (2001b) MALDI-TOF mass spectra of pitch fractions with narrow polydispersity for the calibration of SEC, Proc. 11th Intl Conf. Coal Sci., September 30–October 5, San Francisco, CA, USA; ICCS-Paper 215.

Islas, C.A., Suelves, I., Morgan, T.J., Herod, A.A. and Kandiyoti, R. (2001c) Size exclusion chromatography with 1-methyl-2-pyrrolidinone as eluent: a structure independent method for molecular mass determination, Proc. 11th Intl Conf. Coal Sci., September 30–October 5, San Francisco, CA, USA; ICCS-Paper 217.

Islas, C.A. (2001) PhD Thesis University of London.

Islas C.A., Suelves I., Carter J.F., Herod A.A. and Kandiyoti, R. (2001d) Rapid Commun. Mass Spectrom. 15 845–856.

Islas, C.A., Suelves, I., Carter, J.F., Herod, A.A. and Kandiyoti, R. (2001e) An investigation of Baltic amber and its solvent extracts by various mass spectrometric methods. In Advances in Mass Spectrometry vol 15, ed Emilio Gelpi, Wiley, Chichester, pages 415–417.

Islas, C.A., Suelves, I., Carter, J.F., Herod, A.A. and Kandiyoti, R. (2002a) Rapid Communications in Mass Spectrometry; 16 481–495.

Islas, C.A., Suelves, I., Carter, J.F., Li, W., Morgan, T.J., Herod, A.A. and Kandiyoti, R. (2002b) Rapid Commun. Mass Spectrom. 16 774–784.

Islas, C.A., Suelves, I., Carter, J.F., Apicella, B., Herod, A.A. and Kandiyoti R. (2002c) Paper to Mediterranean Combustion Symposium, Egypt,.

Islas, C.A., Suelves, I., Carter, J.F., Apicella, B., Herod, A.A. and Kandiyoti, R. (2003a) *Comb. Sci. Technol., 175*, 775–791.

Islas, C.A., Suelves, I., Li, W., Morgan, T.J., Herod, A.A. and Kandiyoti, R. (2003b) Fuel 2003; 82:1813–1823.

Islas, C.A., Suelves, I., Millan, M., Apicella, B., Herod, A.A. and Kandiyoti, R. (2003c) *J. Sep. Sci., 26*, 1422–1428.

Jarvis, K.E., Gray, A.L. and Houk, R.S., (1992) Handbook of Inductively coupled plasma mass spectrometry. Chapman & Hall, London UK.

Jiao, C.Q., Phelps, D.K., Lee, S., Huang, Y. and Freiser, B.S. (1993) *Rapid Commun. Mass Spectrom., 7*, 404–408.

John, P., Johnson, C.A.F., Parker, J.E., Smith, G.P., Herod, A.A., Gaines, A.F., Li, C-Z. and Kandiyoti, R. (1991) *Rapid Commun. Mass Spectrom., 5*, 364–367.

John, P., Johnson, C.A.F., Parker, J.E., Smith, G.P., Herod, A.A., Li, C-Z. and Kandiyoti, R. (1993) *Rapid Commun. Mass Spectrom., 7*, 795–799.

John, P., Johnson, C.A.F., Parker, J.E., Smith, G.P., Herod, A.A., Li, C.-Z., Kandiyoti, R., Humphrey, P. and Chapman, J.R. (1994) Fuel 73 1606–1616.

Johnson, B.R., Bartle, K.D., Herod, A.A. and Kandiyoti, R. (1997) J Chromatography A, 758 65–74.

Johnson, B.R., Bartle, K.D., Cocksedge, M., Herod, A.A. and Kandiyoti, R. (1998) Fuel 77 933–945.

Johnson, B.R., Bartle, K.D., Ross, A.B., Herod, A.A., Kandiyoti, R. and Larsen, J.W. (1999) *Fuel, 78*, 1659–1664.

Joy, W.K., Ladner, W.R. and Pritchard, E. (1970) *Fuel, 49*, 26–38.

Karaca, F., Islas, C.A., Millan, M., Behrouzi, M., Morgan, T.J, Herod, A.A. and Kandiyoti, R. (2004) Energy & Fuels, 18, 778–788.

Karaca, F., Behrouzi, M., Morgan, T.J., Herod, A.A. and Kandiyoti, R. (2005a) Energy & Fuels 19, 164–169.

Karaca, F., Behrouzi, M., Morgan, T.J., Herod, A.A. and Kandiyoti, R. (2005b) Fuel 84, 1805–1811.

Karaca, F., Morgan, T.J., Behrouzi, M., Herod, A.A., Bull, I.D. and Kandiyoti, R. (2005c) Size Exclusion Chromatography for the unambiguous detection of aliphatics in petroleum liquid fractions and standard materials, in the presence of aromatics. paper in preparation.

Kendrick, M.J., May, M.T., Plishka, M., Robinson, K.D. (1992) Metals in Biological Systems; Ellis Horwood Ltd., Chichester UK.

Kershaw, J.R. and Smart, P.J. (1994) *Fuel Proc Technol.*, 38, 1–15.

Kim, S.H., Fletcher, R.A. and Zachariah, M.R. (2005) Environ. Sci & Technol. 39, 4021–4026.

Lachas, H., Richaud, R., Jarvis, K.E., Herod, A.A., Dugwell D.R. and Kandiyoti, R. (1999) The Analyst, London, 124, 177–184.

Lachas, H., Richaud, R., Herod, A.A., Dugwell, D.R. and Kandiyoti, R.; (2000) Rapid Commun. Mass Spectrom., 14, 335–343.

Lafleur, A.L. and Nakagawa, Y. (1989) Fuel 68 741–752.

Lakowicz, J.R., (1986) Principles of Fluorescence Spectroscopy: Plenum Press, New York.

Larsen, J.W. and Wei, Y-C. (1988) Energy & Fuels, 2 344–350.

Larsen, J.W., Lapucha, A.R., Wernett, P.C. and Anderson, W.R. (1994) *Energy & Fuels, 8*, 258–265.

Lazaro, M-J., Herod, A.A., Cocksedge, M.J., Domin, M. and Kandiyoti, R. (1997) Fuel 76 1225–1233.

Lazaro, M.J., Herod, A.A., Domin, M., Zhuo, Y., Islas, C.A. and Kandiyoti, R. (1999a) Rapid Commun. Mass Spectrom. 13 1401–1412.

Lazaro, M.J., Islas, C.A., Herod, A.A. and Kandiyoti, R. (1999b) *Energy & Fuels, 13*, 1212.

Lazaro, M.J., Domin, M., Herod, A.A. and Kandiyoti, R. (1999c) J. Chromatogr A 840 107–115.

Lazaro, M-J., Herod, A.A. and Kandiyoti, R. (1999d) Fuel 78 795–801.

Lazaro, M-J., Moliner, R., Suelves, I., Herod, A.A. and Kandiyoti, R. (2001) Fuel 80 179–194.

Leenheer, J.A., Rostad, C.E., Gates, P.M., Furlong, E.T. and Ferrer, I. (2001) Anal. Chem. 3 1461–1471.

Lehrle, R.S. and Sarson, D.S. (1995) Rapid Commun. Mass Spectrom. 9 91–92.

Li, C-Z., Bartle, K.D. and Kandiyoti, R. (1993a) Fuel 72 3–11.

Li, C-Z., Bartle, K.D. and Kandiyoti, R. (1993b) Fuel 72 1459–1468.

Li, C-Z., Madrali, E.S., Wu, F., Xu, B., Cai, H-Y., Güell, A.J. and Kandiyoti, R. (1994a) Fuel 73 851–865.

Li, C.-Z., Herod, A.A., John, P., Johnson, C.A.F., Parker, J.E., Smith, G.P., Humphrey, P., Chapman, J.R., Rahman, M., Kinghorn, R.R.F. and Kandiyoti, R. (1994b) Rapid Commun. Mass Spectrom. 8 823–828.

Li, C-Z., Wu, F., Xu, B. and Kandiyoti, R. (1995) *Fuel, 74*, 37–45.

Li, W., Morgan, T.J., Suelves, I., Carter, J.F., Herod, A.A., Rahman, M. and Kandiyoti, R. (2002) Characterisation and liquefaction of Type 1 kerogens of different maturities. unpublished work.

Li, W., Morgan, T.J., Herod, A.A. and Kandiyoti, R. (2004) *J. Chromatogr* A, *1024*, 227–243.

Lloyd, P.M., Suddaby, K.G., Varney, J.E., Scrivener, E., Derrick, P.J. and Haddleton D.M. (1995) Eur. Mass Spectrom. 1 293–300.

Lyons, P.C., Hercules, D.M., Morelli, J.J., Sellers, G.A., Mattern, D., Thompson-Rizer, C.L., Brown, F.W. and Millay, M.A. (1987) *Int. J. Coal Geology, 7*, 185–194.

Lyons, P.C., Morelli, J.J., Hercules, D.M., Lineman, D., Thompson-Rizer, C.L. and Dulong, F.T. (1990) *Fuel, 69*, 771–775.

Madrali, E.S., Wu, F., Rahman, M., Kinghorn, R.R.F., Herod, A.A and Kandiyoti, R. (1994) Fuel 73, 1829–1835.

Malawer, E.G. (1995) Chapter 1, Introduction to size exclusion chromatography. In Handbook of Size Exclusion Chromatography, ed Chi-san Wu, Marcel Dekker Inc, NY, pages 1–24.

Malhotra, R., McMillen, D.F., Watson, E.L. and Huestis, D.L. (1993) *Energy & Fuels, 7*, 1079–1087.

Maroto-Valer, M.M., Andrésen, J.M. and Snape, C.E. (1998) Fuel 77, 783–785.

Masuda, K., Okuma, O., Knaji, M. and Matsumara, T. (1996) Fuel 75 1065–1070.

Mathis, M., Mohr, M. and Kaegi, R. (2005) Environ. Sci. &Technol. 39, 1887–1892.

McKelvy, M.L., Britt, T.R., Davies, B.L., Gillie, J.K., Lentz, L.A., Leugers, A., Nyquist, R.A and Putzig, C.L. (1996) Anal. Chem. 68 93R.

McKelvy, M.L., Britt, T.R., Davies, B.L., Gillie, J.K., Graves, F.B. and Lentz, L.A. (1998) Anal. Chem. 70, 119R.

McLafferty, F.W. and Turecek, F. *Interpretation of mass spectra*, 4th ed, University Science Books, Mill Valley, Ca, USA 1993.

Meehan, E. (1995) Chapter 2 Semirigid polymer gels for size exclusion chromatography. In Handbook of Size Exclusion Chromatography, ed Chi-san Wu, Marcel Dekker Inc, NY, pages 25–46.

Menéndez, R., Blanco, C., Santamaria, R., Bermejo, J., Suelves, I., Herod, A.A. and Kandiyoti, R. (2001) Energy & Fuels 15 214–223.

Menéndez, R., Blanco, C., Santamaría, R., Domínguez, A., Blanco, C.G., Suelves, I., Herod, A.A., Morgan, T.G. and Kandiyoti, R. (2002) *Energy & Fuels, 16*, 1540–1549.

Merdrignac, I., Truchy, C., Robert, E., Guibard, I. and Kressman, S. (2004) Petroleum Sci Technol. 22(7&8) 1003–1022.

Messenböck, R.C. (1998) PhD Thesis, University of London.

Millan, M. (2005) PhD Thesis University of London.

Millan, M., Morgan, T.J., Behrouzi, M., Karaca, F., Galmes, C., Herod, A.A. and Kandiyoti R. (2005a) Rapid Commun. Mass Spectrom. 19, 1867–1873.

Millan, M., Behrouzi, M., Karaca, F., Morgan, T.J., Herod A.A. and Kandiyoti, R. (2005b). Catalysis Today 109, 154–161.

Millan, M., Behrouzi, M., Karaca, F., Morgan, T. J., Herod, A.A. and Kandiyoti, R. (2005c) Solid state ^{13}C NMR of fractions of coal liquids; unpublished work.

Mills, I., Cvitaš, T., Homann, K., Kallay, N., Kuchitsu, K. (1993) Quantities, units and symbols in Physical Chemistry, 2nd Ed. IUPAC Phys,. Chem. Div., Blackwell, Oxford UK.

Montaudo, G., Montaudo, M.S., Puglisi, C. and Samperi F. (1994) Rapid Commun. Mass Spectrom. 8 1011–1015.

Montaudo, G., Montaudo, M.S., Puglisi, C. and Samperi, F. (1995). Rapid Commun. Mass Spectrom., 9 453–460.

Monte, M.B.M., Coelho, R.R. and Middea, A. (2004) Petroleum Sci Technol 22(&7*) 991–1001.

Mori, S. (1983) Anal. Chem, 55:2414.

Morgan, T.J., Millan, M., Behrouzi, M., Herod, A.A. and Kandiyoti, R. (2005a) Energy & Fuels 19, 164–169.

Morgan, T.J., Morden, W.E., Al-Muhareb, E., Herod, A.A. and Kandiyoti, R. (2005b) Essential oils investigated by size exclusion chromatography and gas chromatography-mass spectrometry. Energy & Fuels 2006 in press ef050364i.

Morgan, T.J., Herod, A.A., Brain, S., Chambers, F. and Kandiyoti, R. (2005c) 1095, 81–88. J Chrom A 1095, 81–88.

Mössbauer, R.L. (1957) Z. Physik. 151, 124–

Müller, S. and Müllen, K. (2005) Chem. Commun. 4045–4064.

Neal, A.C. (1995) HPLC and column liquid chromatography. Chapter 12 in *Chromatography in the Petroleum Industry*. Ed Adlard ER, Elsevier, Amsterdam.

Niessen, W.M.A. and Tinke, A.P. (1995) *J Chromatogr* A, *703*, 37.

Nishioka, M. and Larsen, J.W. (1990) Energy and Fuels, 4 100–106

Oldham, P.B., McCarroll, M.E., McGown, L.B. and Warner, I.M. (2000) Anal Chem 72 197R.

Pacey, P.D. and Glasier, G.F. (2002) Amer. Chem. Soc. Fuel Chem. Div. Preprints 47 no 2, 742–743.

Parker, J.E., Johnson, C.A.F., John, P., Smith, G.P., Herod, A.A., Stokes, B.J. and Kandiyoti R. (1993) Fuel, 72 1381–1391.

Pindoria, R.V., Megaritis, A., Chatzakis, I.N., Vasanthakumar, L.S., Lazaro, M.J., Herod, A.A., Garcia, X.A., Gordon, A. and Kandiyoti, R. (1997a) Fuel, 76, 101–113.

Pindoria, R.V., Lim, J-Y., Hawkes, J.E., Lazaro, M-J., Herod, A.A. and Kandiyoti, R. (1997b) Fuel 76 1013–1024.

Pindoria, R.V., Megaritis, A., Herod, A.A. and Kandiyoti, R. (1998) Fuel 77 1715–1726.

Pindoria, R.V., Chatzakis, I.N., Lim, J-Y., Herod, A.A., Dugwell, D.R. and Kandiyoti, R. (1999) Fuel 78 55–63.

Pipatmanomai, S., Islas, C.A., Suelves, I., Herod, A.A. and Kandiyoti, R. (2001) J. Anal Appl. Pyrol. 58 299–313.

Pipatmanomai, S., Herod, A.A., Morgan, T.J., Paterson, N., Dugwell, D.R. and Kandiyoti, R. (2004) Energy & Fuels 18 68–76.

Poirier, N., Derenne, S., Rouzaud, J-N., Largeau, C., Balesdent, J. and Maquet, J. (2000) *Organic Geochemistry, 31*, 813.

Podzimek, S. (2005) Int. J. Polymer Anal. Characterization 9, 305–316.

Pokorna, L., Gajdosova, D. and Havel, J. (1999) Characterisation of humic acids by capillary zone electrophoresis and matrix assisted laser desorption/ ionisation time of flight mass spectrometry. In Ghabbour EA and Davies G (eds) Understanding humic substances: advanced methods, properties and applications. Royal Society of Chemistry, London, special publication no.247, pages 107–119.

Polman, J.K. and Quigley, D.R. (1991) Energy & Fuels, 5 352–353.

Pretorius, W., Foulkes, M., Ebdon, L. and Rowland, S. (1993) *J. High Res. Chromatogr., 16*, 157–160.

Purevsuren, B., Herod, A.A., Kandiyoti, R., Morgan, T.J., Avid, B. and Gerelmaa, T.Ya. (2004a) Fuel, 83: 799–805.

Purevsuren, B., Herod, A.A., Kandiyoti, R., Morgan, T.J., Avid, B. and Davaajav, Ya. (2004b) Eur J Mass Spec, 10, 101–108.

Qian K.N., Rodgers, R.P., Hendrickson, C.L., Emmett, M.E. and Marshall, A.G., (2001a) *Energy & Fuels* 15, 492 (2001).

Qian, K.N., Robbins W.K., Hughey C.A., Cooper H.J., Rodgers R.P., Marshall A.G., (2001b) *Energy & Fuels*, 15, 1505 (2001).

Rahman M., Herod AA and Kandiyoti R (2000) Fuel 79, 201–205.

Ralph, J.P. and Catcheside D.E.A. (1996) J. Chromatogr A 724 97–105.

Reemstma, T. and These, A. (2003) Anal. Chem. 75 1500.

Reerink, H. and Lijzenga, J., (1975) Anal. Chem. 37, 2160.

Rendall, D. (1987) Fluorescence and Phosphorescence Spectroscopy; J Wiley & Sons, Chichester UK.

Richaud, R., Lachas, H., Collot, A-G., Mannerings, A.G., Herod A.A. and Kandiyoti R.; (1998) Fuel 77, 359–368.

Richaud, R., Lazaro, M-J., Lachas, H., Miller, B.B., Herod, A.A., Dugwell, D.R. and Kandiyoti R. (2000a) Rapid Commun. Mass Spectrom., 14 317–328.

Richaud, R., Lachas, H., Lazaro, M-J., Clarke, L.J., Jarvis, K.E., Herod, A.A., Gibb, T.C. and Kandiyoti, R. (2000b) *Fuel, 79*, 57–67.

Richaud, R., Lachas, H., Healey, A.E., Reed, G.P., Haines, J., Jarvis, K.E., Herod, A.A., Dugwell D.R. and Kandiyoti, R.;(2000c) Fuel, 2000, 79, 1077–1087.

Richaud, R., Herod, AA. and Kandiyoti, R. (2004) Fuel, 83, (14–15), 2001–2012.

Robinson, N., Evershed, R.P., Higgs, W.J., Jerman, K. and Eglinton, G. (1987) The Analyst 112 637–644.

Sato, S., Takanohashi, T. and Tanaka, R. (2005) Energy & Fuels 19, 1991–1994.

Sheu, E.Y., De Tar, M.M. and Storm, D.A.. (1994) Surface activity and dynamics of asphaltenes. In Asphaltene Particles in Fossil Fuels, Exploration, Recovery, Refining and Production Processes; eds M.K. Sharma and T.F. Yen, Plenum Press New York, p118.

Schulten, H.-R. (1982) Fuel 61 670–676.

Sheng, L-S., Covey, J.E., Shew, S.L., Winger, B.E. and Campana, J.E. (1994) Rapid Commun. Mass Spectrom. 8 498–500.

Sherma, J. (1996) Anal. Chem. 68, 1R.

Sherma, J. (1998) Anal. Chem. 70, 7R.

Sherma, J. (2000) Anal. Chem. 72, 9R.

Shimada, K., Lusenkova, M.A., Sato, K., Saito, T., Matsuyama, S., Nakahara, H. and Kinugasa, S. (2001) Rapid Commun Mass Spectrom. 15 277–282.

Shui, H.l., Norinaga, K. and Iino, M. (2002) *Energy & Fuels, 16*, 69.

Sirota, E.B. (2005) Energy & Fuels 19, 1290–1296.

Snape, C.E., Axelson, D.E., Botto, R.E., Delpuech, J.J., Tekely, P., Gerstein, B.C., Pruski, M., Maciel, G.E. and Wilson, M.A. (1989a) Fuel 68, 547–560.

Snape, C.E., Kenwright, A.M., Bermejo, J., Fernandez, J. and Moinelo, S.R. (1989b) Fuel 68, 1605–1608.

Socrates, G. (2001) Infra red and Raman Characteristic Group Frequencies: Tables and Charts 3rd ed. Wiley.

Somsen, G.W., Morden, W. and Wilson, I.D. (1995) *J Chromatogr* A, *703*:613.

Stenson, A.C., Landing, W.M., Marshall, A.G. and Cooper, W.T. (2002) Anal. Chem. 74 4397–4409.

Stock, L.M. and Obeng, M. (1997) *Energy & Fuels, 11*, 987–997.

Strausz, O.P., Peng, P. and Murgich, J. (2002) *Energy & Fuels, 16*, 809–822.

Strlic, M. and Kolar, J. (2003) J Biochem. Biophys. Methods, 56, 265–279.

Suelves, I., Herod, A.A. and Kandiyoti, R. (2000) Abstr. Pap. Am. Chem. Soc., vol 219; p88-GEOC.

Suelves, I., Islas, C.A., Herod, A.A. and Kandiyoti, R. (2001a) Energy & Fuels 15 429–437.

Suelves, I., Lazaro, M-J., Begon, V., Morgan, T.J., Herod, A.A., Dugwell, D.R. and Kandiyoti, R. (2001b) Energy & Fuels 15 1153–1165.

Suelves, I., Islas, C.A., Millan, M., Galmes, C., Carter, J.F., Herod, A.A. and Kandiyoti, R. (2003) Fuel; 82:1–14.

Takanohashi, T., Iino, M. and Nakumara, K. (1994) *Energy & Fuels, 8*, 395.

Tanaka, R., Sato, S., Takanohashi, T., Hunt, J.E. and Winans, R.E. (2004) Energy & Fuels 18 1405–1413.

Tei, A., Herod, A.A., Dugwell, D.R., Kandiyoti, R., Vitolo, S. and Tartarelli, R. (1997) Hydrocracking of biomass-derived tars. Project Report.

These, A. and Reemstma, T. (2003) Anal. Chem., 75, 6276–6281.

These, A., Winkler, M., Thomas, C. and Reemstma, T. (2004) Rapid Commun. Mass Spectrom., 18, 1777–1786.

Thiyagarajan, P. and Cody, G.D. (1997) *Amer Chem Soc Div Fuel Chem, 42*(1) 253–257.

Torregrosa-Rodriguez, P., Martinez-Escandell, M., Rodriguez-Reinoso, F., Marsh, H., de Salazar, C.G., and Palazon, E.R 2000 Carbon 38 535.

Touchstone, J.C. (1992) Practice of Thin Layer Chromatography 3rd ed. John Wiley & Sons, New York

Vandecasteele, C. and Block, CB, (1993), Modern Methods for Trace Element Determination, John Wiley & Sons, Chichester, UK.

Vander Hayden, Y., Popovici, S.T. and Schoenmakers, P.J. (2002) *J. Chromatogr. A, 957*, 127–137.

Vasanthakumar, L.S., Hewitt, G.F., Lazaro, M-J., Herod, A.A., Dugwell, D.R., Kandiyoti, R., Fernando, N. and Paterson, N. (ed) (1998) Fluidised bed gasdification and pyrolysis of biomass. Final Report, EU Contract No CI1*-CT93–0343(DG12HSMU). January

Wang, J. and Takarada, T. (2003) *Fuel Proc. Technol., 81*, 247–258.

Winans, R.E. (1991) *J Anal Appl Pyrol., 20*, 1–13.

Winans, R.E., McBeth, R.L., Hunt, J.E. and Melnikov, P.E. (1991) *Proc. 1991 Int. Conf. Coal Sci.*, 16–20 Sept. Newcastle upon Tyne, UK, p44.

Wolfbeis, O.S. ed (1993) Fluorescence Spectroscopy, New Methods and Applications, Springer-Verlag, Berlin.

Wang, Z.L. and Cheng, R.S. (1998) Chem. J. Chin. Univ.-Chin 19 1522.

Wong, K.K.Y. and de Jong E., (1996) J. Chromatogr A 737 193–203.

Wu, Chi-san Ed (1995) Handbook of Size Exclusion Chromatography, Chromatographic Science Series, Vol. 69; Marcel Dekker: New York.

Wu, Z.G., Jernstrom, S., Hughey, C.A., Rodgers, R.P. and Marshall, A.G., (2003) *Energy Fuels* 17, 946 (2003).
Wu, Z.G., Rodgers, R.P. and Marshall, A.G., (2004) *Anal. Chem.* 76, 2511 (2004).
Wu, Z., Rodgers, R.P. and Marshall, A.G. (2005) Fuel 84 1790–1797.
Yau, W.W., Kirkland, J.J. and Bly, D.D., (1979) Modern Size Exclusion Liquid Chromatography, John Wiley & Sons Inc., NY.
Yevlampieva, N.P., Biryulin, Yu.F., Melenevskaja, E.Yu., Zgonnik, V.N. and Rjumtsev, E.I. (2002). *Colloids and Surfaces A: Physicochemical and Engineering Aspects, 209*, 167–171.
Zhang, S-F., Xu, B., Herod, A.A. and Kandiyoti R., (1996a) Energy & Fuels, 10 733–742.
Zhang, S-F., Xu, B., Moore, S., Herod, A.A. and Kandiyoti R., (1996b) Fuel, 75 597–605.
Zhang, S-F., Xu, B., Herod, A.A., Kimber, G.M., Dugwell, D.R. and Kandiyoti R., (1996c) Fuel 75 1557–1567.
Zhang, S-F., Herod, A.A. and Kandiyoti R., (1997a) Fuel 76 39–49.
Zhang, S-F., Shearman, J., Domin, M., Lazaro, M-J., Herod, A.A. and Kandiyoti R., (1997b) Fuel 76 207–217.
Zhuo, Y., Herod, A.A. and Kandiyoti, R., (2003) "The thermochemical reactions of middle rank coals" Chapter 3 in "Natural and Laboratory-simulated Thermal Geochemical Processes" (R.Ikan, Ed.) Kluwer Academic Publishers, Dordrecht, pages 53–151.
Ziechmann, W., Hubner, M., Jonassen, K.E.N., Batsberg, W., Nielsen, T., Hahner, S., Hansen, P.E. and Gudmundson, A-L. (2000) Humic substances and humification. In Ghabbour EA and Davies G (eds) Humic Substances, versatile components of plants, soils and water. Royal Society of Chemistry, London, special publication no. 259, pages 9–20.

CHAPTER 9

Concluding Remarks: Where to With Solid Fuels?

9.1 Characterizing Solid Fuels and Heavy Hydrocarbon Liquids

The focus on the thermochemical reactions of solid fuels during much of this book had a twofold aim. First, it seemed worthwhile to underline the care required in designing bench-scale experiments capable of producing data that are relevant to basic fuel behavior and useful in process design and operation. Second, when due care and attention is paid to reactor design, similarities and divergences emerge between pyrolysis and liquefaction that provide vital clues about key parameters governing thermal breakdown in coals. The final section on structural characterizations outlines work at the threshold of exciting new advances in the chemistry of fossil fuels. The prizes on offer are nothing less than an improved understanding of fuel structures and compositions and pointers regarding their organic geochemistry – as well as for potential improvements in technologies that must be developed for processing heavy hydrocarbon liquids.

In discussing pyrolysis reactors, we have found that wire-mesh reactors provided data and reaction products less affected by reactor related parameters than most other types of apparatus. However, the construction and commissioning of wire-mesh reactors is expensive and time consuming. Furthermore, only small samples of chars and tars can be generated for further characterization. Entrained flow ("drop-tube") reactors provide an alternative, so long as a measure of scatter in the data can be tolerated. As explained, tar samples generated in this type of apparatus are thermally degraded prior to exiting from the reactor. Fluidized-bed reactors provide another alternative for acquiring background pyrolysis data in gasification or combustion work at temperatures up to about 1,000°C. Once again, however, considerable cracking and thermal alteration of the tar product needs to be tolerated.

Many coals melt upon heating. The generally accepted view of coal plasticity is that of a transient hydrogen donor stage where the solvating and hydrogen-donating vehicle is the hydroaromatic component of the coal itself. The extent of softening and the magnitude of tar yields are perceived to be "directly proportional" to the hydroaromatic hydrogen content. Liptinite group macerals, which nearly always melt on heating, are similarly thought to contain greater proportions of hydroaromatic component. Unfortunately, methods available for determining hydroaromatic contents in coals are neither direct, nor reliable or easy to use.

The plastic phase in coals thus appears related to the release and survival within the particles of molecules detached from the solid matrix, as "extractables", prior to removal from the parent coal particle. These extractables may be subsequently dissolved in a solvent medium, as liquefaction extracts, or remain within the particles as tar precursors during heating in a gaseous medium. When temperatures much above 400–450°C are reached during "dry" heating, the "extractables" undergo cracking reactions leading to the release of tars and gases and the deposition of secondary char.

We have also presented data from electron spin resonance spectroscopy (ESR) suggesting that pre-pyrolysis phenomena leading to the accumulation of extractables within coal particles are actually initiated at quite low temperatures. For the set of coal samples tested, temperatures for the *onset* of co-valent bond scission reactions ranged from 310 to 340°C. It is likely that many bonds need to rupture before large tar precursor molecules can be detached from the solid matrix as "extractables". Both pyrolysis and liquefaction derived data suggest that the process of detachment from the solid matrix gains speed around 350–375°C. It appears therefore that the *onset* of covalent bond cleavage and the actual depolymerization stage, leading to the *release* of tar/extract precursors, are largely sequentially occurring events. It would be useful to formulate a new ESR study with improved instrumentation to revisit and expand some of these findings.

We have also observed that the boundary between softening and non-softening coals is not a rigid one. Many coals and some lignites may soften if heated rapidly, at about 1,000°C s^{-1} or faster. Wire-mesh reactors have the ability to operate over a wide range of controlled heating rates. They have enabled studies of the effect of changes in heating rates on pyrolysis product distributions. In general, low to middle-rank bituminous coals appear more sensitive to changes in heating rates. Vitrinites appear as the more sensitive of the maceral groups, while product distributions from some liptinites and inertinites also show limited sensitivity.

The observed increases in tar yields with increasing heating rate have often been ascribed to greater tar survival through rapid ejection of tar precursors from coal particles. We have also found evidence clearly showing that far greater amounts of extractables can build up inside coal particles during rapid (compared to slow ~1°C s^{-1}) heating, prior to extensive volatile evolution. Both effects tend to contribute to enhanced tar release from coal particles. Coal combustion has been largely excluded from our discussions; it may be noted, however, that greater volatile and tar releases during the devolatilization step would contribute to flame stability and the overall extent of reaction during pulverized fuel combustion. We have also observed that, the presence of larger proportions of extractables within coal particles appears to improve the coking characteristics of coals.

In attempting to explain the larger amounts of extractables observed during rapid heating, we have proposed that sample-derived hydrogen might be assimilated more efficiently into the pyrolyzing mass during fast heating. "Good" coking coals presumably already contain sufficient proportions of the hydroaromatic component and are less sensitive to rates of heating. Similarly, yields from coals liquefied in a sufficiently large excess of hydrogen-donor solvent do not appear to change with heating rate. To confirm these findings, it would be useful to expand the range of heating rates tested during liquefaction. It seems likely from our results, however, that low-to-middle rank coals, transitional between high volatile coals and coking coals, are those marginally deficient in donatable hydrogen – and that the deficiency is made up during rapid heating by more efficient hydrogen transfer from the hydroaromatic component.

We have inferred from existing data that retrogressive char forming reactions compete, during heatup, with free radical quenching/stabilization due to the contribution of hydrogen by the hydroaromatic component. Char forming reactions are observed to proceed to greater extents during slow heating. They appear to be partially suppressed during fast heating. The data suggest these reactions to be rapid and distinct from another class of demonstrably slow recombination-repolymerization reactions observed during both pyrolysis and liquefaction experiments. The latter experiments include the observation of slow char formation in a mini-bomb reactor, where the solvent used was essentially a non-donor liquid, 1-methylnaphthalene. On reflection, this is not a difficult conclusion to understand. In effect, we are proposing a two-speed (fast and slow) regime for the progress of retrogressive repolymerization reactions. This may be more simply rephrased as acknowledging the presence of some more reactive free radicals alongside less reactive and more stable free radicals. It would be useful to test these ideas on a wider selection of samples.

Coal liquefaction work carried out in a flowing-solvent reactor has helped reinforce some of these conclusions. The design of this reactor allows products to be diluted and removed rapidly from the reaction zone. Anywhere up to 80–90% of suitable coals may be liquefied, not just with hydrogen donor solvents, but also when merely using strong solvents (e.g. quinoline) that have little or no hydrogen-donor ability. However, high conversions cannot be achieved in the presence of *all* liquid media. For instance, in the presence of hexadecane, which is neither a hydrogen-donor *nor* a good solvent for coal-derived materials, the mass transfer barrier between the solid and liquid phases appears nearly as difficult to overcome as in the case of "dry" pyrolysis. The major distinguishing feature between pyrolysis and liquefaction sequences thus turns out to be whether and how products chemically already released by the coal matrix *into the particle* are removed from the particle and from the reaction environment.

We have additional difficulties to overcome, if we next wish to find out something of the molecular masses and chemical structures of these tars and extracts. This type of information is valuable not only in providing indications regarding original sample structures but also in helping to reveal the processes by which the molecules were actually released. In Chapter 7, we have reviewed how molecules amenable to analysis by gas chromatography (GC) and the more usual mass spectroscopic techniques (GC-MS, probe-MS) may be analyzed. The upper limit for this type of analysis is normally confined to about 300–350 u for aromatic materials and about 550–600 u for aliphatics. We have also seen how these limits may be extended by the use of supercritical fluid chromatography and liquid chromatography.

However, most coal derived tars and extracts and many petroleum-derived asphaltenes contain significant proportions of high mass material, which are not amenable to analysis by GC-MS, liquid chromatography or probe-MS techniques. Above that molecular mass level, it is no longer relevant to talk about "analysis", in the sense of measuring the concentration of individual molecules or molecular types. The fractions become too complex to think in terms of identifying specific molecular types. Low-resolution mass spectra tend to show ions at every mass while high-resolution methods show several components of different isotopic composition at every mass. The *characterization* methods outlined in Chapter 8 are usually initiated by fractionating the sample, broadly, in terms of molecular mass, using solvent separation or planar or column chromatography. Molecular masses of separated fractions may then be characterized by MALDI (matrix assisted laser desorption/ionization) mass spectroscopy and by size exclusion chromatography (SEC) using 1-methyl-2-pyrrolidinone (NMP) as eluent.

In SEC, the polystyrene calibration has been shown to function with an essentially size dependent mechanism, for part of the way to high mass components. Quantitative agreement has been found between SEC and MALDI-MS up to a mass just above 3,000 u. Attempts at extending this range of agreement seem well worth pursuing. *Structural* characterizations and comparisons of the fractions may be carried out, by using any appropriate analytical technique. UV-fluorescence spectrometry, infrared spectroscopy, solid-state ^{13}C-nmr and pyrolysis-GC-MS provide useful structural clues. We have cited examples of how this battery of tests can be used in conjunction with pyrolysis experiments to characterize naturally occurring samples ranging from amber and kerogens to peats and coals, as well as examining process streams such as petroleum vacuum residues and bitumen fractions.

When we attempted using pyrolysis-GC-MS on the two heaviest coal tar pitch fractions, sparse spectra were obtained showing only a few aliphatic fragments, likely to have evolved from the break-up of aliphatic/alicyclic bridging structures between PAC ring systems. However, no aromatic material could be identified in the mass spectra of these heavy fractions. It would appear that the PAC groups were either embedded in structures that charred readily or were too large to pass through the chromatographic column. A coal liquefaction extract and a low temperature tar both contained more complex alkyl-aromatic structures than those found in the thermally more intensely treated coal tar pitch. However, when subjected to pyrolysis-GC-MS, the two heaviest fractions of these samples gave only a few components, which could mostly be attributed to contaminants (such as phthalates and residual solvent) rather than to molecular fragments from the coal-derived material.

Taken together, these data suggest that within larger molecules of coal-derived samples, increasingly large PAC ring systems are being held together by a variety of aliphatic and alicyclic bridging structures. We know very little regarding either the size (greater than 8–10 rings?) and conformation of the PACs or the aliphatic/alicyclic groups that we presume bind them together. Differential pulse voltammetry (DPV) and oxidation suggest linked 2-, 3- and 4-ring structures. However, the DPV results come the study of asphaltenes of solvent extracts of coal (Figure 2.6) and do not necessarily represent the content of the preasphaltenes of the samples that might be equivalent to the high mass materials described in Chapter 8. Interestingly, extracts and tars from a (geologically much younger) sample of Baltic amber showed no discernible differences in structural features (*e.g.* PAC sizes) between fractions separated on the basis of increasing molecular mass. These findings raise a number of interesting questions, regarding whether and how resinites cross-link with increasing maturity. The results provide indications that, during maturation, larger PACs in *coals* grow through crosslinking reactions around smaller PACs, which serve as nuclei.

We clearly need many fresh lines of enquiry. For example, in SEC, fullerenes *appear* to be of much larger mass than we know them to be, because of their particular conformations. It seems worthwhile to explore the possibility that, in SEC, above the mass ranges indicated by the later eluting "retained peak", above 5,000–8,000 u, larger hydrocarbon molecules may also change (curl?), in solution in NMP, into conformations which cause them to elute at earlier times than would be expected from their molecular masses. It is possible that the normal shape of these high mass molecules is three-dimensional, but this has not been considered previously. The indication that molecules may be three-dimensional comes from the comparison of results from SEC and MALDI-MS. One of the major spurs to the development of mass spectrometry in the last century was the need to analyze petroleum fractions. In the last 30 years, the thrust to the development of mass spectrometry has been its application to biological molecules. The requirement in future may be to develop further the methods used in biological mass spectrometry and apply them to these large molecules in petroleum and coal. In any case, the molecular masses, structures and conformations of material appearing under the excluded peak in size exclusion chromatography require further careful study.

Taken together, these methods are able to detect and partially characterize molecules far larger than the ranges of materials amenable to analysis by conventional chromatographic and mass spectrometric techniques. However, many problems, common to the characterization of coal derived liquids, petroleum residues, kerogens and humic substances remain to be studied. Some of the molecular masses identified under the excluded peak in size exclusion chromatography appear to be very large indeed – as evaluated by polystyrene calibration. Lumping all semblances of large molecules under the "aggregates" label has not proved to be productive for the petroleum industry. To date, that particular line of thinking has not solved any of the problems associated with demonstrably heavy petroleum fractions and residues. Indeed, no one has been able to demonstrate that these materials could be disaggregated.

During the decades when light petroleum feedstocks were cheap and widely available, the heavier fractions could be disposed of by coking, to recover some hydrogen and the petroleum-coke could be sold or discarded as convenient. However, as available feedstocks get progressively heavier and more expensive worldwide, larger molecular mass materials will need to be treated with perhaps a little more attention. The work outlined above suggests that some of these "heavy" materials are not aggregates but large molecules. To achieve adequate economic returns from heavier and more expensive crudes, the petroleum industry will require careful evaluations of their molecular weights and structures. This is only a necessary first step in developing catalysts and catalytic processes that are appropriate to upgrading these materials.

Many of the concepts and methods described in this book could readily find applications outside the vast orbit of solid fuel utilization. The methods described for the characterization of heavy hydrocarbon liquids provide a good example for such wider applicability. While this has been part of the attraction of the work, it is nevertheless important not to forget solid fuels themselves. Before closing, it is useful to revisit the prospects for solid fuel conversion and utilization. Are the concepts and techniques devised for improving their exploitation, in fact, responses to needs and technologies of the past? Will we ever need them again?

9.2 Solids Fuel Conversion to Gases and Liquids

In June 2004, one of the present co-authors (RK) began his talk at the ASME-IGTI meeting in Vienna by stating, "Fossil fuels are cheap! Synthetic fuels cannot compete on economic grounds." Fifteen months later, the price of a barrel of NYMEX crude reached the high sixties before retreating to the low to middle *sixty* dollars level. On the same day, the "New York City gate" natural gas price was above the US$ 12 mark per million BTU (New York close of business 16 September, 2005; http://www.bloomberg.com/markets/commodities/energyprices.html).

Clearly, some fossil fuels are no longer quite so cheap.

With the exception of the long decade after 1973, oil and natural gas prices have remained reasonably low during much of the post-World War II period. There have also been several cycles of scarcity and glut since the mid-1980s, tending to discourage industry from investing in synthetic fuels in a sustained manner. Investors need to have confidence that oil prices will remain above their costs long enough, for them to recoup their considerable investments *and* make a profit. The prospects for making synthetic fuels from coal have thus floundered in past decades, not just through competition from cheaper oil and gas, but also from the perception that oil and gas *might* get cheaper. There have been exceptions. The large Canadian tar-sand reserves are being commercially exploited despite processing costs of about US$ 20 per barrel and recent petroleum price rises will do this operation no harm at all.

The surge in prices, which has prevailed during much of 2004–5, has generally been explained in terms of greater imports by "countries such as India and China". When matched against actual import figures, the perceived impact of imports by these two countries on oil prices appear to have been amplified by uninformed repetition. Chinese imports have reached the 130 million ton level over a period of 11–12 years. It is difficult to see how year on year increases in imports of about 250,000 barrels a day (less than 0.5 percent of total world production) could possibly have such a dramatic impacts on the market. The Indian share of increased imports during 2000–2004 grew from about 70 million tons pa to about 90 million tons pa. At other times, an increase in Indian imports of about 20 million tons *over five years* could have been viewed as reasonably "stable".

The present high-price excursion is being driven by structural elements that will not quickly disappear. A combination of cost cutting exercises by the oil majors and the drive for higher prices by OPEC have left world markets with short inventories, coupled to little shut-in (reserve) production *and* even less reserve refinery capacity. Recent step-up in prices has followed on the back of this close tracking between production, refining and consumption levels, where every minor disruption in supply has driven prices to panic levels. Furthermore, present price levels are attractive for both producers and refiners. Partly, the current dearth of refining capacity has resulted from recent excess capacity leading to fierce competition between refiners that squeezed refinery profit margins to the limit. For a spell, it no longer seemed worthwhile to build oil refineries – until consumption rose to swamp the excess capacity and various local events (notably hurricane Ivan and the Nigerian oil worker strikes of 2004, hurricane Katrina in 2005) lead us to the latest crisis. At present, it is difficult to see oil producers and the oil majors rushing into large-scale investment decisions that would result in limiting their own profits. In any case, constructing new refineries and sinking new production wells are medium term projects. At least three, more probably five to six years are needed for present investment decisions to translate into new refining plant and oil production facilities coming on stream.

Returning to our subject, the latest surge in oil prices – and it is expected to last – will have altered the economic outlook for coal gasification and liquefaction processes substantially. Energy strategists the world over should now be re-examining the economics of synthetic gas-from-coal and oil-from-coal projects, as potentially viable routes for making transport and other fuels. In the UK, coal liquefaction does not form part of the new Department of Trade and Industry strategy for clean coal technology research. Instead, official strategy is directly concerned with the development and implementation of carbon abatement technologies.

Coal liquefaction: In 1996, it was estimated that an oil price of US $32 per barrel would justify building coal liquefaction plants for making transport fuels. At present, there is a clear case for re-considering the economic prospects of this process route. In the 1980s and early 1990s, British Coal had successfully developed a coal liquefaction process with substantial funding from the UK Government. The work led to the construction and operation of a pilot scale plant at Point of Ayr in North Wales. An experimental program of work was completed to plan and the results were of sufficient technical merit to justify the construction of a larger plant. Subsequent developments saw the UK coal industry privatized, with the intellectual property rights of the liquefaction process being transferred to the Government [Paterson, 2005]. Ironically, one of the reasons for abandoning oil-from-coal projects was the high cost of hydrogen. How our brave new world will find it economically sensible to adopt the consumption of pure hydrogen as a basic fuel will need explaining, but clearly, that is in the future.

In the United States, natural gas prices have nearly doubled within a fifteen-month period (since early summer 2004). There is no "spot" market for natural gas in the rest of the world. Liquefied natural gas (LNG) prices may move more rapidly, but in Europe and elsewhere, the impact of price fluctuations on gas prices will be felt more slowly. This would apply particularly to supplies contracted through existing transnational pipelines, which are priced according to formulae that *include* changes in the price of oil as *one* of the elements used for calculating the sales price. Nevertheless, the economics of power generation processes based on coal gasification will now also need to be examined.

Numerous technologically proven high-temperature gasifier systems are commercially available. These include the Texaco slurry feed and the Shell dry feed, oxygen blown gasifiers. PRENFLO technology has been used in the Puerto Llano IGCC (Integrated Gasification Combined Cycle) power plant. In addition, there are attractive new intermediate – to high temperature gasifier designs under development in China [Ren *et al.*, 2004; Peralta *et al.*, 2004]. The main constraint on the proliferation of these systems is, in the first instance, the risk associated with the possible future decline of natural gas prices in the medium term. In the past, the DEMCOLEC coal fired IGCC plant in Buggenum (the Netherlands) has had to sell electricity to the grid on the basis of peak load supply only. This was because the Dutch grid took cheaper gas fired electricity to cover its base-load requirements. Apart from the loss of income, this gave rise to technical difficulties, because coal-fired IGCC power plants are not designed for variable output. Furthermore, lead times for constructing coal fired IGCC plants are never shorter than 3–4 years. The difficult problem for potential investors is to decide whether they can rely on natural gas prices remaining high for the duration and beyond, long enough to return an adequate profit for their investment. So far, the response has been negative.

In theory, it would be desirable to generate electricity from biomass and wastes as well as from coal. Two basic points are crucial to the economics of converting waste to energy. First, the availabilities and calorific values of waste materials fluctuate daily as well as seasonally. Process energy-balances also depend crucially on the moisture contents of incoming wastes. The calorific values of middle rank coals are usually greater than those of biomass/waste materials and far more uniform. Co-processing these materials with coal is one way of smoothing variations in supply and in calorific value. Second, most biomass and waste materials have low mass densities, as well as low energy densities and are therefore more expensive to transport over long distances compared to fossil fuels. In order to cut transport costs, feedstocks must be carried over short distances. With small catchment areas, the capacities of processing units need to be equally restricted, ranging from 4–5 MWe up to perhaps 30–50 MWe, depending on local fuel availability and transport costs.

In the UK, one of the reasons given by Border Biofuels for "reviewing" plans to build pyrolysis plants in Cumbria and Scotland was the high transport costs of collecting and shipping lumber waste. One major exception has been the 240 MWe (550 MWt) Pietarsaari Plant (Finland), equipped with a circulating fluidized bed boiler and entirely fueled with wood chips. Notwithstanding the Finnish experience, which relied on an atypically plentiful and concentrated supply of timber, energy production from biomass/waste does not normally appear to be economically viable without gate fees paid for removing the waste. In view of the non-fossil-fuel obligations of the Kyoto Protocol, and the more mundane obligatory cost of waste disposal, such incentives may not be as unlikely as they may sound at first. It costs nearly US $ 65 to have a ton of waste removed in most of the industrialized world. The thermochemical utilization of biomass/waste for energy production would help reduce net CO_2 emissions as well as diminish overall dependence on fossil fuels.

Of the thermochemical alternatives available for processing biomass/waste with coal, combustion/incineration is the more mature technology at up to the 50 MWe level. Many modern plants with flue gas cleaning are in operation or under construction. Practically all these installations produce electricity *and* supply district heating schemes. Primary research still appears necessary for reducing toxic trace element releases (*e.g.* arsenic, mercury, lead, cadmium) and volatile organic compounds (VOCs). Recent findings also suggest that halogenated dioxins may survive by adhering to fly ash. Furthermore, the commercial difficulties of the 50 MWe Tacoma City (WA, USA) co-combustion unit, which utilized coal with up to 50% refuse-derived fuel plus wood waste, suggest that the economics of these units may require some revision (http://www.energyproducts.com/Tacoma.htm).

Clearly, in the selection of process routes, co-pyrolysis and co-gasification schemes would have to compete with combustion/incineration. In contrast to incinerators, most recent major biomass (or coal/biomass) gasification developments have involved singly constructed units. At present, there are few commercially available turnkey pyrolysis/gasification units available for biomass processing. The FERCO project in Burlington, Vermont and the ARBRE 8 MWe wood chip gasifier constructed with TPS Technology (Eggborough in North Yorkshire, UK), have run into technical as well as financial difficulties. In the longer term, however, gasification schemes appear better suited for suppressing toxic emissions and for providing higher theoretical cycle efficiencies. The availability of proven turnkey gasification technology would much enhance the interest of prospective buyers across the world.

The near absence of turnkey gasification systems on the market serves as testimony that technical as well as economic and administrative problems remain to be resolved, before the widespread use of small-medium scale gasification and co-gasification may be contemplated. There are clear advantages in examining many of these technical problems using bench scale experimental methods; they provide far greater flexibility and speed. They are also usually less expensive by a wide margin. In this book, we have presented several examples of how bench-scale experimentation may be improved to study primary product trends and the fundamental thermal behavior of solid fuels. We have also discussed examples where bench scale experimentation was utilized to provide basic diagnostics, to assist in process design and in solving operating problems in pilot or plant scale equipment. The key to the work that has been accomplished to date was the use of a relatively small number of thermochemical reactors, which are able to simulate wide ranges of process conditions, alongside several novel product-characterization methods.

9.3 Energy Demand – Energy Supplies: The Big Questions

In the fashion conscious world of energy experts and politicians fond of futuristic visions, the present spell of high prices is a time to ponder some difficult policy options. The urgent re-examination of the economics of ethanol as a motor fuel will no doubt come up with positive reports. Ideas about energy farms may yet be revived. Biorefineries, processing trees for chemicals and transport fuels will once again be discussed. However, recent optimistic accounts have widely overlooked the cost of such programs. Subsidies are indispensable for the wider use of cultivated energy crops. The Brazilian government support for automobile-fuel grade ethanol from maize and sugar cane was one of the more expensive examples of its kind. With the recent rise in the price of petroleum-derived gasoline, they might break even for a while.

However, some things do not change rapidly.

Energy crops compete with food production: The concept of energy farms has been much in evidence since Melvin Calvin suggested in the late 1970s that he could extract hydrocarbons directly from one of the smaller rubber plants, the *euphorbia tirucallis*. Dr. Calvin was awarded the Nobel Prize in 1961 for elucidating the chemistry of the photosynthetic process. His excursion into the global energy debate, with the prospect of putting photosynthesis to use as a driving force for making motor fuels must be viewed as creative. The idea has lived on without much reference to the fact that rubber plant extracts would have needed to be processed – cracked, reformed – before any sort of useful hydrocarbons can be made. That nearly 70 percent of the biomass went unused and that, furthermore, rubber plants deplete the soil quite disastrously were ignored amidst the general acclamation.

What is clear, however, is that no one went out to establish plantations of any sort of rubber plant, certainly not in competition to oil or coal or gas production. Instead, sensible farmers continue to plant sugar cane and cereals and vegetables, opium for pharmaceuticals (and occasionally for contraband), cotton for textiles and they do all this without any sort of direction from us, the energy visionaries. Furthermore, in doing so, they do not habitually benefit from any sort of subsidies, apart from farmers in high-cost economies and that now is under discussion.

However, focusing solely on relative price levels misses the more significant point that large-scale energy plantations are unsustainable in the long-term. The arguments are simple. Most crops deplete the soil. Sugar beet and sugar cane, oily plants such as sunflower and cotton have more severely deleterious effects on soil quality. In the industrialized world, the more usual response to soil depletion is to use fertiliser. In the rest of the world, if past practice in Brazil is anything to go by, desertification would inexorably accelerate. Yet, fertiliser grade phosphate rock is also a finite resource, with reserves estimated to be good for another ~400–500 years at present rates of consumption [Bungay, 1985].

In terms of augmenting the supply of fertilisers, there are many possible intermediate solutions. In the UK domestic sewage sludge is actually being used as a source of fertiliser or soil conditioner to make up for soil depletion, even for food crops. Toxic trace element contents of these materials are high. Even though botanists tell us that most plants reject *most* trace metals such as lead, cadmium and mercury (which we have been able to confirm), the ordinary consumer may well cringe. The application will probably be dispensed with, sooner rather than later, since it pollutes clean agricultural land. Alternatively, composting of domestic waste biomass could provide a source of soil conditioner for biomass crops. Here again, transport costs require the sewage supply, composting sites and biomass croping areas to be close to each other.

The second major obstacle to extensive energy crop cultivation is the direct competition with food production. Examples from two projects suggest that between 2,500 and 3,300 hectares of arable land are required to produce sufficient crops for an energy output of 10 MWe by combustion or gasification. These data are based on the experience of the SIGAME Project in Bahia State in North-Eastern Brazil [Worldenergy, 2005] and calculations for the ARBRE-Eggborough [Renewonline, 2002] gasification plant in North Yorkshire (UK). By contrast, in Uzbekistan, one particular commune in Kharezm Province owns 2740 hectares of average-to-poor quality irrigated land, employing 2550 workers in 1700 households, with an average of five persons per household. Thus the amount of land that could produce energy crops sufficient for generating 10 MWe, may alternatively feed approximately 7500 people plus about 850 head of cattle [Kandiyoti, 2004].

Clearly, 10 MWe is a rather small amount of power. We would need 2.5–3.3 *million* hectares of arable land, to cultivate enough energy crops to produce, say 10 GWe of electric power. The total UK electric power requirement is a little more than 70 GWe, so 10 GWe is still a relatively small amount of energy. On the other hand, 2.5–3.3 million hectares represent *all* the arable land of a country such as Angola or Zimbabwe or Mozambique or 60% of all the arable land of Zambia or of Uganda, or 20% of all arable land in agriculturally fabulously rich South Africa. It appears, therefore, that cultivated energy crops would not be allowed to make more than a marginal contribution to base load power production on a global scale. Thus, unless widespread nuclear power based programs are put in place in the very near future, we must consider that, in the near to medium term, a large proportion of power production in the world will continue to be based on fossil fuels.

Not all is lost on the biomass front. Biomass utilization is perceived as CO_2 neutral and there are obvious advantages in generating power from forestry, farming and municipal wastes. As already explained, we would need to develop a new generation of small and efficient gasifiers, to reduce transport costs for these low-density feedstocks. We would need gate fees – to make up for the fundamentally inefficient process, compared to fossil fuel utilization. This is not unrealistic. We have already mentioned that disposing of a ton of waste costs nearly US $ 65 anywhere in the industrialized world. Finally, we need long-term (~15–25 year) supply contracts. Experience suggests that the gate-fee becomes vulnerable, as soon as waste-to-power projects show a profit. This particular route would not provide the universal quick fix we all seem to have been searching for. However, the reuse of waste renewables may actually reduce global CO_2 emissions by a few percent. It is a beginning.

Where will the bulk of our energy come from? At present, we live in a world powered and fueled with coal, gas, oil and alas, nuclear energy. Received wisdom would have us believe we must recover the CO_2 generated by fossil fuel utilization and store it somewhere and that we must burn hydrogen instead of these "dirty" fuels and make up the balance with renewables. Nuclear energy has become politically undesirable and with decommissioning costs finally in the frame (why were they not calculated earlier?), it turns out to be commercially far less justifiable than it had seemed earlier. Thinking ahead for once, it might be instructive to compare the decommissioning costs of nuclear power stations with the potential costs of carbon dioxide capture and storage.

In a conference on using high-sulfur coals in 1989 the then Secretary of Energy of the USA J.D. Watkins was reported as saying "Environmental pressures are putting new pressures on our ability to use our most abundant domestic fuel, coal. Electricity reserve margins are shrinking across the country. Voltage reductions have already been required in the Northeast, and right here in Washington. Meanwhile, New York State officials are rushing to pull fuel rods from a completed, safe nuclear power plant. Our country needs a clear energy blueprint to take the United States into the next century." [Markuszewski & Wheelock, 1990] In this respect, there is no evidence that anything has changed in the USA in the intervening 15 years.

During the same period in the UK, the "dash for gas" has reduced the generation of electricity from coal combustion. We are told, however, that we may have new alternatives. In public discussions of our energy future, phrases like "low carbon economy" and the "hydrogen economy" have become commonplace. The UK Energy White Paper of 2003 [U.K. Government, 2003] implied – but never stated – that most of the *desired* 10% switch to renewables by 2010 could be achieved with the use of wind power.

Clearly, however, there are difficulties. For example, it is generally considered that a wind power contribution greater than 2–3% of total power input would be likely to introduce instabilities into large electricity distribution grids. The UK Energy White Paper itself acknowledges this [UK Government, 2003]. Turning to hydrogen, in the short to medium term, it would have to be produced from fossil fuels and that would generate much CO_2. We are asked to believe that we will capture and store CO_2. Others have argued for hydrogen from the electrolysis of water (i.e. decomposing water into molecular hydrogen and oxygen), but at present, electricity comes from fossil fuel fired power stations or from nuclear power stations. We gain nothing by using up good electricity to make hydrogen. Could that electricity come from solar power? That would be "clean" energy, indeed. It is early days yet. We know we can draw small currents from solar power. However, nearly forty years of research has not given us a sufficiently high voltage source from solar energy. At present, the technology seems far from producing the magnitude of electrical power required for the electrolysis of water at anything approaching the required scale. On the other hand, had electricity from solar power been available, there would have been little reason to waste it on one further process step. After all, electricity is the cleanest form of energy.

The dangers of global warming and consequent climate change may be real. It is incontrovertible that global temperatures are rising in parallel with atmospheric levels of carbon dioxide and other gases such as methane. However, the current consensus that global warming is linked to emissions of these "greenhouse gases" through human agency is clouded by claims that volcanic emissions of carbon dioxide and methane far surpass those attributable to human activity. It is also likely we are in a natural global warming phase. It is possible, but not certain, that the two effects are moving climate change in the same direction. Clearly, more research is required. Meanwhile, there are other factors, which would dwarf the small economies in emissions of the kind envisaged in the UK, such as plans by China to double its coal-fired generator capacity to over 650 GWe by 2020. Many other developing countries including the other Asian giant, India, also plan to increase their fossil fuel utilization. In any case, before the main offender, the United States with nearly 40% of total world emissions joins the consensus, precious little of value can be achieved.

The foregoing must be read as an agenda for doing our homework in respect of energy and environmental matters more carefully. It is difficult not to observe that the energy community has for some years been looking for rapid solutions in places where no quick fixes are to be found. Meanwhile, during the "climate change" debate of the past decade or more (1990–2003), global fossil fuel utilization and CO_2 emissions by the OECD countries have increased by an additional 20–25% [IEA, 2005]. Our lively debate on CO_2 notwithstanding, the share of coal in United States total energy generation has risen from about a quarter to a third since 1993; coal now accounts for more than 50% of electricity production [Gellici, 2005]. Nothing but massive self-deception should make us believe that we are, or that somebody is, doing something about climate change at present.

One clear, credible and practical route to damping down CO_2 emissions would be to economize, to use *less* energy, *less* fossil fuels for non-essential purposes. This requires a consensus based on public awareness. We could do this as much for fear of global warming as for slowing down the rate of resource utilization. Furthermore, it may help remind us, we should not use the Blue Planet as an immense rubbish tip. The second route would be to examine our *real* (i.e. feasible) renewables options and to re-examine the possibilities of nuclear energy. In late 2005, it is no longer a secret that all major energy consuming countries, ranging from the Europeans to Japan and the USA are taking another look at nuclear energy. Even oil and gas rich Iran is working to develop nuclear energy and Russia is willing to sell them the necessary equipment, but that takes us outside our remit. In any case, to supply our main energy requirements, it seems we are back to examining our renewables, nuclear and fossil fuel options. At current oil prices, these could now include the gasification and liquefaction of coal and biomass.

References

Bungay, H., (1985) Energy, the biomass options, Wiley Interscience New York 1985

Gellici, J., (2005) *Energeia* 16; No. 4 p. 4

IEA (2005) Energy Statistics of OECD Countries 2002–2003, OECD/IEA, Paris

Kandiyoti, D., (2004) School of Oriental & African Studies, London, *unpublished field notes.*

Markuszewski, R. and Wheelock, T.D. Eds (1990) Processing and Utilization of High-Sulfur Coals III. Elsevier Science Publishers, Amsterdam; Plenary Lecture by H. Feibus, p3

Paterson, N., (2005) unpublished notes

Peralta, D., Li Xiaoyu, Xu Shisen, Paterson, N., Dugwell, D. R., Kandiyoti, R., (2004) Proc. *"2004 Intl. Hi-Tech Symposium on Coal, Chemical Industry and Coal Conversion", 30–31 Oct. (2004), Shanghai, China*, pp. 136

Ren, Y., Xu, S., Xu, Y., Chen, C. and Xia, J., Proc. 2004 Intl. Hi-Tech Symposium on Coal Chemical Industry and Coal Conversion, 30–31 Oct. (2004), Shanghai, China, p. 172

Renewonline (2002), http://eeru. open. ac. uk/ natta/ renewonline/rol38/6.htmL

UK Government (2003) Energy White Paper: "Our Energy Future – creating a low carbon economy" February 2003 www.dti.gov.uk/energy/whitepaper

Worldenergy (2005); http://www. worldenergy. org/wec-geis/publications/reports/rural/case_studies/annII_brazil.asp

Subject Index